2023 IEEE 10th Workshop on Wide Bandgap Power Devices & Applications (WiPDA 2023)

Charlotte, North Carolina, USA
4-6 December 2023

IEEE Catalog Number: CFP23WBP-POD
ISBN: 979-8-3503-3714-3

**Copyright © 2023 by the Institute of Electrical and Electronics Engineers, Inc.
All Rights Reserved**

Copyright and Reprint Permissions: Abstracting is permitted with credit to the source. Libraries are permitted to photocopy beyond the limit of U.S. copyright law for private use of patrons those articles in this volume that carry a code at the bottom of the first page, provided the per-copy fee indicated in the code is paid through Copyright Clearance Center, 222 Rosewood Drive, Danvers, MA 01923.

For other copying, reprint or republication permission, write to IEEE Copyrights Manager, IEEE Service Center, 445 Hoes Lane, Piscataway, NJ 08854. All rights reserved.

*** This is a print representation of what appears in the IEEE Digital Library. Some format issues inherent in the e-media version may also appear in this print version.

IEEE Catalog Number: CFP23WBP-POD
ISBN (Print-On-Demand):   979-8-3503-3714-3
ISBN (Online):   979-8-3503-3713-6
ISSN: 2641-8274

Additional Copies of This Publication Are Available From:

Curran Associates, Inc
57 Morehouse Lane
Red Hook, NY 12571 USA
Phone: (845) 758-0400
Fax: (845) 758-2633
E-mail: curran@proceedings.com
Web: www.proceedings.com

TABLE OF CONTENTS

Packaging of 15-KV Silicon Carbide Half-Bridge Module Enabled by a Nonlinear Resistive
Polymer Nanocomposite Field-Grading Coating 1
  Zichen Zhang, Shengchang Lu, Carl Nicholas, Justin Lynch, Adam Morgan, Woongje Sung,
  Khai D. T. Ngo, Guo-Quan Lu

Comparison of the Static Characteristics of GaN HEMTs with Different Gate Technologies and the
Impact on Modeling 5
  Xiaomeng Geng, Nick Wieczorek, Carsten Kuring, Oliver Hilt, Mihaela Wolf, Sibylle
  Dieckerhoff

Innovations in GaN Four Quadrant Switch Technology 11
  Geetak Gupta, Carl Neufeld, Davide Bisi, Yulu Huang, Bill Cruse, Peter Smith, Rakesh Lal,
  Umesh Mishra

Cost-Effective Test Setup for Measuring Threshold Voltage Shift of GaN-HEMTs Under Long-
Term Drain-Voltage Stress.. 15
  Daniel Breidenstein, Benedikt Kohlhepp, Thomas Dürbaum

Reduced GaN Half-Bridge IC Switching Loss on Biased Si P-N Junctions 21
  Stefan Mönch, Richard Reiner, Michael Basler, Patrick Waltereit, Rüdiger Quay

Accurate Prediction of Incomplete Zero-Voltage Switching Dynamics and Losses 25
  Mike Zäch, Szymon Beczkowski, Asger Bjørn Jørgensen, Stig Munk-Nielsen

Thermal Performance Investigation of a High-Current & High-Power Density GaN-Based Motor
Drive for All Electric Aircraft Applications... 31
  Armin Ebrahimian, Seyed Iman Hosseini Sabzevari, Waqar A. Khan, Nathan Weise

The Effect of Cryogenic Temperature on Subthreshold Hysteresis of Commercial SiC Power
MOSFETs.. 38
  Monikuntala Bhattacharya, Michael Jin, Jiashu Qian, Limeng Shi, Hengyu Yu, Marvin H.
  White, Anant K. Agarwal

Etch Depth Study for Step-Etched Junction Termination Extensions in Vertical GaN Devices 42
  Andrew T. Binder, Jeffrey Steinfeldt, Andrew A. Allerman, Brian D. Rummel, Caleb Glaser,
  Luke Yates, Robert J. Kaplar

Study of GaN HEMTs Robustness to Application-Like, Software-Controlled Overshoots Emulating
Different Gate Routings in Original 50 Ohms Environment 46
  Ludovic Roche, David Trémouilles, Emmanuel Marcault, Corinne Alonso

Wide Bandgap Semiconductors for LVDC Solid State Circuit Breaker Applications...................... 52
  George Govaerts, Urmimala Chatterjee, Johan Driesen, Wilmar Martinez

Reduction of DC/DC Converters EMI Emission Using Bi- And Unidirectional QR-ZVS Topologies.............. 58
  Abdelmoumin Allioua, David Krause, Andrea Zingariello, Gerd Griepentrog

Impact of Process Variations on Back-Bias Effect in 100V p-GaN Gate AlGaN/GaN HEMTs........................ 64
  M. Cioni, G. Giorgino, A. Chini, G. Marletta, C. Miccoli, M. E. Castagna, G. Luongo, M.
  Moschetti, C. Tringali, F. Iucolano

Investigation on ESD Robustness of 20-V GGNMOS and GDPMOS in 4H-SiC Process with 100-Ns TLP Pulse 69
    Chao-Yang Ke, Ming-Dou Ker

Single and Double-Sided Jet Impingement Cooling for SiC-Based Power Modules 74
    Himel Barua, Shajjad Chowdhury, Jon Wilkins, Burak Ozpineci

An Effective Screening Technique for Early Oxide Failure in SiC Power MOSFETs 80
    Limeng Shi, Jiashu Qian, Michael Jin, Monikuntala Bhattacharya, Hengyu Yu, Marvin H.
    White, Anant K. Agarwal, Atsushi Shimbori, Zhuxian Xu

Investigation of the Constant Current Stress for Charge-To-Breakdown Extraction in Commercial SiC Power MOSFETs 84
    Jiashu Qian, Limeng Shi, Michael Jin, Monikuntala Bhattacharya, Hengyu Yu, Marvin H.
    White, Anant K. Agarwal, Atsushi Shimbori, Zhuxian Xu

Pulse-Voltage Time-Dependent Dielectric Breakdown of Commercial 1.2 kV 4H-SiC Power MOSFETs 88
    Michael Jin, Limeng Shi, Jiashu Qian, Monikuntala Bhattacharya, Hengyu Yu, Marvin H.
    White, Anant K. Agarwal, Atsushi Shimbori, Zhuxian Xu

Common Mistakes in Practical Power Supply Design with Wide Bandgap Devices 92
    Sheng-Yang Yu, Fei Yang

Thermo-Mechanical Analysis of a 650 V/150 a e-GaN HEMT Sandwiched Between a PCB and DBC Substrate 99
    Carl Nicholas, Filip Boshkovski, Emmanuel Arriola, Zichen Zhang, Guo-Quan Lu

Avalanche Capability of SiC MOSFET Under High Current 104
    Xuning Zhang, Ehab Tarmoom, Ali Shahabi, Linda Starr, Dennis Meyer

Scaled Projections of Empirically Verified Hybrid Edge Terminated Vertical GaN Diodes to 20 kV 109
    Tolen Nelson, Prakash Pandey, Daniel G. Georgiev, Raghav Khanna, Michael R. Hontz, Alan
    G. Jacobs, James C. Gallagher, Andrew D. Koehler, Karl D. Hobart, Travis J. Anderson

Investigation of the Impact of Low Thermal Conductivity on Gallium Oxide Power Module Packaging 114
    Mohammad Dehan Rahman, Xiaoqing Song

A Partial Soft-Switching SiC-Based ANPC Single-Phase Inverter with Low THD for Grid-Tied PV Systems 119
    Wenjie Ma, Hu Li, Shan Yin, Xiaohu Pang, Jiayue Fang

A Physical-Based 3rd-Quadrant Behavioral Model for Power SiC MOSFET 125
    Yuzhi Chen, Chi Li, Yifan Wu, Zedong Zheng

Performance Comparison of GaN-Based Multilevel Converters for Electric Vehicle Powertrain Application 131
    Seyed Iman Hosseini Sabzevari, Armin Ebrahimian, Nathan Weise

Improved Non-Destructive Mutual Inductance Estimation Method for Multi-Chip Power Modules 138
    Arthur Boutry, Sergio Jimenez, Andrew Lemmon

SiC MOSFET Device for Radio Frequency Power Conversion 142
    Amaury Gendron, Dumitru Sdrulla, Nathaniel Barr, Tetsuya Takata, Dick Frey, Su-Wen Chen,
    Albert Gu

Edge Termination Design Considerations for 1.2kV 4H-SiC MOSFETs While Utilizing Room Temperature Ion Implantations 148

    Stephen A Mancini, Seung Yup Jang, Zeyu Chen, Dongyoung Kim, Balaji Raghothamachar, Michael Dudley, Woongje Sung

Scalable Test System for Long Term Reliability Assessment of SiC MOSFET Stability in Extreme dV/dt Stress Conditions 154

    Lisi Zhu, David C. Sheridan, Kiran Chatty, Zhan Liu, Arash Salemi, Jin Zhang, Madhur Bobde

Enhanced Conduction and Switching Performance of 1.2 kV 4H-SiC MOSFETs Through High JFET Doping Concentration 159

    Dongyoung Kim, Skylar Deboer, Seung Yup Jang, Adam J. Morgan, Woongje Sung

Co-Optimization Design and Analysis of WBG and UWBG Power Diodes with Operational Regimes 163

    Lee Gill, Jonah Shoemaker, Jack Flicker, Stephen Goodnick, Robert Kaplar, Alan Michaels

Monitoring Current of a GaN HEMT at Ultra-High Magnetic Fields 169

    Brett Setera, Aristos Christou, Natalia Gudino

Switching Loss Reduction on Cascaded H-Bridge Converter with Diode Clamped Transformer Grounding Scheme 174

    Zihan Gao, Ruirui Chen, Dingrui Li, Fred Wang

SiC MOSFETs Performance Modeling in Simulink Simscape Environment 179

    Jacopo Ferretti, Giacomo-Piero Schiapparelli, Enrico Sangiorgi, Andrea Natale Tallarico

Exploring the Impact of Implant Temperature and a Novel Aluminum Ion Source on the Electrical Performance of 4H-SiC PiN Diodes 185

    Justin Lynch, Ryota Wada, Nobuhiro Tokoro, Takashi Kuroi, Woongje Sung

Multi-MHz Auto-Resonant Power Oscillator in a 650 V GaN-On-SOI Technology for Compact Wireless Power Transfer Systems 190

    Manuel Rueß, Dominik Koch, Ingmar Kallfass

Ultra-Wideband Surface Current Sensor Topology for Wide-Bandgap Power Electronics Applications 195

    Ali Parsa Sirat, Hossein Niakan, Babak Parkhideh

Electro-Thermal Design of MV SiC JFET Based Solid State Circuit Breakers 201

    Sima Azizi Aghdam, Mohammed Agamy, Zhongda Li, Peter Losee

Highly-Integrated, Low-Noise, Dual-Output GaN DC/DC for GaN Solid State Power Amplifier Supplies in Space Applications 207

    Dominik Koch, Jeremy Nuzzo, Michael Bosch, Manuel Rueß, Dominik Wrana, Benjamin Schoch, Ingmar Kallfass

Gate Driver with Dynamic Drive Strength on High-Temperature CMOS Process for Heterogeneous Integration Inside the SiC Power Module 213

    Asif Faruque, Ayesha Hassan, Yuyang Wang, H. Alan Mantooth

Analysis of 10 kV SiC MOSFET Module Baseplate Parasitic Capacitance Impact on Switching Loss 219

    Ruirui Chen, Min Lin, Dingrui Li, Zihan Gao, Fred Wang, Hua Bai, Leon M. Tolbert

A Si IGBT Circuit Breaker for Protection of 10 kV SiC MOSFET Power Module 225
Ruirui Chen, Min Lin, Dingrui Li, Zihan Gao, Fred Wang, Hua Bai, Leon M. Tolbert

Planar Implantation Edge Termination for Vertical GaN Power Devices 231
*Yifan Wang, Ming Xiao, Matthew Porter, Ruizhe Zhang, Qihao Song, Albert Lu, Nathan Yee,
Hiu Yung Wong, Yuhao Zhang*

Output Capacitance Loss in Wide-Bandgap and Superjunction Power Transistors: Impact of
Switching Voltage and Current 236
Qihao Song, Qiang Li, Yuhao Zhang

Comprehensive Investigation on Effects of Anti-Parallel Diodes in GaN-Based Converters 240
Kazuma Sakamoto, Yosuke Kato, Kenji Natori, Yukihiko Sato

Finite Control Set Model Predictive Control Based on In-Situ Junction Temperature for Reliability
Enhancement of Power Converters 245
*Jiale Zhou, Ali Parsa Sirat, Chondon Roy, Qiang Mu, Zaheen Mustakin, Luocheng Wang,
Babak Parkhideh, Tiefu Zhao*

P-Type Doping Control of Magnetron Sputtered NiO for High Voltage UWBG Device Structures 251
Matthew A. Porter, Yunwei Ma, Yuan Qin, Yuhao Zhang

Gate Lifetime of P-Gate GaN HEMT Under DC and Switching Overvoltage Stress 258
Bixuan Wang, Ruizhe Zhang, Qihao Song, Qiang Li, Yuhao Zhang

Fabrication AlGaN/GaN Fin-HEMTs with Hexagon Nano-Scale Fin Channel 263
Yu-Hsuan Lu, Yu-Cheng Chang, Wei-Ju Lu, Feng-Ting Lin, Bo-Hsun Xu, Chao-Hsin Wu

Short-Circuit Ruggedness Characterization of State-Of-The-Art 3.3 kV SiC MOSFETs 266
Yizhou Cong, Peiwen Jiang, Ke Wang, Pengyu Fu, Jin Wang, Ashish Kumar, Kraig Olejniczak

Author Index

Packaging of 15-kV Silicon Carbide Half-Bridge Module Enabled by a Nonlinear Resistive Polymer Nanocomposite Field-Grading Coating

Zichen Zhang
CPES, the Bradley Department of Electrical and Computer Engineering
Virginia Tech
Blacksburg, US
zichen2013@vt.edu

Shengchang Lu
CPES, the Bradley Department of Electrical and Computer Engineering
Virginia Tech
Blacksburg, US
lsheng1@vt.edu

Carl Nicholas
CPES, the Bradley Department of Electrical and Computer Engineering
Virginia Tech
Blacksburg, US
carl176@vt.edu

Justin Lynch
College of Nanoscale Science and Engineering
SUNY Polytechnic Institute
Albany, US
lynchjm@sunypoly.edu

Adam Morgan
College of Nanoscale Science and Engineering
SUNY Polytechnic Institute
Albany, US
morganaj@sunypoly.edu

Woongje Sung
College of Nanoscale Science and Engineering
SUNY Polytechnic Institute
Albany, US
sungw1@sunypoly.edu

Khai D.T. Ngo
CPES, the Bradley Department of Electrical and Computer Engineering
Virginia Tech
Blacksburg, US
kdtn@vt.edu

Guo-Quan Lu
CPES, the Bradley Department of Electrical and Computer Engineering
Virginia Tech
Blacksburg, US
gqlu@vt.edu

Abstract—**Emerging medium-voltage silicon carbide devices offer the potential to achieve more efficient and compact power electronics for grid-tied applications. However, the lack of an effective insulation solution for packaging the devices has slowed their widespread adoption. In this work, a 15-kV silicon carbide MOSFET half-bridge module was designed and fabricated. The module was built by silver-sintering the device chips on an aluminum nitride direct-bond-copper substrate with 1.0-mm thick ceramic. The insulation of the module was enhanced by coating a nonlinear resistive polymer nanocomposite along the electrode edges on the substrate prior to silicone-gel encapsulation. The field-grading effect of the coating was tested on aluminum nitride direct-bond-copper substrates. The coated substrates showed an average partial discharge inception voltage of 20.2 kVpeak, which is 85% higher than the uncoated ones and 33% higher than the devices' rated voltage. Finally, prototypes of the modules with functional 15-kV dice were fabricated and tested. The blocking capability of the devices were unchanged after packaging. The fabricated modules in this work will serve as testbed for evaluating the coating's effectiveness under fast pulse-width modulation excitations during switching tests.**

Keywords—*medium-voltage power module, insulation design, nonlinear resistive polymer nanocomposite, field grading, partial discharge*

I. INTRODUCTION

To propel medium-voltage (MV) silicon carbide (SiC) power devices and modules in grid-tied applications [1-5], innovative packaging solutions are needed to improve the module insulation without sacrificing its thermal performance [6, 7]. The escalating blocking voltages of MV SiC devices subject dielectric materials—encompassing encapsulants and ceramic substrates—to significantly intensified electric field (E-field) stresses within the MV SiC power modules, engendering partial discharges, surfacing treeing, and dielectric breakdown [8, 9]. Such dielectric failures abbreviate the device lifespan and provoke reliability issues over extended operational periods.

Recent advances in nonlinear resistive field-grading offer a promising solution [6, 10-14]. Diverging from alternative approaches such as thickening [9] or stacking [15-18] direct-bond-copper (DBC) substrates, the use of a nonlinear resistive field-grading coating at the triple-point (TP) inside a module can significantly reduce the electric field stress without increasing its thermal resistance[6]. Contrasted with more intricate and costly methods centered around TP shape modification [9, 19, 20], the coating process demands a negligible raw material input and can be implemented via an automated dispensing system. We have recently conducted an evaluation of a polymer nanocomposite (PNC) demonstrating nonlinear, field-dependent electrical conductivity. The assessment has established its efficacy in reducing the local electric field (E-field) at the TP by 40%, which consequentially improves the partial discharge inception voltage (PDIV) of the MV power modules' substrate by a substantial margin of over 80% [12, 21]. The PNC-coated substrates with 0.5-mm ceramic thickness were used to package 10-kV SiC diodes and enabled an at least 30% junction-to-case thermal resistance reduction and 84% PDIV improvement [6]. More module design and prototyping with higher voltages and more complex layouts were also done to fully evaluate its practicality in MV power modules [10, 22].

In this study, we applied the coating technology to fabricate a 15-kV SiC MOSFET in a half-bridge module. The module, incorporating a 1.0-mm Aluminum Nitride (AlN) DBC, was

---

This work was supported by the Advanced Research Projects Agency-Energy of the Department of Energy under Grant DE-AR0001008, Virginia Innovation Partnership Corporation under CCF23-0136-HE, US Department of Energy under DE-SC0022676, and the Center for Power Electronics Systems High Density Integration Industry Consortium at Virginia Polytechnic Institute and State University.

979-8-3503-3714-3/23 $31.00 © 2023 IEEE

coated with the PNC field-grading coating. A scaled-down substrate test coupon was prepared and subjected to PDIV measurement to validate the insulation design. Multiple modules were prototyped, and their static blocking characteristics were checked. This module will function as a testbed to assess the coating's effectiveness during subsequent switching tests in the near future.

II. MODULE DESIGN AND SUBSTRATE INSULATION EVALUATION

A. Module Design

Fig. 1 shows the layout of the 15-kV SiC MOSFET half-bridge module. It is made of a 50 mm by 50 mm AlN DBC substrate with an etched pattern for die-attach and electrical routing. The AlN thickness is 1.0 mm, which is considerably thinner than that of the state-of-the-art 10 kV module [3, 15], thanks to its PNC-coated TP edges. All the TP edges on the substrate are coated by the PNC to reduce the localized E-field and improve the PDIV of the module. For heat sinking, the DBC substrate is attached to an aluminum baseplate. For each module, two SiC MOSFETs are attached to the substrate by silver-sintering, and their Source and Gate pads are wire-bonded. Drawing a parallel to the design in [22], two trench widths – 1.0 mm and 3.0 mm – were incorporated into the substrate. The narrower trench aims to isolate signal voltages, such as the voltage differential between SS1 and Gate G1, whereas the wider trench is devised to isolate higher voltage potentials, like the voltage differential between D1 and S1. A current sensor is integrated within the module for ease of current sensing during the dynamic characterization in the future.

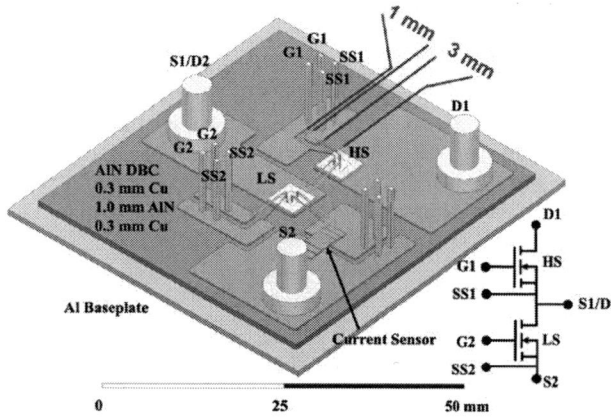

Fig. 1. Package design of the 15-kV SiC MOSFET half-bridge module and its equivalent circuit schematic. All the pins and terminals were labeled accordingly.

B. Substrate Insulation Evaluation

The insulation performance of the package design was evaluated on scaled-down AlN DBC coupons with an etched pattern shown in Fig. 2. The trench widths in this test coupon follow the same design rule used in Fig. 1. Ten samples were prepared.

Five of them were coated with PNC as described in [12] and encapsulated in a silicone gel (SilGel 612). The others were not coated with PNC and were only encapsulated in the silicone gel. They were subject to the same heating profile 180 °C for 30 min to cure the PNC and 150 °C to cure the silicone gel. Then, their PDIVs were measured on the test setup presented in [23].

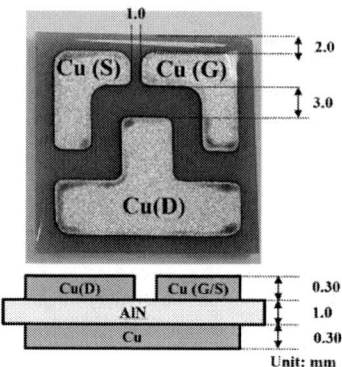

Fig. 2 Photo and a cross-sectional schematic of the scaled-down DBC test coupon with critical dimensions and terminals labeled. The shown sample's TP edges were coated with the polymer nanocomposite.

During the PDIV measurements, a high-voltage 60 Hz AC source was applied to the Drain pad while the Gate/Source pads were grounded. The baseplate was also grounded. The PDIVs were recorded when PD level exceeded 10 pC by a PD checker (MPD-600). Fig. 3 are the measured PDIVs of the coupons with and without the PNC coating. The coated coupons showed an average PDIV of 20.2 kVpeak, which is 85% higher PDIV than that of the uncoated. This improvement matches the reported data in [21, 22] on similar substrates. Detailed E-field analyses can be found in [21, 22].

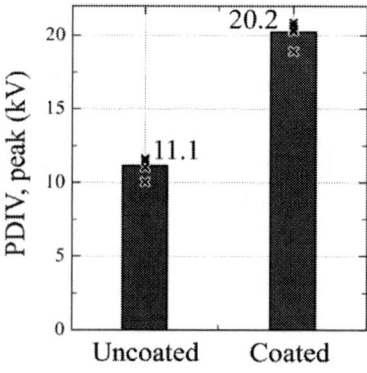

Fig. 3 Measured PDIVs of uncoated and coated AlN DBC test coupon in Fig. 2.

III. MODULE FABRICATION AND TESTING

During the fabrication process of the module, the patterned AlN DBC substrate was initially painted with the PNC solution and cured at 180 °C for 30 minutes. Subsequently, 15-kV SiC dice were attached onto the substrates through silver sintering, followed by wire-bonding. Power terminals, signal pins, and a current sensor were then soldered onto the substrate using a belt reflow furnace. Afterwards, the substrate was attached to the aluminum baseplate and glued within a 3D-printed housing. The final step involved encapsulation with SilGel-612. Figure 4 depicts a fully fabricated 15-kV SiC half-bridge module, featuring the coated AlN DBC substrate.

Fig. 4 Photo of the fabricated 15-kV SiC half-bridge module with AlN DBC substrates coated by the polymer nanocomposite.

Multiple prototype modules were fabricated. The blocking voltages of these modules were measured employing a curve tracer, following their packaging. Prior to this stage, the blocking characteristics of each die were also measured in a fluorinert liquid (FC-70) on a probe station for comparison. Fig. 5 illustrates the measured blocking characteristics of two dices in a prototype module. Remarkably, post-packaging, the devices demonstrated the capability to block in excess of 15 kV, a performance metric that aligns cohesively with their on-wafer characteristics.

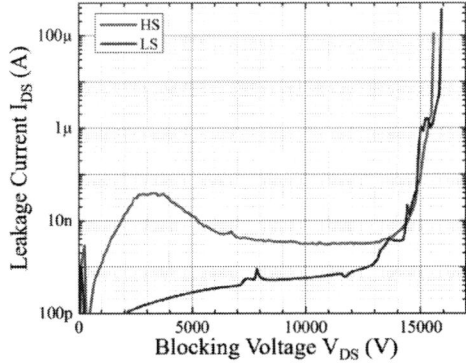

Fig. 5 Blocking characteristics of the packaged 15-kV SiC MOSFETs for both high switch (HS) and low switch (LS).

IV. CONLUSIONS

This study has successfully designed, fabricated, and tested a 15-kV SiC MOSFET half-bridge module with a single 1.0-mm AlN DBC substrate by implementing a PNC TP-coating. The PNC coating, painted at the TP-edges of the DBC substrates, was found to effectively improve PDIV by 85% on a scale-down test coupon. Without utilizing thicker or stacked substrates, this method also reduces the material cost and thermal resistance of the module. Prototype modules were fabricated and tested for their blocking capability. After packaging, the SiC MOSFETs were capable of blocking identical voltages to on-wafer characteristics. This work is the first demonstration of the coating technology for packaging an MV transistor on a DBC substrate with thinner ceramic and a more complex etched pattern. The findings of this work further validate the packaging concept of using nonlinear resistive field-grading PNC for MV power module packaging. Further efforts on the dynamic characteristics of the module are ongoing.

ACKNOWLEDGMENT

The authors thank Qingrui Yuchi and Filip Boshkovski at Virginia Tech for help with the PDIV measurements. The authors also acknowledge Rogers for providing AlN DBC substrates and NBE Tech for the nanosilver paste and the polymer nanocomposite solution.

REFERENCES

[1] M. Xiao, Y. Ma, K. Liu, K. Cheng, and Y. Zhang, "10 kV, 39 mΩ· cm 2 Multi-Channel AlGaN/GaN Schottky Barrier Diodes," *IEEE Electron Device Letters,* vol. 42, no. 6, pp. 808-811, 2021.

[2] N. Yun *et al.,* "Developing 13-kV 4H-SiC MOSFETs: Significance of Implant Straggle, Channel Design, and MOS Process on Static Performance," *IEEE Transactions on Electron Devices,* vol. 67, no. 10, pp. 4346-4353, 2020.

[3] B. Passmore and C. O'Neal, "High-voltage SiC power modules for 10-25 kV applications," *Power Electronics Europe Mag,* no. 1, pp. 22-24, 2016.

[4] S. Ji, Z. Zhang, and F. Wang, "Overview of high voltage SiC power semiconductor devices: Development and application," *CES Transactions on Electrical Machines and Systems,* vol. 1, no. 3, pp. 254-264, 2017.

[5] A. Q. Huang, Q. Zhu, L. Wang, and L. Zhang, "15 kV SiC MOSFET: An enabling technology for medium voltage solid state transformers," *CPSS Transactions on Power Electronics and Applications,* vol. 2, no. 2, pp. 118-130, 2017.

[6] Z. Zhang *et al.,* "Packaging of a 10-kV Double-Side Cooled Silicon Carbide Diode Module With Thin Substrates Coated by a Nonlinear Resistive Polymer-Nanoparticle Composite," *IEEE Transactions on Power Electronics,* vol. 37, no. 12, pp. 14462-14470, 2022.

[7] B. Zhang, M. Ghassemi, and Y. Zhang, "Insulation materials and systems for power electronics modules: A review identifying challenges and future research needs," *IEEE Transactions on Dielectrics and Electrical Insulation,* vol. 28, no. 1, pp. 290-302, 2021.

[8] B. Wang *et al.,* "Chip size minimization for wide and ultrawide bandgap power devices," *IEEE Transactions on Electron Devices,* vol. 70, no. 2, pp. 633-639, 2023.

[9] C. F. Bayer, U. Waltrich, A. Soueidan, E. Baer, and A. Schletz, "Partial discharges in ceramic substrates-correlation of electric field strength simulations with phase resolved partial discharge measurements," *Transactions of The Japan Institute of Electronics Packaging,* vol. 9, pp. E16-003-1-E16-003-9, 2016.

[10] Z. Zhang, C. Nicholas, K. D. Ngo, and G.-Q. Lu, "Package Design of a Double-Side Cooled 20-kV Gallium Nitride Diode Module

With Improved Insulation by Nonlinear Resistive Polymer-Nanoparticle Coating," in *2023 IEEE Applied Power Electronics Conference and Exposition (APEC)*, 2023: IEEE, pp. 1622-1626.

[11] J. Li, Y. Liang, Y. Mei, X. Tang, and G.-Q. Lu, "Packaging design of 15 kV SiC power devices with high-voltage encapsulation," *IEEE Transactions on Dielectrics and Electrical Insulation*, vol. 29, no. 1, pp. 47-53, 2022.

[12] Z. Zhang, K. D. Ngo, and G.-Q. Lu, "Characterization of a nonlinear resistive polymer-nanoparticle composite coating for electric field reduction in a medium-voltage power module," *IEEE Transactions on Power Electronics*, vol. 37, no. 3, pp. 2475-2479, 2021.

[13] K. Li, B. Zhang, X. Li, F. Yan, and L. Wang, "Electric field mitigation in high-voltage high-power IGBT modules using nonlinear conductivity composites," *IEEE Transactions on Components, Packaging and Manufacturing Technology*, vol. 11, no. 11, pp. 1844-1855, 2021.

[14] L. Donzel and J. Schuderer, "Nonlinear resistive electric field control for power electronic modules," *IEEE Transactions on Dielectrics and Electrical Insulation*, vol. 19, no. 3, pp. 955-959, 2012.

[15] C. M. DiMarino, B. Mouawad, C. M. Johnson, D. Boroyevich, and R. Burgos, "10-kV SiC MOSFET power module with reduced common-mode noise and electric field," *IEEE Transactions on Power Electronics*, vol. 35, no. 6, pp. 6050-6060, 2019.

[16] B. Passmore *et al.*, "The next generation of high voltage (10 kV) silicon carbide power modules," in *2016 IEEE 4th Workshop on Wide Bandgap Power Devices and Applications (WiPDA)*, 2016: IEEE, pp. 1-4.

[17] C. F. Bayer, U. Waltrich, A. Soueidan, E. Baer, and A. Schletz, "Stacking of insulating substrates and a field plate to increase the pdiv for high voltage power modules," in *2016 IEEE 66th Electronic Components and Technology Conference (ECTC)*, 2016: IEEE, pp. 1172-1178.

[18] O. Hohlfeld, R. Bayerer, T. Hunger, and H. Hartung, "Stacked substrates for high voltage applications," in *2012 7th International Conference on Integrated Power Electronics Systems (CIPS)*, 2012: IEEE, pp. 1-4.

[19] A. Deshpande, F. Luo, A. Iradukunda, D. Huitink, and L. Boteler, "Stacked DBC cavitied substrate for a 15-kV half-bridge power module," in *2019 IEEE International Workshop on Integrated Power Packaging (IWIPP)*, 2019: IEEE, pp. 12-17.

[20] H. Reynes, C. Buttay, and H. Morel, "Protruding ceramic substrates for high voltage packaging of wide bandgap semiconductors," in *2017 IEEE 5th Workshop on Wide Bandgap Power Devices and Applications (WiPDA)*, 2017: IEEE, pp. 404-410.

[21] Z. Zhang, Q. Yuchi, F. Boshkovski, K. D. Ngo, and G.-Q. Lu, "Field-Grading Effect of a Nonlinear Resistive Polymer-Nanoparticle Composite Triple-Point Coating on Direct-Bond Copper Substrates for Packaging Medium-Voltage Power Devices," in *2022 IEEE Electrical Insulation Conference (EIC)*, 2022: IEEE, pp. 439-442.

[22] Z. Zhang *et al.*, "Packaging of a 15-kV Silicon Carbide MOSFET With Insulation Enhanced by a Nonlinear Resistive Polymer-Nanoparticle Coating," in *2022 IEEE Energy Conversion Congress and Exposition (ECCE)*, 2022: IEEE, pp. 1-4.

[23] Z. Zhang *et al.*, "Packaging of an 8-kV silicon carbide diode module with double-side cooling and sintered-silver joints," in *2021 IEEE Electric Ship Technologies Symposium (ESTS)*, 2021: IEEE, pp. 1-7.

Comparison of the static characteristics of GaN HEMTs with different gate technologies and the impact on modeling

Xiaomeng Geng[a], Nick Wieczorek[a], Carsten Kuring[a], Oliver Hilt[b], Mihaela Wolf[b], and Sibylle Dieckerhoff[a]

[a] Chair of Powe Electronics Technische Universität Berlin, Berlin, Germany
[b] Leibniz-Institut für Höchstfrequenztechnik, Ferdinand-Braun-Institut, Berlin, Germany
Email: xiaomeng.geng@tu-berlin.de

Abstract—**This paper comprehensively studies the impact of different gate technologies on the static characteristics of GaN-HEMTs by comparing three transistors: 1) a GaN-on-SiC transistor fabricated by Ferdinand-Braun-Institut, which has an Ir-based Schottky gate, 2) a GaN-on-Si transistor from GaN Systems with a Schottky contact p-GaN gate and 3) a GaN-on-Si Gate Injection Transistor (GIT) from Infineon with an ohmic contact p-GaN gate. These three types of GaN transistors are further compared with a SiC-MOSFET and a Si-MOSFET. In addition, the impact of different gate technologies on modeling is studied.**

Keywords—**Gallium Nitride (GaN), Device characterisation, HEMT, p-GaN, Schottky-type**

I. INTRODUCTION

Gallium nitride (GaN)-based high-electron-mobility transistors (HEMTs) enable high power density and efficiency in power electronic applications due to their superior material characteristics compared with Si-counterparts. GaN HEMTs utilize a two-dimensional electron gas (2DEG) with high electron concentration forming at the AlGaN/GaN interface as a channel to conduct current. Different gate technologies can be employed to modulate the conductivity of the devices (Fig. 1). A metal layer can be deposited on top of the AlGaN layer as the gate electrode and a Schottky diode is consequently formed beneath the gate. This results in a Schottky gate, and the 2DEG will be depleted if a negative gate-source voltage V_{gs} is applied. This gate technology builds a normally-on GaN transistor commonly used in radio frequency (RF) communication applications. As for power electronic applications, normally-off operation is more desirable for safety considerations, thus various technologies have been studied and successfully applied to GaN HEMTs to shift the threshold voltage of GaN HEMTs to a positive value, such as recessed gate, implanted gate, p-GaN gate, and cascode hybrid [1]. The p-GaN gate technology is currently widely used by many GaN HEMT manufacturers such as EPC, GaN Systems, Infineon, and STMicroelectronics, and exhibits a good balance in performance, reliability, and manufacturability [2]. An extra p-GaN layer is deposited beneath the metal layer on the gate (Fig. 1) to lift the conduction band upwards and thus deplete the 2DEG [3]. Depending on the doping of the p-GaN and the metal used, the metal/p-GaN interface can be either a Schottky contact or an ohmic contact [4]. As for a Schottky contact, a metal/p-GaN Schottky diode is formed in reverse to the p-GaN/AlGaN/GaN pin diode resulting in a back-to-back diode model, while no significant barrier exists between metal/p-GaN for an ohmic contact resulting in a single diode model (Fig. 1).

Several publications have demonstrated that the gate contact affects the characteristics of GaN transistors such as the threshold voltage [4], [5]. Meantime, many studies have been conducted on the modeling of GaN HEMTs to simulate the transistor behavior prior manufacturing, including physics-based models [4], [6], and equivalent-circuit models based on measurement results [7], [8]. The gate technology impacts essential model parameters required for an accurate device simulation, a defining step within the device development process. Therefore, in this paper, comprehensive static characteristics of different gate technologies are studies and an equivalent circuit shown in Fig. 2 is used as transistor model in order to emphasize their importance for device modeling and simulation. The model consists of diodes, parasitic capacitors, channel resistors, and a voltage-controlled current source. The study is based on experimental results of static characterization, and modelling parameters are extracted by curve-fitting, showing the influence of different gate modules in the device model.

Three GaN transistors with GaN buffer and Si/ SiC substrate but different gate technologies, listed in TABLE I. are chosen to conduct the study: 1) a normally-on GaN-on-SiC transistor fabricated by Ferdinand-Braun-Institut (FBH), which has an Ir-based Schottky gate without a p-GaN layer beneath the metal layer, a device layout and an epitaxial stack dedicated for RF applications. The Ir-based Schottky gate technology can also be adapted on high-voltage platform for power device development [9] hence it has been considered for the current gate technology comparison. 2) GS66504B [10] from GaN Systems with a Schottky contact p-GaN gate and 3) GaN GIT IGOT60R070D1 [11] from Infineon with an ohmic contact p-GaN gate. By comparing these three device types, the impact of the p-GaN layer and the contact type of the metal/p-GaN interface on static characteristics of GaN transistors and their modeling have been evaluated. In addition, these three gate types of GaN transistors are compared with the SiC-MOSFET IMW65R107M1H [12] and the Si-MOSFET IPA65R400CE [13] from Infineon, providing deeper insight into the particularity of the static characteristics of GaN transistors, compatibility of the modeling, and their special application requirements.

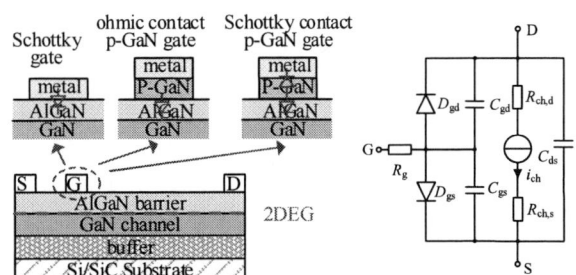

Fig. 1. Schematic cross-section of GaN HEMTs with three different gate types.

Fig. 2. Equivalent circuit of GaN HEMT model.

979-8-3503-3714-3/23 $31.00 © 2023 IEEE

Transistors	Electrical rating		
	Drain-source voltage	On-state resistance	Gate-source voltage
FBH	100 V	130 mΩ	-10 V to \
GS66504B	650 V	100 mΩ	-10 V to 7 V
IGOT60R070D1	600 V	55 mΩ	-10 V to \
IMW65R107M1H	650 V	107 mΩ	0 V to 18 V
IPA65R400CE	650 V	400 mΩ	-20 V to 20 V

TABLE I. ELECTRICAL RATING OF TRANSISTORS

II. REUSLTS AND DISCUSSION

A. Static on-resistance $R_{ds,on}$

The static on-resistance $R_{ds,on}$ of GaN HEMTs is known to be strongly temperature-dependent due to reduced mobility of electrons in the 2DEG at increased device temperature. The $R_{ds,on}$ is measured inside a temperature cabinet while the transistor is turned on and a constant drain current I_d=100 mA is sourced into the device under test (DUT). The measurement results (Fig. 3) show that the GaN transistors exhibit a temperature coefficient much higher than the SiC-MOSFET, but similar to the Si-MOSFET. The temperature dependence of the on-resistance in ohmic contact p-GaN HEMTs is less pronounced compared to Schottky contact p-GaN HEMTs. The reason could be that the barrier height of the Schottky metal/p-GaN interface is higher as the temperature rises, leading to less hole injection and decreased channel conductivity. The channel resistance $R_{ch,on}=R_{ch,d}+R_{ch,s}$ is modeled using $R_{ds,on}$. The transistor profiles are modeled using the formula

$$R_{ds,on} = a \cdot T_j^2 + b \cdot T_j + c, \qquad (1)$$

where T_j is the junction temperature. This modeling approach shows satisfactory agreement with the measured results (Fig. 3).

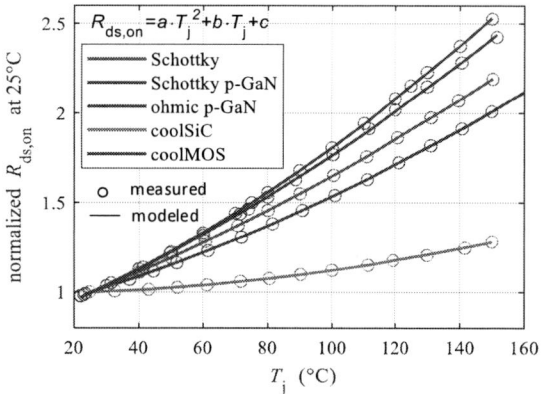

Fig. 3. Measurement results of normalized static on-resistance dependent on temperature for different transistors and modeling with a function of 2nd order.

B. Intrinsic parasitic capacitances

The parasitic capacitances significantly affect the switching losses, time, and transient behavior of transistors, since the capacitances need to be charged or discharged during the switching transitions. Therefore, it is important to characterize the intrinsic parasitic capacitances and model them accurately. Fig. 4 shows the input capacitances $C_{iss}=C_{gs}+C_{gd}$, the output capacitance $C_{oss}=C_{gd}+C_{ds}$ and the reverse transfer capacitance $C_{rss}=C_{gd}$ of the GaN transistors as well as of the SiC- and the

Si-MOSFET acquired via small signal measurements with a Keithley parameter analyzer. The capacitances of all commercial transistors are compared with datasheet values (grey lines), and the small deviation between them demonstrates the reliability of the measurement results.

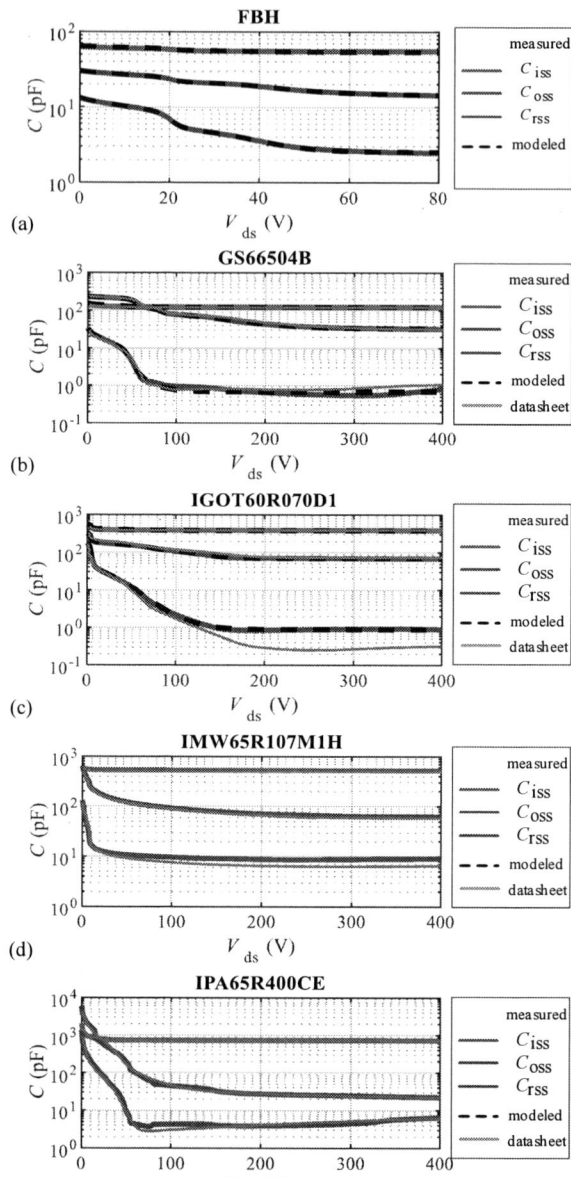

Fig. 4. Capacitance voltage profile at different V_{ds} for (a) Schottky gate GaN transistor (b) Schottky p-GaN gate GaN transistor, (c) ohmic p-GaN gate GaN transistor, (d) SiC-MOSFET, (e) Si-MOSFET, and the comparison between measurements, modeling and datasheet values.

The input capacitance C_{iss} is almost constant for the three gate types GaN HEMTs as well as for the SiC- and Si-MOSFET with varied drain-source voltage V_{ds}. In contrast, C_{gd} and C_{ds} show an obvious non-linear dependency on V_{ds}, since the width of the depletion region increases and thus the capacitances decrease with applied V_{ds}. The nonlinear behavior of the capacitances can be modeled in SPICE with an expression charge. The formula used by GaN Systems is chosen (equation (2) [14]) and achieves a satisfactory agreement (Fig. 4) for all three gate types GaN HEMTs. The amount of charge stored in the capacitors can then be calculated by integration of C over voltage (equation 3).

$$C = a_0 + \sum_{k=1}^{n} a_k \cdot b_k \cdot \frac{e^{b_k(-V+c_k)}}{1+e^{b_k(-V+c_k)}} \qquad (2)$$

$$Q = a_0 + \sum_{k=1}^{n} a_k \cdot \ln(1+e^{b_k(-V+c_k)}) \qquad (3)$$

The normalized capacitances by on-resistance in pF·Ω (Fig. 5) confirm that GaN HEMTs achieve much lower capacitances compared with Si-MOSFET and SiC-MOSFET counterparts with similar electrical ratings and have the potential for high-frequency operation. As for the three GaN HEMTs, the C_{iss} of the Schottky contact p-GaN gate transistor is lower than that of the ohmic contact p-GaN gate transistor. This could be attributed to the additional junction capacitance at the Schottky metal/p-GaN interface in series with the p-GaN/AlGaN junction capacitance, which reduces the total gate capacitance. The FBH transistor with a Schottky gate shows the lowest C_{iss}, which is possibly due to the different junction capacitance caused by the different interface materials.

The impact of the gate-source voltage V_{gs} on the intrinsic capacitances of the transistors in the off-state is studied as well. Fig. 6 shows that the capacitances are not only dependent on V_{ds} but also on V_{gs}, especially the C_{iss}. However, the impact of V_{gs} differs for GaN transistors with different gate technology. For the Schottky contact p-GaN gate and Schottky type gate transistor, V_{gs} has only minimal effects, while V_{gs} shows significant impact on the ohmic contact p-GaN gate transistor. For the SiC-MOSFET, there is no visible impact from V_{gs}, while for the Si-MOSFET, the influence of V_{gs} is significant.

C. Gate diode characteristics

For a Schottky gate and an ohmic contact p-GaN gate, the Schottky diode at the metal/AlGaN interface and pin diode at p-GaN/AlGaN/GaN interface (Fig. 1) are forward biased and conduct current if the gate-source voltage V_{gs} exceeds its threshold voltage, resulting in a non-insulating gate (Fig. 7a). It indicates that a continuous current in mA range must be sourced into the gate during on-state. Therefore, conventional gate driver circuits consisting only of gate resistors cannot be reasonably used, special design considerations are necessary and are discussed in [15]. The gate diode behavior directly affects the control of the transistors and thus should be included in the transistor model. In both cases, the gate diode can be modeled satisfactorily using a standard Berkeley SPICE semiconductor diode model [16]

$$V_D = R_S \cdot I_D + N \cdot V_T \cdot \ln(\frac{I_D}{I_S}), \qquad (4)$$

where V_D and I_D are voltage and current of the diode, I_s is the saturation current, N is the emission coefficient, R_S the parasitic resistance and V_T is the thermal voltage which is defined as $V_T=k \cdot T/q$. In contrast, for a Schottky contact p-GaN gate, two diodes in opposite direction are formed beneath the gate (Fig. 1) and the Schottky metal/p-GaN diode blocks the gate current at forward gate bias, resulting in an insulating gate. A standard driver circuit can be used to drive these gate type GaN HEMTs. However, the gate is still not completely insulating and exhibits a forward leakage current in the μA range that increases with V_{gs}. The current-voltage behavior is similar to a diode but with a much lower conduction current (Fig. 7b).

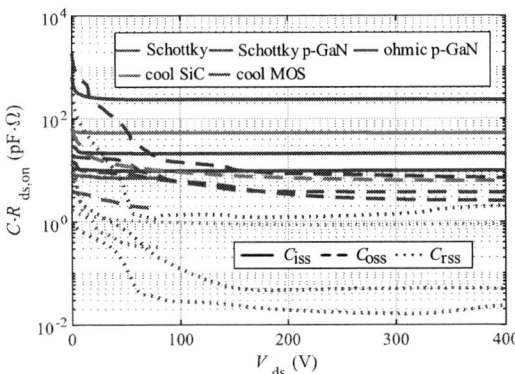

Fig. 5. Comparison of normalized capacitance voltage profiles by static on-resistance for different transistors.

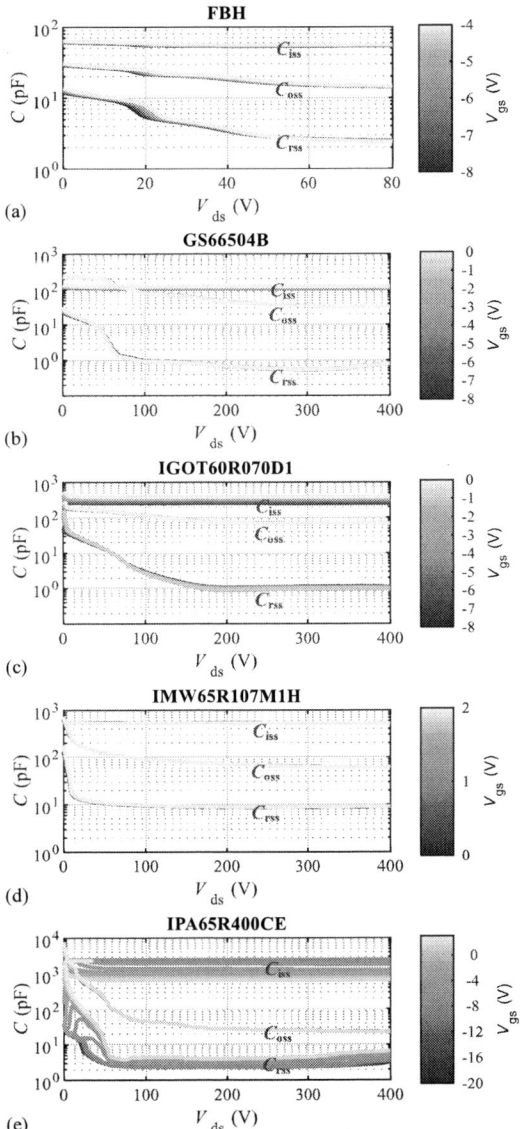

Fig. 6. Capacitance voltage profile at different gate source voltage V_{gs} for (a) Schottky gate GaN transistor (b) Schottky p-GaN gate GaN transistor, (c) ohmic p-GaN gate GaN transistor, (d) SiC-MOSFET, (e) Si-MOSFET.

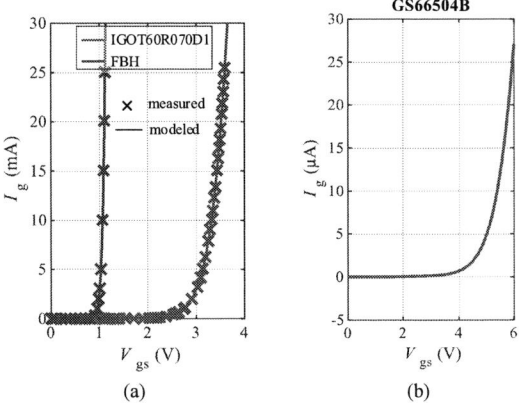

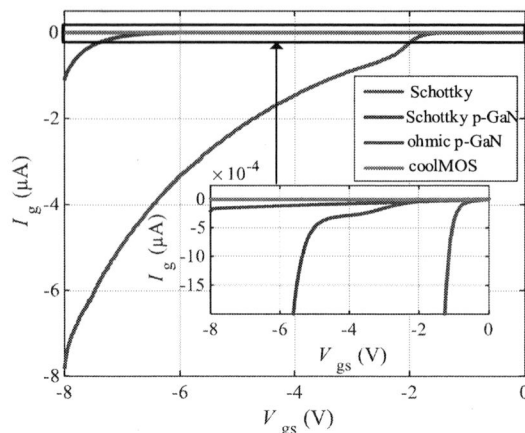

Fig. 7. (a) Measurement and modeling results of forward gate characteristics for Schottky gate and ohmic contact p-GaN gate transistors. (b) Measurement results of forward gate characteristics for Schottky contact p-GaN gate transistor.

Fig. 8. Comparison of reverse gate leakage current between different gate GaN transistors and Si-MOSFET.

D. Reverse gate leakage current

Considering that GaN HEMTs are inherently normally-on devices, and even if the threshold voltage is lifted to a positive value through the insertion of the p-GaN layer, it is relatively low, a negative gate voltage is usually mandatory to safely turn off the devices. A high reverse gate leakage current could reduce breakdown voltage, cause current collapse effects, and degrade the reliability of the devices [17] and thus it is important to analyze it. The reverse gate leakage current of GaN HEMTs can be attributed to different mechanisms, such as Pool–Frenkel emission (PF), Fowler–Nordheim tunneling (FN), trap-assisted tunneling (TAT) emission and sidewall leakage [17], [18], [19]. The reverse gate leakage current of the three GaN HEMTs and the Si-MOSFET is measured using a Keithley parameter analyzer, and the results are shown in Fig. 8. The measurement of the SiC-MOSFET is omitted since a negative gate voltage is not recommended for this transistor [12]. The reverse gate leakage current of GaN HEMTs is much larger than that of the Si-MOSFET. Among the GaN HEMTs, the Schottky gate transistor shows a relatively high gate leakage current in the range of a few µA, while both p-GaN gate transistors show a much lower leakage current. The leakage current of the Schottky gate transistor remains low at first and then increases significantly at V_{gs}=~ −1 V. For the present case, this behavior is as expected as the device semiconductor structure has an RF optimized epitaxial stack. The ohmic contact p-GaN gate transistor shows an obvious increase at V_{gs}=~ −2.5 V for the first time and then presents a saturation and rises significantly again at V_{gs}=~ −5 V. The leakage current of the Schottky contact p-GaN gate transistor remains relatively low and increases only slightly with a negative gate voltage of up to −8V. The results indicate that the aforementioned mechanisms are bias-dependent and different mechanisms dominate for different gate technology, device epi-structure and at different bias voltages. Furthermore, the difference between the Schottky contact p-GaN gate and the ohmic contact p-GaN gate suggests that the Schottky metal/p-GaN barrier affects reverse gate leakage current, and a decrease of reverse gate leakage current by improving the Schottky contact metal/p-GaN interface quality was observed in [20]. However, to distinguish the mechanism of reverse gate leakage current, which is mainly responsible for different gate technologies, more detailed investigations need to be done in the future.

E. Drain leackage current

The drain leakage current determines the breakdown voltage and is characterized for varied V_{ds} and V_{gs} with a Keithley parameter analyzer. All three GaN transistors show drain leakage currents higher than SiC- and Si-MOSFET (Fig. 9). The Schottky gate transistor has the highest value in the 10 x µA range, while both p-GaN gate transistors show drain leakage currents of smaller than two orders of magnitude. However, for both p-GaN gate transistors, the leakage current exhibits a considerable increase when V_{ds} exceeds 300 V, and the increase is more significant for the ohmic contact p-GaN gate transistor. Furthermore, for p-GaN gate GaN HEMTs, the V_{gs} has an obvious impact on the drain leakage current. The drain leakage current increases considerably with decreased V_{gs}. The reason could be that the gate-drain leakage current increases and contributes to the drain leakage current.

F. IV characteristics

The static I-V characterizations are acquired in a pulse test bench introduced in [8]. The gate-source voltage V_{gs} is kept constant in each measurement step, and a drain-source voltage V_{ds} is applied as short pulses with a pulse length of 2 µs to minimize self-heating effects. Similar to MOSFETs, GaN HEMTs exhibit an ohmic behavior and then come to a saturation region as V_{gs} exceeds the threshold voltage V_{th} and V_{ds} increases (Fig. 10). Although the Schottky gate and ohmic contact p-GaN gate are non-insulating and show diode behavior (Fig. 7a), the transistor is still voltage-controlled, and the drain saturation current rises linearly with V_{gs} (Fig. 11). Besides, the saturation current of all three GaN transistors and SiC-MOSFET increases with V_{gs} up to the rated value recommended by the manufacturers, although with different slope, while the saturation current of the Si-MOSFET does not increase further at V_{gs}=10 V. Moreover, the saturation current at a fixed V_{gs} changes with varied drain-source bias especially at low V_{gs} for Schottky gate and Schottky contact p-GaN gate GaN HEMTs as well as of the SiC-MOSFET, while it remains almost constant for the ohmic contact p-GaN gate GaN HEMT and the Si-MOSFET(Fig. 10). This could be attributed to threshold voltage shift [5], [21], which will be presented in the following section.

979-8-3503-3714-3/23 $31.00 © 2023 IEEE

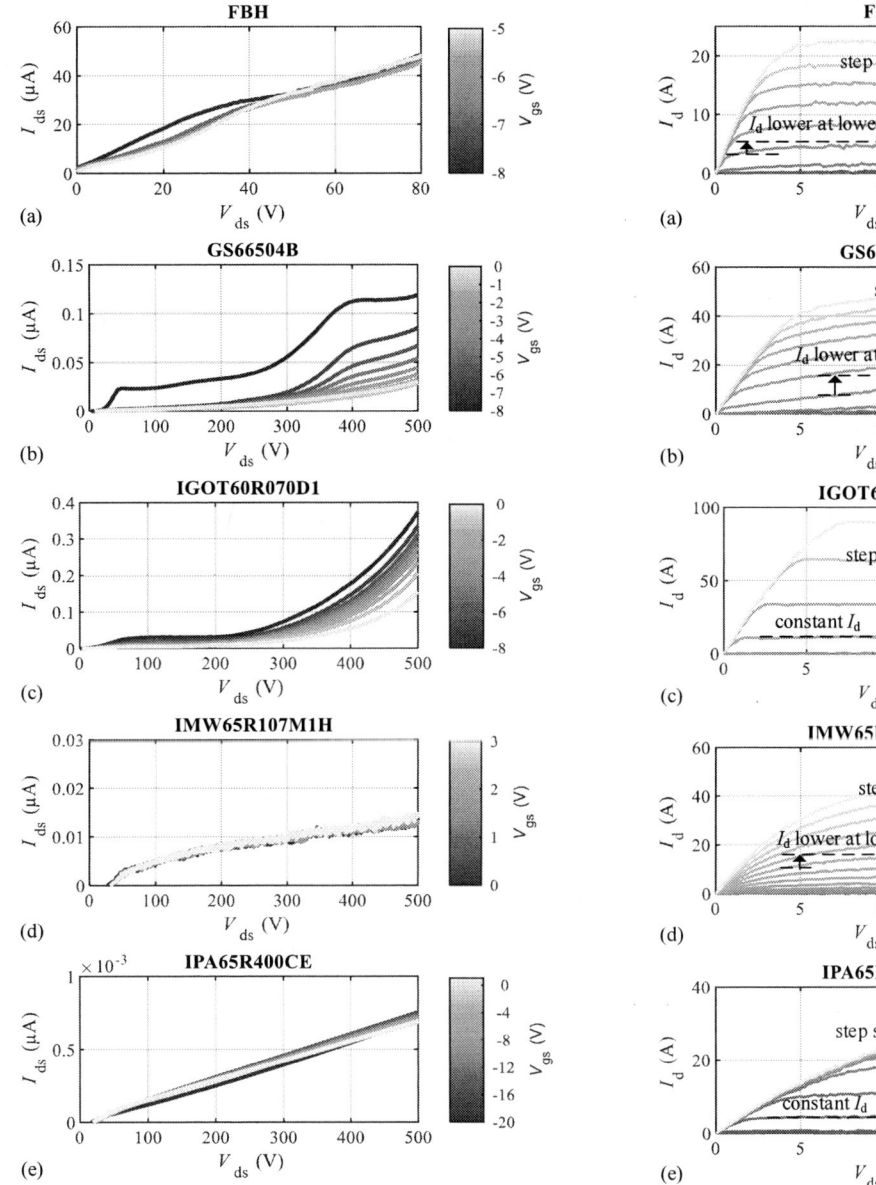

Fig. 9. Measurement results of drain leakage current at different V_{ds} and V_{gs} for (a) Schottky gate GaN transistor (b) Schottky p-GaN gate GaN transistor, (c) ohmic p-GaN gate GaN transistor, (d) SiC-MOSFET, (e) Si-MOSFET.

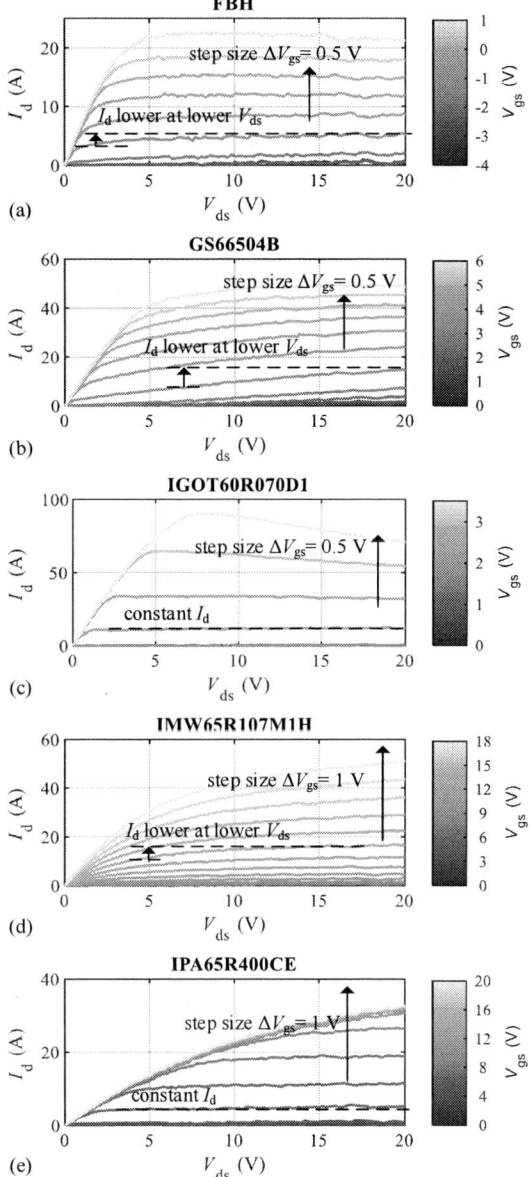

Fig. 10. Static I-V curve of the 1st quadrant for (a) Schottky gate GaN transistor (b) Schottky p-GaN gate GaN transistor, (c) ohmic p-GaN gate GaN transistor, (d) SiC-MOSFET, (e) Si-MOSFET.

G. Threshold voltage

The threshold voltage V_{th} is measured at different drain-source voltage V_{ds} (Fig. 12). During the measurement, the device under test (DUT) is first kept off and a pre-bias voltage V_{ds} is applied to the DUT for 10 s, then the gate-source voltage V_{gs} is increased step by step until the drain current I_d reaches 1 mA. The acquired V_{gs} at I_d=1 mA is defined as V_{th}. As shown in Fig. 12, the Schottky gate GaN HEMT is normally-on and shows a V_{th} of −2.3 V at V_{ds}=5 V. With the insertion of a p-GaN layer, V_{th} is lifted to a positive value, and the two p-GaN gate GaN HEMTs achieve V_{th} of 1.2 V (Schottky contact) and 1.35 V (ohmic contact), respectively. However, V_{th} is still relatively low compared with the SiC-MOSFET (4.1 V) and Si-MOSFET (3.1 V). Besides, a negative threshold voltage

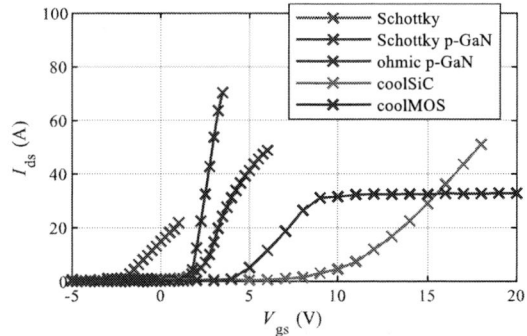

Fig. 11. Transfer curve for different transistors at V_{ds}=20 V.

shift ΔV_{th} is observed for all GaN transistors with increased V_{ds} (Fig. 13), indicating that they are more sensitive to false turn-on in high voltage operation. The ΔV_{th} of the ohmic contact p-GaN gate transistor is relatively low (<0.2 V up to V_{ds}=400 V), while the Schottky gate and Schottky contact p-GaN gate transistors show higher ΔV_{th}. The difference suggests that the p-GaN layer plays a role in a negative ΔV_{th}. A gate/drain coupled barrier lowering (GDCBL) effect could be used to explain the phenomenon [22]. The potential of the p-GaN layer of an ohmic contact p-GaN gate is fixed, while the p-GaN layer is floating for the Schottky contact p-GaN gate due to the reverse biased Schottky metal/p-GaN diode. The diode junction capacitance is in series with the channel capacitance. The potential of the p-GaN layer is raised by increased V_{ds}, and the required V_{gs} to turn on the device is in turn decreased. The Schottky gate configuration is similar, a Schottky metal/AlGaN diode is reverse biased in off-state and its junction capacitance is in series with the transistor channel capacitance. Furthermore, the RF Schottky devices have Fe-dopped buffer with weak back-barrier properties in contrast with C-dopped buffer which allows for very sharp back-barriers, an epi-stack typically used for high-voltage device development. Such designs affect device's susceptibility to punch-through effects which further affects the V_{th} shift. As shown in Fig. 13, The threshold voltage shift can be modeled satisfactorily using

$$\Delta V_{th} = a \cdot \frac{V_{ds}}{1 + b \cdot V_{ds}}. \tag{5}$$

Besides, the Si-MOSFET presents a low ΔV_{th} similar to the ohmic contact p-GaN gate transistor, while the SiC-MOSFET shows the highest ΔV_{th}.

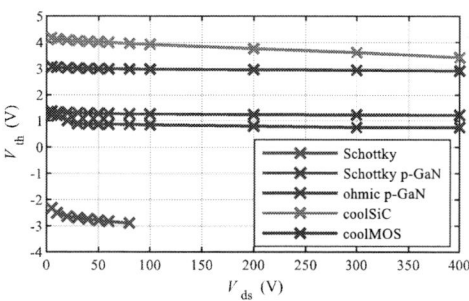

Fig. 12. Measurement results of threshold voltage for different transistors.

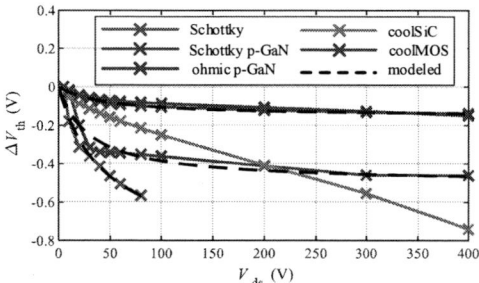

Fig. 13. Comparison of measurement results of threshold voltage shift for GaN HEMTs (solid lines), SiC- and Si-MOSFET; and the modeled results of GaN HEMTs (dashed lines).

III. CONCLUSION AND FUTURE WORK

This paper shows that the different gate technologies used for GaN HEMTs affect their static characteristics. Schottky gate and ohmic contact p-GaN gate are non-insulating and show a diode behavior at forward gate-source bias. The characteristics of the leakage current differ as well. The Schottky contact

p-GaN gate transistor shows the lowest gate leakage current, and the impact of V_{gs} on the drain leakage current is most significant. The threshold voltage shift resulting from the drain bias of the Schottky contact p-GaN gate is much larger than that of an ohmic contact p-GaN gate, resulting in different static I-V behavior. However, further investigations need to be done in the future to understand the physical mechanisms.

REFERENCES

[1] Lidow, Alex, et al. GaN transistors for efficient power conversion. John Wiley & Sons, 2019.

[2] Cheng, Yan, et al. "Observation and characterization of impact ionization-induced OFF-state breakdown in Schottky-type p-GaN gate HEMTs." Applied Physics Letters 118.16 (2021).

[3] He, Jiaqi, et al. "Recent advances in GaN‐based power HEMT devices." Advanced Electronic Materials 7.4 (2021).

[4] L. Sayadi, G. Iannaccone, S. Sicre, O. Häberlen and G. Curatola, "Threshold Voltage Instability in p-GaN Gate AlGaN/GaN HFETs," in IEEE Transactions on Electron Devices, vol. 65, no. 6, pp. 2454-2460, June 2018.

[5] T. Oeder and M. Pfost, "Gate-Induced Threshold Voltage Instabilities in p-Gate GaN HEMTs," in IEEE Transactions on Electron Devices, vol. 68, no. 9, pp. 4322-4328, Sept. 2021.

[6] S. Khandelwal, N. Goyal and T. A. Fjeldly, "A Physics-Based Analytical Model for 2DEG Charge Density in AlGaN/GaN HEMT Devices," in IEEE Transactions on Electron Devices, vol. 58, no. 10, pp. 3622-3625, Oct. 2011.

[7] A. Endruschat, et al., "A Universal SPICE Field-Effect Transistor Model Applied on SiC and GaN Transistors," in IEEE Transactions on Power Electronics, vol. 34, no. 9, pp. 9131-9145, Sept. 2019.

[8] J. Böcker, H. Just, O. Hilt, N. Badawi, J. Würfl and S. Dieckerhoff, "Experimental analysis and modeling of GaN normally-off HFETs with trapping effects," EPE'15 ECCE-Europe, Geneva, Switzerland, 2015, pp. 1-10.

[9] O. Hilt, et al., "10 A/950 V switching of GaN-channel HFETs with non-doped AlN buffer," ISPSD2023, Hong Kong, 2023, pp. 374-377.

[10] GaN Systems, "Bottom-side cooled 650 V E-mode GaN transistor," GS66504B datasheet.

[11] Infineon, "600V CoolGaN™ enhancement-mode Power Transistor," IGT60R07D1 datasheet, Jan. 2021.

[12] Infineon, "650V CoolSiC™ M1 SiC Trench Power Device," IMW65R107M1H datasheet, Dec. 2019.

[13] Infineon, "650V CoolMOS™ CE Power Device," IPA65R400CE datasheet, April. 2017.

[14] GaN Systems, "SPICE model for GaN HEMT usage guidelines and example," Application note 006, Rev. 9, Aug.2016.

[15] X. Geng, C. Kuring, O. Hilt, M. Wolf, J. Wuerfl and S. Dieckerhoff, "Design and Optimization of the Driver Circuit for Non-Insulating Gate GaN-Transistors Enabling Fast Switching and High-Frequency Operation," CIPS 2022, Berlin, Germany, 2022, pp. 1-6.

[16] T. Quarles et al., SPICE interactive user guide. [Online]. Available: http://bwrcs.eecs.berkeley.edu/Classes/IcBook/SPICE/UserGuide/elements_fr.html

[17] X. Jiang, et al. "Reverse gate leakage mechanism of AlGaN/GaN HEMTs with Au-free gate." Chinese Physics B 32.3 (2023):

[18] D, Yan, et al. "On the reverse gate leakage current of AlGaN/GaN high electron mobility transistors." Applied Physics Letters 97.15 (2010).

[19] H. Mojaver and P. Valizadeh, "Reverse Gate-Current of AlGaN/GaN HFETs: Evidence of Leakage at Mesa Sidewalls," in IEEE Transactions on Electron Devices, vol. 63, no. 4, pp. 1444-1449, April 2016.

[20] A. Stockman et al., "On the origin of the leakage current in p-gate AlGaN/GaN HEMTs," 2018 IEEE International Reliability Physics Symposium (IRPS), Burlingame, CA, USA, 2018, pp. 4B.5-1-4B.5-4.

[21] Y. Jia, Z. Wen, Y. Chen, C. -C. Xie, Y. -X. Guo and Y. Xu, "A Threshold Voltage Model for Charge Trapping Effect of AlGaN/GaN HEMTs," in IEEE Access, vol. 7, pp. 120638-120647, 2019.

[22] M. Nuo et al., "Gate/Drain Coupled Barrier Lowering Effect and Negative Threshold Voltage Shift in Schottky-Type p-GaN Gate HEMT," in IEEE Transactions on Electron Devices, vol. 69, no. 7, pp. 3630-3635, July 2022.

Innovations in GaN Four Quadrant Switch technology

Geetak Gupta, Carl Neufeld, Davide Bisi, Yulu Huang, Bill Cruse, Peter Smith, Rakesh Lal, Umesh Mishra

Transphorm Inc., Goleta, CA, USA, ggupta@transphormusa.com

Abstract— A Four Quadrant Switch (FQS) provides bidirectional current carrying and bidirectional voltage blocking capability. Integrated FQS can enable significant size, complexity, and cost savings compared to traditional implementations that use multiple components resulting in limited performance and high cost. We demonstrate GaN based integrated FQS technology. The lateral GaN HEMT technology combined with a common drain configuration allows us to share the high-voltage region and results in a 40% reduction in the die size compared to two discrete GaN switches. The integrated FQS is a 60mOhm switch assembled in a TO247 package with a floating tab. We show excellent bidirectional current conduction and voltage blocking with symmetric current-voltage and capacitance-voltage behavior. The $R_{ON}.Q_g$ is 80% lower and $R_{ON}.Q_{rr}$ is 30% lower than state-of-art SiC MOSFETs resulting in 60% lower switching losses.

Keywords—GaN, power, bidirectional

I. INTRODUCTION

GaN power switches have enabled significant improvements in efficiency and system cost for power conversion applications [1][2]. 650V GaN-on-Si switches have achieved automotive qualification [3] and significant market acceptance in adapters, computing and industrial power supplies [4] due to their superior performance compared to Si or SiC [5] at lower cost. 900V GaN-on-Si switches have also been released [6]. 1200V GaN is under development targeting next-generation EV and higher power industrial applications in the near future [7].

GaN Four Quadrant Switch (FQS) is an innovative technology with potential to disrupt the future power conversion market. The GaN FQS is capable of bidirectional current conduction and bidirectional voltage blocking (Figure 1). This capability enables many converter topologies (e.g., current source inverter, matrix converter, and cyclo-converter) which are used in applications ranging from PV inverters to motor drives to on-board chargers [8,9,10]. However, traditional implementations of such a bidirectional switch require use of back-to-back MOSFETs (2 components) or IGBTs with reverse blocking diodes (4 components) [8]. This makes the implementation very complex with limited performance benefit. Integrated SiC bidirectional devices have also been demonstrated recently but also have very limited benefit over a traditional implementation using back-to-back MOSFETs since the blocking region is not shared for the integrated bidirectional device [11].

The lateral nature of the GaN HEMT allows us to design a bidirectional switch where the high-voltage blocking region is shared thus reducing the $R_{ON,sp}$. The high-mobility of the GaN HEMT is key to realizing low stored charges for fast switching, resulting in compact, low-cost, high performance converters.

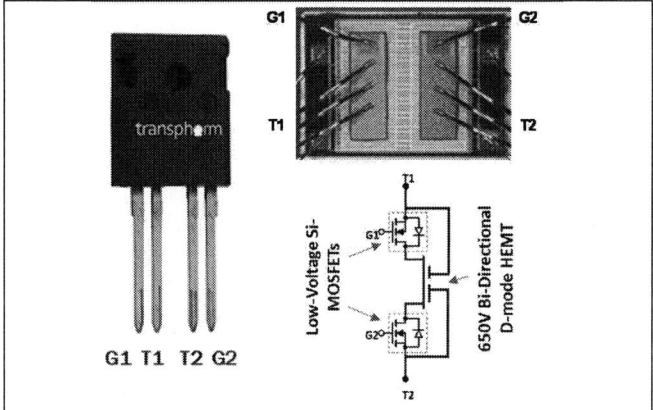

Fig. 1. Integrated GaN FQS in TO247 package (left), microscope image of die (top right), and cascode circuit schematic (bottom right).

The GaN FQS technology is based on Transphorm's qualified and commercialized GaN-on-Si platform. The substrate potential is floating which allows for high-voltage blocking in both directions without the need for complex substrate biasing schemes which are required by p-GaN-based bidirectional devices [12].

II. DESIGN AND CHARACTERIZATION

A. Device Design

The GaN FQS technology utilizes a cascode architecture with a high-voltage GaN HEMT die for voltage blocking and low-voltage Si MOSFETs for normally-off operation.

The high-voltage GaN die consists of a single die with two back-to-back normally-on GaN HEMTs connected in a common drain configuration and sharing the high-voltage blocking region (Figure 2). The design of the high-voltage GaN HEMT is based

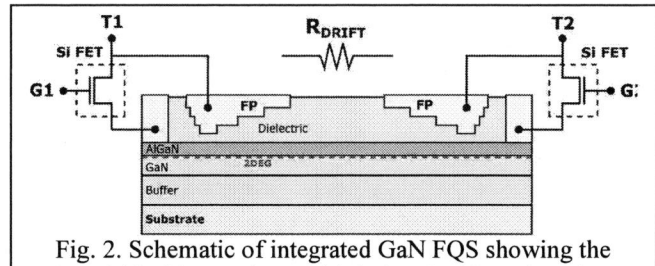

Fig. 2. Schematic of integrated GaN FQS showing the shared blocking region which helps significantly reduce $R_{ON,sp}$.

979-8-3503-3714-3/23 $31.00 © 2023 IEEE

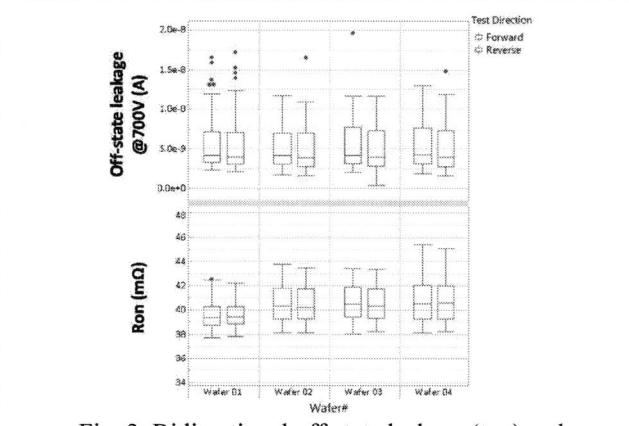

Fig. 3. Bidirectional off-state leakage (top) and bidirectional Ron (bottom) of GaN HEMT die.

on our existing GaN-on-Si technology platform and does not require any special process modules. As mentioned before, the substrate potential is floating, resulting in bidirectional voltage blocking capability without the need to externally control the substrate potential.

Two low-voltage Si MOSFETs are used to achieve normally-off behavior (Figure 2). The Si MOSFETs have maximum gate voltage rating of +/- 20V and Vth ~ 4V resulting in a robust input interface. The Si MOSFETs are selected for easy drivability with standard gate drivers. The two Si MOSFETs are integrated with the GaN HEMT die inside the package using a very low inductance connection. The package

used here is TO-247 which has high power handling capability and good heat dissipation. The thermal tab of the package is electrically isolated from the package terminals. The FQS device presented here and $R_{DS,on}$ and V_{DS} rating of 60mΩ and 650V respectively.

B. Characterization

The GaN HEMT die on-wafer bi-directional leakage and $R_{DS,on}$ are shown in Figure 3. The on-wafer data shows that the measured off-state leakage and $R_{DS,on}$ are same in both directions therefore validating the bidirectional capability. The output characteristics of the fully packaged device also show excellent symmetric behavior (Figure 4). The leakage of the packaged device is < 5µA up to 1500V in both directions (Figure 4). This shows that the device has adequate voltage margin for a 650V blocking voltage rating.

III. SWITCHING PERFORMANCE AND BENCHMARKING

A. Switching in Half-bridge

To validate switching behavior, a half-bridge hard-switched topology is used. Both high-side and low-side devices are GaN FQS switches. In total there are four gates (two for high-side FQS switch and two for the low-side FQS switch) that are driven independently. For this test two of the gates are always kept in ON state. The other two gates are switched in a similar way as traditional uni-directional devices. The resulting switching waveforms are well behaved at 400V and 16A (Figure 5). The switching losses extracted from measured waveforms are also well behaved and low for the GaN FQS (Figure 6).

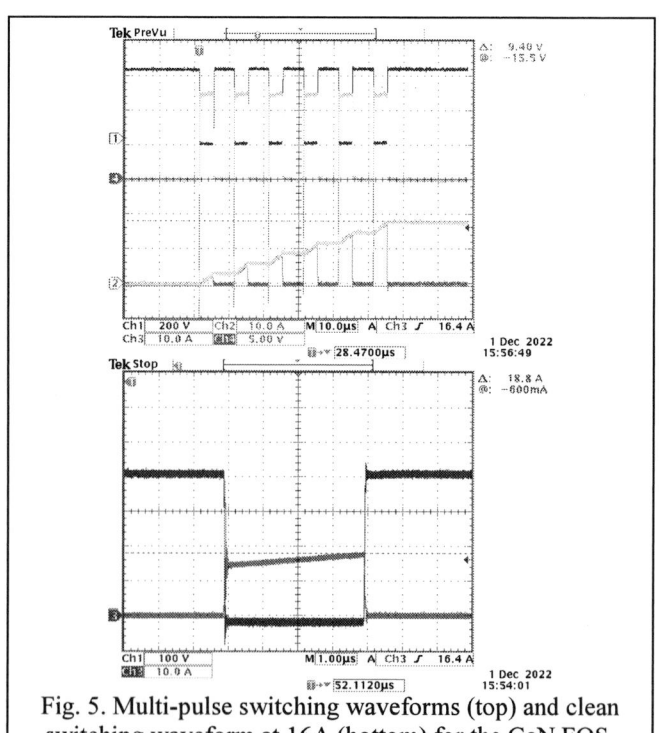

Fig. 5. Multi-pulse switching waveforms (top) and clean switching waveform at 16A (bottom) for the GaN FQS. Only one side of the FQS is switched for this measurement.

Fig. 4. Output characteristics showing symmetric bidirectional current conduction (top) and symmetric bidirectional leakage upto 1500V (bottom).

979-8-3503-3714-3/23 $31.00 © 2023 IEEE

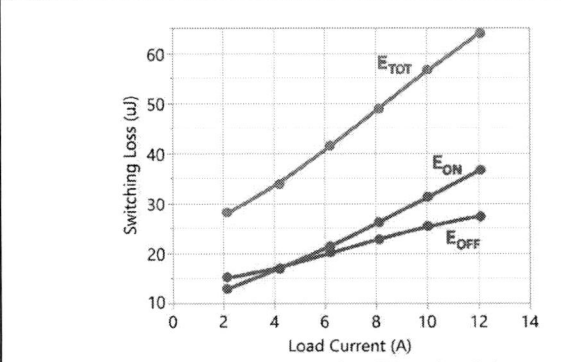

Fig. 6. Switching losses (Eon/Eoff/Etot) of the GaN FQS at 400V. Only one side of the FQS is switched for this measurement.

B. Performance Benchmarking

Next, we benchmark the key figures-of-merit: $R_{ON}*Q_g$, $R_{ON}*Q_{rr}$, and switching loss (E_{tot}) vs SiC MOSFETs and SiC bidirectional device (Figure 7). GaN FQS has excellent $R_{ON}*Q_g$ that is 80% lower than SiC MOSFETs. The cascode architecture is key to achieving low Q_g and translates to fast switching speed of the GaN FQS. The $R_{ON}*Q_{rr}$ is 30% lower than SiC MOSFETs. Cascode GaN devices only have a low-voltage body-diode with negligible Q_{rr}. The Q_{oss} of GaN HEMT based power devices is also low due to the high mobility of the 2DEG. As a result, the overall Q_{rr} (including Q_{oss}) is significantly smaller for the GaN device than the SiC MOSFETs.

The switching losses of GaN FQS are 40% lower than SiC MOSFETs as shown in Figure 7. The advantage of GaN FQS technology comes from three key aspects, high-mobility 2DEG, shared blocking region, and cascode architecture. These fundamental advantages result in very low stored charges and hence low switching loss. In addition, the cost of GaN FQS is expected to be lower than SiC due to low cost of the GaN-on-Si platform, lower part count, and the improvement in die size that is achieved because of blocking region sharing.

IV. BIDIRECTIONAL SWITCHING

The final validation for GaN FQS technology is true bi-directional switching. The GaN FQS has a total of eight states of operation categorized into four categories as listed in Figure 8. For typical implementation in half-bridge topologies, care needs to be taken to ensure that the switch is not in a fully blocking state (both gates are off) when there is a load current

	SiC MOSFET	SiC bidirectional	GaN FQS	GaN advantage
Part Count	2x	1x	1x	Integrated GaN FQS
Blocking voltage	650V	1200V	650V	650V FQS today, 1200V achievable
Ron	50 mΩ	50 mΩ	60 mΩ	
Ron*Qg (single side)	5.4 Ω.nC		1.1 Ω.nC	High mobility of GaN HEMT → Low stored charge
Ron*Qrr (single side, 400V)	15.4 Ω.nC		10.8 Ω.nC	
Etot (400V, 12A)	170 µJ	440 µJ	64 µJ	GaN FQS outperforms SiC

Fig. 7. GaN FQS comparison with state-of-art SiC integrated bidirectional switch.

State			Conditions (current, Ron, etc.) [Junction maintained at 25 °C]	Comments
Gate 1	Gate 2	V₁₂₁		
OFF	OFF	V₁₂₁>0	I = I₂ ≈ –I₁ < 5 µA for 0 V < V₁₂₁ < 520 V	Switch is OFF; blocks in both directions
OFF	OFF	V₁₂₁<0	–I = –I₂ ≈ I₁ < 5 µA for –520 V < V₁₂₁ < 0 V	
ON	OFF	V₁₂₁>0	I = I₂ ≈ –I₁ >0; V₁₂₁ ~ 0.07*I + V₍f₎(I) for 0 A < I < 15 A	Switch conducts in one direction (V₁₂₁>0) with the body-diode drop (V₍f₎(I)) of the OFF MOSFET and blocks in the other direction
ON	OFF	V₁₂₁<0	–I = –I₂ ≈ I₁ < 5 µA for –520 V < V₁₂₁ < 0 V	
OFF	ON	V₁₂₁>0	I = I₂ ≈ –I₁ < 5 µA for 0 V < V₁₂₁ < 520 V	Switch blocks in one direction and conducts in the other direction (V₁₂₁<0) with the body-diode drop of the OFF MOSFET (V₍f₎(I))
OFF	ON	V₁₂₁<0	I = I₂ ≈ –I₁ <0; V₁₂₁ ~0.07*I – V₍f₎(I)) for –15 A < I < 0 A	
ON	ON	V₁₂₁>0	I = I₂ ≈ –I₁ > 0; Ron ~ 0.07Ω for 0 A < I < 15 A	Switch is fully ON in both directions
ON	ON	V₁₂₁<0	I = I₂ ≈ –I₁ < 0; Ron ~ 0.07Ω for –15 A < I < 0 A	

Fig. 8. Switching states of the GaN FQS including fully-ON, fully-OFF, and freewheeling modes.

that needs to be supported. Several strategies can be used to ensure a freewheeling path and will be described in future publications.

To validate bi-directional switching, we use an AC-AC converter circuit that can be used for voltage regulation applications (Figure 9). The circuit is very similar to a half-bridge converter where both the input and outputs are AC signals. Switching control strategy is implemented to ensure that the switches can cycle between states while always ensuring a freewheeling path for inductor current. An input of 180V rms (50Hz) is successfully converted to an output of 230V rms (Figure 9). The switching frequency used is 48kHz. The waveforms show true bidirectional switching of the GaN FQS.

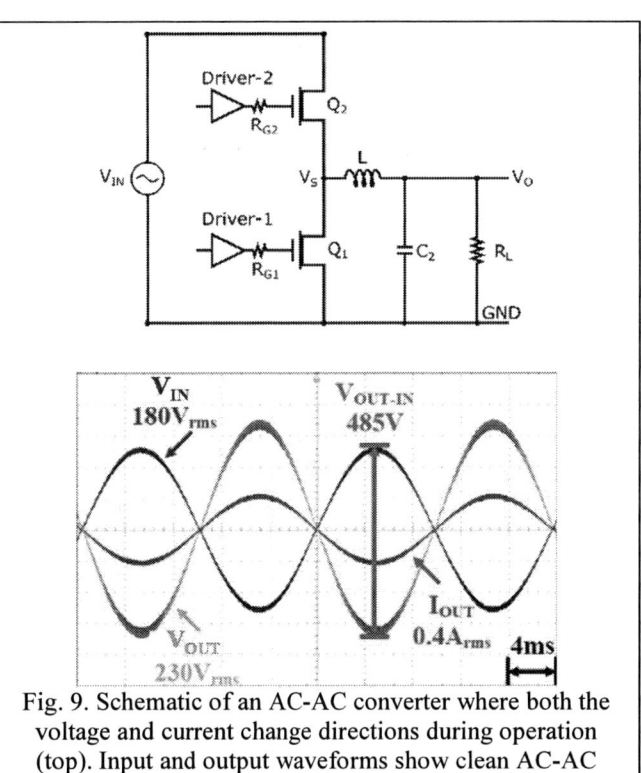

Fig. 9. Schematic of an AC-AC converter where both the voltage and current change directions during operation (top). Input and output waveforms show clean AC-AC conversion (bottom) with line frequency of 50Hz and switching frequency of 48kHz.

V. CONCLUSION

We have demonstrated GaN FQS, a bidirectional GaN switch ideally suited for applications ranging from PV inverters to motor drives and on-board chargers. The GaN FQS is an integrated switch with shared high-voltage region enabling small die size, low cost, and high performance. The switch also has excellent figures of merit including low Q_g, low Q_{rr}, and very low switching losses compared to state-of-the-art SiC. Bidirectional switching capability is also demonstrated using an AC-AC converter. We believe these results show that GaN FQS is a potentially disruptive technology for a wide range of power conversion applications.

ACKNOWLEDGEMENT

We gratefully acknowledge the support of ARPA-E (Program Manager: Dr. Isik Kizilyalli) in funding part of the work demonstrated here.

REFERENCES

[1] Y.F. Wu, J Gritters, L Shen, RP Smith, J McKay, R Barr, R Birkhahn, "Performance and robustness of first generation 600-V GaN-on-Si power transistors", The 1st IEEE Workshop on Wide Bandgap Power Devices and Applications, 6-10, Oct, 2013.

[2] P. Parikh, Y. F. Wu, L. K. Shen, "Commercialization of High 600V GaN-on-Silicon Power Devices", Materials Science Forum, Vols. 778-780, pp. 1174-1179, October 2014.

[3] https://www.transphormusa.com/en/news/supergan_geniv_aecq101/

[4] P. Parikh et al., "650 Volt GaN commercialization reaches automotive standards", ECS Trans., 80, 17, 2017.

[5] https://www.transphormusa.com/en/gan-technology/

[6] https://www.transphormusa.com/en/news/transphorm_900v_geniii_gan_fet/

[7] G. Gupta et al., "1200 V GaN switches on sapphire substrate," 34th IEEE ISPSD, Vancouver, BC, Canada, May 2022, pp. 349–352.

[8] J. Huber and J. W. Kolar, "Monolithic Bidirectional Power Transistors," in IEEE Power Electronics Magazine, vol. 10, no. 1, pp. 28-38, March 2023

[9] R. Amorim Torres, H. Dai, W. Lee, B. Sarlioglu and T. Jahns, "Current-Source Inverter Integrated Motor Drives Using Dual-Gate Four-Quadrant Wide-Bandgap Power Switches," in IEEE Transactions on Industry Applications, vol. 57, no. 5, pp. 5183-5198, Sept.-Oct. 2021.

[10] U. Raheja et al., "Applications and characterization of four quadrant GaN switch," in Proc. IEEE Energy Convers. Congr. Expo. (ECCE), Cincinnati, OH, USA, Oct. 2017, pp. 1967–1975.

[11] K. Han et al., "Monolithic 4-terminal 1.2 kV/20 A 4H-SiC bi-directional field effect transistor (BiDFET) with integrated JBS diodes," 32nd IEEE ISPSD, Vienna, Austria, Sep. 2020, pp. 242–245.

[12] T. Morita et al., "650 V 3.1 mΩcm2 GaN-based monolithic bidirectional switch using normally-off gate injection transistor," 2007 IEEE IEDM, Washington, DC, USA, 2007, pp. 865-868

Cost-Effective Test Setup for Measuring Threshold Voltage Shift of GaN-HEMTs under Long-Term Drain-Voltage Stress

Daniel Breidenstein*, Benedikt Kohlhepp† Thomas Dürbaum‡

Electromagnetic Fields, Friedrich-Alexander-Universität Erlangen-Nürnberg

Erlangen, Germany

Email: *daniel.breidenstein@fau.de, †benedikt.kohlhepp@fau.de, ‡thomas.duerbaum@fau.de

Abstract—**GaN-HEMTs offer low on-resistances and low capacitances, making them attractive candidates for high efficient power electronic converters. In this field of application, the normally-off transistor is the preferred structure but can exhibit threshold voltage instability problems. Undesired threshold-shifts can lead to increased switching losses or unintended turn-on events. Typically, the dedicated and expensive semiconductor-device characterization equipment is unavailable in application-oriented power electronics labs. Therefore, this paper proposes a cost-effective test setup based on standard laboratory equipment to measure the threshold voltage shift due to static long-term drain voltage stress. The test procedure alternates between stress periods and short measurement pulses, during which the transfer characteristic is recorded.**

Index Terms—**GaN-HEMTs, static stress, drain voltage stress, threshold voltage instability**

I. INTRODUCTION

The widespread use of wide-bandgap (WBG) devices in power applications requires reliable and robust WBG semiconductors. Therefore, an accurate characterization is necessary to meet the requirements for reliable and high efficient converters. For Gallium-Nitride High-Electron-Mobility-Transistors (GaN-HEMTs), a dynamic on-state resistance due to charge trapping [1, 2] as well as an instability in the threshold voltage [3, 4] are widely known in literature. Thus, characterizing those effects is necessary to use such components in applications.

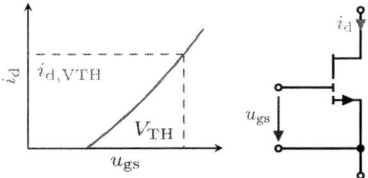

Fig. 1. Transfer characteristic

The threshold voltage V_{TH} delimits the gate voltage below which the device is assumed to be off. Before saturating, the drain current i_{d} of a semiconductor device has a voltage-controlled current source characteristic, where i_{d} is given by the gate voltage u_{gs} according to the transfer characteristic, exemplified in Fig. 1. V_{TH} is the corresponding gate voltage

for a small arbitrary current limit $i_{\mathrm{d,VTH}}$, typically in the range of micro- or milliamperes.

Datasheets lack information on threshold voltage shift, and papers discuss only selected devices, typically not the ones intended for the application at hand. Both negative and positive shifts are possible, depending on factors like stress type, time, and temperature [3], making own measurements for the intended device inevitable. Furthermore, a decrease in threshold voltage enlarges the risk of unintended miller-induced turn-on and increased turn-off and reverse conduction losses. An increased V_{TH} can lead to larger on-resistance and increased turn-on losses. Most studies in literature use costly equipment to evaluate the V_{TH}-shift like an Agilent B1505A power device analyzer [4–6]. However, the power electronic circuit designer requires an appropriate test setup to be able to study V_{TH}-shifts using conventional equipment already available.

For drain-source stress in [7], a similarity between the $i_{\mathrm{d}}\left(u_{\mathrm{gs}}\right)$ transfer characteristic and the $i_{\mathrm{d}}\left(u_{\mathrm{ds}}\right)$ reverse conduction characteristic is used to study V_{TH}-shift of short-term and long-term static stress, recovery after stressing as well as different conditions like temperature. In [8, 9] the output of a current source flows through the channel of a GaN-HEMT and an external gate capacitance, which is in parallel, so it charges the gate capacitance until the whole current is taken over by the channel. The resulting gate voltage corresponds to the threshold voltage for the injected current. The authors of [10] provide a similar approach where a capacitor in series with the DUT is charged by it's drain-current, reducing the gate-source voltage until the threshold limit is reached. All those approaches have in common that the DUT is either part of a modified half-bridge that can apply or remove drain-voltage stress, or an external half-bridge is used.

For gate-source stress in [11–13], the transfer characteristic is measured transiently with a shunt resistor.

In this paper, the ohmic load approach is modified to record the transfer characteristic after long-term drain-voltage stress. Section II discusses the measurement setup and measuring procedure in detail, along with considerations necessary for implementation. The setup is verified in Section III by characterizing a traditional Silicon MOSFET. In Section IV, measurement results for a commercially available $200\,\mathrm{V}$ GaN-HEMT are presented.

II. MEASUREMENT SETUP

To acquire the shift of the threshold voltage, long drain-stress phases have to be interrupted by short measurement intervals. To implement this, Fig. 2 shows the proposed setup's basic structure, which is discussed in this section.

The setup can be divided into two functional parts: one for applying and removing the drain-voltage stress and another for recording the transfer characteristic during the measurement phases. Drain-stress is applied with a half-bridge and a voltage source that can stress the DUT by charging the stabilization capacitor C_{in} to the stress voltage U_{stress} through R_{dis} and discharge it by outputting $0\,V$ before recording the transfer characteristic. Due to the comparable long stress periods, widely used bootstrapping is not sufficient for driver supply. Thus, an additional supply must be used for the high-side switch. To keep the delay from stressing to measuring small, a small time constant of C_{in} and R_{dis} is necessary.

During a measurement phase, D_1 clamps u_{Cin} to the measurement voltage U_c. Using a separate voltage source for this purpose makes the recording process independent of the stress voltage, allowing the study of different stress levels. A third high impedance state of the half-bridge is optional in this phase to avoid a constant current through R_{dis}. After discharging C_{in}, a conventional gate driver suitable for GaN-HEMTs turns on the DUT, whereby the switching process is slowed down by the RC-filter consisting of R_{g1} and C_g. C_g is not the input capacitance of the DUT, but an additional capacitor parallel to the input capacitance. Proper selection of R_{g1} and C_g results in a time constant in the microsecond range, which allows accurate acquisition by an oscilloscope. The waveforms of the gate-source voltage u_{gs} and the drain current i_d result in the transfer characteristic when plotted against each other. The drain current is measured by the shunt resistor R_{shunt}, which is placed at the source connection of the DUT. This provides a common reference point M (ground) for the oscilloscope while limiting the maximum occurring

voltage to U_c to utilize the full vertical resolution of the oscilloscope. The shunt resistor should be selected in such a way that the maximum drain-current $i_{d,max}$, given by (1), is larger than $i_{d,VTH}$. However, it should also not be larger than a few milliamperes to accurately extract V_{TH} with the limited vertical resolution of the oscilloscope.

$$i_{d,max} = \frac{U_c}{R_{shunt}} \tag{1}$$

When $i_{d,max}$ is reached in the measurement phase, a further increase of the gate voltage doesn't provide additional information and is therefore not helpful because it leads to unintended gate-stress, which is not desired during drain-voltage stress characterization. So the driver output can be turned off far before u_{gs} comes close to the driver voltage U_{dr}. For the same reason of unwanted gate-stress, a fast path to discharge the gate is provided by D_2 and $R_{g2} = 0\,\Omega$. This also allows to reapply drain-stress quickly. As an alternative, a second measurement of the transfer characteristic during turn-off is possible with $R_{g2} > 0\,\Omega$, which can lead to a shift between both measurements, like seen in [13, 14] due to short-term effects. Because the driving voltage difference between the gate driver and u_{gs} is lower than during turn-on, only R_{g1} would lead to a significantly longer switching process, so a parallel path with R_{g2} is needed.

A. Measuring procedure

Fig. 3 shows the relevant waveforms during the experiment. An initial measurement pulse is applied, starting at t_0, before which no drain-stress was present, which gives the V_{TH} without drain stress. Therefore the driver is turned on, resulting in an exponential rise of u_{gs} during which the rising edge transfer characteristic is recorded. After $i_{d,max}$ is reached, the driver is turned off at t_1, and the falling edge transfer characteristic can optionally be recorded. When u_{gs} is safely below the threshold voltage, drain stress is applied for an adjustable stress time between t_2 and t_3. After the stress

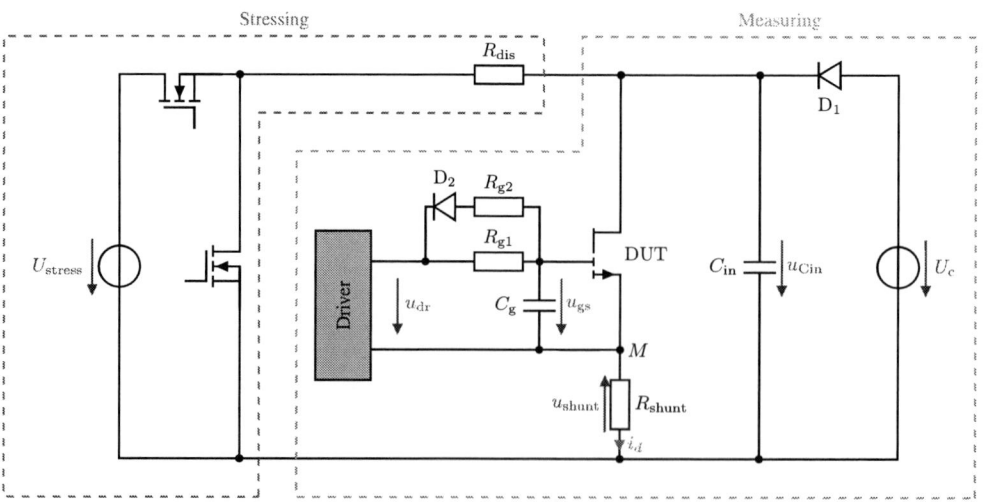

Fig. 2. Structure of measurement setup

period, the lowside switch of the half-bridge discharges C_{in} until the voltage is clamped by D_1 to U_c. After a short blanking time (t_3 until t_4), the measurement process can be repeated as described before.

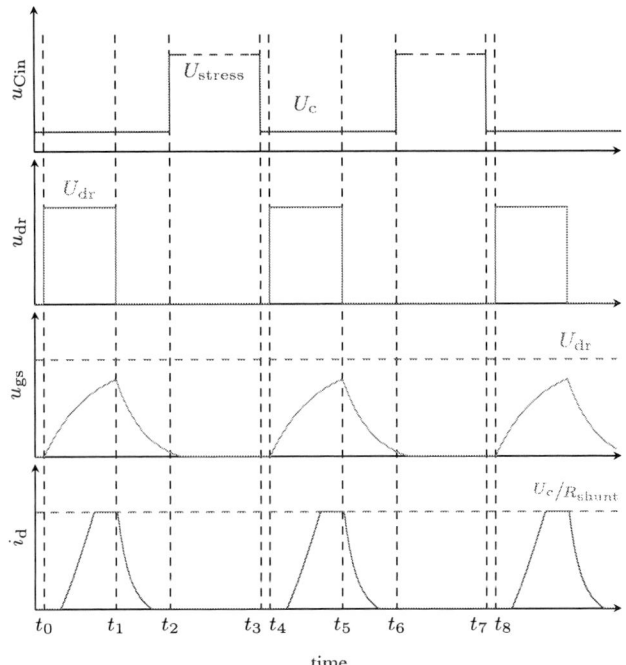

Fig. 3. Waveforms of the measurement process

B. Distortions due to capacitive charging currents

During the turn-on process of the DUT in the saturation region of the output characteristic, the channel current i_{ch} is given by the gate voltage corresponding to the transfer characteristic. As long as u_{ds} is constant, the channel current is equal to the externally measurable drain current i_{d}. However, during a switching transition, where u_{Cds} is changing over time, the effect of the drain-source capacitance C_{ds} has to be included, as shown in Fig. 4.

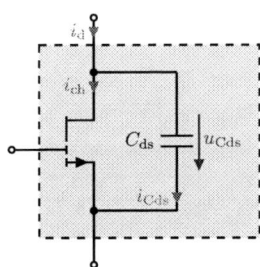

Fig. 4. Capacitive charging and discharging currents

The externally measurable drain current is composed of the actual channel current i_{ch} given by the transfer characteristic and the parasitic charging/discharging current i_{Cds} of C_{ds}. With (2) the actual channel current can be calculated from the external drain current and the drain-source voltage:

$$i_{\text{ch}} = i_{\text{d}} - i_{\text{Cds}} = i_{\text{d}} - C_{\text{ds}} \cdot \frac{du_{\text{Cds}}}{dt} \qquad (2)$$

To obtain the transfer characteristic directly from one turn-on process without distortion through capacitive currents, i_{Cds} must be approximated and subtracted, or i_{Cds} must be kept comparatively small, so the measurable drain current approximately equals i_{ch}. C_{ds} is fixed for a given device, so only a large dt, equivalent to a slow and thus long turn-on process, and a small du_{ds}, implementable with a small measurement voltage U_c, can prevent distortions. That's another reason why an extra voltage source is used during measurement, instead of reusing the already available stress voltage source for this purpose. The same applies to the turn-off process.

C. Parasitic capacities of voltage sources

The reference point M for measuring the gate voltage and shunt current is automatically connected to the protective earth through the oscilloscope when using single-ended probes or coaxial cables. The shunt resistor is placed below the DUT to take advantage of its full horizontal resolution, so the voltage sources U_{stress} and U_c are not referenced to M. Parasitic capacitances of those voltage sources to protective earth, and thus M, which are combined in an equivalent capacity C_{p} in Fig. 5, cause currents flowing through the shunt resistor, which are then visible at the scope.

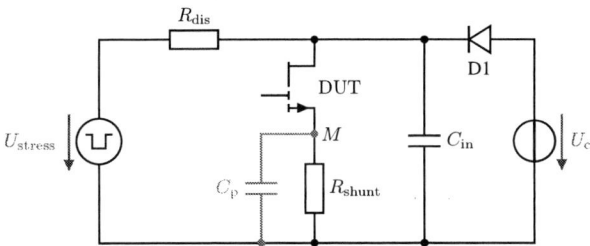

Fig. 5. Parasitic capacities to protective earth

During turn-on, a part of the DUT's source current has to charge C_{p} up from $0\,\text{V}$ to U_c. This portion of the shunt current is not causing a voltage drop over the shunt resistor, so the measured current is lower than the real source current. In the same way, C_{p} is slowly discharged through R_{shunt} during turn-off, resulting in an additional current when the GaN-HEMT is turned off. An even slower switching process would be a solution to reduce the impact, but in contrast to C_{ds}, the C_p can be reduced by e.g. the use of isolating transformers between both voltage sources and the power grid.

D. Capacitive-induced negative gate-voltage stress

To remove the drain-stress voltage before measurement (t_3), not only the input capacitor C_{in} has to be discharged from U_{stress} to U_c, but also the parasitic capacitances C_{ds} and C_{gd} of the DUT. The discharging current of C_{gd} flows as well through the parallel connection of R_{g1}, C_{g} and C_{gs} which is shown in Fig. 6. In case of a large resistance value of R_{g1}, the main path of the discharging current is through C_{g} and C_{gs}.

979-8-3503-3714-3/23 $31.00 © 2023 IEEE

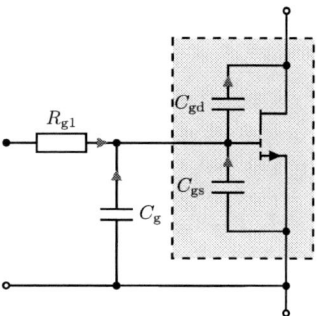

Fig. 6. Cause of negative gate-voltage while removing drain stress

The current induces a negative gate voltage. This is not only a kind of additional stress but also increases the necessary measuring time (t_4 to t_5) without positive influence. And thus, the time at which no stress is applied. To minimize the effect for a given time constant $\tau_{RC,g}$, a small resistor in combination with a large capacitance is recommended. At the rising edge of U_{stress} when reapplying stress (t_6) after measurement, the same effect increases the gate voltage. This could lead to an unintentional turn-on event (miller-induced turn-on) while stress voltage is present, destroying the shunt resistor or the DUT. The previous recommendation to use a large C_g in combination with a small R_g also helps to avoid this failure.

III. VERIFICATION OF THE MEASUREMENT SETUP

To verify the test setup, a conventional $100\,\text{V}$ Si-MOSFET (BSC070N10NS5), for which no threshold voltage shift is expected, is tested for reference.

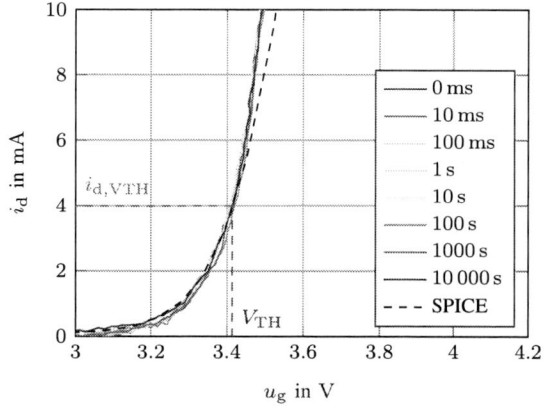

Fig. 7. Measured transfer characteristics of a Si-MOSFET (BSC070N10NS5) afer different drain-voltage stress periods for turn-on process with $U_{stress} = 35\,\text{V}$

Like in all the following measurement results, $i_{d,VTH} = 4\,\text{mA}$ is used to define the threshold voltage. Fig. 7 shows the recorded transfer characteristics after different stress times with $U_{stress} = 35\,\text{V}$. The denoted times shown in the legend refer to the starting point $t_0 = 0\,\text{ms}$ of the first measurement pulse. So for example t_4 is $10\,\text{ms}$, t_8 is $100\,\text{ms}$, and so forth.

Furthermore, the Spice model of the MOSFET, delivered by the manufacturer, is used to extract the transfer characteristic (dashed line in Fig. 8). It shows good correlation with those recorded by the test setup. As it can be seen in the extracted threshold voltages in Fig. 8, no significant shift of the threshold voltage over time can be determined within the measurement accuracy. The initial measurement at $t_0 = 0\,\text{ms}$ is shifted in Fig. 8 to $1\,\text{ms}$ so it appears in the logarithmic scale. Together with the good match with the simulation, the measurement setup is capable of recording transfer characteristics and therefore extract the corresponding V_{TH}.

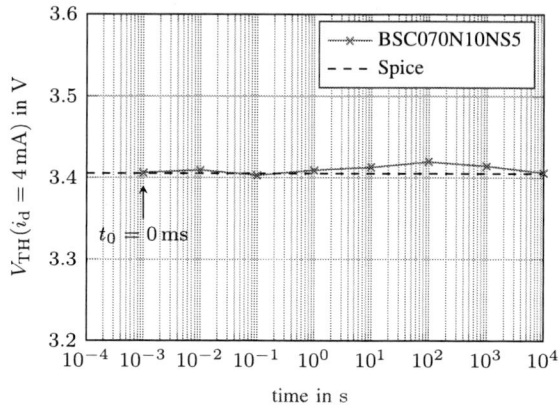

Fig. 8. Threshold voltage of conventional MOSFET after different stress times with $U_{stress} = 35\,\text{V}$

IV. MEASUREMENT RESULTS

For the device under test, a commercial available $200\,\text{V}$ enhancement-mode GaN (EPC2307) is used as an example. The parameters for the test setup (Fig. 2) are summarized in Tab. I. For the selected setup, the transfer characteristic is recorded during falling and rising edge of the gate voltage for each pulse.

TABLE I
CIRCUIT PARAMETERS OF MEASUREMENT SETUP

Parameter	Value
R_{g1}	$4.7\,\text{k}\Omega$
R_{g2}	$1\,\text{k}\Omega$
C_g	$10\,\text{nF}$
$\tau_{RC,g}$	$47\,\mu\text{s}$
R_{shunt}	$100\,\Omega$
R_{dis}	$5\,\Omega$
C_{in}	$100\,\text{nF}$
$\tau_{RC,dis}$	$500\,\text{ns}$
U_c	$2.4\,\text{V}$

Fig. 9 shows the recorded gate voltage and the shunt voltage for the second measurement pulse after the first stressing period, starting at $t_4 = 10\,\text{ms}$. The drain-stress voltage is $75\,\text{V}$. The device was resting without any stress for one week for regeneration from previous stress-tests to ensure a virgin device. Previous tests have shown, that long time revovery constants come along with drain-stress. For the selected ratio of R_g and the external C_g, a negative gate voltage before t_4 and a miller-induced increase of the gate voltage at t_6 are visible but uncritical. The charging current of C_{gd} and C_{ds}, when drain-stress is reapplied at t_6, is visible as a spike.

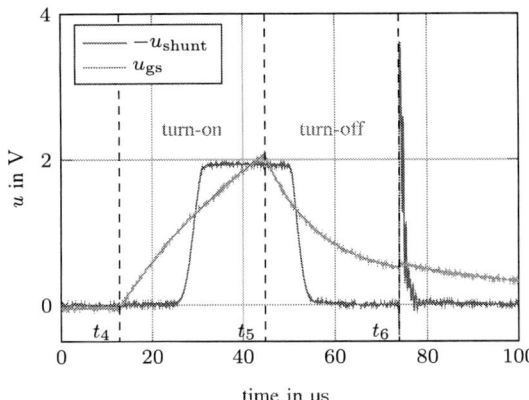

Fig. 9. Recorded waveforms during measurement phase for second pulse with $U_{stress} = 75\,V$

Based on the waveforms, the transfer characteristics for both edges of u_{go} are plotted in Fig. 10 for the fourth measurement pulse as an example. A clear shift is visible between rising and falling edge. Capacitive charging currents from the device's capacitances cannot be the cause, as those would lead to a shift in the opposite direction (During turn-on i_d is reduced by the discharging current of C_{ds}). Furthermore, these have a comparably small impact on the waveforms due to the slow turn-on process in combination with a small U_c.

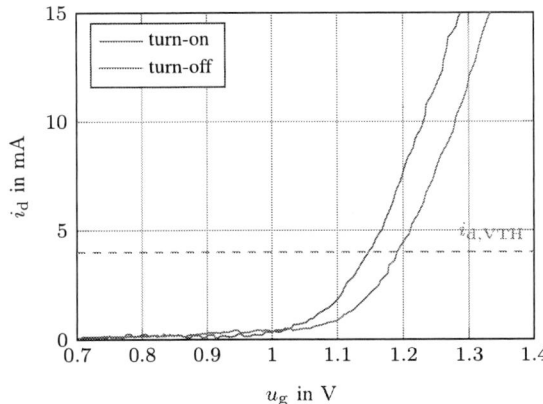

Fig. 10. Comparison of transfer characteristics recorded during turn-on and turn-off with $U_{stress} = 75\,V$ for the fourth measurement pulse

The shift during both edges indicates the presence of short-term effects like the superposed gate-stress during measurement or the regeneration of the drain-voltage stress. Due to the measuring method, such influences are not completely separable.

In Fig. 11 the recorded turn-on transfer characteristics are presented. After the first measurement pulse an increased V_{TH}

is visible for the second pulse. For all following measurements, the transfer characteristic shifts to the left, which is equivalent to a decreasing threshold voltage. Thus, threshold voltage becomes smaller than the one of the initial pulse after $100\,s$.

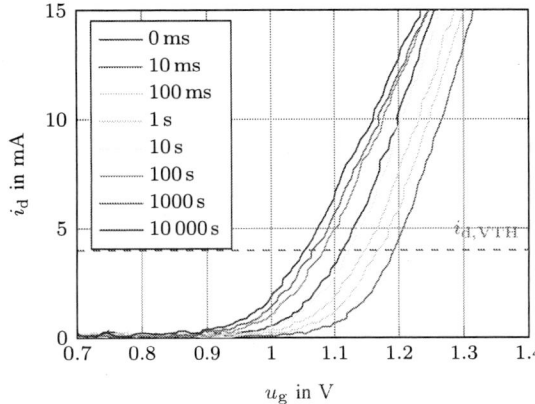

Fig. 11. Measured transfer characteristics of a $200\,V$ enhancement-mode GaN (EPC2307) afer different drain-voltage stress periods for turn-on process with $U_{stress} = 75\,V$

Finally, Fig. 12 shows the extracted threshold voltages for all recorded transfer-characteristics for both edges. For the turn-off transfer characteristics, V_{TH} increases during the first four measurements with a decrease afterward.

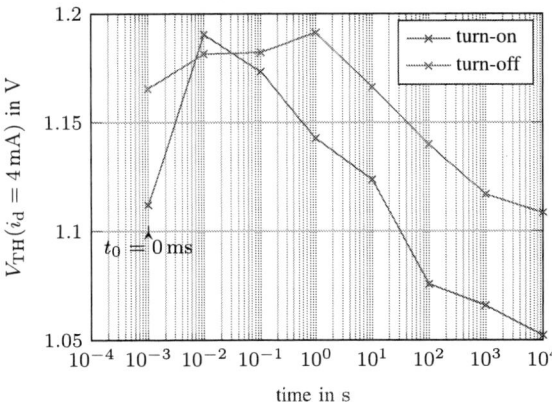

Fig. 12. Threshold voltage of f a $200\,V$ enhancement-mode GaN after different stress times with $U_{stress} = 75\,V$

V. CONCLUSION

In this work, a measurement setup for characterizing the threshold voltage shift of GaN-HEMTs due to long-term static drain voltage stress based on standard laboratory equipment is presented. Measurements were performed for a conventional MOSFET for verification and a $200\,V$ GaN-device. For the last, a shift in threshold voltage is present long-term as well as short-term due to other effects like gate-voltage stress, which is necessary for measurement and therefore, not completely separable. The resulting threshold voltage shift of the GaN device calls for further improvement of the devices and, thus,

979-8-3503-3714-3/23 $31.00 © 2023 IEEE

makes accurate characterization of the devices necessary. For real SMPS applications, where drain-voltage and gate-voltage stresses are alternately present, further research has to be done to characterize the threshold voltage shift.

REFERENCES

[1] B. Kohlhepp, C. Kuring, S. Peller, and D. Kubrich, "Measurement of Dynamic On-State Resistance of High-Voltage GaN-HEMTs under Real Application Conditions," in *2020 22nd European Conference on Power Electronics and Applications (EPE'20 ECCE Europe)*, Lyon, France: IEEE, Sep. 2020, pp. 1–10. DOI: 10.23919/EPE20ECCEEurope43536.2020.9215744.

[2] B. Kohlhepp, D. Kubrich, T. Durbaum, M. Tannhauser, and A. Hoffmann, "Measuring of Dynamic On-State Resistance of GaN-HEMTs in Half-Bridge Application under Hard and Soft Switching Operation," in *2019 IEEE Electrical Power and Energy Conference (EPEC)*, Montreal, QC, Canada: IEEE, Oct. 2019, pp. 1–6. DOI: 10.1109/EPEC47565.2019.9074792.

[3] M. Meneghini *et al.*, "GaN-based power devices: Physics, reliability, and perspectives," *Journal of Applied Physics*, vol. 130, no. 18, p. 181 101, Nov. 14, 2021. DOI: 10.1063/5.0061354.

[4] J. Chen *et al.*, "OFF-State Drain-Voltage-Stress-Induced V_{TH} Instability in Schottky-Type p-GaN Gate HEMTs," *IEEE Journal of Emerging and Selected Topics in Power Electronics*, vol. 9, no. 3, pp. 3686–3694, Jun. 2021. DOI: 10.1109/JESTPE.2020.3010408.

[5] L. Efthymiou, K. Murukesan, G. Longobardi, F. Udrea, A. Shibib, and K. Terrill, "Understanding the Threshold Voltage Instability During OFF-State Stress in p-GaN HEMTs," *IEEE Electron Device Letters*, vol. 40, no. 8, pp. 1253–1256, Aug. 2019. DOI: 10.1109/LED.2019.2925776.

[6] X. Li *et al.*, "Observation of Dynamic V_{TH} of p-GaN Gate HEMTs by Fast Sweeping Characterization," *IEEE Electron Device Letters*, vol. 41, no. 4, pp. 577–580, Apr. 2020. DOI: 10.1109/LED.2020.2972971.

[7] K. Zhong, H. Xu, S. Yang, Z. Zheng, J. Chen, and K. J. Chen, "A Bootstrap Voltage Clamping Circuit for Dynamic V_{TH} Characterization in Schottky-Type p-GaN Gate Power HEMT," in *2021 33rd International Symposium on Power Semiconductor Devices and ICs (ISPSD)*, Nagoya, Japan: IEEE, May 30, 2021, pp. 39–42. DOI: 10.23919/ISPSD50666.2021.9452232.

[8] F. Yang, C. Xu, and B. Akin, "Characterization of Threshold Voltage Instability Under Off-State Drain Stress and Its Impact on p-GaN HEMT Performance," *IEEE Journal of Emerging and Selected Topics in Power Electronics*, vol. 9, no. 4, pp. 4026–4035, Aug. 2021. DOI: 10.1109/JESTPE.2020.2970335.

[9] F. Yang, C. Xu, and B. Akin, "Impact of Threshold Voltage Instability on Static and Switching Performance of GaN Devices with p-GaN Gate," in *2019 IEEE Applied Power Electronics Conference and Exposition (APEC)*, Anaheim, CA, USA: IEEE, Mar. 2019, pp. 951–957. DOI: 10.1109/APEC.2019.8722163.

[10] R. Kumar, S. Samanta, and T.-L. Wu, "Threshold Voltage Instability Measurement Circuit for Power GaN HEMTs Devices," *IEEE Transactions on Power Electronics*, vol. 38, no. 6, pp. 6891–6896, Jun. 2023. DOI: 10.1109/TPEL.2023.3247569.

[11] S. Yang, Y. Lu, H. Wang, S. Liu, C. Liu, and K. J. Chen, "Dynamic Gate Stress-Induced Vth Shift and Its Impact on Dynamic R_on in GaN MIS-HEMTs," *IEEE Electron Device Letters*, vol. 37, no. 2, pp. 157–160, Feb. 2016. DOI: 10.1109/LED.2015.2505334.

[12] P. Lagger, A. Schiffmann, G. Pobegen, D. Pogany, and C. Ostermaier, "Very Fast Dynamics of Threshold Voltage Drifts in GaN-Based MIS-HEMTs," *IEEE Electron Device Letters*, vol. 34, no. 9, pp. 1112–1114, Sep. 2013. DOI: 10.1109/LED.2013.2272095.

[13] A. Kerber *et al.*, "Characterization of the Vt-instability in SiO2 HfO2 gate dielectrics," in *2003 IEEE International Reliability Physics Symposium Proceedings, 2003. 41st Annual.*, Dallas, TX, USA: IEEE, 2003, pp. 41–45. DOI: 10.1109/RELPHY.2003.1197718.

[14] T. Oeder and M. Pfost, "Gate-Induced Threshold Voltage Instabilities in p-Gate GaN HEMTs," *IEEE Transactions on Electron Devices*, vol. 68, no. 9, pp. 4322–4328, Sep. 2021. DOI: 10.1109/TED.2021.3098254.

Reduced GaN Half-Bridge IC Switching Loss on Biased Si p-n Junctions

Stefan Mönch, Richard Reiner, Michael Basler, Patrick Waltereit, Rüdiger Quay

Fraunhofer Institute for Applied Solid State Physics IAF, Tullastr. 72, 79108 Freiburg, Germany (stefan.moench@iaf.fraunhofer.de)

Abstract—Substrate-biasing effects in monolithic GaN half-bridge ICs are suppressed on Si substrates with p-n junction isolation and local buried substrate-to-source terminations, similar to GaN-on-SOI. However, the vertical depletion capacitances of the p-n junctions add to the switch-node capacitance (in contrast to discrete half-bridges) and slightly increase switching times, losses and energies. This work's approach is to intentionally bias the substrate towards negative voltages to reduce the effective vertical depletion capacitance. After C-V and I-V characterization of the vertical Si p-n junction in a commercial GaN half-bridge IC, experimentally a switching loss reduction of up to -6% at 1 MHz, 32 V switching is demonstrated by applying up to -40 V average substrate voltage via a LR-type bias tee. The work provides insight into p-n junction isolation for monolithic GaN power converter ICs.

Index Terms—Gallium nitride, DC-DC converters, substrates, bridge circuits, switching loss, power integrated circuits.

I. INTRODUCTION

Monolithic integration of low-voltage HEMTs as GaN-on-Si half-bridge [1], [2], [3], [4] improves the switching behavior and efficiency by reduced parasitic loop inductances and interconnection resistances. However, the common Si substrate can cause back-gated on-resistance increase, trap-related dynamic issues [5], [6] and capacitive coupling [7]. To avoid on-chip device to device coupling, additional isolation methods can be used, similar to as it was done in the past already for Si-based power ICs [8]: For Si-based substrates, additional vertical isolation is achieved either by silicon on insulator (SOI) [9], Si p-n junction (back surface) isolation [10] as investigated in this work (Fig. 1a) or the similar reverse p-n junction (EBUS) [11] isolation. These approaches eliminate substrate effects locally (if local substrate-to-source terminations of the power transistors are used). Other approaches such as floating Si substrates can also partly reduce substrate effects, increasing the fault-free operation voltage [12].

While a thin buried oxide under the half-bridge transistors in SOI adds a high constant additional capacitance to the switch node (increasing the switching energies and loss [13], the p-n junction can add significantly less capacitance by adjusting the doping concentration, because it defines the p-n junction depletion width and thus capacitance.

From the capacitance data in monolithic half-bridge GaN IC datasheets the existence of vertical p-n junction diodes can be deduced, and is also measured in this work. For symmetrical half-bridge devices the difference between the high-side and

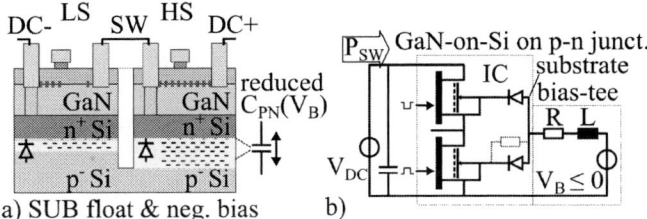

Fig. 1. a) Reduced p-n junction capacitance by increased negative substrate biasing. b) Bias-tee for experiments and observed substrate resistance (dotted).

low-side output capacitance can be assigned to the p-n junction depletion capacitance (further analyzed in [14]).

Commercially, monolithic low-voltage GaN-on-Si ICs which use (most probably) Si p-n junction isolation are available [15], [16] (from EPC Co.). Beside patents, however, only very little information on the p-n junction isolation for GaN power ICs is found in literature [11], [17], [14]. Even though these low-voltage monolithic half-bridges are widely used today it is surprising that the existence of the p-n junction isolation and its effects on switching behavior is rarely discussed or investiated - even though it is key to achieve the fault-free operation by isolation (functionally similar to SOI) for monolithic GaN power topologies (with integrated low and high-side devices) also at higher voltages),

This work investigates a commercial GaN half-bridge on Si substrate with p-n junction isolation. This work's idea is to reduce the lateral isolation capacitance of the Si p-n junction below lateral and monolithic half-bridge transistors (Fig. 1a) by intentionally increased negative substrate biasing, to increase the p-n junction's depletion width and thus to decrease the depletion capacitance.

II. EXPERIMENTAL RESULTS

A. Setup

A commercial 60 V GaN on Si p-n junction symmetrical half-bridge IC is analyzed in this work (EPC2102) in a dc-dc buck converter board without inductor/load (Fig. 2a, using the evaluation board EPC9038). To measure and adjust the substrate voltage, an additional pin was attached to the floating substrate (backside of the IC) by conductive ink after backside thinning and roughening (Fig. 2b). The purpose of the thinning and roughening is to expose the conductive silicon substrate,

979-8-3503-3714-3/23 $31.00 © 2023 IEEE

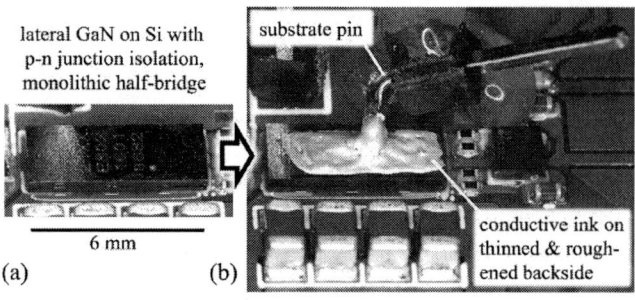

lateral GaN on Si with p-n junction isolation, monolithic half-bridge

substrate pin

conductive ink on thinned & rough-ened backside

6 mm

(a)   (b)

Fig. 2. a) Conventional GaN-on-Si half-bridge EPC2102. b) Attachment of a pin to the substrate by conductive ink after backside thinning and roughening.

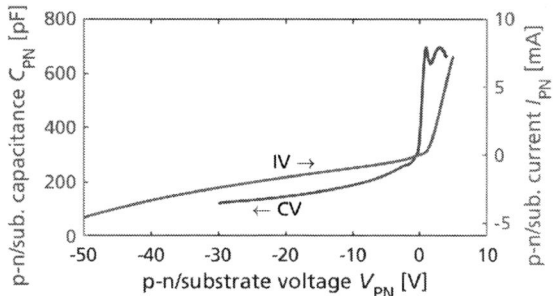

Fig. 3. Capacitance-voltage (CV) and current-voltage (IV) measurements of the Si p-n junction as a function of substrate voltage.

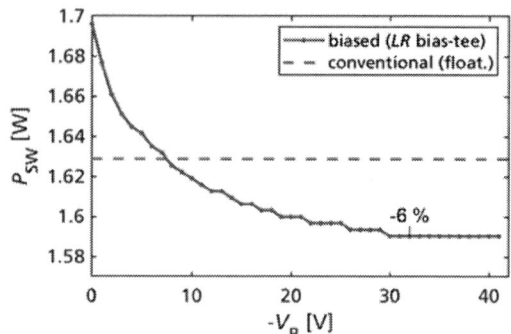

Fig. 4. Measured switching loss at 1 MHz, 32 V dc-link, no inductor or load, for conventional (dashed) and V_B biased substrate.

which visually seems to be otherwise covered by some kind of surface coating with unknown electrical properties.

B. IV and CV Measurement of p-n Junction

The vertical capacitance-voltage characteristic (CV) under the integrated low-side transistor is measured between SUB and DC-, shielding SW with ac-guard of an LCR meter at 100 kHz. The non-linear CV (Fig. 4) confirms a Si p-n junction, and can also be extracted from the datasheet by subtracting the high-side and low-side output capacitance data, which consists of the p-n junction capacitance [14]. The zero-bias capacitance of approx. 350 pF is significantly reduced to below 125 pF already at -30 V substrate bias. This work's later experiments aim to bias the substrate towards such negative voltages and to thereby reduce the switching energies (and thus hard-switching switching losses or resonant switching times). A conductive vertical substrate current path at reverse p-n voltage is observed from the IV data, approx. 12 kΩ at 40 V, Fig. 3. Probably, this resistance was added and designed by the manufacturer to discharge the otherwise floating substrate towards zero during switching, to ensure well-defined voltages on the otherwise only capacitively coupled Si substrate. In this work however, the vertical substrate resistance in addition to the p-n junction is not desired because it will add static conduction loss if the substrate is intentionally biased. It should be noted that in general it is possible to fabricate similar GaN ICs on Si p-n junction isolation substrates also without the observed substrate resistance. This would avoid the

substrate conduction losses present in this work and allow a even lower loss operation of monolithic half-bridges on biased p-n junction substrates.

The p-n junction capacitance $C_{PN}(V_{PN})$ is highest at low bias voltage, and decreases approximately proportional to $1/\sqrt{-V_{PN}}$ during reverse blocking operation $(-V_{PN} \gg 0)$. The non-linear depletion capacitance of a Si p-n junction (with highly asymmetrical doping) is well known from semiconductor device physics. Quantitative examples of depletion-widths, voltage-blocking capability and doping concentration for GaN-on-Si half-bridges are found in [18], [14].

forms a depletion capacitance with non-linear voltage dependent capacitance

C. p-n Junction Biasing for Reduced Switching Loss

This work proposes to bias the substrate towards negative voltages to reduce the depletion capacitances below the low-side and below the high-side transistors as visualized in Fig. 1a. While for a pure p-n substrate a highly-resistive bias-tee would be sufficient, in this work the presence of the observed integrated substrate resistance requires a more complex bias-tee. This work uses a an LR bias-tee (55 Ω, 12.5 mH, Fig. 1b) during switching experiments. The inductance L allows high-frequency capacitive coupling from the switch-node to the substrate while providing an average substrate voltage and the current required for the observed integrated substrate resistor (in parallel to the p-n junction).

The bias voltage V_B and the dc-dc converter input voltage V_{DC} are provided by two independent power supplies such that a power loss breakdown is possible where the switching loss P_{SW} in the power stage, provided by V_{DC}, can be measured separately from the substrate resistance (biasing) loss P_B, which is provided by V_B.

D. Hard Switching Loss Reduction

The half-bridge is operated at 32 V in synchronous mode with 50% duty-cycle and without load and without power inductor. Stepping the substrate bias V_B from zero to negative values, the hard-switching loss P_{SW} during 1 MHz hard-switching is measured. The measured switching loss shown in

979-8-3503-3714-3/23 $31.00 © 2023 IEEE

Fig. 4 (solid) is then compared to the conventional operation (without the bias tee), also shown in Fig. 4 (dashed). The measured switching loss is reduced up to 6% for below -30 V substrate bias voltage. This measurement experimentally verifies this work's idea that a biased p-n junction reduces the substrate capacitance and thereby the switching loss.

This switching loss data was just the power stage loss, neglecting the (avoidable, but in the particular GaN IC used in this work present) substrate resistance loss, which is further analyzed: The additional power loss of the biasing supply was measured and increases nearly quadratic with the biasing voltage. The measured power loss was fitted to a substrate resistance of around $7.2\,\mathrm{k\Omega}$, which is in the range of to the statically measured, but non-linearly voltage-dependent, value of around $12\,\mathrm{k\Omega}$ from the IV measurements. While the switching loss in the power stage reduces monotonic with the substrate bias and then reaches a minimum value, (see Fig. 4), the substrate resistance adds additional power loss, quadratic with the substrate bias. Consequently, a minimum of overall losses as a function of substrate biasing voltage exists. For the 1 MHz operation at 32 V, this optimum is at $V_{\mathrm{B}} = -8\,\mathrm{V}$, where, considering the sum of the additional substrate losses and saved switching losses, still a reduction of 3% overall losses was measured. This is half of the improvement which would be possible if the substrate resistance would not be resent in the GaN IC.

Further investigation of the measured switching loss (Fig. 4) shows that at zero-voltage biasing $V_{\mathrm{B}} = 0\,\mathrm{V}$ and increased loss compared to the conventional operation (without bias tee) was measured. This effect can be explained by the fact that the p-n junction is clamping the capacitive-coupled substrate voltage to around 0 V (GND/DC-). The avoidable substrate resistance is neglected for the following analysis: At very negative substrate voltages the substrate is only capacitively coupled to the dc-terminals (DC-/DC+) and switch-node (SW), and both p-n junctions (under the high and low-side transistors) are effectively series-connected, which has reduced effective capacitance compared to each of the two p-n junction capacitances. At substrate voltages close to zero however, depending on the direction of switch-nod voltage transition, one p-n junction starts to clamp the substrate voltage, such that only one of the two p-n junction capacitances is effective, which has a higher effective capacitance compared to the series connection with another p-n junction. If the RL bias-tee is connected to the substrate it discharges the substrate faster to zero (GND/DC-) than just the integrated substrate resistance. Therefore, at fast-switching half-bridge operation, the initial substrate voltage prior to a low-to-high switch node transition is already closer to zero (compared to operation without bias tee). Figuratively speaking, without bias tee the substrate is capacitive more loosely coupled to the switching nodes than operation with a bias tee around zero bias voltage. Measurements of the substrate voltage during continuous switching operation is carried out to provide more insight into this coupling mechanisms.

Fig. 5 shows the measured switching transients. All mea-surements have in common the 32 V, 1 MHz hard-switching switch-node operation. First, the substrate voltage is measured by connecting a high-impedance voltage probe to the substrate pin (without bias tee), shown as dashed line (conventional). Then, the substrate voltage is also measured for $V_{\mathrm{B}} = 0\,\mathrm{V}$ to -30 V biasing in steps of -10 V (with bias tee). Comparison of the conventional operation with the biased operation at $V_{\mathrm{B}} = 0\,\mathrm{V}$ shows different transient substrate voltages. The biased substrate is discharged quicker (during low switch-node state) to zero compared to the conventional operation. As a consequence, during high switch-node state, the p-n junction conducts earlier and clamps for a larger portion of the switch-node transition. This explains that the proposed solution with a bias tee at $V_{\mathrm{B}} = 0\,\mathrm{V}$ has higher effective substrate capacitance and thus loss than the conventional case.

E. Further Discussion

Even though this work experimentally demonstrated a up to 6% reduction of switching loss in the power stage, this im-provement seems still small, especially given that the measured p-n junction capacitance was more than halved (reduction by over 50%). The total switch-node capacitance which is relevant for hard-switching losses consists not only of the p-n junction capacitance, but also the other terminal capacitances of the lateral GaN HEMTs.

This work investigated low-voltage lateral GaN HEMTs. It is well known that higher voltage HEMTs require more chip area (for the same on-resistance). Therefore, the p-n junction (substrate) capacitance for low-voltage half-bridges is small compared to the conventional transistor capacitances. Even though high-voltage GaN on p-n junction half-bridges are not yet commercially available, at higher voltage the contribution of the substrate capacitance to overall device capacitance increases significantly.

In [14] a more detailed theoretical analysis of monolithic half-bridges with different voltage-ratings was carried out, cal-culating that for 30 V operation (as in this work) the substrate capacitances only contribute below 17% to the switch-node capacitance. Biasing of the p-n junction can only reduce the effective capacitance, but not eliminate it completely. Thus, the reduction of switching loss by substrate biasing is expected to be below the substrate capacitance ratio. This work's measured 6% improvement is well below 17%. Also in [14], for a hypothetical high-voltage half-bridge (600 V rated) it was calculated that at 480 V operation the p-n junction isolation already contributes 35% of the switch node capacitance. Therefore, at higher voltage operation the contribution of p-n junction capacitance is higher, and the biasing approach can then again cause a higher switching loss reduction.

It should be mentioned (see [14]) that due to the non-linear capacitance characteristic of p-n junctions, in contrast to a fixed capacitance of oxide isolation (as in SOI), a monolithic high-voltage half-bridge on p-n junction isolation can result in lower switching loss (and thus higher efficiency) compared to a high-voltage half-bridge on SOI substrate. Since neither of both have been demonstrated so far at high voltage, future

979-8-3503-3714-3/23 $31.00 © 2023 IEEE

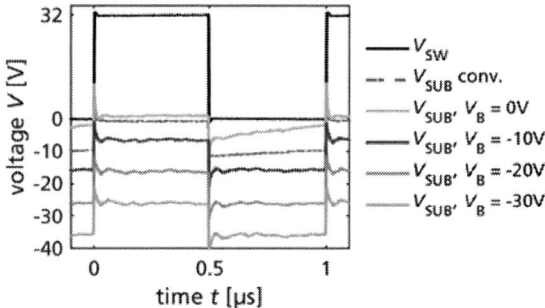

Fig. 5. Measured switch-node and substrate voltage during 32 V, 1 MHz hard-switching, for conventional and biased substrate.

work's should experimentally investigate the substrate capacitance contribution for different isolation methods (p-n junction vs. SOI) and the benefit of substrate biasing, and possible limitations not discussed by this work at higher voltages.

III. CONCLUSION

A p-n junction isolation solves substrate-biasing issues of monolithic GaN power converter topologies and enables efficient and fault-free hard-switching operation. Intentional negative biasing of the substrate allows the reduce the effect of the non-linear pn depletion capacitance, which allows a slight reduction of switching energies.

ACKNOWLEDGMENT

This work was partially supported by the German Federal Ministry of Education and Research (BMBF) through the project GreenICT@FMD (FKZ: 16ME0496); partially by the Fraunhofer Society in the Fraunhofer lighthouse project "ElKaWe - Electrocaloric Heat Pumps" (www.ElKaWe.org); and partially by the Federal Ministry for Economic Affairs and Climate Action (BMWK) in the project "GaN4EmoBiL" (grant ID 01MV23003A).

The author thanks Dirk Meder for modifying the GaN ICs.

REFERENCES

[1] W. Chen, K.-Y. Wong, and K. J. Chen, "Single-Chip Boost Converter Using Monolithically Integrated AlGaN/GaN Lateral Field-Effect Rectifier and Normally Off HEMT," *IEEE Electron Device Letters*, vol. 30, no. 5, pp. 430–432, 2009. [Online]. Available: https://doi.org/10.1109/LED.2009.2015897

[2] S. Ujita, Y. Kinoshita, H. Umeda, T. Morita, S. Tamura, M. Ishida, and T. Ueda, "A compact GaN-based DC-DC converter IC with high-speed gate drivers enabling high efficiencies," in *2014 IEEE Healthcare Innovation Conference (HIC)*. Piscataway, NJ: IEEE, 2014, pp. 51–54. [Online]. Available: https://doi.org/10.1109/ISPSD.2014.6855973

[3] R. Reiner, P. Waltereit, B. Weiss, M. Wespel, M. Mikulla, R. Quay, and O. Ambacher, "Monolithic GaN-on-Si Half-Bridge Circuit with Integrated Freewheeling Diodes: PCIM Europe 2016; International Exhibition and Conference for Power Electronics, Intelligent Motion, Renewable Energy and Energy Management," *PCIM Europe 2016; International Exhibition and Conference for Power Electronics, Intelligent Motion, Renewable Energy and Energy Management*, 2016.

[4] S. Moench, S. Müller, R. Reiner, P. Waltereit, H. Czap, M. Basler, J. Hückelheim, L. Kirste, I. Kallfass, R. Quay, and O. Ambacher, "Monolithic Integrated AlGaN/GaN Power Converter Topologies on High-Voltage AlN/GaN Superlattice Buffer," *physica status solidi (a)*, vol. 218, no. 3, p. 2000404, 2021. [Online]. Available: https://doi.org/10.1002/pssa.202000404

[5] B. Weiss, R. Reiner, P. Waltereit, R. Quay, and O. Ambacher, "Operation of PCB-Embedded, High-Voltage Multilevel-Converter GaN-IC," *2017 IEEE 5th Workshop on Wide Bandgap Power Devices and Applications (Wipda)*, pp. 398–403, 2017. [Online]. Available: https://doi.org/10.1109/WiPDA.2017.8170580

[6] C.-L. Tsai, Y.-H. Wang, M.-H. Kwan, P.-C. Chen, F.-W. Yao, S.-C. Liu, J.-L. Yu, C.-L. Yeh, R.-Y. Su, W. Wang, W.-C. Yang, K.-Y. Wong, Y.-S. Lin, M.-C. Lin, H.-Y. Wu, C.-M. Chen, C.-Y. Yu, C.-B. Wu, M.-H. Chang, J.-S. You, T.-M. Huang, S.-P. Wang, L. Y. Tsai, C.-H. Chern, H. C. Tuan, and A. Kalnitsky, "Smart GaN platform: Performance & challenges," *2017 IEEE International Electron Devices Meeting (IEDM)*, pp. 33.1.1–33.1.4, 2017. [Online]. Available: https://doi.org/10.1109/IEDM.2017.8268488

[7] S. Moench, R. Reiner, B. Weiss, P. Waltereit, R. Quay, O. Ambacher, and I. Kallfass, "Effect of Substrate Termination on Switching Loss and Switching Time using 600 V GaN-on-Si HEMTs with Integrated Gate Driver in Half-Bridges," *2017 IEEE 5th Workshop on Wide Bandgap Power Devices and Applications (Wipda)*, pp. 257–264, 2017. [Online]. Available: https://doi.org/10.1109/WiPDA.2017.8170557

[8] V. Rumennik, "Power devices are in the chips: New power integrated circuits, called PICs, put both power-handling semiconductors and logic on the same IC chip; soon they may be part of every household appliance," *IEEE Spectrum*, vol. 22, no. 7, pp. 42–51, 1985. [Online]. Available: https://doi.org/10.1109/MSPEC.1985.6370755

[9] X. Li, M. van Hove, M. Zhao, K. Geens, V.-P. Lempinen, J. Sormunen, G. Groeseneken, and S. Decoutere, "200 V Enhancement-Mode p-GaN HEMTs Fabricated on 200 mm GaN-on-SOI With Trench Isolation for Monolithic Integration," *IEEE Electron Device Letters*, vol. 38, no. 7, pp. 918–921, 2017. [Online]. Available: https://doi.org/10.1109/LED.2017.2703304

[10] A. Lidow, "Semiconductor devices with back surface isolation, US20120153300A1," Patent US20 120 153 300A1.

[11] J. Wei, M. Zhang, G. Lyu, and K. J. Chen, "Substrate Effects in GaN-on-Si Integrated Bridge Circuit and Proposal of Engineered Bulk Silicon Substrate for GaN Power ICs," *2020 IEEE Workshop on Wide Bandgap Power Devices and Applications in Asia (Wipda Asia)*, 2020. [Online]. Available: https://doi.org/10.1109/WiPDAAsia49671.2020.9360273

[12] B. Weiss, R. Reiner, P. Waltereit, R. Quay, O. Ambacher, A. Sepahvand, and D. Maksimovic, "Soft-switching 3 MHz converter based on monolithically integrated half-bridge GaN-chip," in *WiPDA 2016*. Piscataway, NJ: IEEE, 2016, pp. 215–219. [Online]. Available: https://doi.org/10.1109/WiPDA.2016.7799940

[13] C. Basceri, V. Odnoblyudov, O. Aktas, W. Wohlmuth, K. Geens, A. Vohra, B. Bakeroot, H. Hahn, D. Fahle, M. Heuken, and S. Decoutere, "Propelling the Power Electronics Revolution: 200 mm Diameter, 100 V to 1800 V and Beyond GaN-on-QST High Volume Device Manufacturing Platform," *CS MANTECH Conference*, pp. 223–226, 2022.

[14] S. Mönch, M. Basler, R. Reiner, F. Benkhelifa, P. Döring, M. Sinnwell, S. Müller, M. Mikulla, P. Waltereit, and R. Quay, "GaN Power Converter and High-Side IC Substrate Issues on Si, p-n Junction, or SOI," *e-Prime - Advances in Electrical Engineering, Electronics and Energy*, p. 100171, 2023. [Online]. Available: https://doi.org/10.1016/j.prime.2023.100171

[15] M. A. Briere, "Advanced power devices for many-core processor power supplies," in *2010 International Electron Devices Meeting*, 2010, pp. 13.6.1–13.6.4. [Online]. Available: https://doi.org/10.1109/IEDM.2010.5703357

[16] D. Reusch, "High frequency eGaN monolithic half bridge IC based 12 VIN to 1 VOUT point of load converter," in *WiPDA 2015*. Piscataway, NJ: IEEE, 2015, pp. 371–376. [Online]. Available: https://doi.org/10.1109/WiPDA.2015.7369282

[17] TechInsights, "The EPC 2152 – A Fully Integrated GaN Half-Bridge IC | TechInsights," 2022. [Online]. Available: https://www.techinsights.com/blog/epc-2152-fully-integrated-gan-half-bridge-ic

[18] G. Lyu, J. Wei, Y. H. Ng, Y. Cheng, S. Feng, and K. J. Chen, "Substrate and Trench Design for GaN-on-EBUS Power IC Platform Considering Output Capacitance and Isolation between High-side and Low-side Transistors," *2022 IEEE 34th International Symposium on Power Semiconductor Devices and Ics (Ispsd)*, pp. 185–188, 2022. [Online]. Available: https://doi.org/10.1109/ISPSD49238.2022.9813683

Accurate Prediction of Incomplete Zero-Voltage Switching Dynamics and Losses

Mike Zäch*, Szymon Beczkowski, Asger Bjørn Jørgensen, Stig Munk-Nielsen

AAU Energy
Aalborg University
Aalborg, Denmark
*mrz@energy.aau.dk

Abstract—The switching losses occurring between full soft-switching and hard-switching are described inadequately, and previous attempts of quantifying incomplete zero-voltage switching losses are topology specific and not applicable for any general half-bridge circuit. The ability to quantify these losses is crucial to optimize the converter efficiency across a wide operating range. This paper proposes a general method for accurately estimating the incomplete zero-voltage switching dynamics and losses in half-bridge converters, including the non-linear output capacitance of the semiconductors. The charge balance during the switching transition period is solved to determine the depth of incomplete zero-voltage switching, which can be used to predict the switching losses. The method is verified experimentally, and is shown to be able to predict the depth of incomplete zero-voltage switching accurately, which can be used to calculate the related losses.

I. INTRODUCTION

Wide band-gap (WBG) devices such as silicon carbide (SiC) MOSFETs are extending their application range to various high power industries, such as e.g. electric transportation, due to their favorable properties [1]–[4]. Electric drive trains implementing WBG devices are both lighter and less voluminous due to their superior isolation properties and higher switching frequencies. As e-mobility advances towards larger, high voltage batteries to cover heavy-duty vehicles allowing for goods transportation, the demand for high efficient charging stations covering a wide range of battery voltages increases [5]–[8]. Charging standards such as HPC-350 and upcoming megawatt charging system (MCS) cover a large variety of voltage and current operating points [6], [9]. In order to achieve high-efficiency with fast-switching WBG devices, zero voltage switching (ZVS) topologies are an attractive choice for the power converter [6], [10]–[12]. However, ZVS converters typically need to be operated in a application-specific ZVS area. This can cause limitations in the feasible operating area, which poses design challenges for the demands of wide operating range application [13]. Furthermore, the methods to determine the ZVS area are usually subject to topology-specific assumptions, and often do not account for the non-linearity of the semiconductors [13]–[15]. In [14], the output capacitance of the semiconductor is linearized, which leads to inaccurate results at the initial and final stage of the switching transition. Although Incomplete zero-voltage switching (iZVS) was considered, the assumption that the inductor current would

decrease linearly to zero during the dead time period required a variable dead time dependent on the inductor current. This is topology-specific and rarely the case in common converter topologies. In [15], the non-linearity of the output capacitance is considered, but the method presented is topology-specific for series resonant converters and does not consider iZVS.

This paper proposes a generally applicable method for determining the border between soft and hard-switching of any half-bridge by predicting the voltage remaining across the device under test (DUT) after the switching transition, including the non-linear output capacitance of the semiconductors, as well as the LC resonant interaction in the circuit. This can be used to quantifies the losses of operating a half-bridge with iZVS and increase the feasible operating area. The proposed method is experimentally verified, allowing to determine the remaining voltage during iZVS at any given operating point from only a handful of known circuit and datasheet parameters.

II. QUANTIFICATION OF DEPTH OF INCOMPLETE ZERO-VOLTAGE SWITCHING

In the circuit in Figure 1, the energy stored in the magnetic field of the inductor can be used to achieve soft-switching thus reducing the switching losses in the transistor. However, if the charge provided by the inductor is insufficient to complete the switching transition, iZVS will occur. To quantify the losses during iZVS, the remaining voltage across the DUT after the switching transition period must be determined.

A. Proposed Model

To quantify the magnitude of the remaining voltage across the DUT, the charge balance during the switching transition is solved using four key parameters: The voltages on each side of the inductor v_0 and V_n, the initial current in the inductor at the start of the switching transition i_{L0}, and the non-linear output capacitance of the transistor as a function of voltage $C_{oss}(v_{ds})$. Curve-fitting or a look-up table can be used to obtain the output-capacitance function. Depending on whether Q_H or

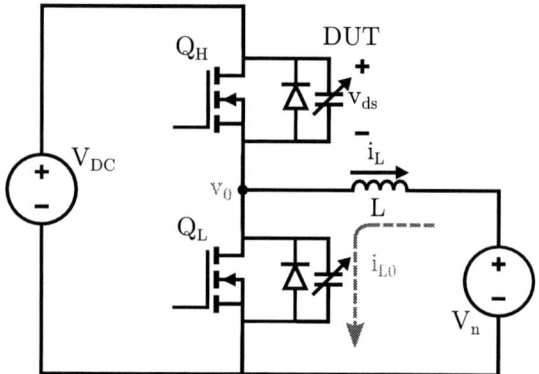

Fig. 1: Half-bridge circuit at the beginning of the switching transition, where Q_L was initially conducting the current i_{L0}, and is to be turned off, while Q_H (DUT) is about to turn on.

Q_L is considered, the parameters v_0 and initial current i_0 have to be adjusted accordingly.

$$v_0 = \begin{cases} 0 & \text{if DUT} = Q_H \\ V_{DC} & \text{if DUT} = Q_L \end{cases} \qquad i_0 = \begin{cases} i_{L0} & \text{if DUT} = Q_H \\ -i_{L0} & \text{if DUT} = Q_L \end{cases}$$
(1)

At the beginning of the dead time period, the voltage across the DUT is equal to the DC voltage, and starts decreasing towards zero as the output capacitance is discharged while the output capacitance of the opposite transistor is charged to the DC voltage. During this period, the circuit shown in Figure 1 can be approximated by replacing the transistors with linearized charge-equivalent capacitors to predict the inductor current.

The output capacitance of a MOSFET is a non-linear function of the V_{ds}, and changes substantially at low voltages [16]. The charge-equivalent capacitance $C_{q,eq}(v_{ds})$ can be calculated by integrating the curve-fitted output capacitance as a function of drain-source voltage from zero to the desired drain-source voltage to get the charge stored at this voltage [14].

$$C_{q,eq}(v_{ds}) = \frac{Q_{oss}(v_{ds})}{v_{ds}}$$
(2)

$Q_{oss}(v_{ds})$ is obtained by integrating $C_{oss}(v_{ds})$ with respect to voltage. By solving the differential equation of the resulting linearized LC-resonant circuit, the time-dependent expression of the inductor current available to discharge the output capacitance of the DUT and charge the opposite transistor can be obtained.

$$i_L(t, v_{ds}) = i_0 \cdot \cos(\omega_0 t) - \frac{v_0 - V_n}{\sqrt{\frac{L}{2C_{q,eq}(v_{ds})}}} \cdot \sin(\omega_0 t)$$
(3)

Where ω_0 denotes the natural frequency of the LC circuit and is given by

$$\omega_0 = \frac{1}{\sqrt{2C_{q,eq}L}}$$
(4)

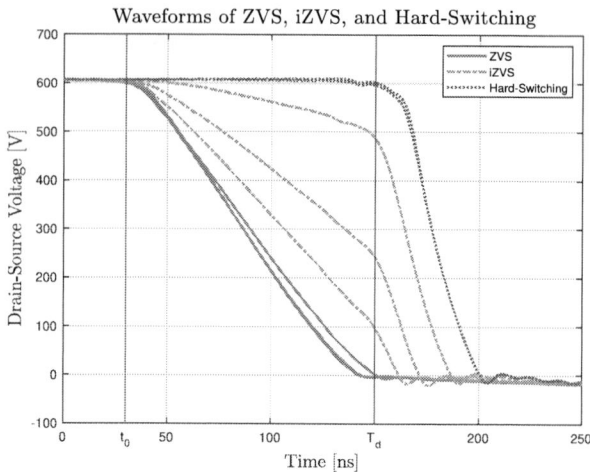

Fig. 2: Drain-source voltage of ZVS, iZVS, and hard-switching. The remaining voltage occurring at T_d must be determined to quantify the iZVS losses.

The charge provided by the inductor during the transition period is expressed as the time-integral of the inductor current from zero to the dead time.

$$\begin{aligned} Q_L &= \int_{t_0}^{T_d} i_L(t)dt \\ &= -\frac{i_0}{\omega_0} \cdot \sin(\omega_0 T_d) + \frac{v_0 - V_n}{\omega_0\sqrt{\frac{L}{2C_{q,eq}}}} \cdot \left(\cos(\omega_0 T_d) - 1\right) \end{aligned}$$
(5)

In order to ensure ZVS, the charge provided by the inductor has to be larger than the charge stored in the output capacitance of the DUT, and the charge required to turn Q_L off, i.e. raise its voltage to the DC voltage. This relation can be used to calculate the remaining voltage across the DUT.

For ZVS to be achieved, the transistor turning off will require a larger amount of charge in the beginning of the switching transition to increase the voltage due to the non-linearity of the output capacitance. Likewise, the opposing transistor will need to lose additional charge to completely discharge by the end of the dead time period. Integrating the output capacitance from zero to the DC voltage yields the charge stored in the output capacitance at a given voltage. The charge required to change the voltage across the DUT by some voltage v can be quantified as

$$Q_{HB}(v) = \int_0^v C_{oss}(v) + C_{oss}(V_{DC} - v) \, dv$$
(6)

Where $Q_{HB}(v)$ is the charge required to change the voltage across the DUT by the voltage v, and V_{DC} is the initial voltage before turn on. The voltage remaining across the DUT can then be calculated by inserting (5) into (6), and solving for V_{rem}. If the remaining voltage is between zero and the DC voltage, iZVS will occur.

$$Q_L = \int_0^{V_{rem}} C_{oss}(v) + C_{oss}(V_{DC} - v) \, dv$$
(7)

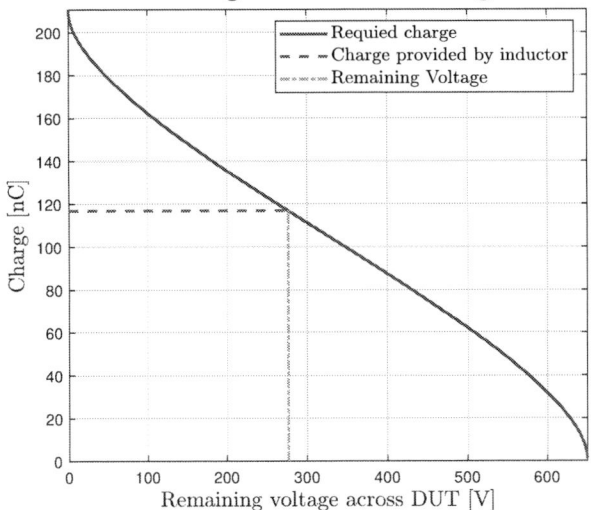

Fig. 3: Example of charge balance in half-bridge during iZVS. The transistors used are C2M0080120D from Wolfspeed [16]. The initial current is 1 A with an inductance of 120 μH. The dead-time is 110 ns and the DC voltage is set to 650 V.

Figure 3 shows an example of the required charge to change the voltage across the DUT by v, and how the remaining voltage across the DUT can be determined by applying (5) and (7). From Figure 3, it can be observed that the charge required to achieve ZVS during the switching transition is equal to 210 nC, which is twice the output charge of the applied transistor with a DC voltage of 600 V. Thus, the initial current to achieve ZVS can be calculated by rearranging (5).

$$i_{\text{zvs}}(v_{\text{ds}}) = -\frac{\omega_0}{\sin(\omega_0 T_{\text{d}})} \cdot$$
$$\left(2Q_{\text{oss}}(V_{\text{DC}}) + \frac{V_{\text{n}} - v_0}{\sqrt{\frac{L}{2C_{\text{q,eq}}}} \cdot \omega_0} \cdot (\cos(\omega_0 T_{\text{d}}) - 1) \right) \quad (8)$$

B. Incomplete Zero-Voltage Switching Losses

The losses that occur during iZVS are approximated by calculating the hard-switching losses that would occur if the DUT was hard-switched with the calculated remaining voltage across it. In [17], a method is presented to accurately calculate the hard-switching losses of SiC MOSFETs based on experimental waveforms and device parameters, taking the internal and external circuit parasitics and non-linearities into account. This method is applied to estimate the turn on losses during iZVS.

$$E_{\text{on,iZVS}} = \int_0^{t_{\text{f}}} v_{\text{ds}}(t) i_{\text{ds}}(t) \mathrm{d}t - \frac{1}{2} L_{\text{ds}} I_0^2 + \frac{1}{2} C_{\text{er}}(V_{\text{rem}}) V_{\text{rem}}^2 \quad (9)$$

Where t_{f} is the fall time, defined from the end of the dead time period until the voltage across the DUT reaches zero, L_{ds} is the sum of the parasitic drain and source lead inductances,

Fig. 4: Full-bridge converter to verify the proposed method.

I_0 is the current at the end of the dead time, and $C_{\text{er}}(V_{\text{rem}})$ is the energy related capacitance of the output capacitance of the transistor evaluated from 0 V to the remaining voltage, calculated as

$$C_{\text{er}}(v) = \frac{2}{V_{\text{rem}}^2} \cdot E_{\text{oss}}(V_{\text{rem}}) = \frac{2}{V_{\text{rem}}^2} \cdot \int_0^{V_{\text{rem}}} Q_{\text{oss}}(v) \mathrm{d}v \quad (10)$$

The drain-source voltage in (9) is approximated as linearly decreasing from the remaining voltage towards zero during the fall time. The drain-source current is calculated using (3).

III. EXPERIMENTAL VERIFICATION

A full-bridge circuit with an inductive load is used to verify the proposed method of determining the remaining voltage and iZVS losses at different operating currents. The drain-source and gate-source voltage of the DUT are observed, while the current in the inductor at the start of the switching transition period is adjusted by modulating the transistors as described in [18]. The experimental setup can be seen in Figure 4.

TABLE I: The following parameters are used in the experiment to verify the remaining voltage across the DUT.

DUT Device	Drain-Source Voltage [V]	Dead Time [ns]	Inductor [μH]
Wolfspeed C2M0080120D (SiC)	600	110	170
Infineon IPB65R125C7 (Si)	50	400	170
Infineon IPB65R125C7 (Si)	75	400	170

Both a Si and a SiC MOSFET are tested to verify the model for different types of transistors and output capacitances. The Si MOSFET is tested at low voltages, as the non-linearity will be most significant in these areas. Waveforms for hard-switching, iZVS, and ZVS for the Si MOSFET at 50 V can be seen in Figure 5. The magnitude of the voltage at the turn-on of the DUT shows the converter is hard-switching (a), iZVS (b & c) and ZVS (d) at different switching currents.

Moreover, the drain-source voltage oscillations are of larger magnitude at increased turn-on voltages, due to the increase in dv/dt when the device is turned on. Apart from increased switching losses at higher turn-on voltages, the resulting switching oscillations can cause undesired EMI,

979-8-3503-3714-3/23 $31.00 © 2023 IEEE

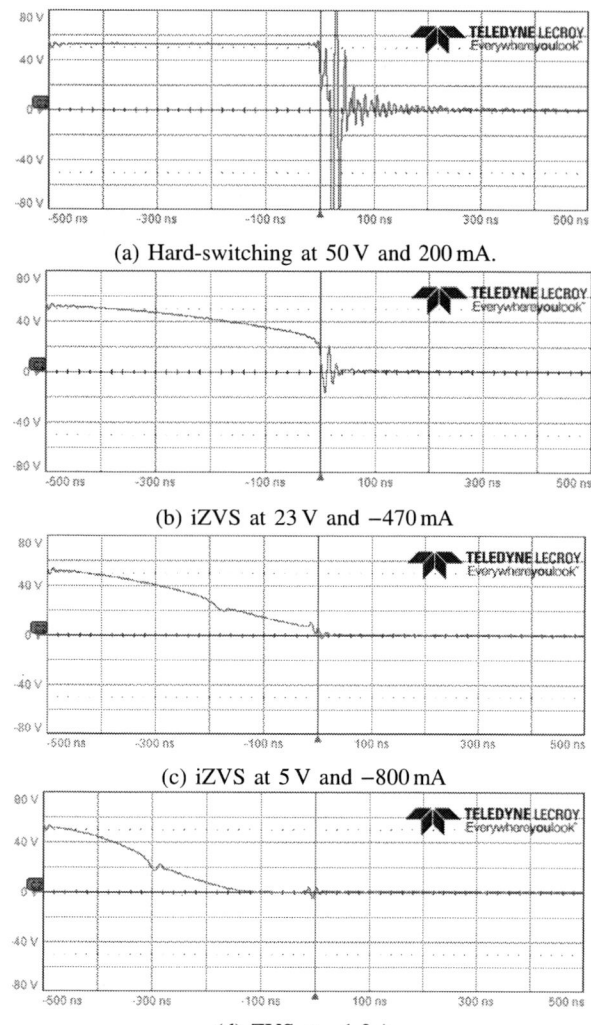

(a) Hard-switching at 50 V and 200 mA.

(b) iZVS at 23 V and −470 mA

(c) iZVS at 5 V and −800 mA

(d) ZVS at −1.3 A

Fig. 5: Experimentally measured drain-source voltage of DUT (Si MOSFET) at different switching currents. The DUT is turned-on at 0 s.

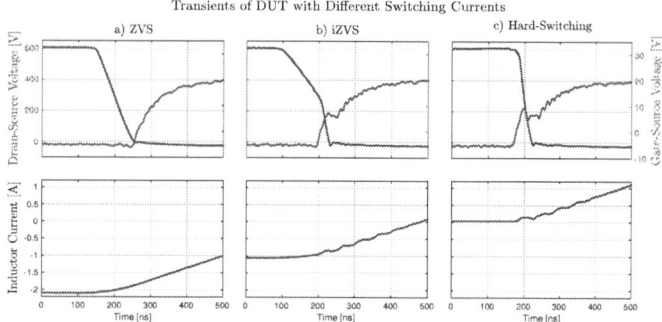

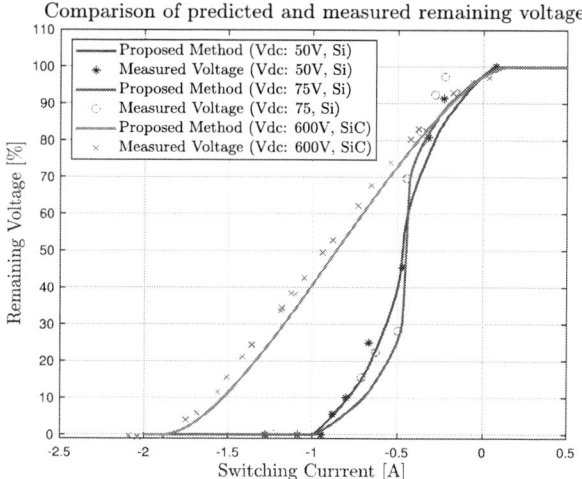

Fig. 6: Experimentally measured switching transients with (a) ZVS at −2.08 A, (b) iZVS at −1.05 A, and (c) hard-switching at 0 A.

Fig. 7: Comparison of remaining voltage predicted by model and measured experimentally as percentage of the DC voltage of the experiments described in Table I.

voltage overshoot, additional power losses, and even shoot-through failures, thus severely compromising the performance of the converter [19]. Figure 6 shows the drain-source and gate-source voltages during the switching transition of a SiC MOSFET operating at a DC voltage of 600 V. The remaining voltage after the switching transition at different inductor currents is used to predict the switching modes of the DUT. It can be observed that the transistor enters iZVS when the inductor current is not sufficiently large to provide the required charge to achieve ZVS during the dead time period. The experimentally measured drain-source voltages at turn-on of the DUT are used to evaluate the proposed model.

A. Evaluation of Proposed Model

The obtained values for the remaining voltage at various switching currents are normalized with respect to the DC voltages and compared to the analytical model in Figure 7. It

can be observed that the non-linearity is most obvious with a Si DUT, which can be attested to the low voltages at which the experiments are conducted. The change in output capacitance is largest at low voltages, which can be observed both at the beginning of the switching transition, when the turn-off transistor is charged from 0 V, as well as at the end of the transition, when the DUT approaches 0 V, as shown in (7). At higher voltages, the linear region of the output charge is reached, which is reflected by the remaining voltage of the SiC device switched at 600 V. To evaluate the accuracy of the proposed model, the root-mean-square deviation (RMSD) from the predicted voltages to the measurements is calculated.

$$\text{RMSD} = \sqrt{\frac{1}{N}\sum_{i=1}^{N}(x_i - \hat{x}_i)^2} \quad (11)$$

Applied to the normalized data seen in Figure 7, the RMSD of the model with respect to the experimental data is 4.7 %. The model does not account for measurement errors or the influence of parasitics. Furthermore, the function used to

Method comparison to predict remaining voltage during iZVS

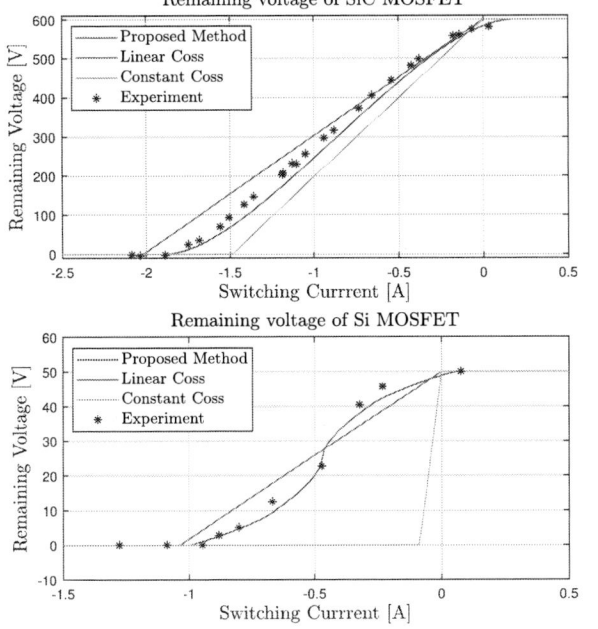

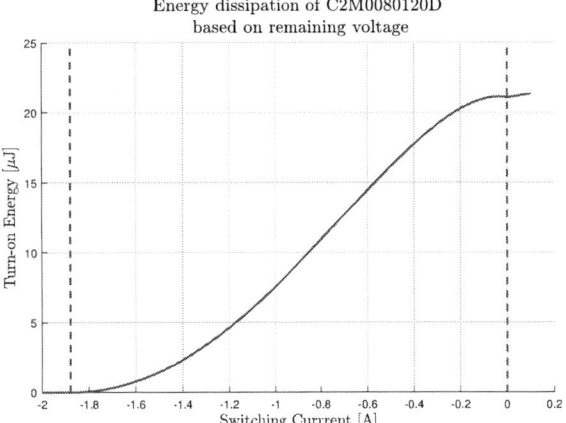

Fig. 9: Calculated turn-on energy based on remaining voltage. iZVS between -1.89 and 0 A.

and the proposed model can predict the ZVS current with high accuracy, while the constant output capacitance method fails to make a precise prediction at low DC voltages, as the significant non-linear part is not considered.

B. Calculation of Turn-on Energy

Finally, the remaining voltage is linked to the turn-on energy. The turn-on energy is calculated for the SiC device as described in Section II-B using (9) as a function of the remaining voltage calculated by the proposed method. The dissipated turn-on energy increases non-linearly between complete ZVS and hard-switching in accordance with the remaining voltage. However, the applied turn-on loss estimation method requires many circuit parameters to be known in order to estimate the fall-time and the drain-source parasitic inductance and can be inaccurate if the parameters are not known precisely. Therefore, further research is required to develop a simplified model tying the remaining voltage to turn-on losses, verifying it with a calorimetric test setup to eliminate deviations caused by inaccuracies of the electrical measurements.

Fig. 8: Comparison of different methods for determining the ZVS current, using Wolfspeed C2M0080120D SiC MOSFET at 600 V.

TABLE II: ZVS current estimation using different methods, compared to experimentally measured

Experiment:	Constant:	Linear:	Proposed:	Measured:
SiC at 600 V	−1.49 A	−2.02 A	−1.90 A	−1.89 A
Si at 50 V	−0.09 A	−1.04 A	−1.00 A	0.95 A

calculate the output charge is a numeric integration of a curve-fit of the output capacitance extracted from the SPICE model provided by the manufacturer, which can also cause deviation of the model from experimentally observed results. Figure 8 shows how the proposed model compares with other common methods for predicting the ZVS current. The constant output capacitance method is widely used to determine the ZVS current and uses the value of the output capacitance listed in the datasheet of the device, usually at the rated voltage of the device [13], [20]. This method entirely disregards the effects of the non-linearity of the output capacitance and is therefore the least precise. The linear output capacitance method uses the charge equivalent capacitance as described in [14]. It provides a more accurate estimation, as the charge contribution from the non-linear part is considered. In the areas dominated by the non-linearity however, the method loses accuracy. The RMSD for the linear method is 10 % compared to the experimental results, which is more than twice the RMSD of the proposed method. The predicted ZVS currents of the different methods are presented in Table II. It can be seen that both the linear output capacitance method

IV. CONCLUSION

The proposed method can be used to accurately determine iZVS losses in any half-bridge configuration with only a handful of circuit and datasheet parameters, allowing the circuit design to be optimized accordingly. Despite many possible sources of error, such as parasitic coupling, measurement errors, and potential inaccuracies during curve-fitting of the output capacitance as a function of drain-source voltage, the proposed resonant charge balance method has a low RMSD below 5 %, outperforming other methods for predicting the remaining voltage after the switching transition, as well as the ZVS current. A qualitative significance of the proposed method can clearly be observed when operating with low DC voltages, where the non-linearity of the output capacitance is dominating.

979-8-3503-3714-3/23 $31.00 © 2023 IEEE

The analytical results determining the depth of iZVS in a half-bridge correlates well with experimental results, and the predicted turn-on energy corresponds to the turn-on energy based on experimental waveforms. The latter should be experimentally verified using a calorimetric test setup, removing the sources of error from electrical measurements. Furthermore, the iZVS losses can be tied to hard-switching losses at a lower DC voltage. To simplify the quantification of switching losses, further research is required to separate and measure turn-on, conduction, and turn-off losses.

REFERENCES

[1] F. Roccaforte, P. Fiorenza, G. Greco, R. L. Nigro, F. Giannazzo, F. Iucolano, and M. Saggio, "Emerging trends in wide band gap semiconductors (sic and gan) technology for power devices," *Microelectronic Engineering*, vol. 187-188, pp. 66–77, 2 2018.

[2] L. Spaziani and L. Lu, "Silicon, gan and sic: There's room for all: An application space overview of device considerations," vol. 2018-May. Institute of Electrical and Electronics Engineers Inc., 6 2018, pp. 8–11.

[3] A. Ghazanfari, C. Perreault, , and K. Zaghib, "Ev/hev industry trends of wide-bandgap power semiconductor devices for power electronics converters," *Hydro-Québec's Center of Excellence in Transportation Electrification and Energy Storage (CETEES)*, 2019.

[4] Y. Berube, A. Ghazanfari, H. F. Blanchette, C. Perreault, and K. Zaghib, "Recent advances in wide bandgap devices for automotive industry." IEEE, 10 2020, pp. 2557–2564. [Online]. Available: https://ieeexplore.ieee.org/document/9254478/

[5] B. Frieske, M. Kloetzke, and F. Mauser, "Trends in vehicle concept and key technology development for hybrid and battery electric vehicles." Institute of Electrical and Electronics Engineers Inc., 10 2014.

[6] C. Suarez and W. Martinez, "Fast and ultra-fast charging for battery electric vehicles - a review," in *2019 IEEE Energy Conversion Congress and Exposition, ECCE 2019*. Institute of Electrical and Electronics Engineers Inc., 9 2019, pp. 569–575, hPC350.

[7] M. Yilmaz and P. T. Krein, "Review of battery charger topologies, charging power levels, and infrastructure for plug-in electric and hybrid vehicles," pp. 2151–2169, 2013.

[8] S. Habib, M. M. Khan, F. Abbas, L. Sang, M. U. Shahid, and H. Tang, "A comprehensive study of implemented international standards, technical challenges, impacts and prospects for electric vehicles," pp. 13 866–13 890, 3 2018.

[9] R. S. K. Moorthy, M. Starke, B. Dean, A. Adib, S. Campbell, and M. Chinthavali, "Megawatt scale charging system architecture." IEEE, 10 2022, pp. 1–8. [Online]. Available: https://ieeexplore.ieee.org/document/9947403/

[10] S. Habib, M. M. Khan, F. Abbas, A. Ali, M. T. Faiz, F. Ehsan, and H. Tang, "Contemporary trends in power electronics converters for charging solutions of electric vehicles," *CSEE Journal of Power and Energy Systems*, vol. 6, pp. 911–929, 12 2020.

[11] Z. Wang and H. Li, "A soft switching three-phase current-fed bidirectional dc-dc converter with high efficiency over a wide input voltage range," *IEEE Transactions on Power Electronics*, vol. 27, pp. 669–684, 2012.

[12] R. Kodoth, H. T, B. K. R., and P. Kanakasabapathy, "Design and development of a resonant converter adapted to wide ouput range in ev battery," *3rd IEEE International Conference on Recent Trends in Electronics, Information & Communication Technology (RTEICT-2018)*, 2018.

[13] T. T. Le, S. Kim, and S. Choi, "A four-phase current-fed push-pull dab converter for wide-voltage-range applications," *IEEE Transactions on Power Electronics*, vol. 36, pp. 11 383–11 396, 10 2021.

[14] M. Kasper, R. M. Burkart, G. Deboy, and J. W. Kolar, "Zvs of power mosfets revisited," *IEEE Transactions on Power Electronics*, vol. 31, pp. 8063–8067, 12 2016.

[15] C.-T. Truong and S.-J. Choi, "A more accurate zvs criterion for resonant converters," 2022, pp. 2264–2267.

[16] C. Inc., "C2m0080120d silicon carbide power mosfet," Cree Inc., Tech. Rep., 2019. [Online]. Available: https://assets.wolfspeed.com/uploads/2020/12/C2M0080120D.pdf

[17] S. K. Roy and K. Basu, "An energy based approach to calculate actual switching loss for sic mosfet from experimental measurement." Institute of Electrical and Electronics Engineers Inc., 6 2021.

[18] J. A. Anderson, C. Gammeter, L. Schrittwieser, and J. W. Kolar, "Accurate calorimetric switching loss measurement for 900 v 10 mω sic mosfets," *IEEE Transactions on Power Electronics*, vol. 32, pp. 8963–8968, 12 2017.

[19] T. Liu, T. T. Wong, and Z. J. J. Shen, "A survey on switching oscillations in power converters," pp. 893–908, 3 2020.

[20] H. Song, D. Xu, and A. J. Zhang, "Re-analysis on zvs condition for llc converter." Institute of Electrical and Electronics Engineers Inc., 6 2021, pp. 1874–1880.

Thermal Performance Investigation of a High-Current & High-Power Density GaN-Based Motor Drive for All Electric Aircraft Applications

Armin Ebrahimian*, Seyed Iman Hosseini Sabzevari*, Waqar A. Khan+, Nathan Weise*

* Department of Electrical and Computer Engineering, Marquette University, Milwaukee, WI, USA
+ Miller Electric, Milwaukee, WI, USA
armin.ebrahimian@marquette.edu, iman.hosseini@marquette.edu, waqar.khan@ieee.org, nathan.weise@marquette.edu

Abstract—Integration of electric machine, drive system, and cooling system in a single structure was proposed as a solution to meet the high power density and efficiency requirements in aviation applications. From the drive perspective, both the efficiency and power density are highly dependent on the switching device selection. Thus, semiconductors with small footprint and low on-state resistance are desirable. Integrating the power converter with the electric machine limits the available space, thus the PCB should be designed accordingly. Considering the space constraint in addition to the efficiency and power density requirements for such an application, the thermal performance of the designed converter is pivotal. In this paper, the thermal performance of a high-current & high-power density GaN-based motor drive is evaluated. At first, using the PLECS software, the performance of the power converter was simulated at rated load conditions. Using PLECS's thermal library, the characteristics of the cooling system are modeled. Based on the simulation results, the cooling system configurations that can satisfy the constraints of the drive system are determined. Finally, the performance of the drive module was evaluated through DC and AC testing.

Index Terms—GaN, Integrated Motor Drive, Thermal Management System

I. INTRODUCTION

There has been a growing interest in research on all-electric aircraft as part of the transition to electrified transportation. Specifically, designing electrified propulsion system is one of the emerging topics in the field [2]. Concerning the aviation industry standards, high power density and efficiency are two important requirements for the propulsion systems [3]. In that regard, 12kW/kg was set as the specific power target for fully integrated motor and drive systems for aviation applications by the United States Department of Energy (US-DOE) [4]. From a high-level perspective, an all electric propulsion system consists of three main parts: electric machine, power electronics interface (drive), and thermal management system (TMS) [5]. As a result, to obtain high specific power, integration of all three parts into a united structure is deemed to be a feasible solution [6]. Such a concept is introduced in the state of the art as integrated motor drive (IMD) [7]–[9]. Since reliability is paramount in aviation applications, the modular design of

The information, data, or work presented herein was funded in part by the Advanced Research Projects Agency-Energy (ARPA-E), U.S. Department of Energy, under Award Number DE-AR0001352. The views and opinions of authors expressed herein do not necessarily state or reflect those of the United States Government or any agency thereof.

IMDs can be considered an added benefit of integrated modular motor drive (IMMD). In IMMDs, reliability is increased since the propulsion system can function at reduced capacity during fault conditions. Also, with modular design comes the ease of maintenance and reduced cost of replacement [10].

From the power electronics standpoint, utilizing semiconductors with a small footprint, low on-state resistance, and high current handling capability is the key to achieving such a compact and efficient design. Furthermore, PCB design and component layout have high importance in an integrated system due to the limited space. Depending on the level of integration and placement of the drive system, with respect to the electric machine, restrictions may be applied to both TMS and the drive system design [1], [11].

Another challenge in IMMDs is designing a TMS that can extract the generated heat from the drive module [12]. Semiconductors are considered to be the main sources of heat due to the conduction and switching losses. In [13], different cooling technologies are introduced and their thermal resistivities are presented. The lower thermal resistance of a cooling system directly relates to its capability of heat extraction from the semiconductors. Thus, a cooling system with the lowest thermal resistance is ideal to obtain the highest efficiency. Creating a simulation platform in which both the electrical and thermal performance of the converter and TMS are modeled can be extremely helpful for the designers to converge to an optimum design [14].

In this paper, the thermal performance of a high-current and high-power density IMMD is investigated. The first goal is to test the designed GaN-based full-bridge converter at the rated current value and observe the junction temperature of the semiconductors. The second target is to extract the heat flux from the experimental results to be used in the TMS design procedure. Although, the final TMS includes heat pipes and liquid cooling, in this paper all the tests are performed using an off-the-shelf heatsink. The rest of the paper is organized as follows: in section II the case study of this research is described. Section III discusses the thermal modeling of the case study in the PLECS software. The simulation and experimental results are provided in sections IV and V, respectively. Lastly, section VI presents the conclusion.

979-8-3503-3714-3/23 $31.00 © 2023 IEEE

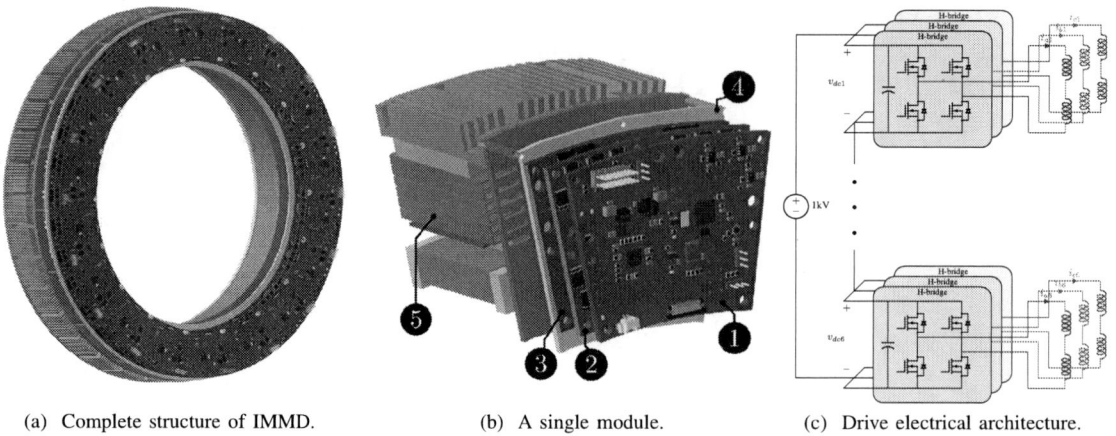

(a) Complete structure of IMMD.   (b) A single module.   (c) Drive electrical architecture.

Fig. 1: (a) Complete structure of the axially stator iron-mounted IMMD. (b) A single module: 1) Communication board; 2) DC capacitor board; 3) GaN PCB; 4) Drive module's heat spreader; 5) Additively manufactured motor coils; (c) Drive topology [1].

TABLE I: Nominal System Parameters

Parameters	Value	Parameters	Value
v_{dc}	1 kV	Coil Resistance (R_s)	46 mΩ
I_{coil}	268.12 Arms	Coil Inductance (L_s)	137 μH
P_{rated}	250 kW	PM Flux (λ_{pm})	60.6 mWb
$f_{o,rated}$	1.25 kHz	Switching Frequency (f_{sw})	20 kHz
Power Factor ($\cos\phi$)	0.8	Target Efficiency (η)	$\geq$ 98%

II. SYSTEM DESCRIPTION

In this research, the case study is a three-phase 250kW IMMD, which is demonstrated in Fig. 1a. The IMMD consists of 18 identical modules, where a single module is shown in Fig. 1b. In other words, there are 6 three-phase sub-modules as illustrated in the electrical architecture of the system in Fig. 1c. Since all the modules are identical, the outcome of studying a single module can be then extended to the whole system. In Fig. 1b, different parts of a single module are labeled as follows: 1) Communication board, 2) DC link capacitor board, 3) GaN-based full-bridge converter, 4) GaN PCB's heat spreader, 5) motor coil.

The primary goal of this research is to test the performance of the designed 16-layer GaN-based full-bridge PCB busbar under the rated load condition while using a temporary off-the-shelf cooling system. Although, the final TMS for both motor coils and the drive is based on heat pipes, at this stage of the project it was decided to conduct the tests using off-the-shelf forced air TMS. Thus, the device under study is only the drive module (machine side TMS considerations are not included in this study), meaning the GaN PCB and DC link capacitor board that are connected to each other through copper bars.

The IMMD parameters are listed in Table I. Considering the modularity and integration aspects of the IMMD, one of the challenges in such a design is to find semiconductors that are compatible with the requirements. Two main requirements

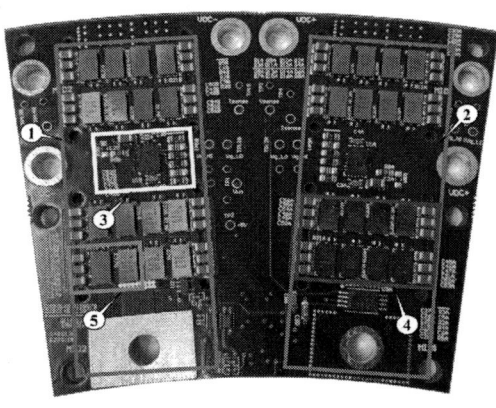

Fig. 2: GaN-based full-bridge module PCB. 1) Left half-bridge; 2) right half-bridge; 3) gate drive circuitry; and 4) a switch block consisting of 4 power poles; 5) power pole and the decoupling capacitors.

in this design are the high power density (which limits the space and weight), and high efficiency. To find semiconductors that fit the application, a figure of merit (FOM) was used, where it was determined that GaN FETs and specifically EPC-manufactured GaN FETs are the best match [1], [11]. Furthermore, due to the high current rating of the application (268.12A), paralleling GaN FETs is required. Also, since the drive module is axially mounted on the stator, the shape of the PCB is dictated based on the stator design. Having the inner and the outer diameter, the board outline is defined, thus the next step is to set the component layout. In a single-phase full-bridge converter, there are four switches (two legs each one having a high side and a low side switch) whereas, in the case study of this research, each switching block represents 8 parallel GaNs.

Fig. 2 demonstrates the GaN-based full-bridge PCB. There

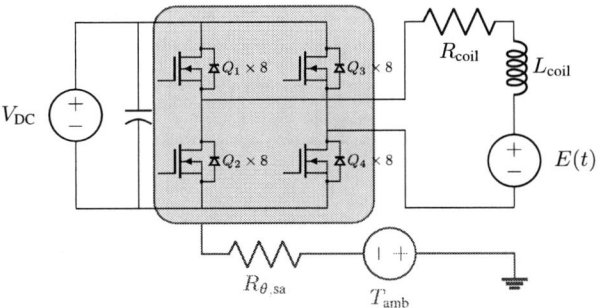

Fig. 3: Simulated system circuitry in PLECS software.

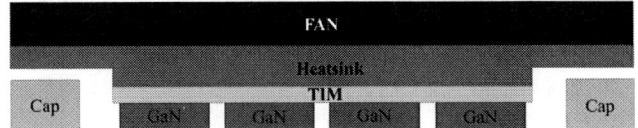

Fig. 4: Side view of a single switch block including the TIM and the heatsink.

are two half-bridges on the left and right sides of the PCB surrounded by red rectangles. Each half-bridge consists of 8 power poles (a combination of a high-side and low-side switch and three decoupling capacitors that are surrounded by the green rectangle) which are controlled by a gate driver (the yellow rectangle in the middle). In other words, there is a total of four switching blocks on the PCB which should be actively cooled down using the TMS. At this stage of the research, the goal is to test the module up to the rated load current, while the temperature of the junction of the semiconductors does not exceed the temperature limit of $125°C$.

In the next section, the thermal modeling approach and the electrothermal simulation setup will be discussed.

III. THERMAL MODELING & HEATSINK SELECTION

In a power converter, switching devices (semiconductors) are the major contributors to the total loss. During the operation, depending on the voltage, current, and turn-on/off characteristics of the switches, their relative switching and conduction losses can be calculated. PLECS is a simulation software in which power electronics converters can be simulated. Also, it has a thermal library which gives the power electronics designer the ability to evaluate the thermal performance of power converters and thermal management systems.

The created model in PLECS consists of two main parts including electrical and thermal components. For the electrical part, a DC voltage source (V_{DC}) is connected to an RLV load through a full-bridge converter. The RLV load represents the motor coil, where R_{coil} and L_{coil} correspond to the resistivity and the inductance of the coil, and the voltage source ($E(t)$) models the generated back EMF by the coil. On the other hand, for the thermal side, the main components are the switch thermal model, thermal interface material (TIM), heatsink, and ambient temperature. Fig. 3 illustrates the general concept of the simulated model in which the blue components represent the TMS.

The side view of a single switch block (number 4 in Fig. 2) is demonstrated in Fig. 4, where the TMS is also shown. A TIM, which is recommended by the GaN manufacturer is used between the heatsink and the GaN switches, which enhances the thermal coupling between the GaNs and the heatsink.

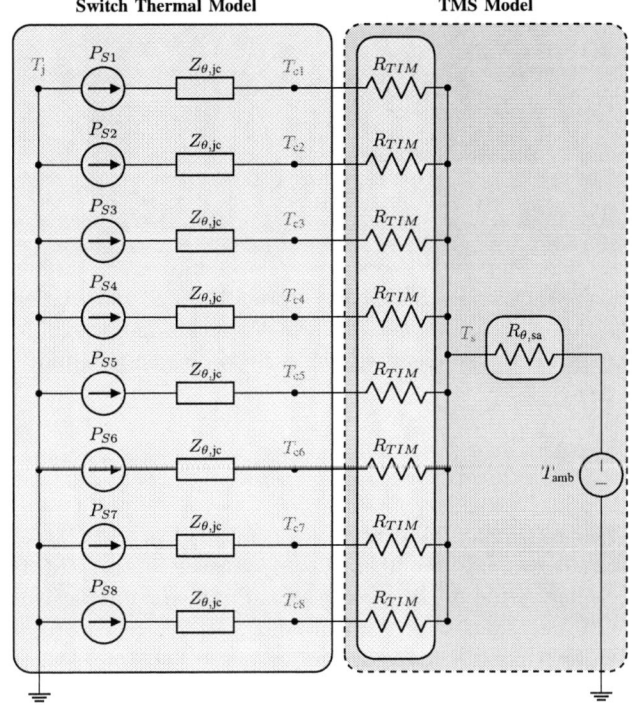

Fig. 5: Loss generation and dissipation network in steady state.

Fig. 5 illustrates the thermal equivalent circuit analogy of the loss dissipation network in the steady state. As it was mentioned at the beginning of this section, the switches are considered the main source of the losses (conduction and switching losses). Each switch dissipates a total loss of P_{Si} ($i \in \{1,2,...,8\}$) through its case to the TIM, heatsink, and finally to the ambient. The switch's thermal model is offered by the manufacturer, which includes loss calculations, and junction-to-case thermal impedance ($Z_{\theta,jc}$). As illustrated in Fig. 4, all the switches are connected to the heatsink through the TIM. Also, in the equivalent circuit analogy (Fig. 5) each switch is connected to the heatsink ($R_{\theta,sa}$) through the TIM ($R_{\theta,TIM}$). The heatsink dissipates all the loss to the ambient which is modeled as a source in Fig. 5. Using the thermal model in the PLECS simulations, the junction temperature (T_j), device's case temperature (T_{ci}, where $i \in \{1,2,...,8\}$), and heatsink temperature (T_s) can be obtained.

In the next section, electrothermal simulation results regarding the selection of a suitable cooling system will be presented.

979-8-3503-3714-3/23 $31.00 © 2023 IEEE   33

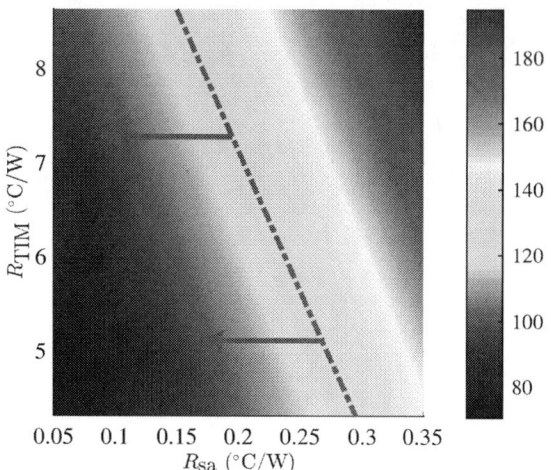

Fig. 6: Maximum junction temperature for different heatsink/TIM combinations from PLECS simulations. The dashed red line and the arrows indicate the permissible heatsink/TIM combinations that satisfy the design limit of $T_j \leq 125°C$.

IV. SIMULATION RESULTS

Electrothermal simulations are conducted in PLECS software to evaluate the thermal performance of the IMMD. Considering the rated operating conditions from Table I, the PWM gate signals for the converter are generated using the open-loop control method. The primary goal is to find the TMS characteristics where both the efficiency lower-limit of 98%, and the junction temperature upper-limit of 125°C are satisfied. To that end, 279 simulations are conducted in which the thermal resistivity of the TIM (R_{TIM}), and the heatsink ($R_{\theta,sa}$) are swept between predefined values. Using the provided information from TIM's datasheet, the values for R_{TIM} are dependent on the applied pressure, ranging from 4.32°C/W to 8.63°C/W (for a switch block consist of 8 GaN FETs). Also, the range for potential heatsink thermal resistance is considered to be between 0.05°C/W to 0.35°C/W covering cooling systems like liquid cooling, heat pipes, and Folded Fin heatsinks.

Analyzing the simulation results revealed that the limiting factor is the junction temperature since the efficiency of above 98% is preserved in all 279 simulations. Fig. 6 demonstrates the highest maximum junction temperature of all the GaNs in a full-bridge converter for different TMS configurations. It is determined that if the heatsink's thermal resistivity is higher than 0.29°C/W, the highest junction temperature will exceed 125°C regardless of the value of R_{TIM}, which is not desirable. As an example, choosing $R_{\theta,sa} = 0.25°C/W$, and $R_{TIM} = 5.39°C/W$, the junction temperature will be 124.35°C. Based on TIM's datasheet and the GaN's surface, the required pressure to obtain the $R_{TIM} = 5.39°C/W$ is 44 PSI.

At this stage of the research, the TMS is only used for

TABLE II: Specifications of the Module Under Test

Part	Description	Part	Description
GaN Switch	EPC 2034C	TIM (1 mm thickness)	TG-A1780
Heatsink & Fan	FH6030MU		

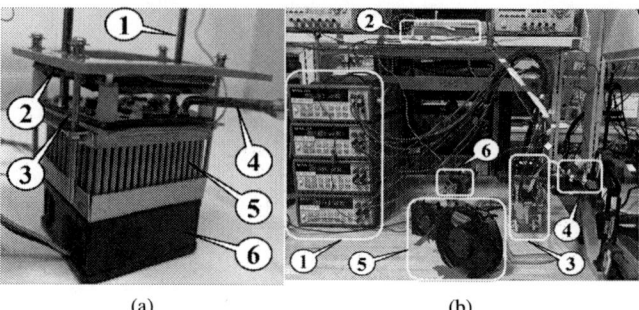

Fig. 7: Experimental setup: a) Module under test: 1)DC busbar; 2) DC capacitor board; 3) GaN PCB; 4) Phase bus bar; 5) Heatsink; 6) Fan. b) DC testing experimental test setup: 1) 5-digit multimeters; 2) RTD data logger; 3) Drive module connected to the DC current source; 4) Microcontroller; 5) Extra fans for advance cooling; 6) LEM current sensor.

continuous testing of the GaN PCB, so it is decided to choose an off-the-shelf forced-air cooling solution. It is worth mentioning that although this is not the final integrated cooling system for the drive system, the obtained data from this study, specifically the heat flux, will be used during the design phase of the integrated TMS.

V. EXPERIMENTAL RESULTS

To assess the performance of the designed 16-layer GaN PCB, numerous DC and AC tests were conducted. Fig. 7a shows the module under test in which the DC link capacitor bank, GaN PCB, and the off-the-shelf cooling system are labeled. General information about the GaN-based converter and the TMS is presented in Table II. The DC test setup is also demonstrated and labeled in Fig. 7b. In the following, different DC tests are explained and their results are discussed. During the DC tests, the injected current to the module under test

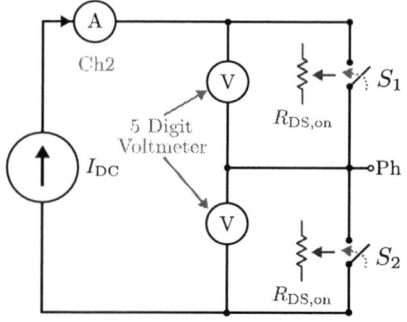

Fig. 8: Half-bridge DC testing circuitry.

979-8-3503-3714-3/23 $31.00 © 2023 IEEE

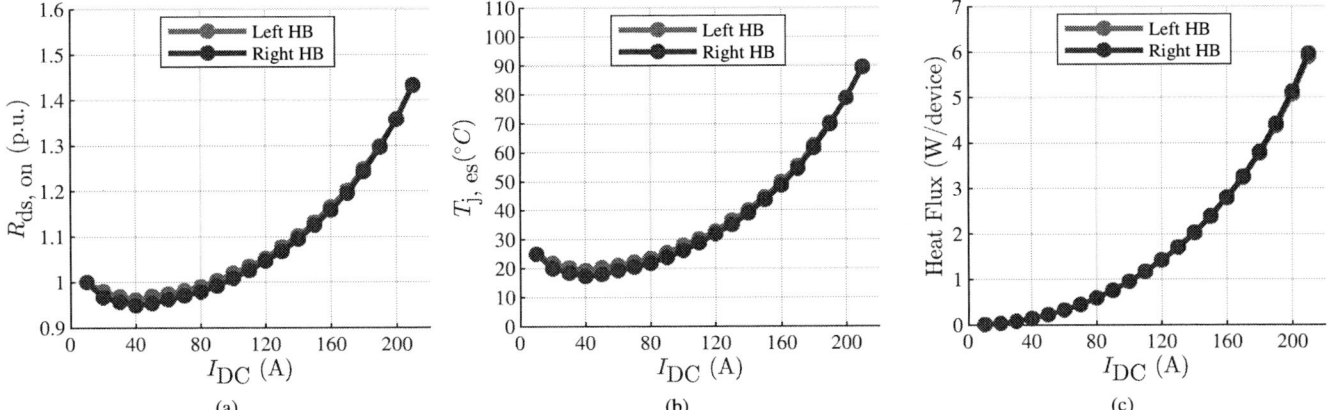

Fig. 9: Half-bridge DC test results for both of the legs: (a) Normalized $R_{ds,on}$. (b) Estimated junction temperature. (c) Dissipated heat flux per device.

increases in steps while in each step the operator waits for the resistance temperature detectors (RTDs), which are placed near the switches, to reach the thermal equilibrium (steady-state), and after recording the data, proceed to the next step.

It is worth mentioning that due to the limited available off-the-shelf heatsinks, the chosen heatsink has a thermal resistivity of 0.35°C/W, which is higher than the desired value. The effect of this discrepancy is observed during the tests.

A. Half-Bridge DC Testing

The test circuit for DC testing of one of the half-bridges is illustrated in Fig. 8. As it is shown in Fig. 8, a DC current source is connected to one of the half-bridges, and the high-side and low-side switches of one of that half-bridge are shorted (while the high-side and low-side switches on the other half-bridge are open circuit). The drain-source voltages of the high-side and low-side switches are measured using 5-digit multimeters to be used for $R_{ds,on}$ calculations. Knowing the injected current from the DC current source that is measured through a LEM current sensor, and measuring the drain-source voltages of the GaNs, the $R_{ds,on}$ can be calculated. A fundamental assumption here is that all the 8 parallel FETs in both high-side and low-side positions are conducting the same current. Considering the confirmed simultaneous turn on/off of the FETs, and also the compact design this is a fair assumption [11]. The final target of the DC test is to inject 190ADC into each half-bridge. This is the RMS value of the switch current while supplying the rated load RMS current (268.12A). However, during the half-bridge testing, the injected DC current was increased up to 210ADC, which is 20A above the target value.

DC testing results for half-bridge topology are presented in Fig. 9. The DC test was performed on both of the half bridges separately. The per unit value of the $R_{ds,on}$ is calculated while considering the recorded $R_{ds,on}$ at 10ADC (where the FETs were at room temperature) as the base and the results are shown in Fig. 9a. Using the GaN's datasheet, the estimated

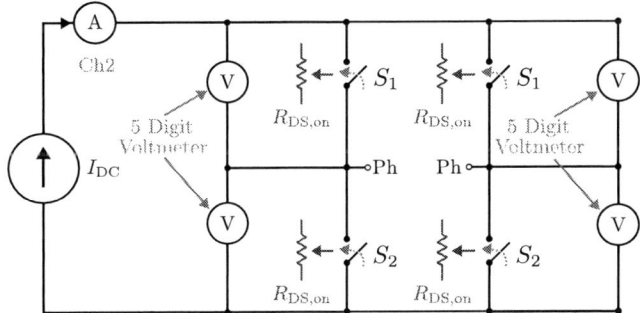

Fig. 10: Full-bridge DC testing circuitry.

junction temperature can be extrapolated based on the calculated per unit $R_{ds,on}$, which is shown in Fig. 9b. Finally, the heat flux for each device can be calculated using the obtained $R_{ds,on}$, and the measured current as follows:

$$\text{Heat Flux} = \frac{R_{ds,on} \times I_{DC}^2}{8} \quad \left(\frac{W}{\text{device}}\right) \quad (1)$$

The obtained heat flux will then be used during finite element analysis (FEA) to design the heatpipe based TMS.

The first thing that can be inferred from Fig. 9 is that the results for both the right and left legs are in good agreement. From Fig. 9a, it can be observed that by increasing the current, especially after 80A, the per unit $R_{ds,on}$ increases in an exponential manner. Based on the increase in the estimated junction temperature shown in Fig. 9b, these results are in accordance with the expectation that GaN's on-state resistance is highly sensitive to the temperature increase.

Although the selected heatsink had a higher $R_{\theta,sa}$ than the desired value, during the half-bridge testing the junction temperature remained below 125°C. At 190ADC, the $R_{ds,on}$ is 1.3 p.u., the estimated junction temperature is 69.75°C, and the heat flux is 4.42W/device. In the next section, the same test will be repeated for the full-bridge topology.

979-8-3503-3714-3/23 $31.00 © 2023 IEEE

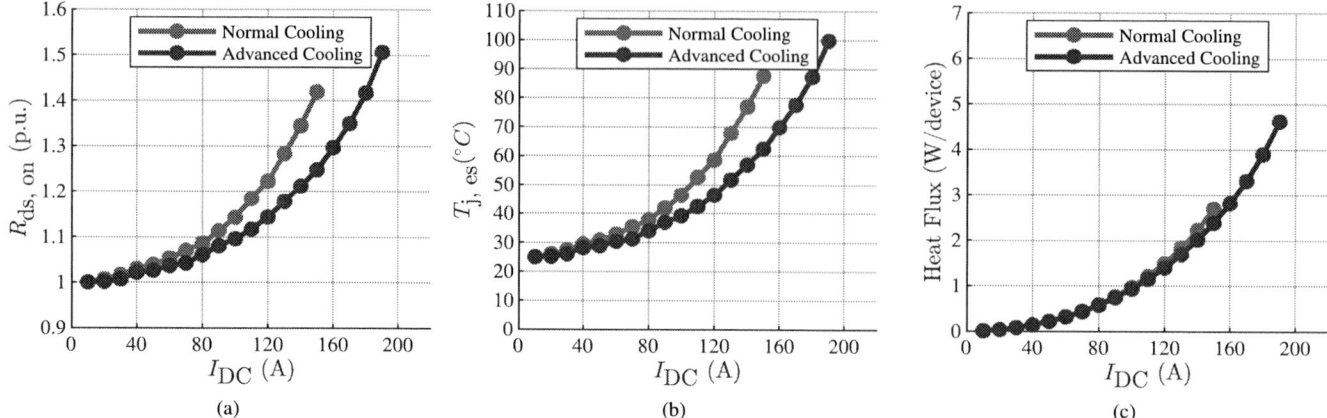

Fig. 11: Full-bridge test results with, and without extra cooling (the x-axis represents the DC current flowing through each half bridge): (a) Normalized $R_{ds,on}$. (b) Estimated junction temperature. (c) Dissipated heat flux per device.

B. Full-Bridge DC Testing

The schematic of DC testing for full-bridge topology is shown in Fig. 10 where all the FETs in both legs are shorted while the current is injected into the module through DC terminals. It is noteworthy that to obtain the 190ADC target for the full-bridge test, the current source should inject twice that (380ADC) to the module under test. Also, one core assumption in the calculation of this section is the current that flows through both legs is equal. Considering the symmetric design and the good agreement of the half-bridge test results for both legs in the previous section, this is a fair assumption. The results presented in Fig. 11 demonstrate two tests. In the first test (the red curves), the utilized cooling solution is the selected off-the-shelf heatsink, while in the second test (the blue curves) two extra fans are added to cool down the DC busbars. The reason is that during the first test, it was observed that the DC busbars were getting hot, and consequently the temperature of the GaN PCB and the DC link capacitor PCB were rising to their limits. By adding extra fans and cooling DC busbars, the DC test on full-bridge topology was performed up to 190ADC.

The results of all DC tests are compared in Table III, in which the $R_{ds,on}$, $T_{j,es}$, and the heat flux are higher in full-bridge topology compared to the half-bridge even with added fan to cool down the busbars. During the first full-bridge test, the TMS showed a deficiency in extracting the generated heat from the GaNs. Considering the obtained simulation results in section IV, this was expected since the selected off-the-shelf heatsink had a higher thermal resistivity than $0.29°C/W$. Consequently, the excess heat flowed through the GaN PCB, DC busbars, and DC capacitor PCB which led to an alarming temperature increase. In the second set of DC tests on the full-bridge topology, adding two extra fans cooled down the DC busbars and made it possible to achieve the predefined target.

The overall results of the DC testing on both half-bridge and full-bridge topologies verify that the improved cooling

TABLE III: Comparison of the Experimental Results

Test	$I_{DC,max}$ (ADC)	$R_{ds,om}$ (p.u.)	$T_{j,es}$ ($°C$)	Heat Flux (W/device)
Half-Bridge DC Test	190	1.31	69.75	4.42
Full-Bridge DC Test (Normal Cooling)	150	1.42	87.57	2.69
Full-Bridge DC Test (Advanced Cooling)	190	1.51	99.85	4.62

system (combination of the off-the-shelf heatsink and extra fans) is capable of cooling the system during equivalent rated conditions. Although the switching loss was not accounted for in these tests, the simulation results confirm that they are negligible compared to conduction loss. The obtained 4.62W/device heat flux from full-bridge DC testing will be used for FEA simulation during the final TMS design. However, a 25% of overdesign margin will be considered as a common practice which means the heat flux per device will be considered 5.78W/device.

C. AC Testing

After gaining confidence in the cooling system, by successfully passing the DC test for full-bridge, AC testing was performed on the module. Similar to the simulated system, an inductive load was driven by the drive module which is connected to a DC voltage source. Same as DC testing, AC testing was performed in steps where the DC voltage is set at a constant value, and using the modulation index, the load current is increased. Despite the DC testing, the $R_{ds,on}$ was not calculated in the AC testing since the exact current going through the switches was not measured. Thus, the estimated junction temperature was not directly calculated. However, the RTD reading results in the AC tests were compared to the ones in the DC testing. The module was tested at 258Arms for 16 minutes, where the RTDs showed 71.6°C as the thermal equilibrium point. Comparing the recorded

979-8-3503-3714-3/23 $31.00 © 2023 IEEE

RTD temperature with DC tests, it can be inferred that the estimated junction temperature is 105°C which is below the 125°C design constraint.

Having the results for both DC and AC testing, it can be concluded that the designed drive module is capable of properly functioning at rated conditions. The obtained results from these tests will be used to design the heat pipe based TMS.

VI. Conclusion

In this paper, the thermal performance of a module of an IMMD is investigated. At first, the drive module and TMS were modeled in PLECS software. Performing electrothermal simulations, the optimum range for TMS system configuration for such a system is determined. In other words, a step-by-step procedure including simulation, experimental tests, and post-processing of the results, was presented to evaluate the thermal performance of the drive module.

Next, a temporary cooling solution was chosen and utilized to perform DC testing and investigate the thermal performance of the drive module. The drive module was tested up to the equivalent rated load current. During these tests, the deficiency of the selected cooling system was revealed, which led to an increase in the DC busbars temperature. To mitigate the issue, two extra fans were added to cool down the busbars. Finally, the module successfully passed the DC tests by operating at the equivalent rated load current continuously until reached the thermal equilibrium point.

After successfully finishing the DC tests, AC testing was performed which the drive module was tested up to 96% of the rated load current, and using the RTD readings the junction temperature was estimated to be 105°C. Thus the drive module successfully passed all the thermal performance investigation tests. The obtained heat flux will be used in FEA simulations to finalize the heat pipe based TMS design.

References

[1] A. Ebrahimian, W. A. Khan, S. Iman Hosseini S, and N. Weise, "Electrothermal design of a gan-based axially stator iron-mounted fully integrated modular motor drive," in *2022 IEEE Transportation Electrification Conference & Expo (ITEC)*, 2022, pp. 733–739.

[2] S. Yin, K. J. Tseng, R. Simanjorang, Y. Liu, and J. Pou, "A 50-kw high-frequency and high-efficiency sic voltage source inverter for more electric aircraft," *IEEE Transactions on Industrial Electronics*, vol. 64, no. 11, pp. 9124–9134, 2017.

[3] W. Lee, S. Li, D. Han, B. Sarlioglu, T. A. Minav, and M. Pietola, "A review of integrated motor drive and wide-bandgap power electronics for high-performance electro-hydrostatic actuators," *IEEE Transactions on Transportation Electrification*, vol. 4, no. 3, pp. 684–693, 2018.

[4] S. Koushan, S. Vahid, and A. El-Refaie, "Study of the current ripple effect of a modular machine drive on torque ripple and losses for an spm machine with additively manufactured hollow conductor coils," in *2022 International Conference on Electrical Machines (ICEM)*, 2022, pp. 1948–1954.

[5] R. Abebe, G. Vakil, G. Lo Calzo, T. Cox, S. Lambert, M. Johnson, C. Gerada, and B. Mecrow, "Integrated motor drives: state of the art and future trends," *IET Electric Power Applications*, vol. 10, no. 8, pp. 757–771, 2016. [Online]. Available: https://ietresearch.onlinelibrary.wiley.com/doi/abs/10.1049/iet-epa.2015.0506

[6] R. A. Torres, H. Dai, T. M. Jahns, B. Sarlioglu, and W. Lee, "Cooling design of integrated motor drives using analytical thermal model, finite element analysis, and computational fluid dynamics," in *2021 IEEE Applied Power Electronics Conference and Exposition (APEC)*, 2021, pp. 1509–1509.

[7] S. Wu, C. Tian, W. Zhao, J. Zhou, and X. Zhang, "Design and analysis of an integrated modular motor drive for more electric aircraft," *IEEE Transactions on Transportation Electrification*, vol. 6, no. 4, pp. 1412–1420, 2020.

[8] F. Tokgoz, O. Gulsuna, F. Karakaya, G. Cakal, and O. Keysan, "Mechanical and thermal design of an optimized pcb motor for an integrated motor drive system with ganfets," *IEEE Transactions on Energy Conversion*, vol. 38, no. 1, pp. 653–661, 2023.

[9] A. H. Mohamed, H. Vansompel, and P. Sergeant, "An integrated modular motor drive with shared cooling for axial flux motor drives," *IEEE Transactions on Industrial Electronics*, vol. 68, no. 11, pp. 10 467–10 476, 2021.

[10] J. Swanke, H. Zeng, D. Bobba, T. M. Jahns, and B. Sarlioglu, "Design and testing of a modular high-speed permanent-magnet machine for aerospace propulsion," in *2021 IEEE International Electric Machines & Drives Conference (IEMDC)*, 2021, pp. 1–8.

[11] W. A. Khan, A. Ebrahimian, S. I. Hosseini S., and N. Weise, "Design of high current, high power density gan based motor drive for all electric aircraft application," in *2022 IEEE 9th Workshop on Wide Bandgap Power Devices & Applications (WiPDA)*, 2022, pp. 247–253.

[12] A. Tenconi, F. Profumo, S. E. Bauer, and M. D. Hennen, "Temperatures evaluation in an integrated motor drive for traction applications," *IEEE Transactions on Industrial Electronics*, vol. 55, no. 10, pp. 3619–3626, 2008.

[13] S. Lee, "Optimum design and selection of heat sinks," *IEEE Transactions on Components, Packaging, and Manufacturing Technology: Part A*, vol. 18, no. 4, pp. 812–817, 1995.

[14] A. H. Mohamed, H. Vansompel, and P. Sergeant, "Electrothermal design of a discrete gan-based converter for integrated modular motor drives," *IEEE Journal of Emerging and Selected Topics in Power Electronics*, vol. 9, no. 5, pp. 5390–5406, 2021.

The Effect of Cryogenic Temperature on Subthreshold Hysteresis of Commercial SiC Power MOSFETs

Monikuntala Bhattacharya, Michael Jin, Jiashu Qian, Limeng Shi, Hengyu Yu, Marvin H. White and Anant K. Agarwal

Dept. of Electrical & Computer Engineering, The Ohio State University, Columbus, Ohio 43210, USA

bhattacharya.119@osu.edu, agarwal.334@osu.edu

Abstract—This work explores the effect of cryogenic temperature on the subthreshold hysteresis for commercially available 1.2kV SiC Planar and Trench MOSFETs. With the reduction in temperature, planar devices show a significant increase in hysteresis, whereas trench devices show the opposite trend. For a particular temperature, the subthreshold slope has been found to be significantly shifted for trench devices in comparison to room temperature, whereas no change in slope has been observed in planar devices. The underlying mechanism has been investigated and presented in this work.

Keywords—*SiC MOSFET, Planar and trench, interface traps, hysteresis, subthreshold voltage shift*

I. INTRODUCTION

Silicon Carbide (SiC) based converters for high-power applications have gained popularity due to their superior material properties like higher breakdown voltage, better thermal conductivity, and faster switching [1]. In recent years, SiC MOSFETs have gained interest for their potential applications at low temperatures, such as cryogenic instrumentation, magnetic energy storage systems, space electronics, etc.[2, 3] The reliability and ruggedness of commercially available 4H-SiC planar double-diffused MOSFET (DMOSFET) and trench MOSFET under room temperature and high temperatures have been studied extensively in the past decades due to their potential industrial applications[4-6], but the low-temperature studies are relatively few[7].

One of the primary challenges with SiC MOSFETs is the extremely high interface state density, which results in inferior channel mobility and threshold voltage instability [4, 8]. In power converters where multiple such MOSFETs are connected in parallel, instability in threshold voltage during operation results in uneven current sharing that leads to long-term reliability issues. Depending on the nature and behaviour, the threshold voltage instability can be categorised into two groups, namely, fully recoverable hysteresis shift and permanent bias temperature instability (*i.e.*, PBTI and NBTI) [5]. Although there are numerous studies on the permanent BTI drift under different gate bias conditions at high temperatures, very few highlighted the hysteresis effect in detail. [9, 10].

In this work, the impact on subthreshold voltage hysteresis due to cryogenic temperature has been presented by comparing commercially available 1.2 kV planar DMOSFET and trench MOSFET.

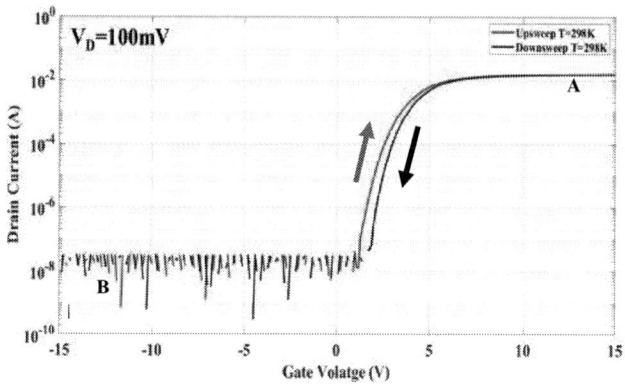

Fig. 1. Transfer characteristic curve of vendor E at room temperature

II. HYSTERESIS PHENOMENA

The hysteresis effect (V_{th} hysteresis) can be seen by utilizing the transfer characteristic of SiC MOSFETs in the subthreshold regime. Fig. 1. displays the transfer characteristics of planar MOSFETs from Vendor E at room temperature (298K) under a constant drain voltage (V_D) of 100 mV. The gate voltage (V_G) is swept from -15V to +15V, as shown in the red curve, called the up-sweep, whereas V_G: +15V to -15V, as shown in the black curve, is known as down-sweep. The change in voltage during hysteresis (ΔV_{th}^{hyst}) has been measured by the change in gate voltage during up-sweep (V_{th}^{up}) and down-sweep (V_{th}^{down}) at a constant drain current of 100nA.

Mathematically, it can be given as, [10]

$$\Delta V_{th}^{hyst} = V_{th}^{down} - V_{th}^{up} \qquad (1)$$

The main contributor to the hysteresis phenomena is the presence of defect states at the SiC/SiO$_2$ interface that undergo trapping and de-trapping of electrons and holes under the gate sweep.[9] One simple way to explain this is by considering points A and B in Fig. 1. At point A, the MOSFET channel is inverted and all the interface states across the bandgap are filled by electrons. At point B, the MOSFET channel is accumulated with holes and all the interface states across the bandgap are empty. By utilizing the shift of threshold voltage, the total number of defect states (N_{it}) across the entire bandgap can be calculated as follows: [10]

$$\Delta V_{th}^{hyst} = \frac{Q_{it}}{C_{ox}} = \frac{qN_{it}}{C_{ox}} = \frac{qN_{it}t_{ox}}{\xi_{ox}\xi_0} \qquad (2)$$

This work was supported by the Block Gift Grant from II-VI (Coherent) Foundation.

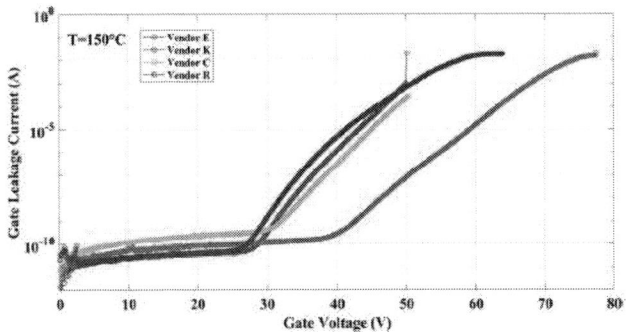

Fig. 2. Ramp-to-Breakdown measurement of all vendors at 150°C

where Q_{it} is the total interface trapped charge, C_{ox} is the gate oxide capacitance, and t_{ox} is the gate oxide thickness.

A. Understanding Threshold Voltage During Upsweep

Under negative gate voltage, i.e., deep accumulation, Fermi level (E_f) gets pinned to the valance band, resulting in positively charged interface traps. When the system is pushed to inversion from accumulation, the Fermi level moves across the entire band gap, resulting in a non-steady state for the interface traps, which comes back to equilibrium again by electron capture from the conduction band and hole emission into the valance band. Although electron capture is an instantaneous process, hole emission time constant (τ_{ep}) depends on temperature and can be given as[10]

$$\tau_{ep} = \frac{1}{v_T \sigma N_v} \exp\left(\frac{E_t - E_v}{KT}\right) \qquad (3)$$

where v_T is the thermal drift velocity, σ is the capture cross section and N_v is the effective density of states in the valance band.

B. Understanding Threshold Voltage during Downsweep

During deep inversion, the Fermi level is pinned to the conduction band edge, and when the gate bias is swept from deep inversion (+15V) to the gate voltage at which V_{th} is measured (V_{th}^{down}), the position of the Fermi level changes slightly. The thermal equilibrium is restored by electron emission to the conduction band. The electron emission time constant (τ_{en}) can be given as[10]

$$\tau_{en} = \frac{1}{v_{th} \sigma N_c} \exp\left(\frac{E_c - E_t}{KT}\right) \qquad (4)$$

The restoration process is almost instantaneous since the trap states are energetically located very close to the conduction band.

III. EXPERIMENTAL PROCEDURE

All the measurements have been carried out in a Lakeshore CCS-400/202 Cryostat. The device under test (DUT) is kept under a vacuum level of 0.1mTorr while cooled by pumping liquid Helium inside the cryostat to avoid condensation. The measurements have been carried out at 90K, 180K and 298K (Room Temperature). The static characterization, i.e., transfer characteristics, have been evaluated using a Keysight B1506A Power device analyzer

Table I. INFORMATION FOR TESTED DEVICES

Vendor	Structure	Device parameters		
		Estimated t_{ox}	*Current Rating*	*On Resistance*
E	Planar	45.5nm	11A	280mΩ
K	Asymmetric Trench	70.4nm	13A	286mΩ
C	Planar	45.7nm	12A	690mΩ
R	Double Trench	58.1nm	17A	208mΩ

connected with the CCS-400/202 Cryostat. The measurement has been performed after the temperature is stabilized within 0.1K of the desired value.

In order to reliably estimate the gate oxide thickness, high-temperature (150°C) ramp-to-breakdown measurements have been done and shown in Fig. 2. Using an oxide electric field of 11MV/cm[11, 12], t_{ox} has been calculated for all the vendors. The detailed information on all the tested devices has been presented in Table I.

IV. RESULTS AND DISCUSSION

Fig. 3(a) shows the transfer characteristics at V_D=100mV and V_G=±15V for 1200V planar MOSFET from Vendor E under all temperature variations (T= 90K,180K and 298K). It is evident that with the decrease in temperature, ΔV_{th}^{hyst} increases significantly, i.e., ΔV_{th}^{hyst} has a negative temperature dependency. On the contrary, the asymmetric trench device from Vendor K (Fig. 3(b)) has shown a positive temperature dependency. A similar effect has been observed in other vendors, namely Vendor C (planar) and Vendor R (double trench), as shown in Fig. 3(c) and (d), respectively. Fig.4 represents the ΔV_{th}^{hyst} shift from all the vendors for T= 90K, 180K and 298K.

The effect of temperature on subthreshold hysteresis for planar and trench MOSFETs can be understood by considering the variation in device structure and fabrication process along with the effect of temperature on threshold voltage.

(a) *Temperature vs. Threshold Voltage*: Temperature plays an important role in determining the Fermi level position in the band gap. At the onset of the strong inversion, where V_{th} is measured, surface potential (φ_s) equals the twice of Fermi potential (φ_F), i.e., φ_s=2 φ_F[13].
With the decrease in temperature, the Fermi level moves closer to the valance band in the p-bulk. Therefore, additional band bending is required to invert the surface (φ_s=2φ_F). In other word, the threshold voltage increases as the temperature goes down. A variation of threshold voltage with temperature, measured in the downsweep curve and I_D=100nA, has been presented in Fig.5.

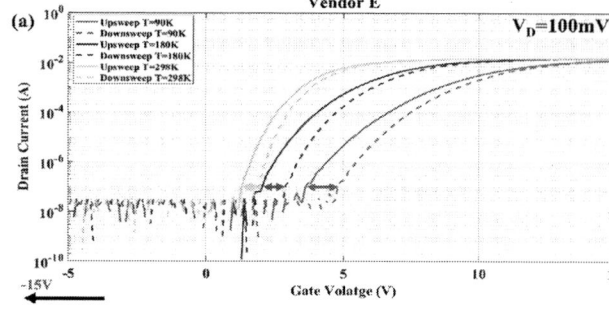

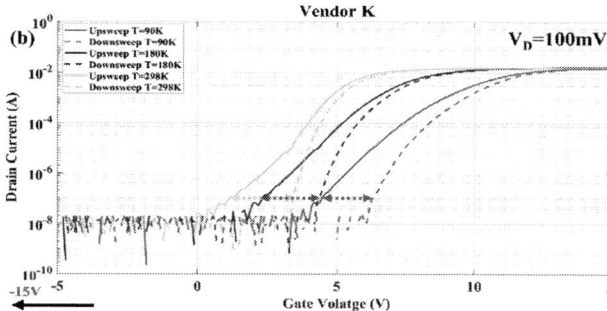

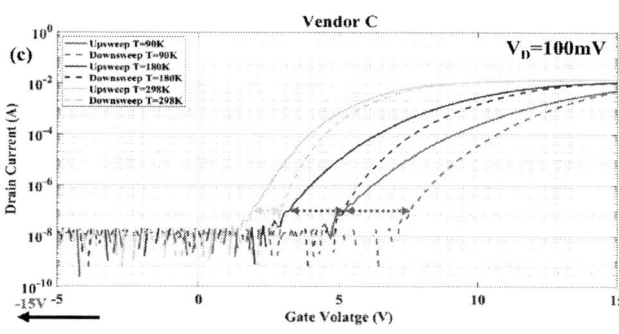

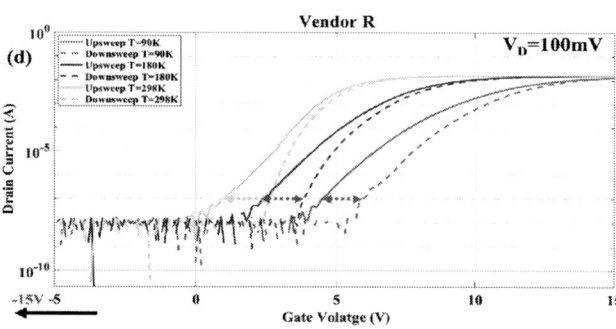

Fig. 3. Transfer characteristic curves of (a) Vendor E, (b) Vendor K, (c) Vendor C and (d) Vendor R at V_D = 100mV and T=90K, 180K and 298K

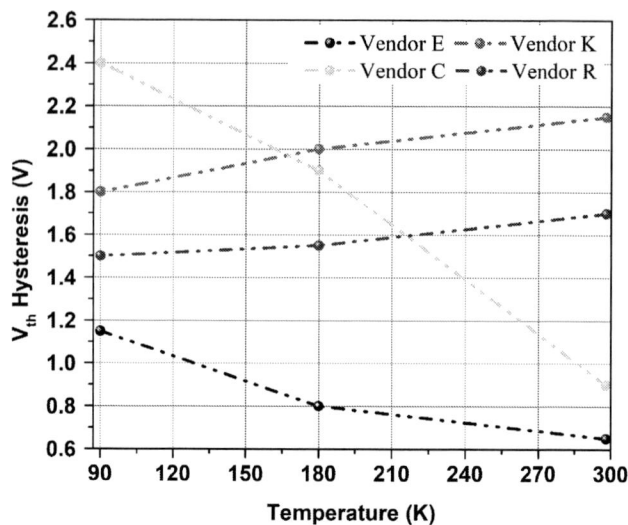

Fig. 4. ΔV_{th}^{hyst} variation with temperature for all tested devices

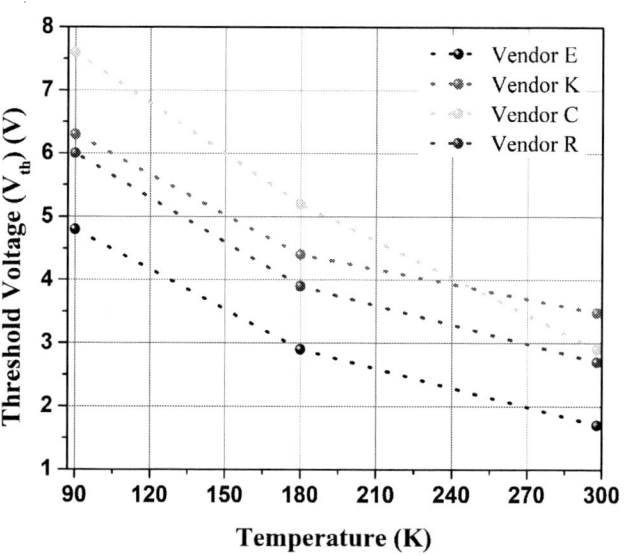

Fig. 5. V_{th} variation with temperature

(b) *Variation in oxide traps due to processing*: Trap states present in the gate oxide are considered as the primary cause behind the subthreshold hysteresis. The total number of near interface traps has been calculated for all the vendors using eq. 2 and at a temperature of 298K and has been shown in Table II. Previous works have also reported a higher density of oxide trap states at or near the interface states for trench devices compared to planar devices[14]. This higher density of traps may have been introduced due to different gate oxide processing techniques, that is, deposition of gate oxide instead of thermal growth[6].

(c) *Device Structure*: Another important factor that can substantially affect the subthreshold hysteresis behavior is the device structure[15]. Considering the trapping and de-trapping mechanism explained in the previous section, during upsweep, V_{th}^{up} depends on electron capture and hole emission process. The gate oxide in planar and trench devices is grown or deposited on different crystal planes

979-8-3503-3714-3/23 $31.00 © 2023 IEEE   40

which may have significant effect on energy dependence of the density of interface traps and their capture cross-sections in different parts of the band gap.

Table II.   CALCULATED N_{it} FOR ALL VENDORS AT ROOM TEMPERATURE (298K)

Vendor	Structure	Parameters	
		Estimated t_{ox}	Calculate N_{it}
E	Planar	45.5nm	3.08×10^{11}/cm³
K	Asymmetric Trench	70.4nm	6.59×10^{11}/cm³
C	Planar	45.7nm	4.25×10^{11}/cm³
R	Double Trench	58.1nm	6.31×10^{11}/cm³

During the downsweep, the V_{th}^{down} primarily depends only on the measurement temperature due to the movement of Fermi level, for both planar and trench devices. However, in case of the upsweep the mechanism alters significantly. Due to the variation in structure and the presence of more hole traps, the increment of V_{th}^{up} is more prominent on trench devices compared to their planar counterparts. As a result, $\Delta V_{th}^{hyst} = V_{th}^{down} - V_{th}^{up}$, goes up as the temperature goes down for the planar device and goes down as the temperature goes down for the trench device.

V.   CONCLUSION

Fully recoverable hysteresis measurement has been proven to be an important method to determine the total number of defect states present at the SiC/SiO₂ interface, distributed over the entire bandgap. This paper has highlighted the effect of temperature on the hysteresis measurement on commercially available planar and trench MOSFETs. Planar MOSFETs (Vendor E and Vendor C) have shown a negative temperature dependency, where as trench MOSFETs (Vendor K and Vendor R) have displayed positive temperature dependency of hysteresis shift. It can be interpreted by considering the device structure and gate oxide quality of both the devices. Presence of higher density of traps increases the threshold voltage during upsweep, for trench devices, which leads to a net reduction of hysteresis threshold voltage, whereas for planar devices no such effect can be seen. As a result, the planar devices and trench devices display opposite temperature dependency.

ACKNOWLEDGEMENT

The authors like to acknowledge Coherent (previously known as II-VI) Foundation for the funding.

REFERENCES

[1] J. Millan, P. Godignon, X. Perpiñà, A. Pérez-Tomás, and J. Rebollo, "A survey of wide bandgap power semiconductor devices," *IEEE transactions on Power Electronics*, vol. 29, pp. 2155-2163, 2013.

[2] K. Rajashekara and B. Akin, "Cryogenic Power Conversion Systems: The next step in the evolution of power electronics technology," *IEEE Electrification Magazine*, vol. 1, pp. 64-73, 2013.

[3] L. J. Woodend, P. M. Gammon, V. A. Shah, A. Pérez-Tomás, F. Li, D. P. Hamilton, *et al.*, "Cryogenic characterisation and modelling of commercial SiC MOSFETs," in *Materials Science Forum*, 2017, pp. 557-560.

[4] X. Zhong, H. Jiang, G. Qiu, L. Tang, H. Mao, C. Xu, *et al.*, "Bias temperature instability of silicon carbide power MOSFET under AC gate stresses," *IEEE Transactions on Power Electronics*, vol. 37, pp. 1998-2008, 2021.

[5] B. Asllani, A. Fayyaz, A. Castellazzi, H. Morel, and D. Planson, "VTH subthreshold hysteresis technology and temperature dependence in commercial 4H-SiC MOSFETs," *Microelectronics Reliability*, vol. 88, pp. 604-609, 2018.

[6] S. Zhu, L. Shi, M. Jin, J. Qian, M. Bhattacharya, H. L. R. Maddi, *et al.*, "Reliability Comparison of Commercial Planar and Trench 4H-SiC Power MOSFETs," in *2023 IEEE International Reliability Physics Symposium (IRPS)*, 2023, pp. 1-5.

[7] K. Tian, A. Hallen, J. Qi, M. Nawaz, S. Ma, M. Wang, *et al.*, "Comprehensive characterization of the 4H-SiC planar and trench gate MOSFETs from cryogenic to high temperature," *IEEE Transactions on Electron Devices*, vol. 66, pp. 4279-4286, 2019.

[8] S. Yu, M. H. White, and A. K. Agarwal, "Experimental determination of interface trap density and fixed positive oxide charge in commercial 4H-SiC power MOSFETs," *IEEE Access*, vol. 9, pp. 149118-149124, 2021.

[9] D. Peters, T. Aichinger, T. Basler, G. Rescher, K. Puschkarsky, and H. Reisinger, "Investigation of threshold voltage stability of SiC MOSFETs," in *2018 IEEE 30th International Symposium on Power Semiconductor Devices and ICs (ISPSD)*, 2018, pp. 40-43.

[10] T. Aichinger, G. Rescher, and G. Pobegen, "Threshold voltage peculiarities and bias temperature instabilities of SiC MOSFETs," *Microelectronics Reliability*, vol. 80, pp. 68-78, 2018.

[11] S. Zhu, T. Liu, M. H. White, A. K. Agarwal, A. Salemi, and D. Sheridan, "Investigation of gate leakage current behavior for commercial 1.2 kV 4H-SiC power MOSFETs," in *2021 IEEE International Reliability Physics Symposium (IRPS)*, 2021, pp. 1-7.

[12] T. Liu, S. Zhu, S. Yu, D. Xing, A. Salemi, M. Kang, *et al.*, "Gate leakage current and time-dependent dielectric breakdown measurements of commercial 1.2 kV 4H-SiC power MOSFETs," in *2019 IEEE 7th Workshop on Wide Bandgap Power Devices and Applications (WiPDA)*, 2019, pp. 195-199.

[13] H. L. R. Maddi, S. Nayak, V. Talesara, Y. Xu, W. Lu, and A. K. Agarwal, "Characterization of Near Conduction Band SiC/SiO 2 Interface Traps in Commercial 4H-SiC Power MOSFETs," in *2022 IEEE 9th Workshop on Wide Bandgap Power Devices & Applications (WiPDA)*, 2022, pp. 22-25.

[14] M. Chaturvedi, S. Dimitrijev, D. Haasmann, H. A. Moghadam, P. Pande, and U. Jadli, "Comparison of commercial planar and trench SiC MOSFETs by electrical characterization of performance-degrading near-interface traps," *IEEE Transactions on Electron Devices*, vol. 69, pp. 6225-6230, 2022.

[15] B. Asllani, A. Castellazzi, O. A. Salvado, A. Fayyaz, H. Morel, and D. Planson, "V TH-hysteresis and interface states characterisation in SiC power MOSFETs with planar and trench gate," in *2019 IEEE International Reliability Physics Symposium (IRPS)*, 2019, pp. 1-6.

979-8-3503-3714-3/23 $31.00 © 2023 IEEE

Etch Depth Study for Step-Etched Junction Termination Extensions in Vertical GaN Devices

Andrew T. Binder*, Jeffrey Steinfeldt, Andrew A. Allerman, Brian D. Rummel, Caleb Glaser, Luke Yates, and Robert J. Kaplar

Sandia National Laboratories, Albuquerque, NM, 87123 USA
*abinder@sandia.gov

Abstract—**This work reports on the optimal dose for a step-etched single-zone junction termination extension by means of a multi-point study on etch depth. Breakdown and device characteristics are reported on over one hundred devices for each dataset to determine a statistically significant representation of the population. Electroluminescence imaging during avalanche breakdown confirms the point at which the JTE switches from full depletion to partial depletion, which corresponds to the maximum breakdown.**

Index Terms—**Vertical GaN, Edge Termination, Junction Termination Extension, Power Devices, Breakdown**

I. INTRODUCTION

Wide-bandgap semiconductors have a significant advantage over conventional Si-based electronics by leveraging materials properties to achieve higher breakdown voltage, lower on-resistance, and high-frequency operation [1], [2]. Gallium nitride (GaN) is of interest for vertical device architectures and specifically for MOSFETs where channel mobility advantages [3], [4], compared to SiC, may position it well for sub-1200 V applications where the advantage of SiC starts to diminish.

One of the principal design challenges for vertical power devices is the management of electric fields at the periphery [5], [6]. For real non-ideal devices, the breakdown is limited by a large surface electric field, yielding a much lower breakdown voltage than for an ideal planar device limited by the peak electric field in the bulk. To address this issue, a junction termination extension (JTE) around the edges of the power device can be formed so that the depletion near the surface is increased, thereby reducing field crowding. Compared to SiC devices, limitations in selective-area doping for GaN [7] provides an additional challenge in designing edge termination structures. One common edge termination style for vertical GaN devices is the step-etched JTE, which does not require ion implantation [8]. Due to the etching processes required to form the JTE, and subsequent passivation, surface conditions can have a large impact on JTE effectiveness.

This study reports on the optimal etch depth to achieve peak breakdown voltage by evaluating a range of JTE doses.

This research was supported by the US Department of Energy (DOE) Vehicle Technologies Office (VTO) under the Electric Drive Train Consortium. Sandia National Laboratories is a multi-mission laboratory managed and operated by National Technology & Engineering Solutions of Sandia, LLC, a wholly owned subsidiary of Honeywell International Inc., for the U.S. Department of Energy's National Nuclear Security Administration under contract DE-NA0003525.

Compared to other reports on this topic [9], [10], this study is based on a statistically significant data set, which provides a more robust outlook on optimization compared to limited quantity datasets.

II. EDGE TERMINATION THEORY

Edge termination design is bounded in theory by Gauss's law and the charge enclosed in the Gaussian surface (the JTE) [11], [12]. The principle behind edge termination is that an abrupt termination of the junction at the periphery of the device causes field crowding at the surface leading to premature breakdown resulting from electric field enhancement at the edge. Edge terminations like the JTE are designed to spatially spread the electric field at the surface and thereby reduce field crowding. Gauss's law relates the electric flux through a closed surface to the total charge inside that surface as shown in Eq. 1:

$$\phi = E \cdot A = \frac{\sigma \cdot A}{\varepsilon_r \cdot \varepsilon_0} \tag{1}$$

where ϕ is the electric flux, σ is the charge per unit area, ε is the permittivity, A is the area of the Gaussian surface, and E is the electric field. In this manner, the charge inside the JTE acts as a sink for the electric field (E-field) at breakdown, spreading the field spatially and causing the field lines to terminate into the JTE. A well-designed edge termination minimizes the surface E-field relative to the E-field in the bulk thereby maximizing the breakdown voltage of the device.

Following Gauss's law the charge (σ) per unit area can be represented as the JTE charge (n_{JTE}) times the elementary charge q (Eq. 2).

$$\sigma = n_{JTE} \cdot q \tag{2}$$

Rearranging these equation presents a relationship between JTE charge, which is the design target, and the critical electric field (E_{crit}) (Eq. 3).

$$n_{JTE} = \frac{E_{crit} \cdot \varepsilon_r \cdot \varepsilon_0}{q} = N_A \cdot t_{jte} \tag{3}$$

In the case of a step-etched JTE, the dose (or charge) in the JTE is simply the doping and thickness product in the region. Hence, for a given doping profile, dose can be controlled by varying the etch depth.

979-8-3503-3714-3/23 $31.00 © 2023 IEEE

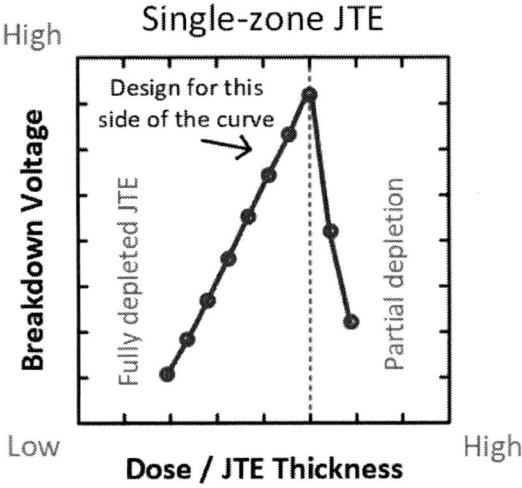

Fig. 1: Theoretical relationship between JTE dose/thickness and breakdown voltage for a single-zone JTE.

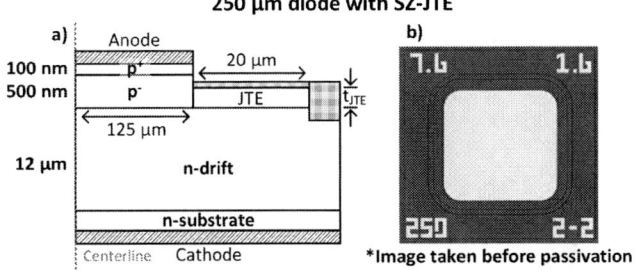

Fig. 2: (a) Axisymmetric schematic representation of devices under test and (b) a microscope image of a representative device as fabricated prior to passivation.

Given these relationships, deviation from the ideal JTE charge will result in increased surface E-field crowding and a reduction in breakdown. If the JTE dose is too high, it will not fully deplete at breakdown, and in the ideal design, the JTE is perfectly depleted exactly when the material's critical electric field is reached. This is represented in Fig. 1, which demonstrates this theoretical relationship between breakdown voltage and JTE dose/thickness. For a single-zone JTE, the JTE efficiency (the ratio of the breakdown voltage in a terminated device to the breakdown voltage of an ideal one-dimensional device) is very sensitive to changes in total charge within the Gaussian surface, and therefore sensitive to the thickness and doping density of the JTE [13], [14]. This sensitivity motivates the need to calibrate the dose by means of a multi-point study.

III. DEVICE STRUCTURE AND FABRICATION DETAILS

For the purpose of an edge termination study a simple pn diode is used as the baseline device. These results are relevant for any pn junction based (GaN) device including MOSFETs and JFETs. The pn diodes are epitaxially grown using a Veeco metal-organic chemical vapor deposition (MOCVD) system on a 2" freestanding c-plane GaN substrate from Mitsubishi Chemical Corporation (MCC). The substrate is n-type Si doped ($n_0 = 1 \times 10^{17} cm^{-3}$) followed by epitaxial layers that form the n-type drift [Si] (12 μm at $N_D \approx 1.5 \times 10^{16} cm^{-3}$) and p-type anode consisting of 500 nm p$^-$ GaN (Mg $\approx 1 \times 10^{18} cm^{-3}$), 100 nm p$^+$GaN (Mg $\approx 3 \times 10^{19} cm^{-3}$), and 10 nm p^{++}GaN contact layer (Mg $\approx 2 \times 10^{20} cm^{-3}$). The pn diode has an area of 250-μm-by-250-μm with a single-zone step-etched JTE that is 20 μm in length. Further details of the device structure and a device image are provided in Fig. 2.

Device processing followed our standard procedure [15]. Etch isolation and JTE etching were performed via BCl$_3$/Cl$_2$-based inductively coupled plasma reactive ion etching (ICP-RIE). Anode metalization consists of a Pd/Au metal annealed at 600 °C. The cathode metalization consists of a non-alloyed Ti/Al/Ti/Ni/Au contact. Device passivation consists of a bi-layer 100 nm thick atomic layer deposited (ALD) Al$_2$O$_3$ plus 2 μm of SiN by plasma enhanced chemical vapor deposition (PECVD).

IV. RESULTS AND DISCUSSION

To investigate the impact of JTE thickness on breakdown voltage, this experiment evaluates three quarter wafers each with varying edge termination etch depth. Devices were evaluated using a Keysight B1505A Power device Analyzer/Curve Tracer. The fabricated pn diodes were immersed in Fluorinert (FC-70) during testing to avoid premature breakdown in air or at the device surface. Measurements were performed under dc bias unless otherwise indicated. For this experiment, forward bias tests to 8 mA were used to screen outliers. After forward bias screening, the data presented here-in represents a minimum of 50 devices per subset (up to 100 devices depending on yield). Reverse I-V characteristics were further screened to present the top 50% performers based on maximum breakdown voltage. It should be noted that the trends presented in this work hold true even if considering 100% of devices after forward I-V screening.

Results for JTE thicknesses of 140 nm, 250 nm, and 295 nm are shown in Fig. 3. Notice that the median breakdown voltage appears to monotonically increase with increasing JTE thickness. Average breakdown from the three quarters (the data from Fig. 3) is shown in Fig. 4 (a) and compared to the simulation model. Experimental results match well to the simulation model, with the simulation model predicting that for t_{JTE} of 140 and 250 nm, the edge termination is fully depleted, and for t_{JTE} of 295 nm, the edge termination is only partially depleted (refer back to Fig. 1). Electroluminescence imaging of the devices during avalanche confirm this theory, showing that for t_{JTE} of 250 nm the location of breakdown is at the inner edge of the JTE [Fig. 4 (b)], which corresponds to a fully depleted JTE. Likewise, for t_{JTE} of 295 nm the breakdown location shifts to the outer edge of the JTE [Fig. 4 (c)] corresponding to a partially depleted JTE.

Optimization of dose in the JTE is a complex topic in that it requires understanding of a variety of non-ideal factors that influence the net dose seen by the JTE. To match the simulation and theory-based optimal dose to the experimental

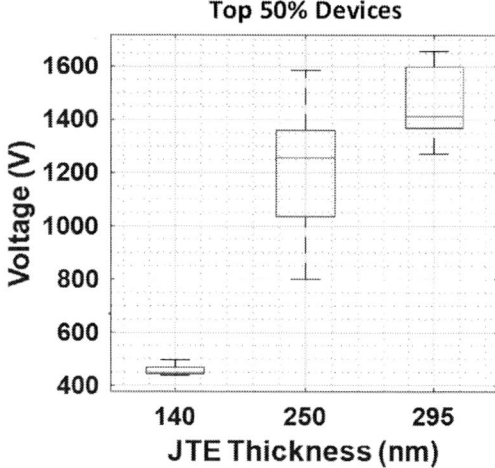

Fig. 3: Statistical breakdown results as a function of JTE thickness for the screened devices. Breakdown data shows a monotonic increase in voltage with increasing JTE thickness.

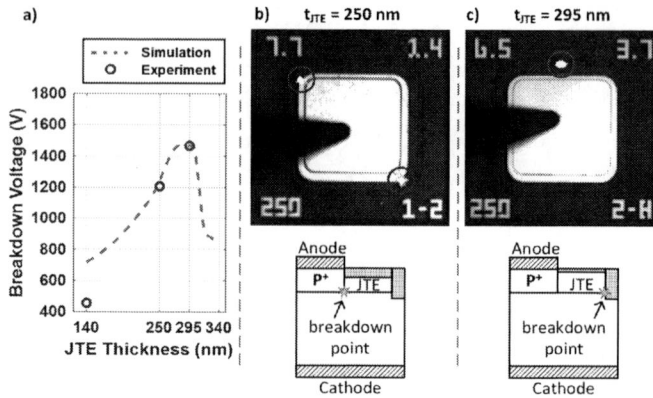

Fig. 4: (a) Breakdown results comparing experimental data to the simulation model. Images showing electroluminiscence during avalanche confirm that at t_{JTE} = 250 nm (b) the JTE fully depletes and breakdown occurs near the inner edge. At t_{JTE} = 295 nm (c) the JTE does not fully deplete and breakdown occurs near the outer edge.

results, a number of factors were considered, which provided the necessary offset and adjustment. The primary factors under consideration that influence JTE dose are from fixed charge in the passivation, plamsa induced etch damage to the surface, and alterations in the p-type doping profile. Further details on these non-ideal effects is expected to be the subject of future work. For more details on the simulation model, see our previous work [8], [16].

V. CONCLUSION

Edge termination optimization requires careful calibration to ensure a high JTE efficiency. A number of factors can influence dose in the JTE and shift the optimal design point considerably from the theoretical baseline. In theory, maximum breakdown voltage can be achieved when the JTE dose matches the critical electric field of the semiconductor, which will occur at the border between partial and full depletion of the JTE. Use of electroluminescence imaging during avalanche breakdown confirmed that the JTE switched from being fully depleted at a thickness of 250 nm to being partially depleted at a thickness of 295 nm. The thickest JTE (t_{JTE} = 295 nm) showed the highest average breakdown voltage, nearing 1500 V. Outliers from both t_{JTE} = 250 nm and t_{JTE} = 295 nm showed peak breakdowns of around 1600 V. Results presented match well against the theoretical model with some adjustment required to account for non-ideal factors, such as charge from the passivation and plasma-induced surface damage. More details on the contribution of these non-ideal factors is expected to be the subject of future work.

VI. ACKNOWLEDGEMENTS

The authors would like to acknowledge the contributions of the following individuals: Hoang Vuong for assistance with microfabrication, Mike Smith for assistance with electrical testing, Richard Floyd for discussions regarding scope and direction of the manuscript, and Paul Sharps for logistics support from his department.

REFERENCES

[1] J. A. Cooper and D. T. Morisette, "Performance Limits of Vertical Unipolar Power Devices in GaN and 4H-SiC," *IEEE Electron Device Letters*, vol. 41, no. 6, pp. 892–895, 6 2020.

[2] T. J. Anderson, S. Chowdhury, O. Aktas, M. Bockowski, and J. K. Hite, "GaN Power Devices – Current Status and Future Directions," *The Electrochemical Society Interface*, vol. 27, no. 4, pp. 43–47, 12 2018. [Online]. Available: http://interface.ecsdl.org/lookup/doi/10.1149/2.F04184if

[3] D. Ji, C. Gupta, S. H. Chan, A. Agarwal, W. Li, S. Keller, U. K. Mishra, and S. Chowdhury, "Demonstrating >1.4 kV OG-FET performance with a novel double field-plated geometry and the successful scaling of large-area devices," in *Technical Digest - International Electron Devices Meeting, IEDM*. Institute of Electrical and Electronics Engineers Inc., 1 2018, pp. 1–9.

[4] H. Otake, S. Egami, H. Ohta, Y. Nanishi, and H. Takasu, "GaN-based trench gate metal oxide semiconductor field effect transistors with over 100cm2/(Vs) channel mobility," *Japanese Journal of Applied Physics, Part 2: Letters*, vol. 46, no. 25-28, p. L599, 7 2007. [Online]. Available: https://iopscience.iop.org/article/10.1143/JJAP.46.L599 https://iopscience.iop.org/article/10.1143/JJAP.46.L599/meta

[5] V. A. Temple and W. Tantraporn, "Junction Termination Extension for Near-Ideal Breakdown Voltage in p-n Junctions," *IEEE Transactions on Electron Devices*, vol. 33, no. 10, pp. 1601–1608, 1986.

[6] W. Sung and B. J. Baliga, "A Comparative Study 4500-V Edge Termination Techniques for SiC Devices," *IEEE Transactions on Electron Devices*, vol. 64, no. 4, pp. 1647–1652, 4 2017.

[7] M. J. Tadjer, B. N. Feigelson, J. D. Greenlee, J. A. Freitas, T. J. Anderson, J. K. Hite, L. Ruppalt, C. R. Eddy, K. D. Hobart, and F. J. Kub, "Selective p-type Doping of GaN:Si by Mg Ion Implantation and Multicycle Rapid Thermal Annealing," *ECS Journal of Solid State Science and Technology*, vol. 5, no. 2, pp. P124–P127, 12 2016. [Online]. Available: http://jss.ecsdl.org/lookup/doi/10.1149/2.0371602jss

[8] J. R. Dickerson, A. T. Binder, G. Pickrell, B. P. Gunning, and R. J. Kaplar, "Simulation and Design of Step-Etched Junction Termination Extensions for GaN Power Diodes," in *4th Electron Devices Technology and Manufacturing Conference, EDTM 2020 - Proceedings*, Penang, 2020.

[9] H. S. Lee, Y. Zhang, Z. Chen, M. W. Rahman, H. Zhao, and S. Rajan, "Design and Fabrication of Vertical GaN p-n Diode with Step-Etched Triple-Zone Junction Termination Extension," *IEEE Transactions on Electron Devices*, vol. 67, no. 9, pp. 3553–3557, 9 2020.

[10] Y. Duan, J. Wang, Z. Zhu, G. Piao, K. Ikenaga, H. Tokunaga, S. Koseki, M. Bulsara, and P. Fay, "Ion-implanted triple-zone graded junction termination extension for vertical GaN p-n diodes," *Applied Physics Letters*, vol. 122, no. 21, p. 11, 5 2023. [Online]. Available: /aip/apl/article/122/21/212104/2892536/Ion-implanted-triple-zone-graded-junction

[11] T. Kimoto and J. A. Cooper, *Fundamentals of Silicon Carbide Technology: Growth, Characterization, Devices and Applications.* Wiley, 2014, vol. 9781118313.

[12] B. J. Baliga, *Fundamentals of power semiconductor devices.* Boston, MA: Springer US, 2008. [Online]. Available: http://link.springer.com/10.1007/978-0-387-47314-7

[13] X. Wang and J. A. Cooper, "Optimization of JTE Edge Terminations for 10 kV Power Devices in 4H-SiC," *Materials Science Forum*, vol. 457-460, pp. 1257–1262, 2009. [Online]. Available: www.scientific.net/MSF.457-460.1257

[14] W. Tantraporn and V. A. K. Temple, "Multiple-zone single-mask junction termination extension - A high-yield near-ideal breakdown voltage technology," *IEEE Transactions on Electron Devices*, vol. 34, no. 10, pp. 2200–2210, 1987.

[15] L. Yates, B. P. Gunning, M. H. Crawford, J. Steinfeldt, M. L. Smith, V. M. Abate, J. R. Dickerson, A. M. Armstrong, A. Binder, A. A. Allerman, and R. J. Kaplar, "Demonstration of >6.0-kV Breakdown Voltage in Large Area Vertical GaN p-n Diodes With Step-Etched Junction Termination Extensions," *IEEE Transactions on Electron Devices*, vol. 69, no. 4, pp. 1931–1937, 4 2022.

[16] A. T. Binder, J. R. Dickerson, M. H. Crawford, G. W. Pickrell, A. A. Allerman, P. Sharps, and R. J. Kaplar, "Bevel edge termination for vertical GaN power diodes," in *2019 IEEE 7th Workshop on Wide Bandgap Power Devices and Applications, WiPDA 2019.* Raleigh: IEEE, 2019, pp. 281–285.

Study of GaN HEMTs Robustness to Application-Like, Software-Controlled Overshoots Emulating Different Gate Routings in Original 50 Ohms Environment

Ludovic Roche*[†‡], David Trémouilles*, Emmanuel Marcault[†], Corinne Alonso*
*LAAS-CNRS, Université de Toulouse, CNRS, UPS, Toulouse, France
[†]CEA Occitanie, Toulouse, France
[‡]ludovic.roche@laas.fr

Abstract—In the field of transient tolerance tests, few studies have been conducted on gate voltage spikes of GaN HEMT components. A parametric generation of such overvoltage either implies precise and tedious hardware gate circuitry design, or require some simplification of the waveform making it less representative of actual and practical cases.

Such tests will be easier to conduct thanks to the original 50 Ohms environment setup proposed in this work. This original test bench improves usual measurement bandwidths. It also allows spatial isolation of a tested device with its driver and power circuits (e.g. over one meter), limiting their complex interactions.

As a first demonstration of the setup capability, the breakdown overvoltage of a p-GaN HEMT is determined. Several gate voltage overshoots as high as twice the DUT nominal rating are then demonstrated not to degrade the tested device. Moreover, allowing such overshoots may enable to significantly reduce switching losses.

Index Terms—GaN, Overshoot, Transient, Spike, Deported Control, 50 Ohm Environment, Robustness, characterization

Symbol Definition

- s is the Laplace variable;
- $\underline{f}(s)$ is the Laplace transform of f(t);
- $H(t)$ is the Heaviside function;
- $H_T(t) = H(t - T)$.

I. Introduction

Wide BandGap (WBG) devices display better behavior than Si devices regarding power and volume efficiency for static converters. GaN High Electron Mobility Transistors (HEMTs) enable higher working frequencies and reduced switching losses. Thus, smaller passive components are needed as filters in power applications. However, fast switching tend to trigger parasitic oscillations in both gate and drain circuitry. These oscillations result from the complex interaction of the inductive and capacitive behaviors of the components, circuits, and parasitics. They can lead to energy losses and failures. Equivalent models are proposed in literature [1], but identification

This work was supported by the LAAS-CNRS PROOF platform, partly financed by the Occitanie region.

of their parameters from actual PCBs is complex. Therefore, precisely predicting oscillations is much more difficult than modeling experimental ones. In particular, overvoltage studies require the generation of reproducible, precisely controlled and measurable OVSs. This entails either tedious PCB design, or the use of simpler waveforms, making them less representative of actual cases.

This work presents a generic characterization-setup design with full control of a power component state and its circuit parasitics. To do so, concepts from the radiofrequency transmission line model, and their application to move away a GaN HEMT from its command and power circuits are presented. A setup achieving high-frequency bandwidth is proposed along with its theoretical analysis. Several OVS robustness tests are conducted as a demonstration of the performances of the setup.

In section II, the origins and typical waveforms of gate OVerShoots (OVSs) are explored and compared to the existing literature of OVS robustness studies. The proposed setup and the underlying theoretical background is presented in section III. Experimental results are discussed in section IV.

II. GaN HEMTs Gate Overvoltages: Origins, Characterization, Consequences

Several phenomena can lead gate-source (or source sensing, if applicable) to hazardous voltage ringing and overshoots. Depending on the amplitude of the oscillations, the component is degraded at varying speeds [2]. An application-scale consequence is the false turn-off; upon switching on, the gate voltage oscillates lower than its threshold, causing undesired, partial switching [3].

Firstly is discussed typical origins and waveforms of gate OVS. Then, in the second part is a review of existing gate OVS robustness characterization methods.

A. Origins and Consequences of Gate Overshoots

GaN transistors feature intrinsic fast switching speed and excellent radiation tolerance. However, while fast transient are beneficial regarding the decrease of switching losses, the

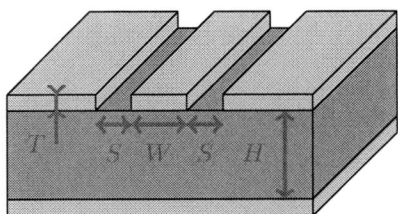

Fig. 1: A typical copper track section on a FR4 power PCB. $H = 1.5\,\text{mm}$, $S = 0.15\,\text{mm}$, $W = 0.2\,\text{mm}$, $T = 35\,\mu\text{m}$. $\epsilon_r = 4.5$, $\epsilon_{r,effective} = 2.20$. S and W are defined accordingly to EPC GaNs gate pads.

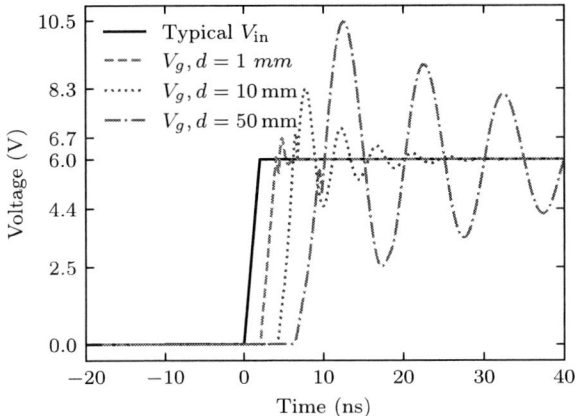

Fig. 2: Simulation: influence of the length of the gate copper track, time shifted for visual clarity. The modeled track is described in Fig. 1. Without parasitics, $V_g = V_{in}$.

parasitic RLC of the surrounding circuits are all the more excited by the high frequency content of the waveforms, creating oscillations jeopardizing the components.

Copper tracks alone provide enough inductance to create oscillations. A typical copper track section is presented in Fig. 1. According to [4], this section creates a linear inductance $L_l = 0.4\,\text{nH}\,\text{mm}^{-1}$.

As an example, the effects of the length of the copper track between a typical driver and a p-GaN HEMT GS-065-004-1-L gate are simulated in Fig. 2; the longer the track, the higher its inductance and the higher the perturbation on the gate signal. Indeed, the track inductance resonates with the transistor gate-capacitance, creating oscillations that can be modeled by a 2nd or 4th order oscillator [1]. This is the reason why it is common to reduce the length of copper tracks to achieve better signal [5].

Furthermore, gate signal distortions may have many other origins like radiated coupling of adjacent tracks, especially power signal tracks. Perturbations from other parts of a circuit can also propagate through power supplies and tracks. These perturbations can in turn resonate with the presented equivalent LC circuitry.

B. Robustness Characterization

Since overshoots can occur in the most common converter designs, as shown in Part. II-A, it is essential to quantify the robustness of GaN HEMTs to those overshoots.

Several setups allow studying GaN gate robustness.

DC studies are the most straightforward to conduct since LC elements does not interfere in this case. The gate DC Breakdown Voltage (BV) of GaN HEMTs is thus generally provided in all device datasheets.

He et al. [6] present a study between stepped DC and pulsed test, which progressively steps up a DC stress applied at $100\,\text{kHz}$, with $50\,\%$ duty cycle, until failure.

A setup approaching the typical waveforms presented in Part. 1 was designed by Wang et al. [2] [7], to find that gate overshoot robustness can be fitted with both lognormal and Weibull distribution. This design lets an inductance be charged upon a controlled level, to then discharge it in the gate of a GaN transistor. The inductance and gate capacitance then produce ringing similar to the expected realistic waveform but around $0\,\text{V}$.

OVS robustness studies are less common than DC ones due to the strong impact of the circuitry on the gate signals. To perform parametric testing, the waveforms applied must be precisely controlled, and there must be no parasitics due to, nor in the measurement path. Since only few millimeters of copper tracks suffice to create parasitics for fast transient signals, the resulting gate and measurement circuitry must be dense; its conception is therefore tedious. With GaN packages of different sizes and shapes, redesigning optimized test-PCBs can be time-consuming.

Moreover, while dense PCB design allows for the reduction of parasitic inductances, it makes impossible to apply an environmental stress to the device only, rising the need for robust and costly circuits for each device, and hampering any efforts to study GaN environmental reliability.

III. PROPOSED 50 OHM SETUP

A common solution for parasitic management at any length of conductors is controlled impedance. This concept is based on transmission lines equations, which is standard in the field of radio-frequency.

The proposed 50 Ohms circuitry is presented in the following parts. The transmission line theory completely encompasses the lumped impedance approach developed in Part. II-A, but is fairly different in its calculations and reasoning. Therefore, it is presented in the following parts independently of the lumped-model approach.

A. Transmission Lines: Selected Theoretical Elements

While lumped elements are a valid model for the propagation of electromagnetic signals in conductors of small length compared to the signal spatial wavelength, the tools provided by the transmission line model are valid for any length of cable.

A set of two conductors part of a current loop is defined by two main parameters in the transmission line model:

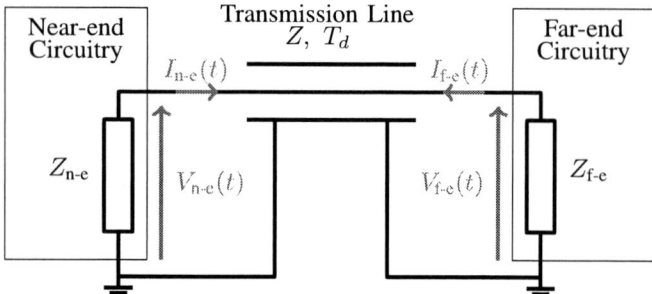

Fig. 3: Parameters of a transmission line and its environment.

- its characteristic impedance Z;
- the travel time of an electromagnetic signal across the length of the conductor T_d.

A circuit containing a transmission line can be described with the following set of parameters, schematized in Fig. 3):

- the transmission line, defined by its Z and T_d;
- the output impedance of the near-end side of the transmission line $Z_{\text{n-e}}$;
- the input impedance of the far-end side of the transmission line $Z_{\text{f-e}}$.

These elements allow calculating the near-end and far-end reflection coefficients:

$$\mathcal{R}_{\text{n-e}}(s) = \frac{Z_{\text{n-e}}(s) - Z}{Z_{\text{n-e}}(s) + Z} \ ; \ \mathcal{R}_{\text{f-e}} = \frac{Z_{\text{f-e}}(s) - Z}{Z_{\text{f-e}}(s) + Z} \quad (1)$$

While Z is usually real, these coefficients can be complex depending on $Z_{\text{n-e}}$ and $Z_{\text{f-e}}$. Note that a reflection coefficient is null only if the termination impedance is equal to the line impedance.

With these parameters, the system can be analyzed as such:

- By replacing the transmission line with its characteristic impedance Z, the voltage across Z is calculated. This voltage is the incident voltage wave $V_{i,\text{n-e}}(t)H(t)$;
- This voltage wave travels across the transmission line. Upon arriving at the far-end, the voltage wave:

$$V_{i,\text{f-e}}(t) = V_{i,\text{n-e}}(t - T_d)H_{T_d}(t) \quad (2)$$

is reflected. The reflected wave is defined by:

$$\underline{V_{r_1,\text{f-e}}} = \underline{V_{i,\text{f-e}}}\mathcal{R}_{\text{f-e}} \quad (3)$$

- This reflected voltage wave then travels back to the near-end and reaches it with a delay:

$$V_{r_1,\text{n-e}}(t) = V_{r_1,\text{f-e}}(t - T_d) \quad (4)$$

This wave is reflected as $V_{r2,\text{n-e}}(t)$, defined by:

$$\underline{V_{r_2,\text{n-e}}} = \underline{V_{r_1,\text{n-e}}}\mathcal{R}_{\text{n-e}} \quad (5)$$

- Subsequent calculations of reflections can be made following this method. If set as in Fig. 3, the same relations apply for current calculations.

- At any time, the apparent voltage is defined by the sum of all the voltage waves defined previously:

$$\begin{cases} V_{\text{tot,n-e}}(t) = V_{i,\text{n-e}}(t)H(t) + \sum_{k=0}^{\infty} V_{rk,\text{n-e}}(t) \\ V_{\text{tot,f-e}}(t) = V_{i,\text{f-e}}(t) + \sum_{k=0}^{\infty} V_{rk,\text{f-e}}(t) \end{cases} \quad (6)$$

- The DC analysis of the currents and voltages of the circuitry are obtained with an infinite number of rebounds that converge to a stationary DC state; thus, they do not depend on T_d. They can be obtained for any T_d, especially $T_d = 0$.

B. Global Description and Functioning of the Setup

The proposed setup is presented in Fig. 4. Following are presented the parts of the setup: driver-side, power-side, measurements; and its advantages for efficient characterization.

1) Driver side: The gate of the DUT is interfaced with a $50\,\Omega$ interconnection, and a voltage probe. The far-end of the gate $50\,\Omega$ coaxial cable is connected to a $50\,\Omega$-output-impedance waveform generator, in order for the reflection coefficient (1) at this end to be null.

As illustrated in Fig. 5, for any incident voltage $V_i(t)$ sent through the line by the generator, one reflection $V_r(t)$ occurs on the DUT side and there is no further reflection on the generator, thanks to the null reflection coefficient on its side.

Applying the results of the previous section (III-A), the following parameters are defined:

- the cable is defined by its impedance $Z_c = 50\,\Omega$ and delay T_c;
- the input impedance of the DUT gate is Z_g, and the output impedance of the generator is $Z_{\text{gen}} = 50\,\Omega$;
- the reflection coefficient on the DUT side of the cable is therefore:

$$\mathcal{R}_{\text{gate}} = \frac{Z_g - Z_c}{Z_g + Z_c} \quad (7)$$

Then, the two voltage waves present at any time at the DUT gate are:

- the generator incident voltage V_i, with a delay T_c created by the cable:

$$V_{i,\text{DUT}}(t) = V_i(t - T_c)H_{T_c}(t) \quad (8)$$

- and the reflected wave V_r, defined by $\mathcal{R}_{\text{gate}}$ and $V_{i,\text{DUT}}$:

$$\underline{V_{r,\text{DUT}}} = \mathcal{R}_{\text{gate}}\underline{V_{i,\text{DUT}}} \quad (9)$$

These waves are represented in Fig. 4. The apparent voltage at the DUT gate is thus:

$$V_g(t) = V_{i,\text{DUT}}(t) + V_{r,\text{DUT}}(t) \quad (10)$$

Then, by combining (9) and (10):

$$\underline{V_g} = \underline{V_{i,\text{DUT}}}\left(1 + \mathcal{R}_{\text{gate}}\right) \quad (11)$$

Hence, any desired gate voltage can be obtained by generating the appropriate V_i defined by (11) and (7):

979-8-3503-3714-3/23 $31.00 © 2023 IEEE

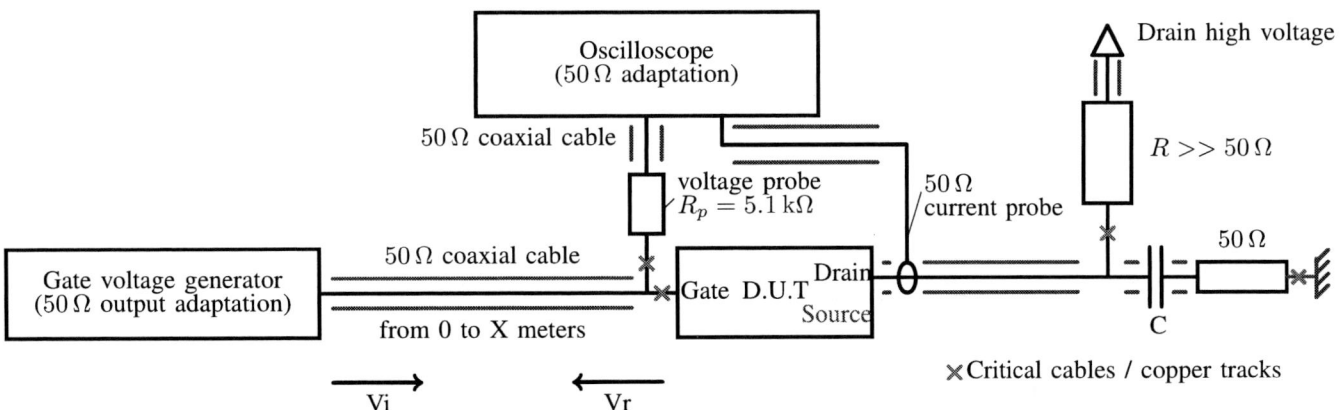

Fig. 4: Proposed $50\,\Omega$ setup. Each $50\,\Omega$ cable can be of arbitrary length, contrarily to the short critical paths.

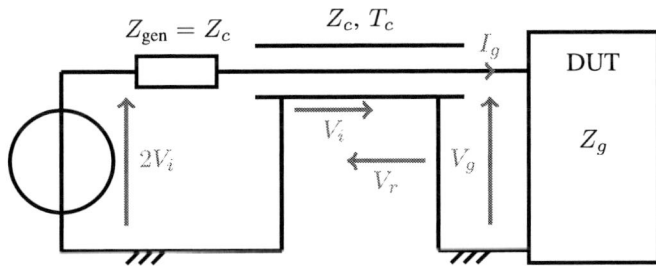

Fig. 5: Gate circuitry model. With a Thevenin model for the waveform generator, V_i is generated with a $2V_i$ voltage source.

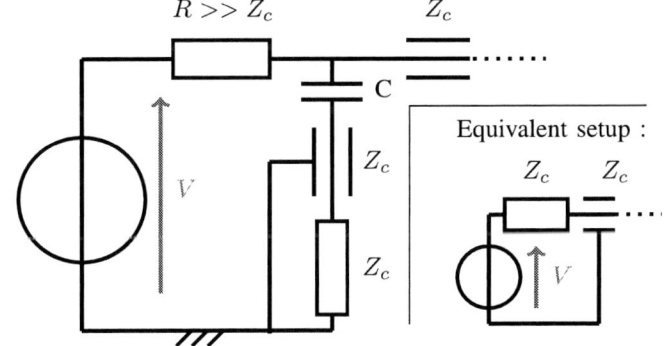

Fig. 6: Drain voltage generator circuitry.

$$V_{i,\text{DUT}} = V_g \frac{Z_g + Z_c}{2Z_g} \qquad (12)$$

Since Z_g is highly non-linear and, in some cases, depends on the power circuitry as well as the DUT intrinsic properties, (12) can be rewritten as:

$$V_{i,\text{DUT}} = \frac{V_g + I_g Z_c}{2} \qquad (13)$$

Since $V_{i,\text{DUT}}$ is only a T_g-shifted V_i, V_i is entirely defined by V_g and I_g. Moreover, with Z_c being real, (14) is similar in time-domain:

$$V_{i,\text{DUT}}(t) = \frac{V_g(t) + I_g(t)Z_c}{2} \qquad (14)$$

The reproduction of an in-application measured (V_g, I_g) is thus possible with the proposed setup. While V_g is usually measurable on PCBs, I_g can be obtained through simulation, knowing V_g.

As displayed in Fig. 5, a Z_c Thevenin voltage source providing $2V_i$ generates the appropriate V_i through the line.

In the following experiments, the theoretical V_i waveforms obtained by this method are synthesized thanks to a home-made generator. This waveform generator is currently in patenting process.

2) Power side: In the same way as in Part. III-B1, the drain of the DUT is connected to a cable of controlled impedance. This cable is terminated by the drain voltage generator, which is constituted of a high voltage source, a high value resistor R, a decoupling capacitor and an adaptation $50\,\Omega$ resistor. The drain voltage generator is represented in Fig. 6. This generator acts as a Z_C-adapted voltage source: the high resistor branch charges the capacitance and ensures the high-frequency stability of the voltage generator. Fast drain transients generated by the DUT are not reflected by the generator thanks to the capacitance and Z_c resistor.

The equivalent drain circuit is actually described by the same schematic as in Fig. 5, with Z_{DS} in place of Z_G. Eq. (14) can thus be rewritten as:

$$V_{i,\text{DUT}}(t) = \frac{V_{\text{DS}}(t) + Z_c I_{\text{DS}}(t)}{2} \qquad (15)$$

Eq. 15 is equivalent to the circuitry in Fig. 7. Correspondence between Fig. 6 and Fig. 7 confirms that $V = 2V_i$. The power side circuitry thus behaves as in Fig. 7. Hence, even tho the generator is deported, the drain-source voltage behaves as if commuting on an adjacent resistive charge.

3) Critical signal paths: In order to eliminate all parasitic, all copper tracks or cable must be of controlled impedance.

979-8-3503-3714-3/23 $31.00 © 2023 IEEE   49

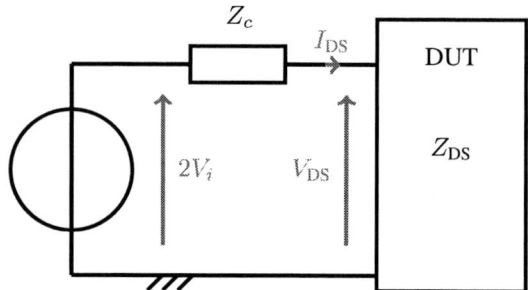

Fig. 7: Power side circuitry equivalent according to Fig. 6. As represented in Fig. 5, $2V_i = V$.

However, some signal paths must be as short as possible to prevent impedance discontinuity and unwanted signal rebounds. Those are marked with red crosses in Fig. 4.

- High value resistors, either for probing (R_p) or injecting small currents (R) must be as close to the main signal path as possible. If not, a signal travelling through the main line would be divided between its main track and this sub-path, with the sub-path termination not being adapted. Thus, rebounds would occur and be visible as parasitics on the main signal.
- Adaptation resistors must be connected directly to the ground. The rebounds and parasitic are as dampened as the ground paths are short.
- For voltage probing, the voltage probe must be as close as possible to the main line termination. If not, the delay δT between the probe and the line termination would be apparent in the measurement. In this later case:

$$V_{\text{probe}} = V_{i,\text{DUT}}(t + \delta T) + V_{r,\text{DUT}}(t - \delta T) \quad (16)$$

4) Experimental implementation and results: Several measurements are made to extract data using the proposed setup.

- The voltage probe allows monitoring V_g. As displayed in Fig. 7 and demonstrated in the corresponding calculations, a cable Z_c adapted at its termination behaves like a Z_c resistor. Thus, the oscilloscope voltage is $V_{\text{oscilloscope}} = \dfrac{50}{50 + R_p} V_{\text{GS}}$.
- The drain current probe is designed to keep the $50\,\Omega$ impedance and thus to avoid parasitic. Contrarily to the gate voltage probe that must be placed close to the gate, the position of the drain current probe on the line does not really matter. V_i is DC, so the time-shifting described by (16) only shifts the measurement in time without distorting it. This time-shift can be measured independently and compensated in the oscilloscope settings.
- The oscilloscope used for this series of experiment is a DSOS104A, with a 1 GHz bandwidth.

5) Advantages of the setup; Parasitics: By controlling the impedance of all interconnections, copper tracks inductances presented in part II-A are integrated in the calculations within the complete model of transmission lines. Therefore, there are no parasitics as long as the line impedance is controlled.

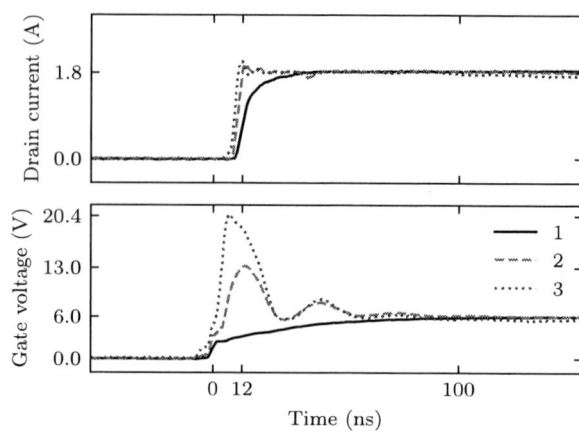

Fig. 8: Typical experimental waveforms applied to the gate to find its spike breakdown (22 V OVS). The drain switching energies are of $3.3\,\mu\text{J}$ (1) and $0.6\,\mu\text{J}$ (2).

However, electromagnetic coupling can still occur from the power side to the command side. This is prevented by only using well-defined, shielded sections, as in Fig. 1. These section have a well-defined impedance and the current loop is very small, due to the proximity of all conductors. $50\,\Omega$ cables are shielded and bring very little to no parasitics. Clean signal paths of controlled impedance are straightforward to route in this application due to the permitted long track length and small number of tracks. The only remaining source of parasitics is the interconnections or welds; these are however as small as the signal path through the welding is short, thus actually creating very little to no parasitics.

IV. ROBUSTNESS TESTING

Measurements were conducted with the proposed circuitry and a homemade gate voltage generator.

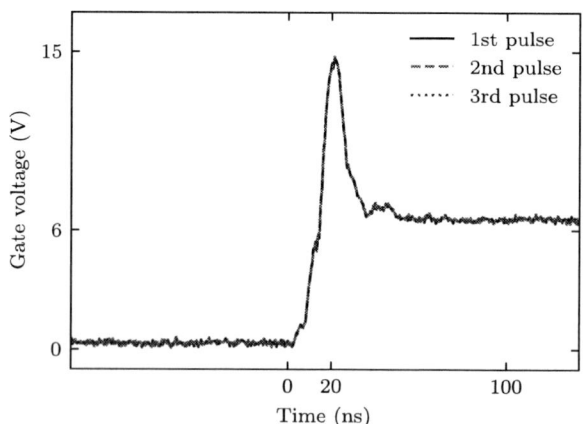

Fig. 9: Experimental stress waveforms. The excellent repeatability and low parasitics level of the generated waveforms are in line with the developped theoretical elements.

979-8-3503-3714-3/23 $31.00 © 2023 IEEE

Firstly, p-GaN HEMTs GS-065-004-1-L were submitted to a wide range of OVS, as displayed in Fig. 8, to find their approximate dynamic Breakdown Voltage BV_{dyn}. Their apparent gate-source resistance was measured with an ohmmeter (Keysight U1241C) before and after each stress. The resistance was not measurable by the ohmmeter for an unbroken device (over $10\,M\Omega$). The breakdown voltage was found to be around $22\,V$ OVS, after which the measured resistance was around $7.5\,M\Omega$.

After determining the approximate 1-pulse BV_{dyn}, the gate of two p-GaN HEMTs GS-065-004-1-L were submitted to a series of three $15\,V$ OVS displayed in Fig. 9. The superposition of the three pulses show the excellent replicability of the stress.

To monitor the gate and its possible degradation, two indicators are used: gate leakage and gate threshold measurements. The corresponding measurements are respectively presented in Fig. 10a and Fig. 10b.

No degradation was detected in Fig. 10, thus showing the robustness of GaN HEMTs to at least few, short gate voltage OVS.

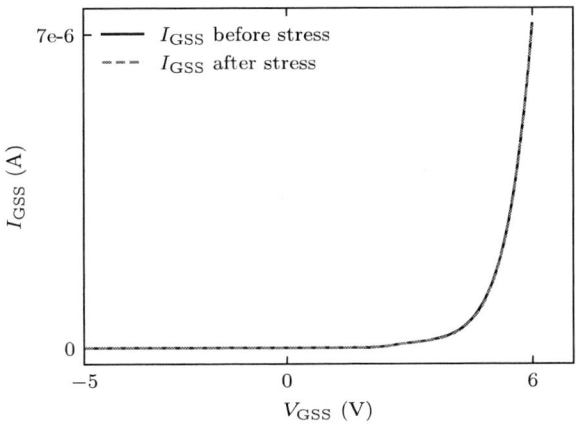

(a) Gate leakage monitoring, before and after stress.

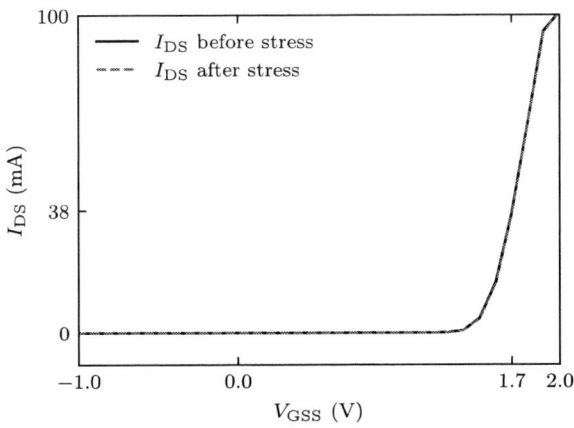

(b) Gate threshold monitoring. The B1505 compliance was set to $100\,mA$.

Fig. 10: Degradation monitoring on the gate. The chosen health monitoring parameters show no sign of degradation.

This result and the measured 1-pulse BV_{dyn} are coherent with Wang et al. [2], although with different gate OVS waveforms. The switching energies are much lower with voltage OVS, as displayed in Fig. 8; this is related to the higher dV/dt with overshot waveforms, or the overshoot itself.

V. CONCLUSION

This work develops theoretical tools to apply any given voltage or current on a power device gate through ordinary $50\,\Omega$ cables. Similarly, the drain side circuit can be moved away (e.g. one meter).

As a demonstration, the spike breakdown voltage of a GaN gate was measured at four times the nominal gate voltage. A few stress waveforms as high as twice the DUT nominal rating have no measurable impact on the gate health.

This demonstration emphasizes the advantages of the setup: the interactions between the instrumentation circuitry and the DUT are suppressed, as well as parasitics.

Thus, this setup can be used for various testings, including long term gate reliability studies (e.g. repeated overshoots using realistic gate waveforms). Easier gate health monitoring is enabled, as well as gate signal optimization to achieve optimal switching in power converters.

Cable and tracks of any length can be used with the proposed setup. Especially, the use of long cables highly simplify environmental, thermal and wafer-level dynamic characterization.

REFERENCES

[1] J. P. Kozak, A. Barchowsky, M. R. Hontz, N. B. Koganti, W. E. Stanchina, G. F. Reed, Z.-H. Mao, and R. Khanna, "An Analytical Model for Predicting Turn-ON Overshoot in Normally-OFF GaN HEMTs," *IEEE Journal of Emerging and Selected Topics in Power Electronics*, vol. 8, no. 1, pp. 99–110, Mar. 2020, conference Name: IEEE Journal of Emerging and Selected Topics in Power Electronics.

[2] B. Wang, R. Zhang, H. Wang, Q. He, Q. Song, Q. Li, F. Udrea, and Y. Zhang, "Gate Lifetime of P-Gate GaN HEMT in Inductive Power Switching," in *2023 35th International Symposium on Power Semiconductor Devices and ICs (ISPSD)*, May 2023, pp. 20–23, iSSN: 1946-0201.

[3] H. Ishibashi, A. Nishigaki, H. Umegami, W. Martinez, and M. Yamamoto, "An analysis of false turn-on mechanism on high-frequency power devices," in *2015 IEEE Energy Conversion Congress and Exposition (ECCE)*. Montreal, QC, Canada: IEEE, Sep. 2015, pp. 2247–2253. [Online]. Available: http://ieeexplore.ieee.org/document/7309976/

[4] R. N. Simons, *Coplanar Waveguide Circuits, Components, and Systems*, 1st ed. Wiley, Mar. 2001. [Online]. Available: https://onlinelibrary.wiley.com/doi/book/10.1002/0471224758

[5] J. Strydom, D. Reusch, S. Colino, and A. Nakata, "Using Enhancement Mode GaN-on-Silicon Power FETs (eGaN® FETs)."

[6] J. He, J. Wei, S. Yang, Y. Wang, K. Zhong, and K. J. Chen, "Frequency- and Temperature-Dependent Gate Reliability of Schottky-Type p -GaN Gate HEMTs," *IEEE Transactions on Electron Devices*, vol. 66, no. 8, pp. 3453–3458, Aug. 2019, conference Name: IEEE Transactions on Electron Devices.

[7] B. Wang, R. Zhang, H. Wang, Q. He, Q. Song, Q. Li, F. Udrea, and Y. Zhang, "Dynamic Gate Breakdown of p-Gate GaN HEMTs in Inductive Power Switching," *IEEE Electron Device Letters*, vol. 44, no. 2, pp. 217–220, Feb. 2023, conference Name: IEEE Electron Device Letters.

Wide Bandgap Semiconductors for LVDC Solid State Circuit Breaker applications

George Govaerts
ESAT, ELECTA
KU Leuven - EnergyVille
Leuven, Belgium
george.govaerts@kuleuven.be

Urmimala Chatterjee
IMEC

Leuven, Belgium
Urmimala.Chatterjee@imec.be

Johan Driesen
ESAT, ELECTA
KU Leuven - EnergyVille
Leuven, Belgium
johan.driesen@kuleuven.be

Wilmar Martinez
ESAT, ELECTA
KU Leuven - EnergyVille
Diepenbeek, Belgium
wilmar.martinez@kuleuven.be

Abstract—**In the quest for energy-efficient Low Voltage Direct Current grids, this paper delves into the characteristics of wide bandgap semiconductors based on GaN and SiC materials. The characteristics are investigated from a solid state circuit breaker design point of view. Characteristics such as on-resistance, junction temperature, current interruption capability, voltage and current rating, switching speed and bidirectionality are discussed. While wide bandgap semiconductors offer great performance improvements across several categories, commercially available devices focus on power electronic applications which have different objectives and consequently are not optimized for protection applications.**

Index Terms—**LVDC, WBG, Protection, SSCB, GaN, Si, SiC**

I. Introduction

The increasing electrification, the nature of modern electric devices and the continuous development of power electronics favour distribution of electrical energy through Low Voltage Direct Current (LVDC) grids. Many benefits are pointed out in literature. Increased energy efficiency, material resource saving, increased reliability, high controllability and reduced total system cost are often mentioned. LVDC is defined in the EU Low Voltage Directive (2014/35/EU) standard and includes DC voltages from 75V up to 1500V. [1], [2]

The design of LVDC grids and the absence of natural voltage and current zero crossing cause different system dynamics during load changes and fault conditions. To maximize the availability of these grids, fast protection systems need to be installed. These devices allow for detection and interruption of fault events before the converter voltage stabilizing capacitors have discharged to the point of instability. For LVDC grids, these dynamics are in the range of microseconds. [3]–[6]

While for Alternating Current (AC) grids, mechanical circuit breakers provide sufficient switching speed and interruption capability, for Direct Current (DC) grids, the mechanical operation prevents fast operation and suffers from arcing during current interruption. the Solid State Circuit Breaker (SSCB) provides a solution for increased fault interruption speed and arcing occurrence. [7]

Traditionally, Si based transistors like MOSFET, IGBT and Thyristor are used in SSCB design. These devices allow for

The work of G. Govaerts was supported by the Ph.D. Grant of Research Foundation Flanders (FWO) with the following reference to the grant number: 1SC6221N.

protection devices with high switching speeds and interruption capabilities. However, a major limitation of using semiconductor devices, is their conducting losses. For mechanical circuit breakers, the conducting losses are negligible. For SSCB, these conducting losses have an impact on the total system efficiency and therefore must be considered. Furthermore, the fast dynamics of LVDC grids cause high transient spikes on the protection circuitry. Additionally, while mechanical switches are inherently bidirectional, semiconductor switches require anti-series or anti-parallel configuration to achieve bidirectionality which increases device losses. [8]

Wide Bandgap (WBG) semiconductors are an upcoming technology with the potential to replace Si based semiconductors in multiple applications. Literature shows that WBG semiconductors have less conduction losses, higher blocking voltages, increased switching speeds and higher temperature ratings. This paper looks into the characteristics of SiC and WBG semiconductor devices with the aim of determining its applicability in SSCB design. [9], [10]

In Section II the following characteristics are discussed:

- On-resistance
- Junction temperature
- Current interruption capability
- Voltage & Current rating
- Switching speed
- Bidirectionality

in Subsection II-D an Unclamped Inductive Switching (UIS) setup is designed to test the overvoltage capability of vertical GaN devices. Section III summarizes the findings and concludes this paper.

II. Relevant properties of semiconductors for LVDC protection

A. On-resistance

Since LVDC grids are considered more energy efficient than Low Voltage Alternating Current (LVAC) grids, it is important that these promises are supported through the entire chain. During steady state operation of the grid, the protection devices will be in a conducting state. Since the protection devices will spend most of the time in this state, the losses that occur here are an important parameter to consider.

Traditionally mechanical circuit breakers are used to protect LVDC grids. Here, the conducting losses are linked to the the electrical parameters of the conducting elements inside this protection device. Since these elements can be designed to minimize the energy losses, these mechanical protection devices are considered to be the most efficient solution.

The need for faster protection devices devices lead to the increasing interest in SSCBs. These components have limited design flexibility and therefore tend to have higher conducting losses compared to mechanical switches. With the development of wide bandgap semiconductors, the efficiency of these solid state switches is on the rise.

Typical values for the on-resistance of mechanical circuit breakers for LVDC applications range from $1e-5\Omega$ - $1e-3\Omega$. The losses occurring within these protection devices can be considered negligible compared to the cable losses within the system. For commercially available semiconductors, the on-resistance varies depending on the used technology. For 650V Si MOSFET an On-resistance of $15m\Omega$ is available. For 650V Silicon Carbide (SiC) an on-resistance of $18m\Omega$ is achieved. While Gallium Nitride (GaN) semiconductors have the lowest theoretical on-resistance, which can be seen in Fig. 1, this technology is the least mature. [2], [11] For a 650V GaN FET, an on-resistance of $35m\Omega$ can be found. However, for SSCBs the voltage suppression circuit and potential bidirectional requirements further increase the on-resistance of these devices. in Section II-F, the potential bidirectionality of GaN Semiconductors and its influence on efficiency are studied. While the efficiency of mechanical circuit breakers and Si based semiconductors have been optimized in recent years, the development of new WBG topologies and production techniques allow this technology to mature and show its full potential in the coming years. [12], [13]

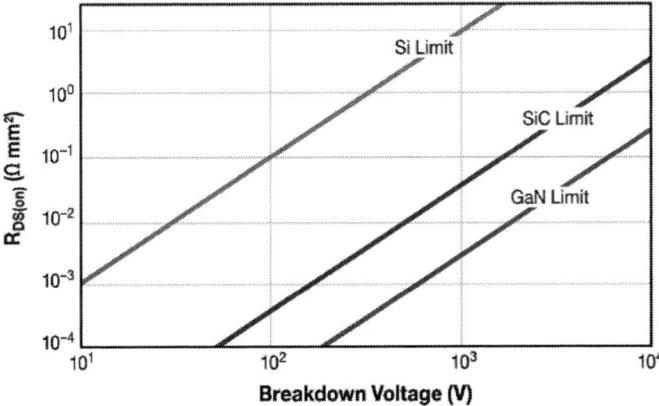

Fig. 1: Theoretical specific ON-resistance for a one square millimeter device versus blocking voltage capability for Si-, SiC-, and GaN-based power devices

B. Junction temperature

The fast interruption of short circuit events in LVDC grids cause high stress on the switching components. While all SSCB circuits feature voltage suppression components such as a MOV or a TVS diode, the change from low impedance to high impedance during interruption causes the switch to absorb some of the fault energy. Due to the high speed at which the thermal energy is generated, the heat cannot be transferred to the cooling fins causing the switch to increase in temperature. Therefore, it is beneficial for SSCB applications that the switch used has a high thermal inertia which is mainly determined by the physical size of the component, a high maximum junction temperature T_j which is determined by the material properties and high electron mobility, which is determined by the doping.

The maximum operating temperature for a Si MOSFET is usually limited to 150°C while WBG devices usually are rated to 175°C. In [14], tests were performed with a Si and a SiC device of similar rating and the same packaging demonstrating stable operation of respectively 190°C and 220°C. It was demonstrated that although the intrinsic temperature capability is much higher, the temperature limit is determined by the fundamental electron mobility temperature dependence. For GaN, similar operating temperatures are expected since the electron mobility is limiting the operation temperature instead of the material properties.

In [15], it is demonstrated that although SiC has a higher temperature rating than Si. Current interruption capability is limited by the fast increase in temperature during transient events caused by the smaller packaging due to the higher power density of the SiC component. This shows a difference between optimal design requirements for SSCB protection and Power Electronics (PE) devices. Smaller chips are beneficial in PE for achieving higher switching frequencies, while for SSCB protection additional reduction in switching speed does not influence the behavior of the protection device, instead it causes the chip to heat up faster resulting in reduced robustness.

C. Current interruption capability

The current interruption capability of the semiconductor device is a key design parameter. The higher the current interruption capability the higher the robustness of the protection circuit. In [15], tests have been performed on the short circuit interruption capability of GaN, SiC and Si devices. The measurements indicate that GaN has the lowest interruption capability failing to interrupt the current even within the linear region of the component. This is caused by the higher di/dt causing larger voltage spikes resulting in higher temperatures in the smaller GaN device. The Si Device Under Test (DUT) showed the highest current interrupting capability by interrupting a fault current while saturated. The Si DUT has an advantage having a larger chip size resulting in a higher thermal inertia, even when the maximum junction temperature of the SiC device was higher.

D. Voltage & Current rating

The higher electric field densities at which WBG semiconductor operate allow them to withstand higher voltage levels compared to Si semiconductors. For LVDC protection,

979-8-3503-3714-3/23 $31.00 © 2023 IEEE

the forward blocking voltage is an important parameter since it determines whether the circuit breaker is able to prevent current flowing through the interrupted branch. Typical Si devices are limited to $0.5MV/cm$, where GaN devices can go up to $2MV/cm$ and SiC even up to $3.5MV/cm$. While this increased electric field density is usually optimised for PE design, where this is used to reduce the thickness of the active region in vertical devices and consequently allowing increased switching speeds and reducing cost, this property can also be exploited to increase the robustness of the semiconductor for SSCB applications. [16], [17]

Fast switching of high fault currents during short circuit events causes voltage overshoot across the semiconductor switch. Although the inductance of the fault path is typically small, the high di/dt causes large voltage spikes. Even though the switch is protected against these overvoltages by the Overvoltage Protection (OVP) components, it is important that the switch has a high robustness against these events.

In [18] tests were performed on a SiC-SIT which have a similar structure to JFETs while having shorter channel lengths allowing very low on resistances while maintaining a high breakdown voltage. Tests showed a higher inductive current interruption capability compared to a Si MOSFET. Furthermore the SiC-SIT features triodelike characteristics enabling great suppression of voltage overshoot during turn-off events.

In [10] and [19] different SiC semiconductors are compared. It was found that the saturation current density for SiC MOSFET decreases slower than for the SiC JFET. This resulted in the MOSFET conducting higher currents and consequently heating up faster. Furthermore the higher current through the MOSFET causes higher overvoltage during switching. The SiC BJT is current limited by the amplification factor beta β and therefore less prone to damage due to short circuit current.

In [20] two short circuit failure modes of SiC MOSFETs are investigated: power strike and gate oxide failure due to gate ringing.

1) Power strike: Power strike occurs when the thermal capability of the semiconductor is exceeded. Usually this occurs during the interruption of a short circuit event due to the high power that is dissipated within the semiconductor. Three contributors are identified, heating caused by the nominal current flow which the device is designed to handle for an unlimited time, heating caused by the uninterrupted short circuit current, which the device should be able to handle for a specified time and heating caused by the interruption of the short circuit event. Here the semiconductor operates at an increased voltage V_{DS} and current I_D causing energy to be dissipated in the device. Due to the transient behavior, all energy is dissipated in the junction without transferring the heat to the case or cooling system. This causes a large increase in junction temperature and potentially in the thermal runaway of the component.

2) Gate oxide failure due to gate ringing: The second failure mode discussed is gate ringing. Gate ringing occurs during the interruption process and is caused by the common source

inductance and high switching speed of the semiconductor. Due to di/dt during interruption, voltage spikes may occur at the gate. These voltage spikes can lead to unwanted triggering of the gate and could cause degradation of the gate oxide in case these voltage spikes exceed the gate voltage limit.

3) Avalanche: Another failure mode is avalanche breakdown. Avalanche breakdown occurs when the forward overvoltage limit of a semiconductor is exceeded. The robustness against avalanche, or the voltage surge robustness is tested using an UIS setup. The result is determined by measuring the avalanche energy that can be dissipated in the DUT. Avalanche Energy is defined as the amount of energy a MOSFET can withstand when it is set into avalanche mode or its breakdown voltage $V_{BR,DSS}$ is exceeded. The higher the avalanche energy, the more robust the device is considered. Typically, this rating is important when the switch faces simultaneous high voltage and high current such as during short circuits or turn-off.

for MOSFETs, the parasitic body diode is responsible for the avalanche rating. During Avalanche conditions, the device behaves like an ideal current source where the current is determined by the series circuit inductance. GaN HEMTs on the contrary do not have this body diode and therefore behave differently during voltage surge events. Their breakdown is expected to be related to a dielectric breakdown caused by the high electric field presented during surge conditions. The GaN device behaves like a voltage source with fixed current during surge events, the drain leakage current remains constant even above the maximum rating of the device.

In order to observe these variations, a JEDEC JESD 24-5 compliant UIS test platform was developed. The setup is shown in Fig. 2 and Fig. 3. The circuit consist of a voltage source, inductor, the DUT and a gate driver. A current is sent through the inductor and the DUT in conducting state. When the threshold current is reached, the gate driver turns off the DUT causing its V_{DS} to increase due to the charged inductance. The voltage V_{DS} and current I_D are measured and can be used to determine the absorbed energy within the DUT.

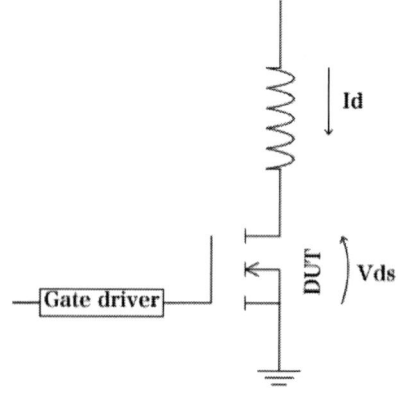

Fig. 2: Schematic of UIS setup

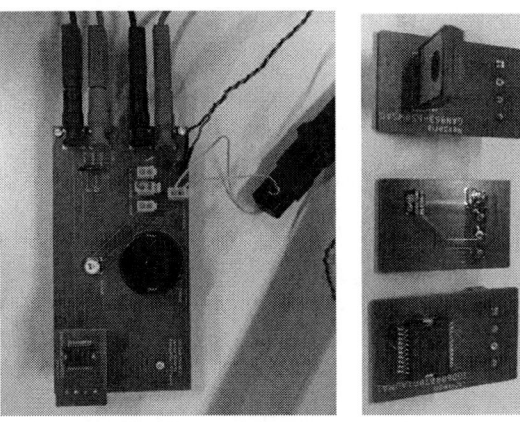

Fig. 3: Picture of UIS setup

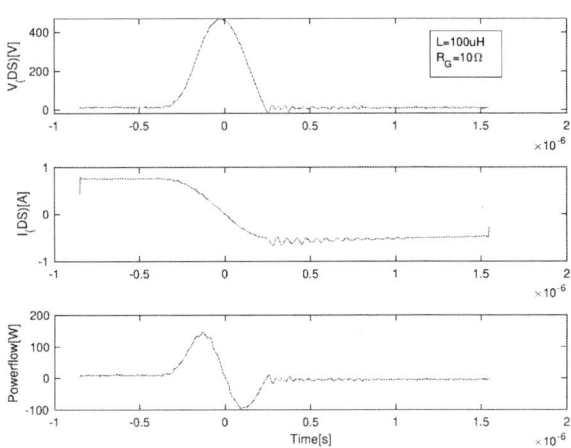

Fig. 4: Infineon GaN - surge voltage within device limits

Three commercially available GaN components where selected and are summarized in table I. Since only the Infineon provided consistent results the other DUT's are not further discussed. In Fig. 4 and Fig. 5 the results of the test performed on the Infineon GaN are shown. Fig. 4 shows that a surge voltage of $500V$ is tolerated by the DUT and that the surge energy is absorbed within the device. When the DUT is turned off, the V_{DS} starts to increase and the drain current consequently drops. The voltage reaches a peak at the time the current goes through zero. Now the current becomes negative while the voltage across the DUT is still positive indicating the DUT operates in the second quadrant. If the energy is calculated that enters the DUT and the energy that is released by the DUT, it can be seen that the DUT did not absorb much of the surge energy. Most of the energy is released again during the second quadrant operation suggesting little heating of the DUT. In Fig. 5 the same test is performed for the case where V_{DS} exceeds the voltage surge capability of the DUT. When the DUT is turned off, the V_{DS} starts increasing until a limit is reached, now the V_{DS} can't be sustained and collapses. The DUT showed voltage blocking capability of 1.5 times the rated breakdown voltage $V_{DS,bd}$ of 800V.

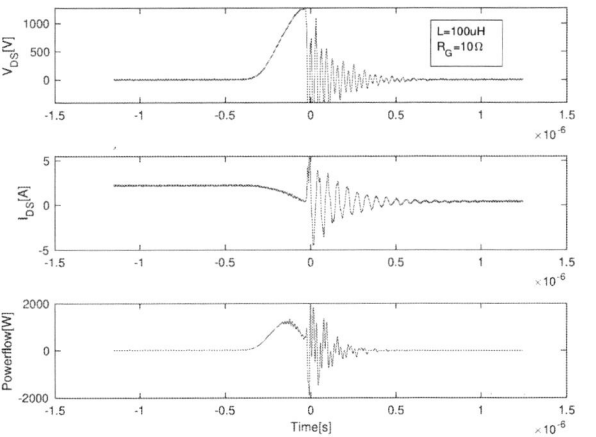

Fig. 5: Infineon GaN - surge voltage exceeding device limits

TABLE I: Specs of tested GaN DUT's

Manufacturer	Infineon	GaN systems	Nexperia
Model	IGO60R070 D1AUMA1	GS66504B	GAN063-650WSAQ
$V_{DS,max}$	600V	650V	650V
$V_{DS,transient}$	/	750V	800V
$R_{DS(on)}$	70mΩ	100mΩ	50mΩ
I_D	31A	15A	34.5A
$I_{D,pulse}$	60A	36A	150A
P_{tot}	125W	120W	143W
T_j	150C	150C	175C
dV/dt	200V/ns	/	40V/ns
R_{thJC}	1C/W	1C/W	/
C_{oss}	72pF	33pF	130pF

In [21], the surge energy ruggedness was also tested on two different GaN transistor, a HD-GIT and a SP-HEMT. It was concluded that GaN transistors do not feature an avalanche mechanism and therefore have a different breakdown mechanism caused by electric field breakdown. This also indicates that comparing avalanche energy for GaN transistors does not provide meaningful insight into the actual surge ruggedness of the device. Although the failure mechanism was the same for both types of GaN transistor, the failure location was different resulting in different breakdown behavior. While the GaN SP-HEMT fails at the drain and gate region resulting in loss of gate control, the GaN HD-GIT failed at the drain where a leakage path was created. Furthermore, it was concluded that failure occurred at voltages much higher than the rated voltage and that failure was independent of the dV/dt but rather determined by the peak overvoltage transient. Finally, it was shown that devices with a higher C_{OSS} are better suited to withstand the surge energy since the capacitance temporarily absorbs the surge energy as is also shown in Fig. 4. This means that larger devices can be considered more robust making it

979-8-3503-3714-3/23 $31.00 © 2023 IEEE

beneficial to over dimension the transistor. However, these devices suffer from slower switching which is not beneficial for application in PE devices but is beneficial for application in LVDC protection circuitry.

Similar conclusions where obtained in [19]. The surge energy was linked to the output capacitance C_{OSS} and the breakdown was linked to electric field breakdown causing the formation of a leakage path between drain and source. Finally, some proposals to improve surge voltage robustness were made.

It can be concluded that for Si and SiC devices, the surge voltage robustness is limited by the thermal capability of the device, whereas for GaN devices, the electric field poses a limit on the voltage surge capability.

E. Switching speed

While the transition from milliseconds (MCB) to microseconds (SSCB) plays a big role in the protectability of LVDC grids, the increased switching speed of WBG semiconductors benefits PE design by allowing smaller passive components and consequently reducing the size of these devices. In SSCB design, the voltage stability benefit for a further increase in switching speed is negligible and mainly influences the peak of the switching transients and the energy absorption within the semiconductor switch. The higher transient peaks following from an increase in switching speed, requires more robust OVP components to absorb these transients and protect the semiconductor device. While the increase in switching speed also allows for a faster interruption of the current, reducing the energy absorption within the semiconductor. Tests show that this benefit is not prominent and device failure due to thermal stress is more common when the device size decreases. [9], [12], [15]

Although the increased switching speed serves little benefit in short circuit protection, in [16] the possibilities of using GaN for OVP is discussed. It was concluded that SiC and GaN semiconductors are capable to be used as E1 protection due to the increased switching speed whereas Si was not.

F. Bidirectionality

A Bidirectional Switch (BDS) is an essential component of a LVDC SSCB for an efficient protection system. A BDS, capable to conduct current and block voltage in both polarities is conventionally made by connecting discrete IGBTs and diodes either back-to-back or in anti-parallel way. [22], [23] Consequently, an LVDC SSCB with discrete components makes the system less efficient, large, and complicated due to the usage of multiple power devices and its drivers. Yet, a high-power on-chip bidirectional switch that offers a high blocking voltage with current-handling capability is not available to date. Lateral GaN technology is believed to allow four quadrant operation by sharing the voltage blocking region. This design cannot be used for vertical power devices and therefore provides unique opportunities for Bidirectional GaN devices. [24] While these new topologies are being researched, monolithic GaN and SiC power ICs provide bidirectional

current and voltage blocking capabilities into a single device. These devices are realised by integrating multiple discrete components to a single device. Integration helps to achieve a compact circuit with less components that reduces the overall power loss, the parasitics and complexity of the circuit. GaN HEMT BDS can be a valuable alternative of the reverse blocking IGBT (RB-IGBT) based BDS due to its low losses and high switching speed operation, which in-turn helps to build a compact, efficient system. By using lateral GaN technology it is possible to realise a high voltage bidirectional switch. Mainly there are two types of BDS which are anti-parallel and anti-serially connected HEMT. The output characteristics I_r, V_{th} and V_{on} can be controlled by changing the AlGaN barrier thickness. [25] In [26], a monololithic GaN BDS is implemented in a 100A SSCB design achieving an R_{on} of $22m\Omega$. In [27], a GaN monololithic BDS was used in a 650V $200m\Omega$ SSCB. Similar to monolithic GaN BDS, monolithic SiC BDS is used to improve SSCB design. In [28] a monolithic integrated SiC BIDFET with $46m\Omega$ R_{on} for a 1.2kV device is reviewed.

III. CONCLUSIONS

This paper looks into the characteristics of WBG semiconductors with the aim to determine the applicability for LVDC protection circuits. WBG semiconductors haven't yet reached the maturity level of Si based transistors but already show improvements across the board. While the increased efficiency of these WBG semiconductors is beneficial, this in term allows for smaller packaging enabling increased switching speeds of WBG semiconductors. While this is useful for PE design, the resulting reduced thermal inertia limits the transient capabilities of the component. Different voltage breakdown mechanisms of GaN are evaluated using an UIS setup. It is shown that GaN is capable of blocking voltage exceeding their maximum rating. Finally, the bidirectional operation of monolithic GaN and SiC devices allow for reduced complexity and switching losses of bidirectional SSCB design.

REFERENCES

[1] M. Marc, D. Roggo, M. Säteri, T. Tuomarmäki, and S. Ranta, "Lvdc vs lvac: A comparison of system losses," in *2023 IEEE 32nd International Symposium on Industrial Electronics (ISIE)*, 2023, pp. 1–4.

[2] H. B. Abdeljawed and L. E. A. Ouni, "Review of low voltage dc technology prospects in smart buildings," in *2022 5th International Conference on Advanced Systems and Emergent Technologies (IC$_A$SET)*, 2022, pp. 319 – −323.

[3] G. Govaerts, L. Hallemans, J. Driesen, and W. Martinez, "Design of a bipolar dc grid fault emulator," in *2020 5th IEEE Workshop on the Electronic Grid (eGRID)*, 2020, pp. 1–6.

[4] X. Yan, Z. Yu, L. Qu, Z. Gan, C. Ren, J. Wu, J. Liu, R. Zeng, and Y. Huang, "Snubber branch design and development of solid-state dc circuit breaker," *IEEE Transactions on Power Electronics*, vol. 38, no. 10, pp. 13 042–13 051, 2023.

[5] R. Kheirollahi, S. Zhao, H. Zhang, and F. Lu, "Complementary commutation-based -type dc sscb," in *2023 IEEE Applied Power Electronics Conference and Exposition (APEC)*, 2023, pp. 514–519.

[6] W. Da, G. George, H. Leonie, E. Glenn, P. V. Tichelen, G. V. D. Broeck, and J. Driesen, "Proposals for grid code of bipolar low-voltage DC (BIDC) grids Author:," 2022.

[7] Z. Ganhao, "Study on dc circuit breaker," in *2014 Fifth International Conference on Intelligent Systems Design and Engineering Applications*, 2014, pp. 942–945.

[8] P. Aditya, S. N. Banavath, A. Lidozzi, A. Chub, and D. Vinnikov, "Bidirectional sscb for residential dc microgrids with reduced voltage and current stress during fault interruption," in *2023 IEEE 17th International Conference on Compatibility, Power Electronics and Power Engineering (CPE-POWERENG)*, 2023, pp. 1–6.

[9] H. Li, X. Lyu, K. Wang, Y. Abdullah, B. Hu, Z. Yang, J. Wang, L. Liu, and S. Bala, "An ultra-fast short circuit protection solution for e-mode gan hemts," in *2018 1st Workshop on Wide Bandgap Power Devices and Applications in Asia (WiPDA Asia)*, 2018, pp. 187–192.

[10] D.-P. Sadik, J. Colmenares, G. Tolstoy, D. Peftitsis, M. Bakowski, J. Rabkowski, and H.-P. Nee, "Short-circuit protection circuits for silicon-carbide power transistors," *IEEE Transactions on Industrial Electronics*, vol. 63, no. 4, pp. 1995–2004, 2016.

[11] A. Lidow, *GaN Power Devices and Applications*, 3rd ed. EL SEGUNDO, 2019.

[12] R. Rodrigues, Y. Du, A. Antoniazzi, and P. Cairoli, "A review of solid-state circuit breakers," *IEEE Transactions on Power Electronics*, vol. 36, no. 1, pp. 364–377, 2021.

[13] S. Zheng, R. Kheirollahi, J. Pan, L. Xue, J. Wang, and F. Lu, "Dc circuit breakers: A technology development status survey," *IEEE Transactions on Smart Grid*, vol. 13, no. 5, pp. 3915–3928, 2022.

[14] S. Pyo and K. Sheng, "Junction temperature dynamics of power mosfet and sic diode," in *2009 IEEE 6th International Power Electronics and Motion Control Conference*, 2009, pp. 269–273.

[15] M. Marwaha, M. H. M. Sathik, K. Satpathi, Y. Yu, F. Sasongko, J. Pou, C. Gajanayake, and A. K. Gupta, "Comparative analysis of si, sic and gan field-effect transistors for dc solid-state power controllers in more electric aircraft," in *2021 IEEE 12th Energy Conversion Congress Exposition - Asia (ECCE-Asia)*, 2021, pp. 592–597.

[16] J. Flicker, E. Schrock, and R. Kaplar, "Reverse breakdown time of wide bandgap diodes," in *2022 IEEE 9th Workshop on Wide Bandgap Power Devices Applications (WiPDA)*, 2022, pp. 26 30.

[17] "Introduction to Wide Band-Gap Semiconductors." [Online]. Available: https://navitassemi.com/introduction-to-wide-bandgap-semiconductors/

[18] Y. Sato, Y. Tanaka, A. Fukui, M. Yamasaki, and H. Ohashi, "Sic-sit circuit breakers with controllable interruption voltage for 400-v dc distribution systems," *IEEE Transactions on Power Electronics*, vol. 29, no. 5, pp. 2597–2605, 2014.

[19] X. Failure analysis of normally-off GaN HEMTs under avalanche conditionsg, G. Wang, Y. Li, A. Q. Huang, and B. J. Baliga, "Short-circuit capability of 1200v sic mosfet and jfet for fault protection," in *2013 Twenty-Eighth Annual IEEE Applied Power Electronics Conference and Exposition (APEC)*, 2013, pp. 197–200.

[20] S. Zhao, R. Kheirollahi, H. Zhang, and F. Lu, "Short circuit fault induced failure of sic mosfets in dc solid-state circuit breakers," in *2022 IEEE 9th Workshop on Wide Bandgap Power Devices Applications (WiPDA)*, 2022, pp. 116–121.

[21] R. Zhang, J. P. Kozak, M. Xiao, J. Liu, and Y. Zhang, "Surge-energy and overvoltage ruggedness of p-gate gan hemts," *IEEE Transactions on Power Electronics*, vol. 35, no. 12, pp. 13 409–13 419, 2020.

[22] H. Umeda, Y. Yamada, K. Asanuma, F. Kusama, Y. Kinoshita, H. Ueno, H. Ishida, T. Hatsuda, and T. Ueda, "High power 3-phase to 3-phase matrix converter using dual-gate gan bidirectional switches," in *2018 IEEE Applied Power Electronics Conference and Exposition (APEC)*, 2018, pp. 894–897.

[23] C. Klumpner and F. Blaabjerg, "Using reverse-blocking igbts in power converters for adjustable-speed drives," *IEEE Transactions on Industry Applications*, vol. 42, no. 3, pp. 807–816, 2006.

[24] N. Flaherty, "Transphorm to develop bidirectional four quadrant GaN switch," 2022. [Online]. Available: https://www.eenewseurope.com/en/transphorm-to-develop-bidirectional-four-quadrant-gan-switch/

[25] D. Bergogne, O. Ladhari, L. S. C. Gillot, R. Escoffier, and W. Vandendaele, "The single reference bi-directional gan hemt ac switch," in *2015 17th European Conference on Power Electronics and Applications (EPE'15 ECCE-Europe)*, 2015, pp. 1–7.

[26] Y. Kinoshita, T. Ichiryu, A. Suzuki, and H. Ishida, "100 a solid state circuit breaker using monolithic gan bidirectional switch with two-step gate-discharging technique," in *2020 IEEE Applied Power Electronics Conference and Exposition (APEC)*, 2020, pp. 652–657.

[27] Z. J. Shen, Z. Miao, A. M. Roshandeh, P. Moens, H. Devleeschouwer, A. Salih, B. Padmanabhan, and W. Jeon, "First experimental demonstration of solid state circuit breaker (sscb) using 650v gan-based monolithic

bidirectional switch," in *2016 28th International Symposium on Power Semiconductor Devices and ICs (ISPSD)*, 2016, pp. 79–82.

[28] A. Kanale, T.-H. Cheng, S. S. Shah, K. Han, A. Agarwal, B. J. Baliga, D. Hopkins, and S. Bhattacharya, "Switching characteristics of a 1.2 kv, 50 m sic monolithic bidirectional field effect transistor (bidfet) with integrated jbs diodes," in *2021 IEEE Applied Power Electronics Conference and Exposition (APEC)*, 2021, pp. 1267–1274.

979-8-3503-3714-3/23 $31.00 © 2023 IEEE

Reduction of DC/DC Converters EMI Emission Using Bi- and Unidirectional QR-ZVS Topologies

Abdelmoumin Allioua, David Krause, Andrea Zingariello and Gerd Griepentrog

Institute for Power Electronics and Control of Drives
Technical University of Darmstadt
Darmstadt, Germany
abdelmoumin.allioua@lea.tu-darmstadt.de

Abstract—**Switched-mode power supplies (SMPSs) operating at high switching frequency (HSF) in the MHz range offer advantages such as improved output voltage regulation, reduced filter size and lower system cost. However, these benefits come with challenges. HSFs can lead to increased total losses, reduced efficiency and severe electromagnetic interference (EMI) emissions, due to transient current and voltage (dv/dt, di/dt). Although wide bandgap (WBG) devices – such as Gallium Nitride (GaN) and Silicon Carbide (SiC) transistors – can afford HSFs, but on the cost of greater switching losses and EMI noise, especially when opting for hard-switching pulse width modulation (PWM) converters. To address the EMI emissions, various methods have been explored, which include the use of EMI filters, the adoption of different spread-spectrum modulation methods, selecting appropriate components along with improved circuit design, and soft-switching techniques. This paper presents the use of the quasi-resonant (QR) zero voltage switching (ZVS) technique for the purpose of reducing the converter EMI noise and enhancing its efficiency, while opting for HSF up to 4 MHz with GaN-HEMT, using fixed LC resonant tank parameter.**

Index Terms—**QR, ZVS, GaN-HEMT, Fixed LC resonant tank, High switching frequency, EMI noise reduction, High efficiency.**

I. Introduction

Accompanying HSFs, two major difficulties with switching devices appears, namely high switching stress and losses due to the presence of parasitic inductances and capacitances. This causes the switch to operate under inductive turn-off and capacitive turn-on conditions. Specifically, through switching-off an inductive load, voltage spikes are induced by the sharp di/dt across the parasitic inductances. On the other hand, when switching-on at a high voltage level, the energy stored in the parasitic output capacitance of the switch C_{oss} is dissipated inside the device, in addition to the Miller effect which is coupled into the drive circuit, leading to significant noise and instability. While not severe in lower switching frequencies (SFs), the capacitive turn-on loss becomes dominant factor when the SF is raised to the MHz range [1]. In this work, the dc-dc converters efficiency including switching losses and EMI noise reduction will be considered at an early design phase, by adopting soft-switching quasi-resonant converters (QRCs) with ZVS topology, which allows for operation at HSF and supports higher input voltages without sacrificing efficiency.

II. Quasi-Resonant Converters

The concept of resonant switches was introduced to operate under favorable condition, either at ZVS or zero current switching (ZCS). This operation is achieved by shaping the switch current and voltage waveform using LC resonant circuit, to mitigate the switching losses and alleviate EMIs. This concept can be directly applied to conventional converters, simply by replacing their switch by a resonant one. However, ZCS is effective in reducing switching losses for power devices with large tail current in the turn-off process, besides its operation is rather limited to low MHz range and cannot solve the problem of high switching losses associated with capacitive turn-on [1]. On the other hand, ZVS eliminates the capacitive turn-on losses. In addition, with the growing interest of WBG switches, which have significantly high turn-on losses compared to their turn-off losses, ZVS satisfies their demand to further cut down the turn-on losses [2]. Therefore, in general ZVS is more favorable than ZCS in dc-dc converters for high frequency (HF) operation.

For ZVS, when a device allowing bidirectional current flow (e.g., Transistor with its body diode D_B) is used, as depicted in Fig. 1(a), the voltage U_{C_R} across the resonant capacitor cannot oscillate below zero, as illustrated in Fig. 2(a). Such a case is referred to as 'half-wave mode'. Conversely, if a unidirectional switch is employed, where a diode D_s is placed in series with the transistor, to prevent reverse current flow through its body diode D_B as illustrated in Fig. 1(b), the resonant capacitor C_R is then placed in parallel with this series combination. As a result, its voltage continues to resonate below zero, as seen in Fig. 2(b). This configuration is termed to as 'full-wave mode'.

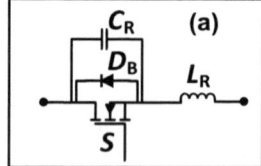 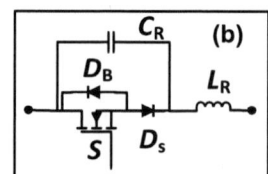

Fig. 1. ZVS M-type switches. (a) Bidirectional resonant switch, (b) Unidirectional resonant switch.

In any case, the transistor is turned on at ZVS condition. To allow for one complete oscillation, the off-time is commonly

979-8-3503-3714-3/23 $31.00 © 2023 IEEE

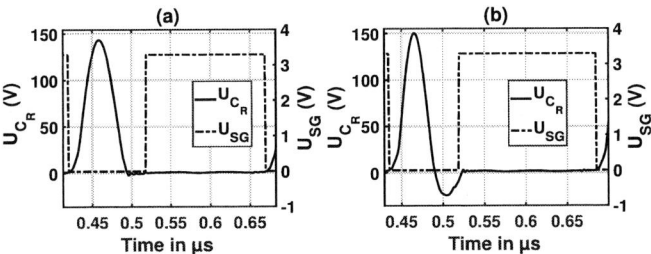

Fig. 2. Simulated waveforms. (a) Half-wave operation mode, (b) Full-wave mode of operation.

kept constant. In QRCs, the output voltage regulation is usually achieved by varying the on-time. Here, a resonant cycle of fixed duration (off-time) is separated by non-resonant stage of variable duration (on-time) [3].

This leads to the less desirable variable switching frequency (VSF) of QRCs, since the variation of the SF broaden the spectrum of the EMI emissions. This can impose challenges when the converter must adhere to specific Electromagnetic Compatibility (EMC) emissions standard. Moreover, the VSF adds more constraints on EMI filter development, increasing its design complexity and makes it more challenging.

For instance, the dc voltage-conversion ratio (VCR) versus the normalized switching frequency of a QR-ZVS buck converter is given by Eq. (1), as elaborated in [1].

$$ A = \frac{2\pi(1 - M)}{[\alpha + \frac{r}{2M} + \frac{M}{r}(1 - \cos\alpha)]}, \; \alpha = \pi + \arcsin\left(\frac{r}{M}\right) \quad (1) $$

Where the resonance characteristic impedance is $Z_\text{R} = \sqrt{L_\text{R}/C_\text{R}}$, the resonance frequency is defined as $f_\text{R} = 1/2\pi\sqrt{L_\text{R} \cdot C_\text{R}}$ and the VCR is referred to as $M = U_\text{out}/U_\text{in}$, while the normalized load resistance and switching frequency are $r = R_\text{L}/Z_\text{R}$ and $A = f_\text{S}/f_\text{R}$ respectively.

The buck QRC steady-state operation, along with switching cycle analysis and waveforms are detailed in [1]. Likewise, this approach can be applied in boost, buck-boost, flyback and bridge topologies.

In general, QRCs employing half-wave mode have a VCR sensitive to load variation as shown in Fig. 5(a) with the analytical curves. However, simply by modifying the resonant switch to full-wave structure the VCR becomes nearly load independent as illustrated in Fig. 5(b) [1]. This behavior comes from how energy moves in the resonant tank. Under heavy loads, the resonant inductor current is mostly offset by the output current, leading to very little energy going back to the source. On the other hand, with lighter loads, more energy goes back to the source. The full-wave mode has a series diode D_s that helps manage and control the energy in the resonant tank. This diode ensures that extra tank energy can be sent back to the source during light load situations, helping to keep the VCR stable. However, the half-wave mode doesn't have this energy-return feature, requiring a change in the SF to maintain

the desired output voltage during different loads, an extensive breakdown on this matter is elaborated in [5]. These findings will be further validated experimentally in Part V of this study.

The full-wave mode appears to be superior. Yet, it has some drawbacks, due to the series diode conduction loss and also the fact that the switch output capacitance C_oss cannot be directly absorbed by C_R. Consequently, C_oss is charged to the peak voltage during resonance as can be observed in Fig. 2(b), and this capacitive energy is dissipated in the transistor during turn-on, resulting in power dissipation [3]. The buck converter in QR-ZVS half-wave and full-wave mode are chosen to illustrate the design and operation of the proposed solution in non-isolated converters. Since isolated converters exhibit substantial differences in operation, characteristics, and design [3], they are not discussed in this paper.

III. QR-ZVS Buck Converter Design

QR-ZVS buck converters brings into the design process more challenges. To overcome the undesirable VSF concern of QRCs, different approaches can be used, such as using tuneable inductors as in [4], or by means of tuneable capacitors (e.g., switched capacitive banks) for the LC tank. However, the tuneable inductors method is valid to extend the ZVS operation range with low SFs < 1 MHz in a relatively small bandwidth i.e., kHz range, due to the magnetic material sensitivity in HFs. In addition, other constraints rise with tuneable capacitors, like power handling issues, plus tuning range and speed.

In the present work, evaluations are performed under an open control loop, utilizing 4 MHz switching signals and fixed LC tank components. This setup should ensure ZVS condition, with output voltage regulation achieved by adjusting the off-time within the permissible ZVS window.

The primary aim of this analysis is to underline the load dependency differences between the half-wave and full-wave topologies, demonstrating how reduced load sensitivity can be advantageous for future closed loop control strategies.

A. Converter Prototypes and Specifications

To demonstrate the feasibility of the aforementioned approach, a 70 W buck converter for 48V/24V conversion was designed with three variants: a hard-switching (HS) benchmark, and two QR-ZVS variants for half-wave and full-wave soft switching as shown in Fig. 3. All converters operate in continuous conduction mode (CCM), to limit the load influence on the voltage ripple and the VCR. The designed converters specifications are mentioned in Table I.

B. Basis for Using GaN-HEMTs

We specifically chose GaN High-Electron-Mobility Transistors (HEMTs) due to their pronounced advantages, especially when paired with soft-switching. GaN-HEMTs have lower on-resistances and capacitances for a given blocking voltage compared to Metal-Oxide-Semiconductor Field-Effect Transistors (MOSFETs) [7]. Their inherent characteristics as low charge devices allow them to operate efficiently at multi-MHz SFs. Remarkably, they can switch voltages up to several 100 V in

979-8-3503-3714-3/23 $31.00 © 2023 IEEE

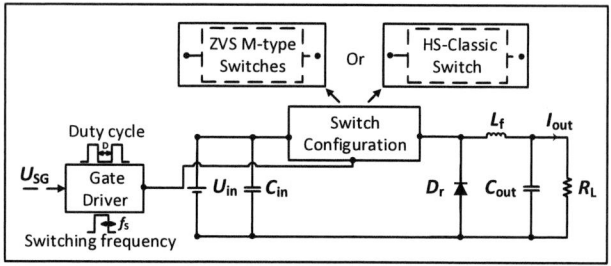

Fig. 3. Buck converter configurations, with HS-classic switch option, and the QR-ZVS M-type possibilities.

TABLE I
PARAMETER OF THE DESIGNED BUCK CONVERTERS

	Buck converters parameters		
	Hard-switching	*Half-wave*	*Full-wave*
C_{in}	3.2 µF	3.2 µF	3.2 µF
C_{out}	110 nF	110 nF	110 nF
L_f	10 µH	10 µH	10 µH
$L_R{}^a$	-	550 nH	550 nH
$C_R{}^a$	-	300 pF	200 pF
$f_R{}^a$	-	8.23 MHz	8.13 MHz
f_S	4 MHz	4 MHz	4 MHz
D_{cycle}	50%	60%	58.5%
R_L	6 Ω	6 Ω	6 Ω
U_{in}	48.55 V	48.11 V	48.40 V
U_{out}	23.26 V	22.44 V	22.48 V
P_{in}	97.42 W	96.08 W	99.75 W
P_{out}	70.62 W	80.17 W	80.10 W
η	72.49%	83.44%	80.30%

^{a}Provided for soft-switching cases only.

a few nanoseconds, due to their fast rise and fall times [8]. Additionally, GaN-HEMTs naturally have minimal turn-off losses [9]. This becomes especially relevant when considering the hard-switching design, where turn-on losses are the main contributor to the converter's overall losses.

C. Derivation of the Resonant Tank Parameters

The values for C_R and L_R depend on both f_R and Z_R. The resonance frequency f_R is selected based on the VCR and the chosen SF, as detailed in Eq. (1). The minimum Z_R to ensure ZVS for different loads is given in [6] by:

$$Z_R \geq \frac{R_{L_{max}}}{M_{min}} \quad (2)$$

IV. IMPLEMENTATION DETAILS AND CHALLENGES

In this section, we explore the challenges associated with QRCs operation and selecting appropriate components.

A. Components Selection for Implementation

1) Input Capacitor: The input capacitor C_{in} is selected mainly because of the desired minimum ripple. To reduce both the Equivalent Series Resistance (ESR) and Equivalent Series Inductance (ESL) effects, two capacitors are used in parallel. This ensures that their combined impedance curve doesn't show a peak, which could influence the noise spectrum. Since

the input capacitor handles HF currents, its placement on the printed-circuit-board (PCB) is crucial for EMI considerations. Therefore, C_{in} should be close to both the switch and D_r. This placement serves to minimize the current loop area and the parasitic inductance associated with the PCB traces, which helps in reducing EMIs. Moreover, having a ground plane beneath those critical components provides a low impedance path for the return current, further assisting in EMI reduction.

2) HF Behavior of the Resonant Inductor: With the converter operating at HSFs, frequency-dependent losses in L_R become a significant factor influencing the buck converter's overall efficiency. In modeling the inductor at these HFs, it's essential to consider its inter-winding capacitance, the skin effect, and the core losses. Among these, the core losses – especially from eddy currents and hysteresis – stand out the most. During the design process, several components were evaluated using simulations, and an inductor, with an iron alloy core material, was the selected choice.

3) Filter Components: Based on the guidelines from [10], the output filter elements were designed for current and voltage ripples of 10% and 0.4%, respectively. The HSF of the converter plays a significant role in this selection process. To make sure the output filter operates correctly, the Self-Resonant Frequency (SRF) of L_f should be significantly above the SF [11]. As a result, the chosen L_f has an SRF of 28 MHz.

B. Operation Challenges in QRCs

1) Half-wave Converter: In the selection of the rectifier diode D_r, its junction capacitance C_{jr} stands out as a key parameter. When the switch is on, this C_{jr} resonates with L_r. One way to minimize this effect is to choose D_r with lower C_{jr}. While the Multi Resonant Converter (MRC) approach offers optimization for the switching conditions of both the switch and the D_r as proposed in [12], it was not explored in this paper, allowing to highlight other aspects of the QRCs.

2) Full-wave Converter: In contrast to the half-wave converter, where the charge from C_{oss} is redirected back to C_{in} during the negative period of the switch current, the series diode D_s in the full-wave mode prevents the reverse current flow through the switch. However, a displacement current still flows due to C_{js} (junction capacitance of D_s) and C_{oss}. As C_{js} gets charged, C_{oss} discharges.

Consequently, for ZVS to be achieved, C_{oss} should not be larger than C_{js}. Otherwise, C_{js} will be fully charged before C_{oss} discharges, preventing U_{C_R} from reaching zero condition. To enhance the switching conditions of the designed full-wave converter, an external 680 pF capacitor was added across D_s.

C. Efficiency and Performance Analysis

The efficiency was determined by measuring the power at the input and output of the converters. Notably, losses from the gate driver were excluded from this analysis.

With the elimination of turn-on losses due to ensuring ZVS condition, the soft-switching converters can enhance efficiency. In contrast, the traditional hard-switching converter only achieves a modest efficiency of 72.49%. In comparison, a noticeable improvement was seen with the soft-switching converters, where the full-wave converter efficiency was 80.30%, which is slightly inferior to that of the half-wave converter with 83.44%, primarily due to the additional conduction losses of D_s. The main source of system losses in the QRCs comes from L_R. To further reduce core losses, an air-core inductor, possibly in the form of a PCB planar coil, offers potential to improve the QRCs efficiency. These results underline the clear benefits of resonant converters, in achieving higher power conversion efficiency, leading to better energy use and possibly longer life for components due to reduced operational stress.

V. Zero-Crossing in QRCs

This section explores the factors that impact the achievement of the ZVS condition in QRCs, particularly focusing on duty cycle variations. Through systematic analysis, we determine the boundaries within which the converter maintains ZVS condition.

In Fig. 4, the dashed lines indicate the boundaries of the ZVS condition based on U_{C_R} measurements. The duty cycle was systematically reduced until ZVS was no longer sustained, thereby establishing these shaded limits.

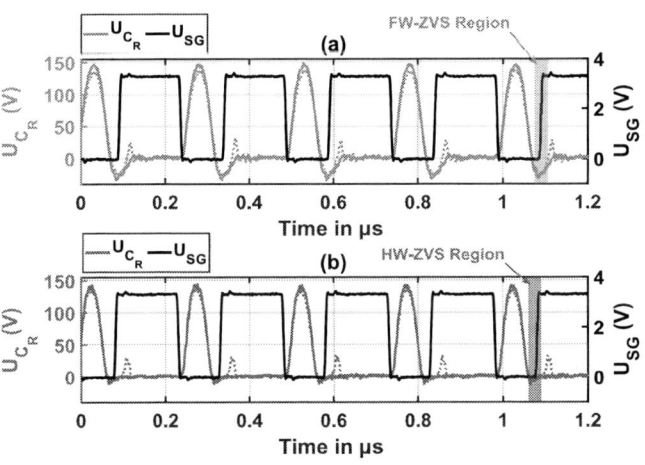

Fig. 4. Measurement of the QR-ZVS buck converter, representing the switching signal U_{SG} and the resonant capacitor voltage U_{C_R}. In (a) full-wave and (b) half-wave operation mode.

The permissible window for duty cycle variation while maintaining the ZVS condition is ±12.6% and ±12.3%, which represent 30.79 ns and 31.51 ns time windows for the full-wave and the half-wave configurations, respectively. Such a range offers more flexibility in operation, allowing the converter to handle variations in input voltage and load without losing the ZVS capability.

VI. Load Regulation Dynamics

In this analysis, utilizing an open control loop, the performance of the QRCs was examined. A signal generator was employed to generate the 4 MHz switching signals. The aim of this investigation was to emphasize the load dependency of the half-wave topology, compared to its full-wave counterpart.

This insight not only deepens our understanding of the full-wave approach but also highlights its practical advantages in real-world applications where load conditions may fluctuate frequently.

A. Load Dependency

Taking into account $C_{oss} = 550\,\text{pF}$, Z_R was calculated to be $\approx 28\,\Omega$. The load values were then carefully chosen to reflect the calculated normalized loads r_{calc} of 0.1, 0.2 and 0.4, aiming to validate the load dependence asserted theoretically to both approaches.

For this evaluation, while maintaining the normalized frequency unchanged, the VCR was measured for the three different load cases, as detailed in Table II. In Fig. 5, these measured values are plotted alongside the analytically calculated curves for comparison.

As depicted in Fig. 5(a), the half-wave structure exhibited a pronounced load dependency. In contrast, the full-wave approach demonstrated good resilience against load changes, and remains relatively stable even with load variations as shown in Fig. 5(b).

TABLE II
VCR Measurement for Load Dependency Analysis

	Half-wave			Full-wave		
R_L	3 Ω	6 Ω	12 Ω	3 Ω	6 Ω	12 Ω
A	0.4857	0.4857	0.4857	0.4917	0.4917	0.4910
r_{meas}	0.1035	0.2071	0.4141	0.1048	0.2096	0.4192
M	0.3325	0.4665	0.5753	0.4004	0.4558	0.5047

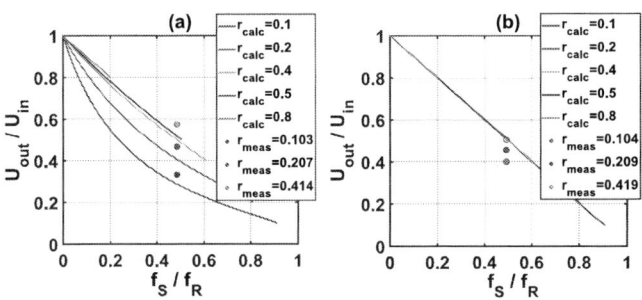

Fig. 5. QR-ZVS buck converter normalized output voltage vs. the normalized switching frequency for different normalized load values in (a) half-wave mode, and in (b) full-wave mode.

This robustness, inherent in the full-wave approach, indicates multiple advantages including enhanced system stability, predictable performance across varied load scenarios, and potentially extended component lifespan due to reduced stress from load variations. Furthermore, this also improve the converter transient response to sudden changes in operating conditions.

B. Full-Wave Benefits in Heavy Loads Conditions

The advantage of the uni-directional switch is highlighted in this part. In fact, I_{L_R} reach faster the initial condition and allows a smaller turn-off time. This advantage can be appreciated better with higher output currents. In Fig. 6(a) and Fig. 6(b), the locus of U_{C_R} and I_{L_R} shows that with the uni-directional switch the period where U_{C_R} has to be zero is reduced.

In Fig. 6(b), the distinctions between full-wave and half-wave modes can be observed from the ideal state locus of I_{L_R} (normalized with Z_R) on the x-axis, and U_{C_R} on the y-axis. When the switch turns off at t_0, I_{L_R} remains equal to I_{out}. Thereafter, U_{C_R} increases linearly towards U_{in} following Eq. (3).

$$U_{C_R}(t) = \frac{I_{out}}{C_R} \cdot t \qquad (3)$$

At point t_1, D_r begins to conduct, initiating the resonant oscillation stage of U_{C_R} and I_{L_R}, as given by Eqs. (4)–(5), respectively.

$$U_{C_R}(t) = U_{in} + Z_R \cdot I_{out} \cdot \sin(\omega_R \cdot t), \ \omega_R = 1/\sqrt{L_R \cdot C_R} \quad (4)$$

$$I_{L_R}(t) = I_{out} \cdot \cos(\omega_R \cdot t) \qquad (5)$$

At t_2 when U_{C_R} returns to zero, the transistor can turn on with ZVS. The inductor current I_{L_R} needs to reach I_{out} prior to the subsequent turn-off. The half-wave behavior of I_{L_R} is detailed by Eq. (6).

$$I_{L_R,\text{half-wave}}(t) = \frac{U_{in}}{L_R} \cdot t + I_{Lt_2} \qquad (6)$$

For high I_{out} scenarios, the advantages of the full-wave mode are evident. In contrast, the half-wave mode often requires more time to return to its initial state, given that the voltage across the inductor is U_{in}. On the other hand, with the presence of D_s in the full-wave converter case, the range of the resonant tank is extended to include voltages where $U_{C_R}(t) < 0$. This change allows ZVS to happen until the moment t_3. At this time, the current I_{Lt_3} is greater than I_{Lt_2}, leading to a quicker return to the initial state where $I_{L_R} = I_{out}$ as shown in Eq. (7) [6].

$$I_{L_R,\text{full-wave}}(t) = \frac{U_{in}}{L_R} \cdot t + I_{Lt_3} \qquad (7)$$

In Fig. 6(a), the measured results are presented. When comparing with the theoretical predictions in Fig. 6(b), due to the chosen value of C_{out}, the load current I_{out} doesn't follow a constant current source approximation in the initial phase, leading to a non-linear charging of C_R. Beyond the time t_1, the resonant tank formed by L_R and C_R operates as described in Fig. 6(b). The advantages of the full-wave approach become evident after t_2.

VII. ELECTROMAGNETIC COMPATIBILITY PERFORMANCE

The EMC test stand used for evaluating the EMI of the QRCs compared to the hard-switching case is shown in Fig. 7. It comprises a regulated dc-power supply, which is connected to the converter through a 2-meter unshielded parallel pair

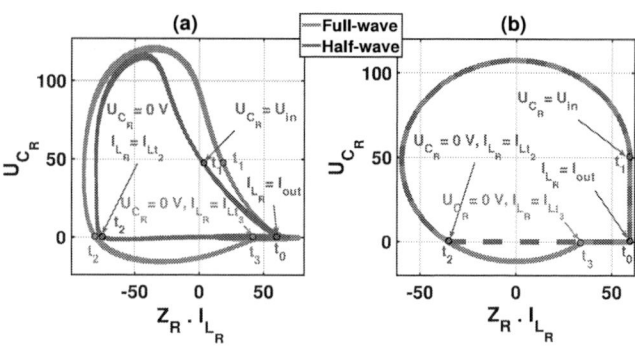

Fig. 6. State locus curve of U_{C_R} and Z_R in half-wave and full wave operation mode. (a) Measured results, (b) Simulated predictions.

cable via a Line Impedance Stabilization Network (LISN). The converter's dc output is then transmitted through a 1-meter unshielded parallel pair cable to a resistive load.

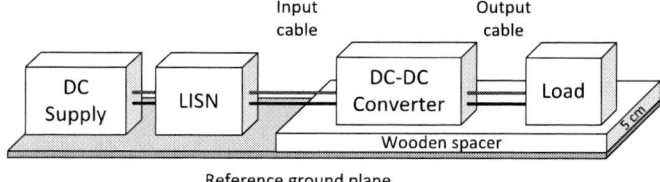

Fig. 7. Schematic of the used test setup for the conducted EMI measurements.

A. EMI Assessment of DC-DC Converters

In this assessment, we conducted in-depth analysis of the EMI behaviour of the converters. The primary objective was to characterize the input noise associated with each converter. For this purpose, the Common Mode (CM) noise was examined. These measurements are essential to understanding the EMI behavior of the converters and offering insights into their interference patterns in real-world application.

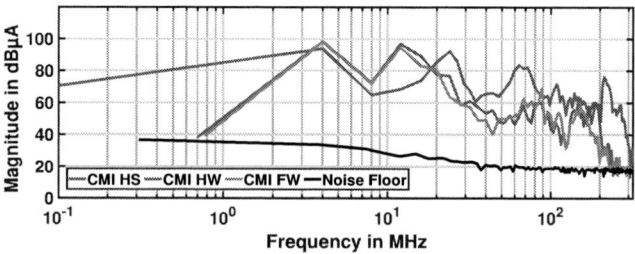

Fig. 8. Frequency spectrum of the converters input CM noise.

As shown in Fig. 8, among three converter topologies – two soft-switching and one hard-switching – it was observed that below 3.5 MHz, the hard-switching converter displayed a more dominant CM noise spectrum. However, between 3.5 MHz to 19 MHz, the soft-switching configurations exhibited increased noise. This increase in noise for soft-switching converters comes from different reasons. First, the QRCs have rapid changes in current and voltage during resonance, which can enhance the noise content near the SF. This is primarily because the U_{C_R} oscillates and returns to zero before the subsequent switch turn-on, introducing new HF content absent in hard-switching configuration. Furthermore, the inclusion of resonant elements in QRCs, brings in their own parasitic inductances and capacitances, potentially contributing to the HF noise. Notably, frequencies between 4 to 18 MHz include the harmonics of the SF and resonant frequency. Beyond 19 MHz, extending up to 320 MHz, the hard-switching converter again surpassed its counterparts in CM noise magnitude. This highlights the advantage of soft-switching over hard-switching in reducing EMI noise at higher frequencies.

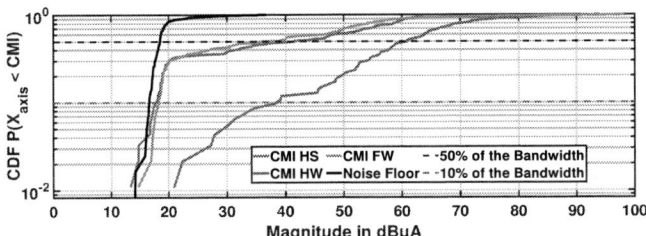

Fig. 9. CDF of the input CM noise frequency spectrum

To provide a more comprehensive understanding of the converters EMI behavior across the frequency spectrum, the CM noise was also represented using a Cumulative Distribution Function (CDF). The CDF makes it possible to quantify the probability that the CM noise will exceed a certain noise level in dBμA, at specific frequency bandwidth. This statistical approach presents a straightforward method to identify the frequency bandwidths where each converter topology might cause significant interference. The CDF plot, shown in Fig. 9, assists in highlighting those frequency bands where the chance of exceeding a certain noise magnitude is the greatest, further showcasing the benefits and limitations of each converter configuration in terms of EMI performance.

To illustrate, within 50% of the bandwidth, the hard-switching topology's noise spectrum exceeds 60 dBμA. In contrast, during the same bandwidth, the soft-switching topologies remain below 40 dBμA. This difference further highlights the better EMI performance of the soft-switching configurations when compared to the hard-switching topology within this frequency range.

B. Applications and Suitability

In the previous EMI analysis, specific EMC norms were not used as a benchmark, and particular applications for the converters were not specified. The primary objective was to demonstrate that employing HSF with soft-switching QR-ZVS topologies can substantially mitigate EMI noise. Notably, with the converter's power rating at 70 W and its input-output voltage specification of 48/24 V, it is well-suited to fit specific applications within the automotive sector, particularly

for heavy-duty vehicles and trucks. These might include systems such as starting, lighting, ignition controls, or auxiliary power supplies for various onboard electronics. Moreover, with more refined EMI filtering, the noise level can be further reduced to better align with the demands of such applications, highlighting its relevance and potential utility in real-world scenarios.

VIII. CONCLUSION AND FUTURE WORK

In this paper, a QR-ZVS converter approach was presented to reduce EMI noise and enhance efficiency compared to its hard-switching counterpart. This was achieved while maintaining a fixed switching frequency of 4 MHz using GaN-HEMT and relying on fixed LC resonant tank parameters. A prototype of the proposed topology was constructed and evaluated in terms of efficiency, load dependency, and EMI performance.

The soft-switching topologies demonstrated decreased EMI noise and exhibited superior efficiency compared to the reference hard-switching design.

Regarding load dependence, measurements confirmed that the full-wave topology exhibits less sensitivity to load variations compared to the half-wave design.

Switching at such a HF introduced additional thermal challenges in the converter design, necessitating careful attention in future prototypes.

The EMI filter design also offers room for refinement, enabling the noise spectrum to meet specific EMC standards, such as CISPR-25.

REFERENCES

[1] K. H. Liu, "High-frequency quasi-resonant converters techniques," Ph.D. dissertation, Virginia Polytechnic Inst. State Univ., Blacksburg, VA.1986.

[2] L. Gu and W. Zhu, "A Family of Zero-Voltage-Switched Resonant Converters: Derivation, Operation, and Design," in IEEE J. Emerg. Sel. Top. Power Electron, vol. 9, no. 2, pp. 2098-2108, April 2021.

[3] W. A. Tabisz, P. Gradzki and F. C. Lee, "Zero-voltage-switched quasi-resonant buck and flyback converters — Experimental results at 10 MHz" in IEEE PESC, pp. 404-413, 1987.

[4] E. Smailus, G. Griepentrog, M. Lutze and M. Pfeifer, "Improvement of ZVS range in Dual Active Bridge Converters using nonlinear inductors by ferrite block insertion," in IEEE PCIM, pp. 1-7, 2020.

[5] F. C. Lee, "High-frequency quasi-resonant converter technologies," in Proceedings of the IEEE, vol. 76, no. 4, pp. 377-390, April 1988.

[6] V. Wuti, A. Luangpol, K. Tattiwong, S. Trakuldit, A. Taylim and C. Bunlaksananusorn, "Analysis and Design of a Zero-Voltage-Switched (ZVS) Quasi-Resonant Buck Converter Operating in Full-Wave Mode," 6th ICEAST, Chiang Mai, Thailand, pp. 1-4, 2020.

[7] Y. Wang, O. Lucia, Z. Zhang, S. Gao, Y. Guan and D. Xu, "A Review of High Frequency Power Converters and Related Technologies," in IEEE OJIES, vol. 1, pp. 247-260, 2020.

[8] Z.-L. Zhang and Y.-F. Liu, High Frequency MOSFET Gate Drivers: Technologies and Applications, Institution of Engineering and Technology, 14 Sept 2017.

[9] T. Yao and R. Ayyanar, "A Multifunctional Double Pulse Tester for Cascode GaN Devices," in IEEE TPEL, vol. 64, no. 11, pp. 9023-9031, Nov. 2017.

[10] Schlienz, Ulrich. Schaltnetzteile und ihre Peripherie: Dimensionierung, Einsatz, EMV. Germany: Springer Fachmedien Wiesbaden, 2020.

[11] L. Crane, "Selecting the Best Inductor for Your DC-DC Converter," Coilcraft. [Online].

[12] W. A. Tabisz and F. C. Y. Lee, "Zero-voltage-switching multiresonant technique-a novel approach to improve performance of high-frequency quasi-resonant converters," in IEEE TPEL, vol. 4, no. 4, pp. 450-458, Oct. 1989.

979-8-3503-3714-3/23 $31.00 © 2023 IEEE

Impact of Process Variations on Back-Bias Effect in 100V p-GaN Gate AlGaN/GaN HEMTs

M. Cioni[1], G. Giorgino[1,2], A. Chini[2], G. Marletta[1], C. Miccoli[1],
M. E. Castagna[1], G. Luongo[1], M. Moschetti[1], C. Tringali[1] and F. Iucolano[1]

1. STMicroelectronics, Stradale Primosole n. 50, 95121 Catania, Italy,
2. Dipartimento di Ingegneria "Enzo Ferrari", University of Modena and Reggio Emilia,
Via P. Vivarelli 10, 41125 Modena, Italy
e-mail: marcello.cioni@st.com

Abstract—In this paper, we investigate the impact of Buffer resistivity and AlGaN barrier design on back-bias stress performed on 100 V p-GaN gate AlGaN/GaN HEMTs. To this end, we compare the results obtained in terms of (i) vertical leakage, (ii) back-bias stress on Transmission Line Measurements (TLM) structures and (iii) back-gating on real transistors. Concerning the latter, a novel test sequence is implemented to monitor the drain current evolution during the stress and evaluate the impact on V_{TH} and R_{ON} parameters after 1000 s stress with V_{SUB}=-50 V. Results indicate that high resistive buffer can significantly reduce the back-bias effect, but also the AlGaN barrier design can affect the parameters drift due to a different two-dimensional electron gas (2DEG) density.

Index Terms - GaN HEMTs, Back-Effect, Vertical Leakage, 2-DEG density, R_{ON}-degradation, V_{TH} drift.

I. INTRODUCTION

100V p-GaN gate AlGaN/GaN HEMTs are particularly important in power applications to realize both discrete and monolithic solutions. Concerning the latter, the monolithic integration of high-side and low-side transistor in a typical half-bridge configuration yields an important issue from the technological point of view [1]. Given the common substrate potential shared by high and low-side device (see Fig. 1), the high-side transistor experiences a voltage difference between its source and substrate terminal during on-state.

This so-called back-gating effect can be responsible for R_{ON}-degradation and V_{TH} drift, making it extremely important to evaluate this type of stress to understand the impact on performances in real applications.

Different approaches can be considered for this scope. In [2, 3], substrate ramping was used to investigate back bias effect, while in [4, 5] current transients on Transmission Line Measurement (TLM) structures were adopted. Even if both techniques are valid, they are commonly applied to TLM structures and not on the ultimate transistor. Goal of this work is to develop a simple technique able to capture the back-gating effect on the real transistor. Moreover, we aim to correlate the results obtained with measurements on simplified

structures to process variations in order to understand the cause for the parameters degradation induced by this kind of stress. This is of paramount importance to provide useful feedback for improving device manufacturing.

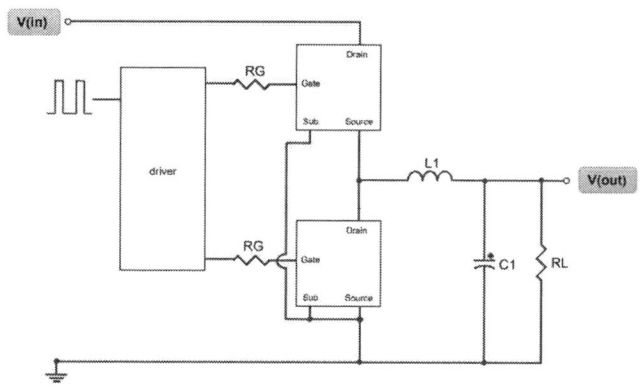

Figure 1. Schematic of a typical Half-Bridge configuration used in a Buck converter topology. In the monolithic integration of high-side and low-side transistors, both devices share the same substrate potential, yielding a voltage difference between source and substrate of the high-side transistor when it is turned on.

The paper is organized as follows: in Section II, we briefly describe the devices tested in this work, providing a first indication on the main differences shown by the samples considered. In Section III we report the preliminary characterization performed on simplified test structures, while in Section IV, back-bias stress/measurement is applied on real transistors to evaluate the different behavior shown by the samples. The correlations with the process variations and related discussion are reported in Section V, leading to the conclusion drawn in Section VI.

II. DEVICES DESCRIPTION

Devices tested in this work were AlGaN/GaN single heterojunction HEMTs grown on Silicon substrate. Four

979-8-3503-3714-3/23 $31.00 © 2023 IEEE

different AlGaN barrier layer designs have been considered, namely #1, #2, #3 and #4. On the other hand, GaN buffer was Carbon (C) to obtain a semi-insulating layer [6]. Particularly, three different GaN Buffer designs have been tested, herby called Sample A, Sample B and Sample C. Devices gate length was < 1 μm while gate-source and gate-drain spacings were < 1 μm and < 2 μm, respectively. This geometry allows operation in the 0 V to 100 V voltage range. In Fig. 2 we reported the typical cross section of the devices under tests.

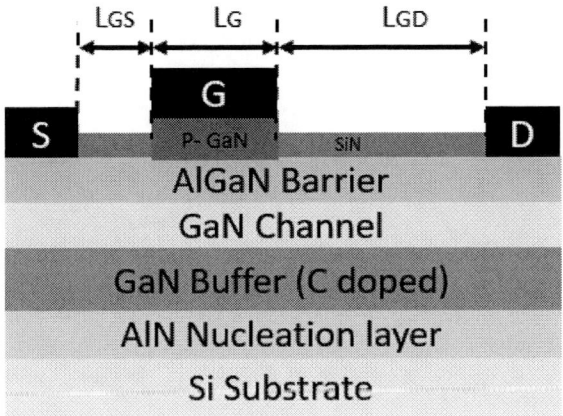

Figure 2. Schematic cross-section of the Device Under Test (DUT). DUTs presented a C-doped buffer layer grown on top of a Silicon (Si) substrate and Aluminum Nitride (AlN) nucleation layer. P-GaN gate is used to achieve normally-off operations, depleting the 2-Dimensional Electron Gas (2DEG) at the AlGaN/GaN heterojunction under the gate terminal even for V_{GS} = 0 V. Four different AlGaN barrier layer designs have been considered, namely #1, #2, #3 and #4. On the other hand, three different GaN Buffer designs have been tested, herby called Sample A, Sample B and Sample C.

The different AlGaN barrier layer designs could significantly impact the I_D-V_{GS} curves of the samples, yielding the differences in terms of triode current (i.e., R_{ON}) and pinch-off voltage shown in Fig. 3. Conversely, no significant differences were shown by samples A, B and C in terms of I_D-V_{GS} characteristics, since all the three samples showed the same AlGaN barrier as #3.

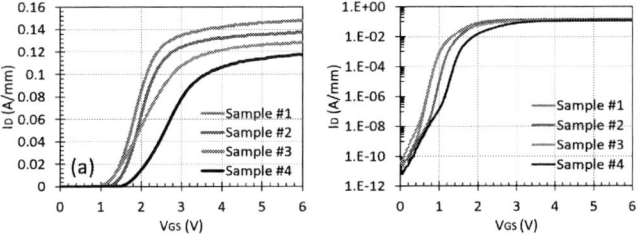

Figure 3. I_D-V_{GS} characteristics acquired on samples #1, #2, #3 and #4: (a) in linear scale, we can appreciate a significant difference in terms of triode current, while in (b) logarithmic scale, we can see a reduced pinch-off voltage for the devices presenting a higher conductivity. For a given AlGaN barrier design, no significant differences are observed by changing the buffer between A, B and C design.

In the following section, we introduce the structures and test conditions employed, along with the preliminary characterization performed.

III. TEST STRUCTURES AND PRELIMINARY CHARACTERIZATION

The test structures used in this work are sketched in Fig. 4.

(i) We first analyze TLM structures (Fig. 4(a)) under Back-bias stress to see the impact of the stress on the 2DEG conductivity.

(ii) Then, to highlight the different resistivity of the three samples considered, vertical leakage measurements (I-V) were performed on the structure reported in Fig. 4(b).

(iii) Finally, tests were performed on real transistors (see Fig. 4(c)), presenting the features described in Section II.

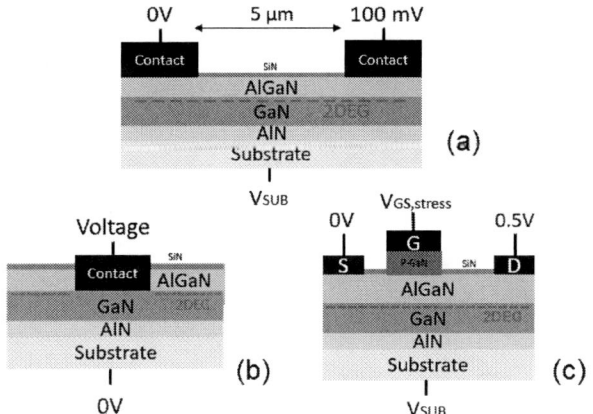

Figure 4. Tested Structures: (a) TLM structure used for back-bias stress; (b) vertical structure used for I-V vertical isolation measurement and (c) actual transistor used for back-bias stress in on-state to emulate high-side configuration in which the Source terminal faces a positive potential difference w.r.t. the substrate.

Concerning the tests performed on TLMs, we firstly measured the current between 5 μm separated pads, with a 100 mV potential applied between pads and V_{SUB}=0 V. This measurement allows to fix a reference fresh value (I_0) for evaluating the Back-Effect degradation. Afterward, the same potential is still applied between pads and the current is monitored for a stress time of 1000 s V_{SUB}=-50 V. The evolution of the current degradation over time (ΔI_{TLM}=I-I_0) is reported in Fig. 5(a) for the samples A, B and C.

On the same samples, vertical leakage measurements have been carried out on the test structure reported in Fig. 4(b). Particularly, four different voltages have been applied to the structure between the top pad and substrate (50 V, 100 V, 150 V and 200 V) yielding the results reported in Fig. 5(b).

As we can see, the Back-Effect degradation is reduced for samples showing a lower vertical leakage, suggesting that a higher vertical resistivity could help in reducing this issue.

979-8-3503-3714-3/23 $31.00 © 2023 IEEE

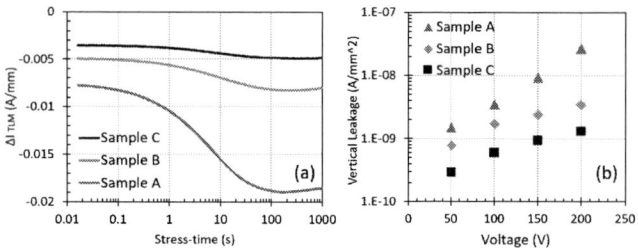

Figure 5. (a) ΔI_{TLM} (I-I$_0$) transients acquired on TLM structure with VSUB=-50V and (b) vertical leakage measurements performed on samples A, B and C. Samples with higher buffer resistivity (lower vertical leakage) are less sensitive to back-bias stress.

Even if this preliminary characterization performed on simplified structures already provides interesting results, it is important to investigate the impact of back-bias stress on the ultimate transistor to see if the trends observed are representative of the final device behavior.

IV. EXPERIMENTAL RESULTS ON REAL TRANSISTORS

The tests on real transistors were carried out by applying the stress/measurement sequence depicted in Fig. 6.

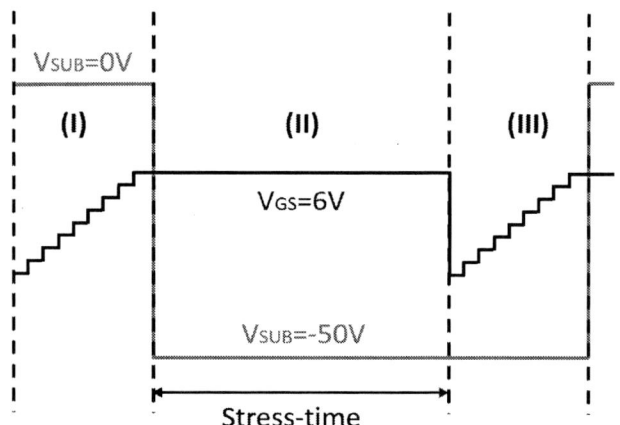

Figure 6. Waveforms employed for the stress/measurement sequence during back-bias stress in on-state. (I) Fresh I_D-V_{GS} is acquired; (II) Negative V_{SUB} (-50 V) is applied in on state (V_{GS}=6 V; V_{DS}=0.5 V) for 1000s; (III) post-stress I_D-V_{GS} is acquired with V_{SUB}=-50 V to reduce current recovery.

The test sequence proposed consists on three main steps.

(I) First, the Fresh I_D-V_{GS} is acquired to set a reference fresh point for the device's parameters.

(II) Then, a negative V_{SUB} (-50 V) is applied in on state (V_{GS}=6 V, V_{DS}=0.5 V) for 1000 s allowing the monitoring of the triode current over several time decades. Particularly, the stress time was set to observe the complete ionization of Buffer traps that have been reported to show time constants of several tens of seconds at room temperature [7].

(III) After the stress, the post-stress I_D-V_{GS} is acquired by still applying V_{SUB}=-50 V to reduce current recovery. The test

is performed on devices presenting different buffer resistivity (samples A, B and C).

The results obtained on these three samples are reported in Fig. 7, Fig. 8 and Fig. 9 respectively.

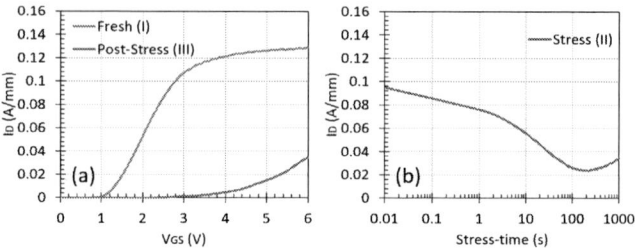

Figure 7. (a) I_D-V_{GS} acquired with the proposed method for Back-bias stress on DUT: Fresh curve (I) and post-stress curves (III) acquired on sample A after a 1000 s stress at V_{SUB}=-50V (II) during which the on-state current transient is monitored (b).

For sample A, we observed a large V_{TH} drift and R_{ON} degradation after the stress phase which is accompanied by a strong dynamic reduction of the triode current during the stress (Step II).

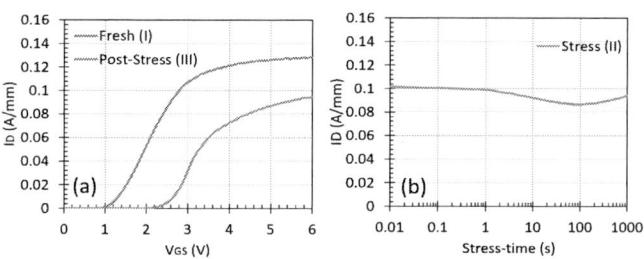

Figure 8. (a) I_D-V_{GS} acquired with the proposed method for Back-bias stress on DUT: Fresh curve (I) and post-stress curves (III) acquired on sample B after a 1000 s stress at V_{SUB}=-50V (II) during which the on-state current transient is monitored (b).

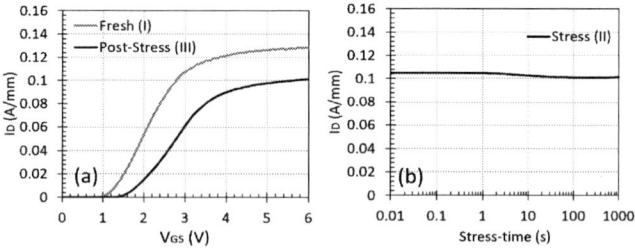

Figure 9. (a) I_D-V_{GS} acquired with the proposed method for Back-bias stress on DUT: Fresh curve (I) and post-stress curves (III) acquired on sample C after a 1000 s stress at V_{SUB}=-50V (II) during which the on-state current transient is monitored (b).

On sample B and Sample C, we observed a reduced degradation on both V_{TH} and R_{ON} parameters, as well as a reduced transient amplitude during the stress at V_{SUB}=-50 V. These results are totally aligned with those observed on TLM

979-8-3503-3714-3/23 $31.00 © 2023 IEEE

for the corresponding samples. The fact that similar results were obtained on both TLM and real transistors (compare Fig. 5(a) and Fig. 7(b), 8(b) and 9(b)) is coherent with previous literature [8] and suggests that the p-GaN gate was not affecting the experiment. In fact, according to the applied stress, the dynamic reduction of the current over time could be associated to the 2DEG charge variation induced by Buffer traps [7, 9] and not associated to gate instabilities.

The same stress/measurement sequence was then applied on devices showing the same buffer design, but different AlGaN barrier. This yielded devices (#1, #2, #3 and #4) with different 2DEG density. The results are reported in Fig. 10 for which we compared the pre- and post-stress I_D-V_{GS}.

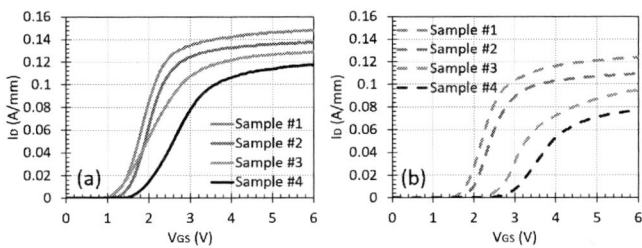

Figure 10. (a) Fresh I_D-V_{GS} curves measured on samples showing different AlGaN barrier (#1, #2, #3, #4) and (b) post-stress I_D-V_{GS} curves acquired after 1000 s back-bias stress in on-state with VSUB=-50 V. Samples presenting a higher fresh current level at V_{GS}=6 V present a reduced current degradation after 1000 s stress time, stemming for a higher carrier density in the 2DEG after the back-bias stress.

In general, a higher fresh current yielded a lower degradation after stress that is totally coherent with a higher carrier density that is less prone to be depleted. To assess the validity of this statement, we report in the following section the correlation observed between 2DEG density and R_{ON}-degradation.

V. CORRELATIONS AND DISCUSSION

In Fig. 11, we reported the correlation found between R_{ON}-degradation and 2DEG density.

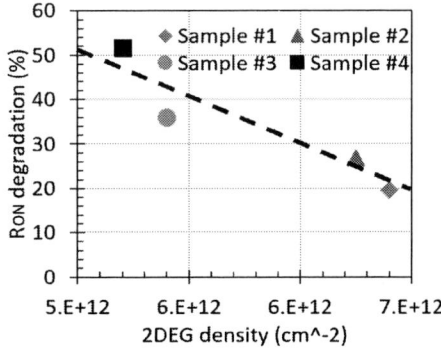

Figure 11. Correlation between 2DEG density (extracted from C-V measurements) and R_{ON} degradation after back-bias stress: samples with more populated 2DEG show a reduced degradation thanks to a higher carrier density that is difficultly depleted.

Essentially, the R_{ON} degradation decreases while increasing the 2DEG density. This is coherent with the fact that a larger carrier availability in the channel is less prone to be depleted, yielding a more conductive channel after the stress and thus a lower R_{ON} degradation.

However, this is just a part of the story. In fact, we have previously observed a strong impact of the Buffer resistivity on the back-bias effect. This is better highlighted in Fig. 12, in which we correlated the R_{ON} degradation with the vertical leakage current.

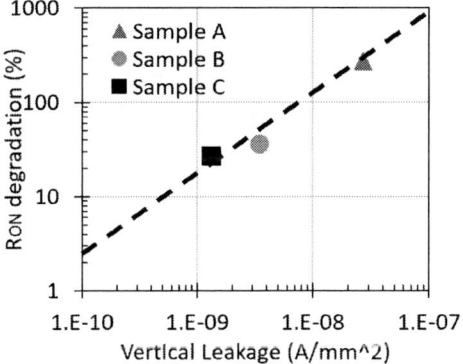

Figure 12. Correlation between vertical leakage and R_{ON} degradation (%) after back-bias stress: samples with more insulating Buffer show a reduced degradation.

As we can see, the degradation increases significantly while increasing the vertical leakage current, suggesting that a poor vertical isolation (i.e., less resistive Buffer) is more likely to cause a strong back-bias effect on the ultimate transistor. Conversely, a higher Buffer resistivity can reduce the degradation and, in principle, prevent the back-bias issue.

VI. CONCLUSIONS

In this work we proposed a novel technique to evaluate the back-bias effect on 100V p-GaN gate AlGaN/GaN HEMTs. Thanks to the method proposed, we evaluated the impact of Buffer resistivity and 2DEG density on the R_{ON}-degradation induced by means of back-bias stress. Essentially, an increase in the 2DEG density or in the buffer resistivity can contribute to prevent back-gating effect, thus allowing the realization of half-bridge topology by through monolithic GaN integration.

ACKNOWLEDGMENT

This work has been carried out in the framework of the European Project GaN4AP. The project has received funding from the ECSEL JU under Grant Agreement No. 101007310. The authors would like to acknowledge Antonino Parisi, Santo Reina and Santo Principato for the technical support.

REFERENCES

[1] J. Wei, M. Zhang, G. Lyu and K. J. Chen, "GaN Integrated Bridge Circuits on Bulk Silicon Substrate: Issues and Proposed Solution," in IEEE Journal of the Electron Devices Society, vol. 9, pp. 545-551, 2021, doi: 10.1109/JEDS.2021.3077273.

[2] M. J. Uren et al., ""Leaky Dielectric" Model for the Suppression of Dynamic RON in Carbon-Doped AlGaN/GaN HEMTs," in IEEE Transactions on Electron Devices, vol. 64, no. 7, pp. 2826-2834, July 2017, doi: 10.1109/TED.2017.2706090.

[3] H. Chandrasekar et al., "Buffer-Induced Current Collapse in GaN HEMTs on Highly Resistive Si Substrates," in IEEE Electron Device Letters, vol. 39, no. 10, pp. 1556-1559, Oct. 2018, doi: 10.1109/LED.2018.2864562.

[4] A. Chini et al., "Experimental and Numerical Analysis of Hole Emission Process From Carbon-Related Traps in GaN Buffer Layers," in IEEE Transactions on Electron Devices, vol. 63, no. 9, pp. 3473-3478, Sept. 2016, doi: 10.1109/TED.2016.2593791.

[5] M. Cioni et al., "Evidence of Carbon Doping Effect on VTH Drift and Dynamic-RON of 100V p-GaN Gate AlGaN/GaN HEMTs," 2023 IEEE International Reliability Physics Symposium (IRPS), Monterey, CA, USA, 2023, pp. 1-5, doi: 10.1109/IRPS48203.2023.10117585.

[6] M. J. Uren, J. Moreke and M. Kuball, "Buffer Design to Minimize Current Collapse in GaN/AlGaN HFETs," in IEEE Transactions on Electron Devices, vol. 59, no. 12, pp. 3327-3333, Dec. 2012, doi: 10.1109/TED.2012.2216535.

[7] M. Cioni, N. Zagni, F. Iucolano, M. Moschetti, G. Verzellesi and A. Chini, "Partial Recovery of Dynamic RON Versus OFF-State Stress Voltage in p-GaN Gate AlGaN/GaN Power HEMTs," in IEEE Transactions on Electron Devices, vol. 68, no. 10, pp. 4862-4868, Oct. 2021, doi: 10.1109/TED.2021.3105075.

[8] F. Iucolano et al., "Correlation between dynamic Rdsou transients and Carbon related buffer traps in AlGaN/GaN HEMTs," 2016 IEEE International Reliability Physics Symposium (IRPS), Pasadena, CA, USA, 2016, pp. CD-2-1-CD-2-4, doi: 10.1109/IRPS.2016.7574586.

[9] M. Meneghini et al., "Temperature-dependent dynamic RON in GaNbased MIS-HEMTs: Role of surface traps and buffer leakage," IEEE Trans. Electron Devices, vol. 62, no. 3, pp. 782–787, Mar. 2015, doi: 10.1109/TED.2014.2386391.

Investigation on ESD Robustness of 20-V GGNMOS and GDPMOS in 4H-SiC Process with 100-ns TLP Pulse

Chao-Yang Ke and Ming-Dou Ker

Institute of Electronics, National Yang Ming Chiao Tung University, Hsinchu, Taiwan

Abstract—The ESD robustness of 20-V GGNMOS and GDPMOS fabricated by the 4H-SiC process was investigated by the 100-ns transmission-line-pulse (TLP) pulse. The experimental results show that, under the test of breakdown mode, there is no correlation between the ESD robustness and the number of fingers. However, under the test of forward mode, the ESD robustness can be enhanced effectively by increasing the number of fingers. When the ESD current is conducted in the forward mode through the body diode of GGNMOS or GDPMOS, the ESD robustness can be effectively enhanced. Hence, the concept and schematic diagram of the whole-chip ESD protection are proposed to achieve sufficient ESD robustness of SiC ICs. Furthermore, the power-rail ESD clamp is an indispensable element to achieve sufficient ESD robustness.

Keywords—SiC, ESD, TLP, GGNMOS, GDPMOS, body diode.

I. INTRODUCTION

SiC devices have become more popular due to the high-voltage and high-temperature characteristics. These two advantages make SiC devices more competitive than Si devices for high-power applications, such as electric vehicles, data centers, green power infrastructures, and rail tractions. Recently, some prior works had demonstrated the successful integration of the high-power VDMOSFET with the low-voltage (10-V or 20-V) drive circuit in a monolithic chip fabricated in the same SiC process [1], [2]. In order to fulfill the commercialization of SiC-based integrated circuits (ICs), the reliability issue needs to be solved. The electrostatic discharge (ESD) event is one of the most challenging issues which is inevitable when it comes to mass production. The human body model (HBM) is an international standard for ESD event, which was specified to verify the ESD robustness of IC products [3]. However, during the whole period of HBM ESD test, the voltage and current of ESD stress on the device under test cannot be clearly quantified. To improve the efficiency of ESD design during the developing phase of IC products, the transmission-line-pulse (TLP) system was therefore invented to effectively quantify the ESD robustness of the device by TLP-measured I-V curves [4].

Although several works on SiC ESD devices were recently conducted [5]-[7], ESD robustness of typical ESD protection structures, which named as the GGNMOS (gate-grounded NMOS) and the GDPMOS (gate-V_DD PMOS) structures, was not clearly investigated in the SiC process. In this work, the ESD robustness of GGNMOS and GDPMOS with different numbers of fingers fabricated in a 4H-SiC process was characterized, respectively. Furthermore, the failure analysis

was also conducted to investigate the failure mechanism. To achieve sufficient ESD robustness, the concept of whole-chip ESD protection with GGNMOS, GDPMOS, and power-rail ESD clamp in the SiC process is demonstrated.

II. TEST STRUCTURE AND MEASUREMENT SETUP

A. Test Structures and DC Characteristics

The device structures of 20-V NMOSFET and PMOSFET used in this study are shown in Fig. 1. In this structure, there was no P-type isolation layer under the body of 20-V NMOSFET and PMOSFET during the SiC chip fabrication. However, when the high-voltage VDMOSFET is integrated into the same monolithic SiC chip, the P-type isolation layer and the corresponding floating guard rings must be added to guarantee that the 20-V devices can sustain the high voltage from the backside of the SiC chip [1], [2]. Their DC I_{ds}-V_{ds} curves with different gate biases (V_{GS}) are shown in Figs. 2(a) and 2(b). For the 20-V NMOSFET with a single finger width (W) of 10 μm and finger length (L) of 1.5 μm, the maximum driving current under V_{GS} of 20 V and V_{DS} of 20 V was around 0.65 mA. For the 20-V PMOSFET with the same W and L values, the maximum driving current under V_{GS} of -20 V and V_{DS} of -20 V was around 0.038 mA.

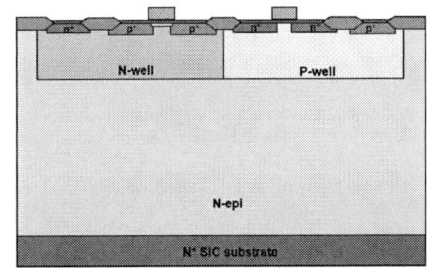

Fig. 1. The cross-sectional view of 20-V NMOSFET and PMOSFET in the SiC process.

To investigate ESD robustness of the GGNMOS (gate-grounded NMOS) and the GDPMOS (gate-V_DD PMOS), the device dimension must be drawn large enough, as well as the layout style of test device must be drawn with ESD protection consideration. The width/length (W/L) of each finger of the GGNMOS and GDPMOS was set to be 50μm /1.5μm, and the finger number was set to be 2, 6, and 12, respectively, in the SiC test chip. The typical layout top views of the 20-V NMOSFET and PMOSFET with multi-finger (12 fingers) structures are shown in Figs. 3(a) and 3(b), respectively.

979-8-3503-3714-3/23 $31.00 © 2023 IEEE

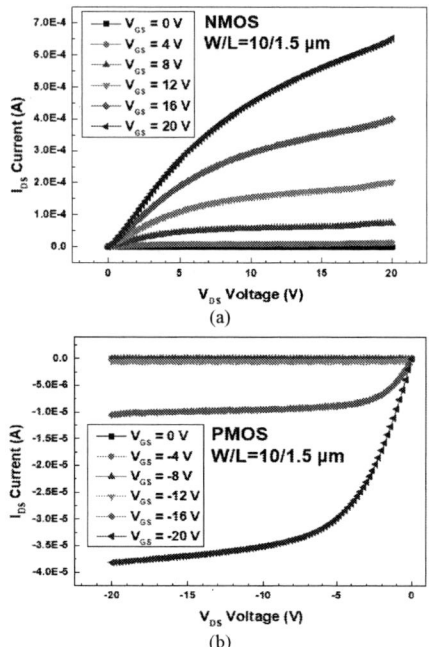

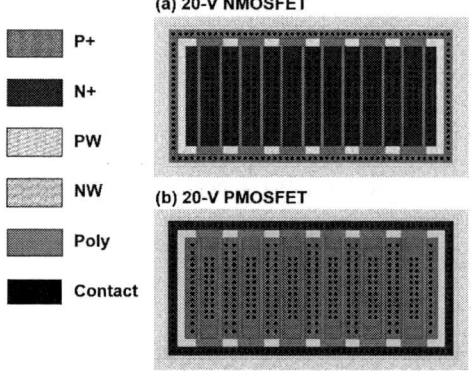

Fig. 2. The DC I_{ds}-V_{ds} curve of (a) 20-V NMOSFET and (b) 20-V PMOSFET under different gate biases.

Fig. 3. The layout top views of (a) 20-V NMOSFET and (b) 20-V PMOSFET on the SiC test chip.

B. Measurement Setups

The ESD events may have positive or negative voltage polarity to the reference ground, and the zapping point may apply at the drain or source side of the GGNNMOS and GDPMOS. The combinations of measurement setup for the device-level TLP test are shown in Figs. 4(a)-4(d). The TLP test can be divided into two modes. One is the breakdown mode, which means that TLP is zapped onto the reverse junction of the device, and the other is the forward mode, which means that TLP is zapped onto the forward junction of the device. The rise time and pulse width of the typical TLP waveform are set to 10 ns and 100 ns, respectively. The parasitic BJT of the silicon device can be triggered during the

breakdown mode test. Hence, this kind of phenomenon will be also worthy of investigation on the SiC devices.

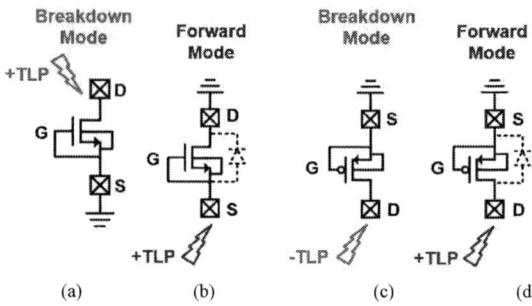

Fig. 4. The measurement setup of device-level TLP test with (a) GGNMOS in breakdown mode, (b) GGNMOS in forward mode, (c) GDPMOS in breakdown mode, and (d) GDPMOS in forward mode.

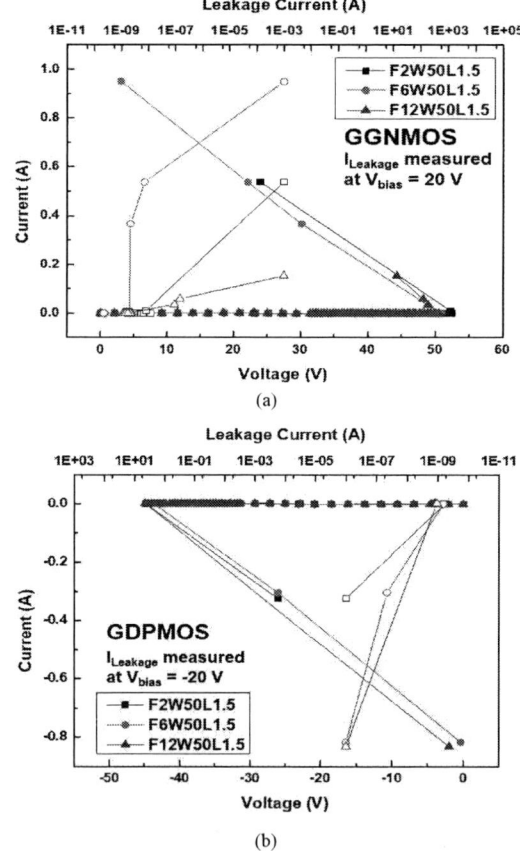

Fig. 5. TLP-measured I-V curves of (a) GGNMOS and (b) GDPMOS in breakdown mode.

III. EXPERIMENTAL RESULTS

A. Breakdown Mode

The TLP-measured I-V curves of GGNMOS and GDPMOS in the breakdown mode with different number of fingers are shown in Figs. 5(a) and 5(b). It can be observed that

979-8-3503-3714-3/23 $31.00 © 2023 IEEE

the parasitic BJT was not triggered until the device failed. Hence, there is no correlation between the number of fingers and the failure current. The failure currents of GGNMOS and GDPMOS in the breakdown mode were all around 0.1 ~ 0.4 mA. Furthermore, the failure voltage is not increased by increasing the number of fingers. Therefore, the ESD robustness of GGNMOS and GDPMOS in breakdown mode cannot be improved by increasing the number of fingers.

B. Forward Mode

However, with respect to the forward mode, the failure current can be substantially enhanced by increasing the number of fingers due to the body-diode characteristics. The TLP-measured I-V curves of GGNMOS and GDPMOS in forward mode with different finger numbers are shown in Figs. 6(a) and 6(b). For GGNMOS, the failure currents of devices with finger number of 2, 6, and 12 are 165 mA, 257 mA, and 344 mA, respectively. For GDPMOS, the failure currents of devices with finger number of 2, 6, and 12 are 499 mA, 1034 mA, and 1470 mA, respectively. As a result, the ESD robustness in both GGNMOS and GDPMOS can be effectively improved by increasing the number of fingers in the forward mode.

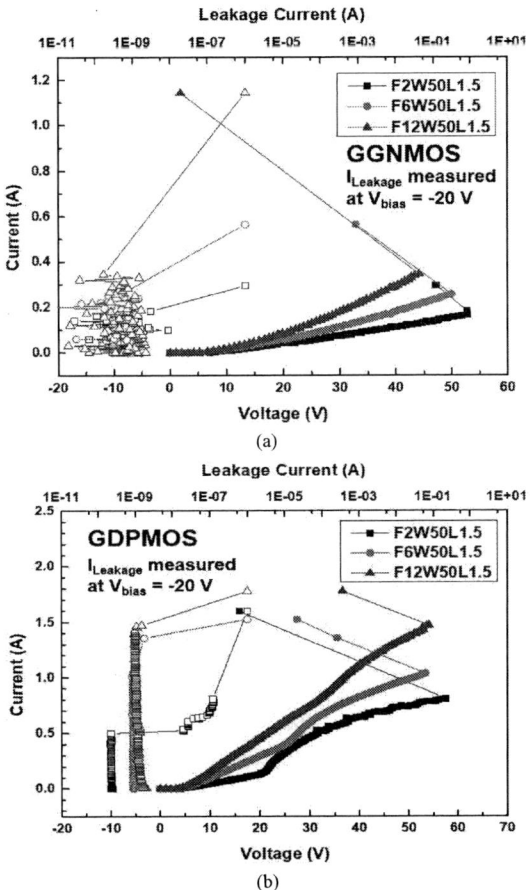

Fig. 6. TLP-measured I-V curves of (a) GGNMOS and (b) GDPMOS in forward mode.

C. Correlation Between Failure Currents and Device Total Widths

The correlations between the failure currents and the total widths of the GGNMOS and GDPMOS devices are clearly shown in Figs. 7(a) and 7(b), respectively. For GGNMOS with a total width of 600 µm, the failure current is 344 mA (0.4 mA) in forward mode (breakdown mode). For GDPMOS with the same total width of 600 µm, the failure current is 1470 mA (0.1 mA) in forward mode (breakdown mode). Moreover, the failure current of GDPMOS in forward mode is larger than that of GGNMOS under the same total width. This can be ascribed to the fact that the sheet resistance of P-well is 10 times higher than that of N-well in the given SiC process.

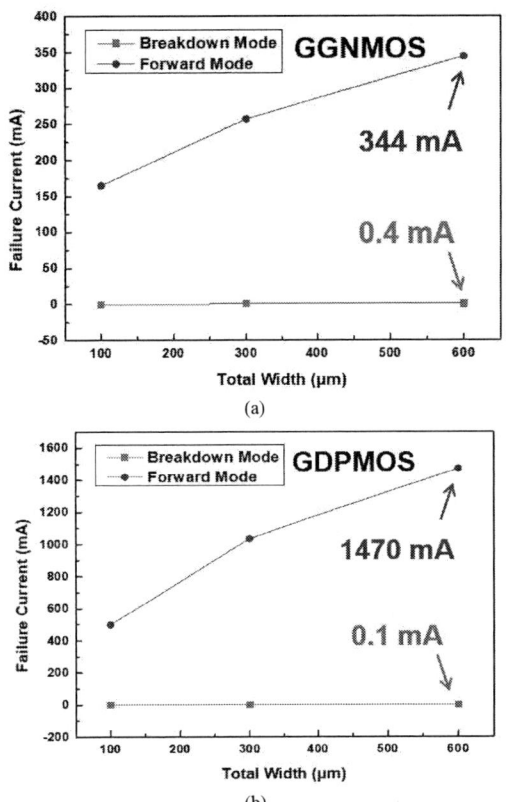

Fig. 7. The correlation between failure currents and device total widths of (a) GGNMOS and (b) GDPMOS.

D. Failure Analysis

In order to further investigate the failure mechanism of GGNMOS and GDPMOS in the breakdown mode, the failure analysis was conducted by using *Optical Beam Induced Resistance Change (OBIRCH)* to find the failure site. The images of the failure samples of GGNMOS and GDPMOS zapping in the breakdown mode with the finger number of 12 are shown in Figs. 8(a) and 8(b), respectively. For GGNMOS, the damage was found at only one site, which was located in the source region. Similarly, for GDPMOS, the damage was

979-8-3503-3714-3/23 $31.00 © 2023 IEEE

also found at only one site, which was located in the region between the source and the drain. Because the failure site was found in only one small region rather than in many locations among the multiple fingers of GGNMOS and GDPMOS, the miscorrelation between the failure current and the finger number can be reasonably interpreted.

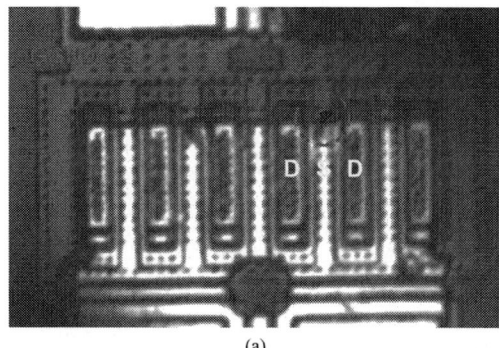

(a)

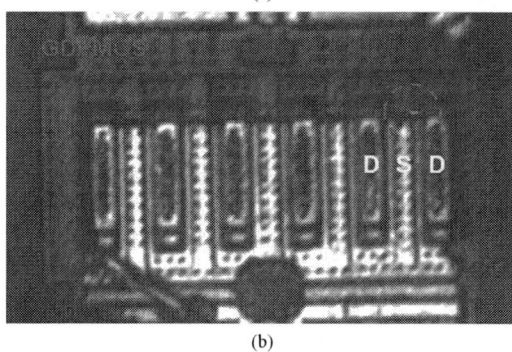

(b)

Fig. 8. The images of the failure analysis by using OBIRCH on the failed sample of 12-finger (a) GGNMOS and (b) GDPMOS, after zapping in the breakdown mode.

IV. Discussion

Based on the ESD robustness of GGNMOS and GDPMOS investigated in this work, if GGNMOS and GDPMOS are used to design the whole-chip ESD protection as what has been commonly used in the silicon CMOS ICs, it is necessary to conduct the ESD current in the forward mode to guarantee the sufficient ESD level. The ESD protection of the input pin can be realized by GGNMOS and GDPMOS, as those shown in Figs. 9(a) and 9(b). To discharge ESD current through the GGNMOS or GDPMOS in the forward mode, the power-rail ESD clamp is indispensable to conduct the ESD current from V_{DD} to V_{SS} [8].

Figs. 9(a) and 9(b) show the schematic diagram of the SiC power IC with whole-chip ESD protection scheme. The PS mode (positive ESD zapping from input to V_{SS}) and the ND mode (negative ESD zapping from input to V_{DD}) ESD stresses are two critical weak conditions when ESD zapping on the input pin [8]. For the PS mode, the ESD current will be first conducted through the body diode of the GDPMOS (in forward mode) to the V_{DD} power line, and then it will be conducted

through the power-rail ESD clamp to the grounded V_{SS} line. For ND mode, the negative ESD current will be first conducted from the input pin through the body diode of the GGNMOS (in forward mode) to the V_{SS} line, and then it will be conducted through the power-rail ESD clamp to the grounded V_{DD} line. Based on these two aforementioned operations of ESD protection, it is confirmed that the power-rail ESD clamp plays an important role in whole-chip ESD protection [8]. Therefore, to enhance the ESD level of SiC ICs, the development of the power-rail ESD clamp in the SiC process is urgently needed.

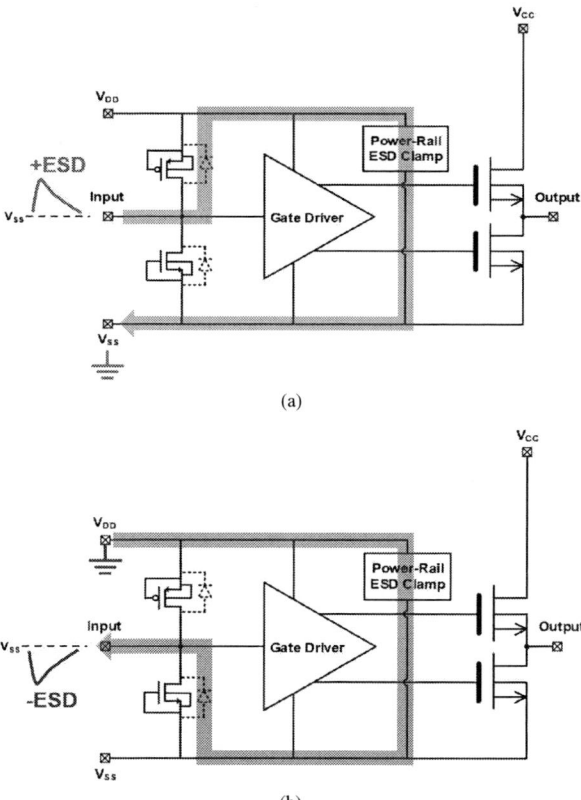

(a)

(b)

Fig. 9. The schematic diagram of the SiC IC with the design of whole-chip ESD protection under the ESD zapping condition of (a) PS mode and (b) ND mode.

V. Conclusions

In this study, the TLP tests have been used to explore the ESD robustness of 20-V GGNMOS and GDPMOS fabricated in a SiC process. The effect of different number of fingers on the TLP failure current has been investigated. Under the breakdown mode, the experimental results show that there is no correlation between the failure current and the number of fingers. The failure analysis has shown evidence that the failure site was found in only one small location on the GGNMOS or GDPMOS after it was zapping under the breakdown mode. However, under forward mode, the ESD robustness in both GGNMOS and GDPMOS can be effectively

979-8-3503-3714-3/23 $31.00 © 2023 IEEE

enhanced by increasing the number of fingers. Finally, the concept of whole-chip ESD protection scheme for SiC IC is demonstrated, where the power-rail ESD clamp can help to conduct ESD current through the GGNMOS or GDPMOS in the forward mode. Thus, the SiC-based power-rail ESD clamp is the key item to be developed and verified in the near future.

ACKNOWLEDGMENT

This work was supported by National Science and Technology Council (NSTC), Taiwan, under Contracts of NSTC 112-2218-E-A49-017 and 110-2622-8-009-017-TP1. The authors thank the research group of Prof. Bing-Yue Tsui for their technical support and Mr. Yu-Xin Wen for the layout of test chip.

REFERENCES

[1] B.-Y. Tsui, T.-K. Tsai, C.-L. Hung, and Y.-X. Wen, "Design and characterization of the junction isolation structure for monolithic integration of planar CMOS and vertical power MOSFET on 4H-SiC up to 300 °C," in *IEDM Tech. Dig.*, 2022, pp. 9.3.1-9.3.4.

[2] B.-Y. Tsui *et al.*, "First integration of 10-V CMOS logic circuit, 20-V gate driver, and 600-V VDMOSFET on a 4H-SiC single chip," in *Proc. IEEE International Symposium on Power Semiconductor Devices and ICs*, 2022, pp. 321-324.

[3] *Electrostatic Discharge (ESD) Sensitivity Testing Human Body Model (HBM)*, EIA/JEDEC Standard EIA/JESD22-A114-A, 1997.

[4] T. J. Maloney and N. Khurana, "Tranmission line pulsing techniques for circuit modeling of ESD phenomena," in *Proc. EOS/ESD Symp.*, 1985, pp. 49-54.

[5] P. Lai, H. Wang, A. Abbasi, S. Roy, A. Rashid, A. Mantooth, and Z. Chen, "Investigation of ESD protection in SiC BCD process," in *Proc. IEEE Workshop Wide Bandgap Power Devices Appl.*, 2019, pp. 405–409.

[6] K.-I. Do, B.-S. Lee, and Y.-S. Koo, "Study on 4H-SiC GGNMOS based ESD protection circuit with low trigger voltage using gate-body floating technique for 70-V applications," *IEEE Electron Device Letters*, vol. 40, no. 2, pp. 283-286, Feb. 2019.

[7] K.-I. Do, J.-I. Won, and Y.-S. Koo, "A 4H-SiC MOSFET-based ESD protection with improved snapback characteristics for high-voltage applications," *IEEE Trans. Power Electronics*, vol. 36, no. 5, pp. 4921-4926, May 2021.

[8] M.-D. Ker, "Whole-chip ESD protection design with efficient VDD-to-VSS ESD clamp circuits for submicron CMOS VLSI," *IEEE Trans. Electron Devices*, vol. 46, no. 1, pp. 173-183, Jan. 1999.

Single and Double-Sided Jet Impingement Cooling for SiC-Based Power Modules

Himel Barua
Building and Transportation
Science Division, *Oak Ridge
National Laboratory*
Oak Ridge, TN, USA
baruah@ornl.gov

Shajjad Chowdhury
Building and Transportation
Science Division, *Oak Ridge
National Laboratory*
Oak Ridge, TN, USA
chowdhuryms@ornl.gov

Jon Wilkins
Building and Transportation
Science Division, *Oak Ridge
National Laboratory*
Oak Ridge, TN, USA
wilkinsjp@ornl.gov

Burak Ozpineci
Building and Transportation
Science Division, *Oak Ridge
National Laboratory*
Oak Ridge, TN, USA
burak@ornl.gov

Abstract—Efficient thermal management of power electronics systems is crucial for higher reliability. With the miniaturization of systems, high-loss-density electronics require cooling systems that can extract a large amount of heat. This study explored a liquid-jet-impingement-based direct substrate cooling system for single-sided and double-sided cooling to improve heat extraction efficiency and improve the power density by reducing the volume and mass. The cooling system was implemented for a SiC-based direct bonded copper substrate. Numerical simulations were performed to determine the effects of nozzle diameter, the number of nozzles, and nozzle array orientation on single-sided cooling and thermal performance gain over double-sided cooling. A novel manifold design was proposed that reduced the volume and mass of the manifold and still achieved the target power density. The performance of the proposed design was compared with the pin-fin-based cooling system used in the BMW I3 module, and a comparative analysis was done.

Keywords—*Jet impingement, double-sided cooling, variable nozzle, and flow rate distribution*

I. INTRODUCTION

With the trend of miniaturizing power electronic systems, the corresponding loss density becomes larger even though these systems are more efficient than conventional ones [1]. This results in thermal challenges that require new thermal management solutions. Jet impingement using liquid jets is a technology that can cool power electronic systems efficiently [2]. Power electronics applications generally use horizontal liquid cooling, as shown in Fig. 1(b), in which the flow path can have different shapes [3], [4]. The average heat transfer coefficient (HTC) of horizontal liquid cooling is 20,000 W/(m²-K) [5], and a high-volume heat sink is required. With jet impingement cooling, HTCs of 115,000 W/(m²-K) are achievable in a smaller volume [5]. In indirect cooling, as shown, the power module is either soldered, sintered, or attached to the heat sink through a thermal interface material. The resulting thermal resistance is high because of additional layers. Direct substrate cooling, removes the baseplate and thermal interface layers, and the liquid jet impinges on the lower Cu layer of the direct bonded Cu (DBC). Removing these two layers would reduce the thermal resistance by ~30% [2].

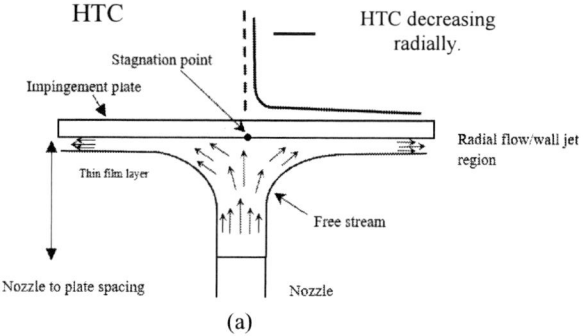

(a)

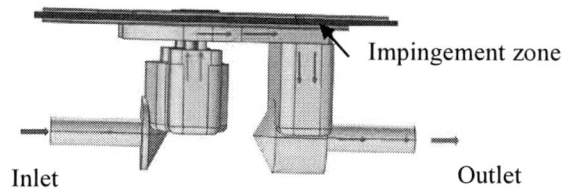

(b)

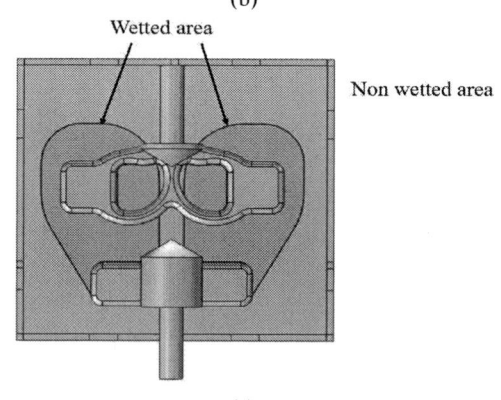

(c)

Fig. 1. (a) Jet structure, (b) cooling manifold, and (c) plan view of manifold.

Jet impingement cooling has been used for applications such as electric motor end-winding cooling [6], [7] and direct cooling of electronic devices [8]. Still, it has not been commonly applied for power electronics applications because of the high pressure drops in the cooling loop. Multiarray nozzles are widely used for jet impingement cooling [9].

This manuscript has been authored by UT-Battelle, LLC, under contract DE-AC05-00OR22725 with the US Department of Energy (DOE). The US government retains and the publisher, by accepting the article for publication, acknowledges that the US government retains a nonexclusive, paid-up, irrevocable, worldwide license to publish or reproduce the published form of this manuscript, or allow others to do so, for US government purposes. DOE will provide public access to these results of federally sponsored research in accordance with the DOE Public Access Plan (http://energy.gov/downloads/doe-public-access-plan).

979-8-3503-3714-3/23 $31.00 © 2023 IEEE

Increasing the number of nozzles for a constant flow rate would reduce the HTC at the impingement point. It would also reduce the pressure drop considerably, reducing the corresponding pumping power. Research has been performed to explore the effect of design parameters such as the jet length to diameter ratio [10], the number of jets and jet diameter [11], and plain and finned surfaces [12]. Smaller jet diameters yield higher HTC values and higher pressure drops, requiring higher pumping power. Finned surfaces yield higher HTC values than plain surfaces because of the larger heat transfer area and flow interaction between the fins. Double-sided cooling has been widely used for horizontal cooling because of its high heat extraction capability because heat is extracted from the device on both the top and the bottom sides [13], [14]. In double-sided cooling, the semiconductor devices are sandwiched between two substrates with the help of metallic spacers to extract the heat from both the top and bottom sides of the module. Although this design enables more heat transfer space, the additional layers and interconnects increase thermal resistance.

US DRIVE 2025 [15] set the power density target for integrated electric drive as 33 kW/L and as 100 kW/L for inverters. In inverter design, the heat sink takes up close to 33% of the inverter volume. Besides volume, in some applications such as aerospace applications, the weight of the components also plays an important role. Conventional horizontal cooling heat sinks are made of either Al or Cu. The thermal conductivity of Cu is close to 1.5 times greater than that of Al, but Cu is also approximately 4 times heavier than Al [16]. Also, thick baseplates are common and spread heat, which increases both volume and weight.

In the present research, the goal was to design a jet-impingement-based cooling system that kept the device temperature within 150°C and the pressure drop within the US DRIVE technical goal of 2 psi [17] and to reduce the volume and mass as much as possible.

In DBC-based substrates, heat coming out from devices spreads at a certain angle depending on the number of layers and the thickness of each layer [17], [18], [19]. So, beyond a certain area under the footprint of a device, heat does not spread far Taking this into account, a novel manifold design was proposed wherein the nozzles are placed below the devices covering an area three times larger than the device footprint instead of covering the whole bottom Cu footprint. This design has two advantages: it reduces the volume and corresponding mass and also enables higher jet speed with a lower flow rate. Increasing the number of nozzles by keeping flow rate constant reduces the jet speed, which reduces corresponding local heat transfer at the impinged point but also reduces the total pressure drop. To have the optimized number of jets to keep both the device temperature and pressure drop within the targets requires careful manifold design.

In this research, though several numerical simulations, an optimized jet-impingement-based cooling design was derived. A novel manifold design was proposed which allowed uniform flow distributions throughout all the nozzles and kept the pressure drop within 2 psi. Manifold design for jet impingement cooling is challenging because it requires that uniform flow through all the nozzles is ensured.

Both single- and double-sided cooling were explored, and thermal performances were compared.

The rest of the article is organized as follows. Section II(A) describes the physical structure of the jet and the layout of the power module and cooling system. Section II(B) shows the model development and thermal properties of different layers of the module. Section III(A) describes the effect of distance between the jet exit and impingement surface, section III(B) discusses the effect of diameter, and section III(C) discusses the effect of the number of nozzles and their orientations on device temperature and system pressure drops. In section III(D), single- and double-sided jet impingement cooling are compared. Finally, section IV provides the conclusion of this study.

II. THEORY AND MODELING

A. Power Module Layout

Fig. 1(c) shows the full-scale power module layout. The model included two SiC devices that were soldered into the top Cu surface of the DBC. Each device had a power loss of 100 W, which corresponded to a heat flux of 316 W/cm². The DBC comprised three layers, the top and bottom Cu layers and the middle AlN, Al₂O₃ layer. Table I lists the design parameters of each component. The jet impingement manifold was attached under the substrate. Fig. 1(b) and (c) shows the manifold design for the module. The height of the manifold was 10 mm, and the top side of the manifold, which was directly attached to the bottom Cu of the substrate, was made of a gasket (polymer-based material) (part number: 8985K422) [20]. The manifold did not cover the whole bottom Cu; rather, as shown in Fig. 1(c), it covered a butterfly-shaped area, which was approximately five times larger than the device footprint. The thickness of this gasket dictated the distance between the jet and the impinged surface. Initially, the distance was 1.5 mm (1/16 in.). The coolant was introduced through a circular channel of 3 mm diameter and enters an array of 6 nozzles per device; the diameter of each nozzle was 1 mm. Later, the nozzle diameter, nozzle number, and the impingement distance were optimized based on the device temperature and acceptable pressure drop from the inlet to the outlet of the manifold.

TABLE I.   DESIGN PARAMETERS OF THE POWER MODULE

Component	Thickness (mm)	Area (mm²)
SiC	0.18	31.65
Top Cu layer	0.45	1,227.48
Middle AlN layer	0.64	3,963.90
Bottom Cu layer	0.45	1,754.20
Solder (Cu–Sn)	0.05	31.65

The nozzles were placed in an in-line shape at the footprint of the device under the bottom Cu layer. The nozzle height was 0.5 mm, and the diameter was 1 mm. The coolant was water, and the thermal conductivity and viscosity of water were considered as functions of temperature. In jet impingement cooling, high-speed coolant impinges on the hot surface from

979-8-3503-3714-3/23 $31.00 © 2023 IEEE

a nozzle perpendicular to the impingement surface. At the impingement point, the flow changes its trajectory as shown in Fig. 1(a) and becomes a radially accelerating flow because of the pressure gradient. This change increases heat transfer because of boundary-layer thinning. At the wall jet zone, the radial flow decelerates, and the heat transfer rate decreases along the flow direction.

B. Model Development

The scope of this study was limited to a numerical study. A heat conduction equation was coupled with the conservation of mass, momentum, and energy to resolve the fluid flow and heat transfer. All the flow rates were kept such that the corresponding Reynolds number in (1) was less than 2,300 to keep the flow pattern laminar. The Mach number of the flow was less than 0.3, and the flow was considered incompressible.

$$Re = \frac{\rho v d}{\mu} \quad (1)$$

where Re is the Reynolds number, ρ is the density of the jet liquid, v is the velocity of the jet, d is the diameter of the nozzle, and μ is the viscosity of the liquid. COMSOL was used for geometry preparation, mesh generation, model development, and to develop solutions. Table II lists the material properties for each layer of the modules.

TABLE II.   MATERIAL PROPERTIES FOR EACH POWER MODULE LAYER

Component	Thermal conductivity (W/(m-K))	Density (kg/m³)
Si	130	2,329
Top and bottom Cu layers	400	8,940
Middle Al₂O₃ layer	27	3,900
Solder	50	9,000

III. RESULTS AND DISCUSSION

Numerical experiments were conducted to understand the behavior of jet-impingement-based cooling. As mentioned earlier, the number of design variables plays an important role in the thermal performance of jet-impingement-based cooling. For the present study, the nozzle diameter, number of nozzles, nozzle orientations, and the distance between the jet and the impingement surface were explored.

A. Effect of Length-to-Diameter Ratio

For the same nozzle diameter, increasing the distance of the impinged surface and the jet exit showed a detrimental effect on thermal performance. The best thermal performance was obtained with a length-to-diameter (L/d) ratio of 1. For a 1 mm nozzle diameter, the distance was 1 mm. For manufacturability, the minimum height was kept as 1 mm. For a constant flow rate and nozzle diameter, increasing the distance would reduce the velocity of flow close to the impingement surface, which would affect the heat transfer. Fig. 2 shows the reduction of jet velocity along the flow path from the jet exit. The velocity is normalized by jet inlet velocity. With a larger gap between the jet exit and the impinged surface, with the same velocity flow starts to decelerate earlier, and the velocity is less near the impingement point. The highest velocity was obtained at a L/d

ratio of 1, which means the jet diameter and the gap between the jet and impinged surface were the same. To seal the coolant, the sealant rubber thickness was required to be at least 1 mm, which is why a L/d ratio less than 1 was not chosen.

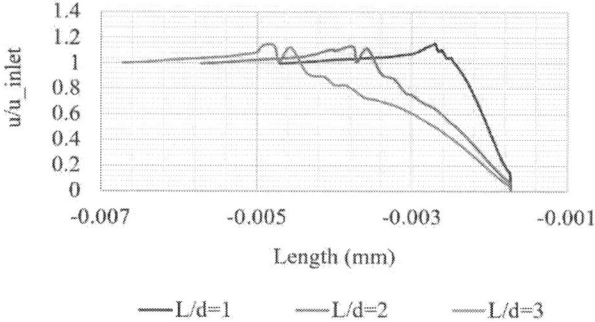

Fig. 2.   Change of normalized jet speed with flow direction.

B. Effect of Nozzle Diameter

To understand the effect of nozzle diameter, two designs were compared. In design A, an array of six nozzles was placed at the footprint of each device. Each nozzle diameter was 1 mm, whereas in design B, the nozzle diameter was 0.5 mm. Because of manufacturing challenges, 0.5 mm was the smallest diameter that could be used.

Fig. 3(a) and (b) shows that with a 50% reduction of jet diameter, the device temperature decreased 20%. But this came at an expense. Fig. 4 shows the pressure drop in the manifold from the inlet to the outlet. With reduction of the jet diameter, the jet exit velocity increased by a factor of four. Fig. 4(b) shows that when the device temperature decreased 20%, the pressure drop increased by a factor of approximately six. Per the US DRIVE 2025 technical goal, the pressure drop for an inverter cooling system should be less than or equal to 2 psi.

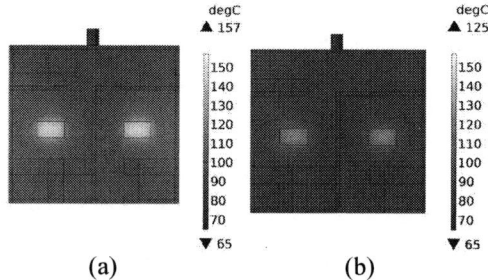

Fig. 3.   Temperature distribution in the module for (a) 1 mm diameter nozzles and (b) 0.5 mm diameter nozzles.

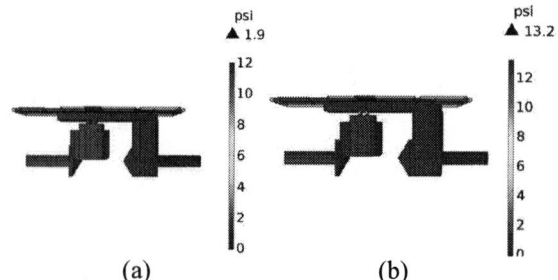

Fig. 4.   Pressure drop for (a) 1 mm nozzles and (b) 0.5 mm nozzles.

C. Effect of Number of Nozzles

In the previous section, it was shown that with a smaller jet diameter, the device temperature decreased because of high jet speed. But the pressure drop was much higher than the goal. One approach to solve this issue would be increasing the number of nozzles. For a constant flow rate, increasing the number of nozzles would reduce the individual jet speed but would also increase the coverage. Fig. 5 shows the two designs considered for studying the effect of the number of nozzles.

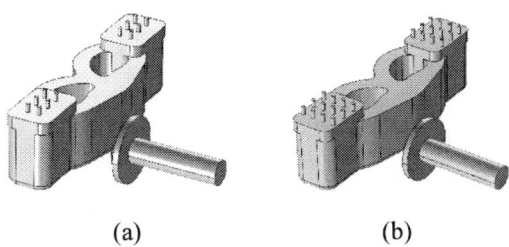

(a)   (b)

Fig. 5. Designs with (a) 6 nozzles and (b) 12 nozzles.

Fig. 6 shows that with an increasing number of nozzles, the device temperature increased. The reason is that with a constant flow rate, increasing the number of nozzles decreases the jet velocity, which corresponds to lower heat transfer at the impingement surface. But increasing the number of nozzles decreased the pressure drop, which allowed for a higher flow rate with the same pumping power. Fig. 6(a) and (b) shows that increasing the number of nozzles twofold did not change the device temperature significantly (it increased only 1°C), whereas the pressure drop decreased threefold.

Fig. 7 shows that with an increasing number of nozzles, thermal performance decreased linearly, but the pressure drop decreased in polynomial fashion as did the pumping power.

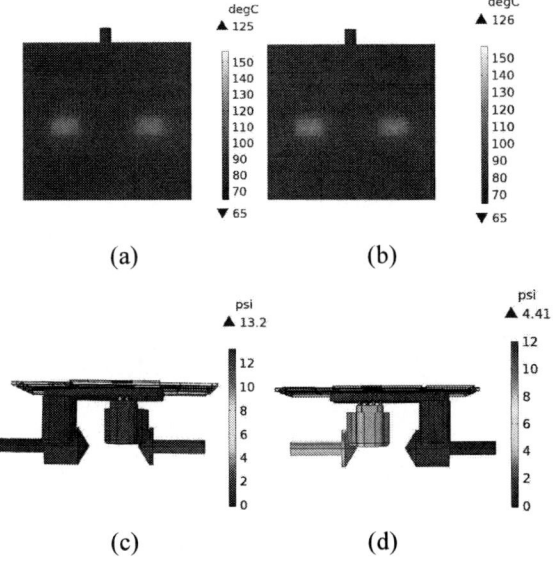

Fig. 6. Temperature distributions for (a) 6 nozzles per device and (b) 12 nozzles per device; pressure drops for (c) 6 nozzles per device and (d) 12 nozzles per device.

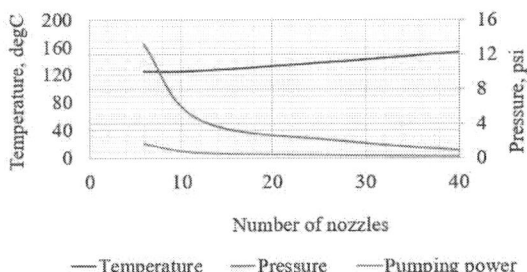

Fig. 7. Relationship between the number of nozzles and device temperature, pressure drop, and pumping power.

D. Effect of Orientation

As the previous section showed, increasing the number of nozzles twofold reduced the pressure drop threefold (12 psi to 4 psi). To reach the pressure drop target set by US DRIVE 2025 (2 psi), the number of nozzles needs to be increased. Two 28-nozzle designs were proposed, one with the nozzles distributed in an in-line orientation and one with the nozzles distributed in a circular orientation. These designs are shown in Fig. 8.

The temperature distributions and the pressure drops from the inlet to the outlet show the device temperature for the circular orientation was 2°C lower than that for the in-line orientation. The pressure drops were very similar for both designs.

Fig. 9(a) shows that increasing the number of nozzles from 12 to 28 per device caused the device temperature to increase 11% but also caused the pressure drop to decrease 57%, making the pressure drop less than 2 psi. This result gives an understanding about how the number of nozzles affects the heat transfer. For a constant flow rate, increasing the number of nozzles reduces thermal performance but also reduces the pressure drop to push the coolant from the inlet to the outlet. This enables more flow to be pushed with the same pumping power because pumping power is the product of flow rate and pressure drop.

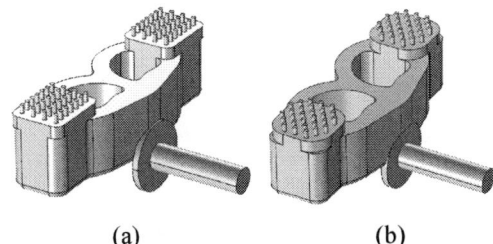

(a)   (b)

Fig. 8. (a) In-line and (b) circular nozzle distributions.

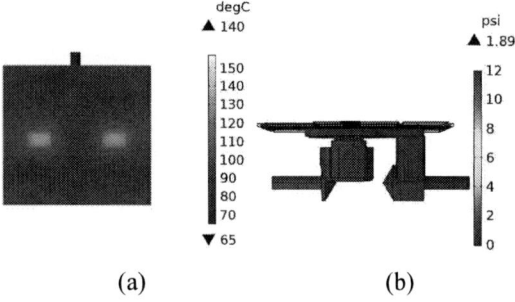

(a)   (b)

Fig. 9. (a) Temperature in the module and (b) pressure drop from inlet to outlet for staggered design.

E. Double-Sided cooling

Previous sections have discussed how the design variables affected the thermal performance for jet-impingement-based cooling. Thus far, the results indicate that reducing diameter increases jet velocity, which improves heat transfer but increases the pressure drop. Increasing the number of nozzles balances out the pressure drop because it reduces exit jet velocity but also decreases the heat transfer. To approach this challenge, double-sided cooling was proposed. Double-sided cooling allows heat to be extracted from devices from both the top and bottom directions. A 3 mm Cu post was soldered to the device, and another DBC substrate was soldered to the top of the post. The top and bottom DBC have exact geometric dimensions.

For double-sided cooling, there are two inlets and outlets. The main feeder divides the flow in two inlets. The main feed inlet has a 1 L/min flow rate, and the top and bottom manifolds have a 0.5 L/min inlet flow rate.

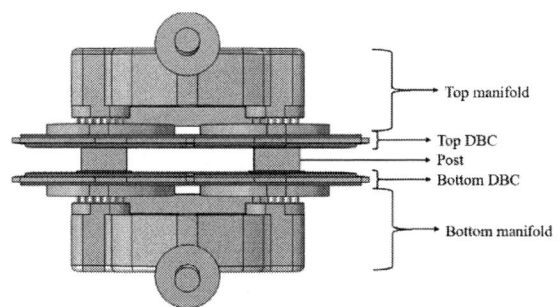

Fig. 10. Double-sided packaging for jet impingement.

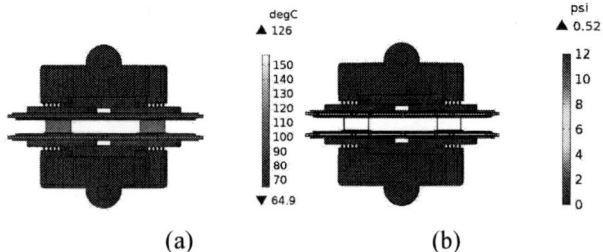

(a)   (b)

Fig. 11. (a) Temperature distribution and (b) pressure drop across the manifolds for a 1 L/min flow rate.

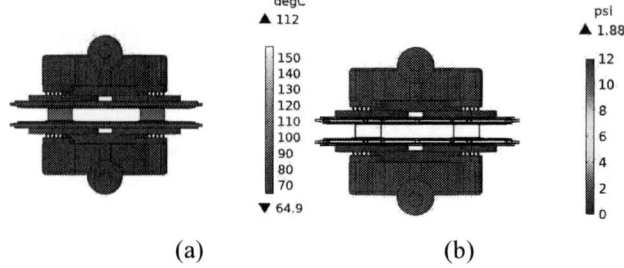

(a)   (b)

Fig. 12. (a) Temperature distribution and (b) pressure drop across the manifolds for a 2 L/min flow rate.

Fig. 9(a) and 12(a) show that for the same flow rate, double-sided cooling reduced the device temperature 10% (from 140°C to 126°C), but the pressure drops for each manifold decreased 50%. This reduction of pressure drop allows double the flow rate for the same pumping power. Fig. 12(a) shows that for a flow rate of 2 L/min, the corresponding device temperature decreased to 112°C, and the pressure drop was 1.88 psi, which is lower than 2 psi. For a constant pumping power, double-sided cooling produced a 20% reduction of device temperature compared with single-sided cooling. A flow rate higher than 2 L/min created a pressure drop higher than the 2 psi target.

F. Benchmarking

To evaluate the thermal performance of the proposed design, the results were compared with the BMW I3 module (FS800R07A2E3) heat sinks. The module has a pin-fin-based heat sink that comprises 50 cylindrical pin fins (see Fig., 13 (a)). Each pin fin is 8 mm tall, and the heat sink has a 4 mm thick Al baseplate. The pin radius is 2.28 mm, and the spacing between the pins is available in [21].

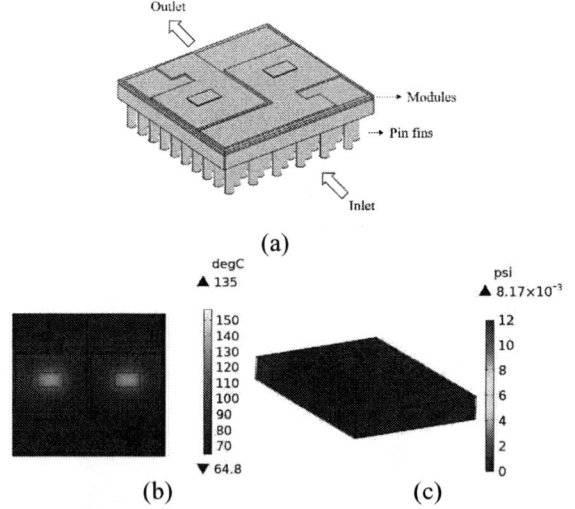

(a)

(b)   (c)

Fig. 13. (a) Pin-fin geometry, (b) temperature distribution in the modules, and (c) pressure drop in the manifold.

For a 1 L/min flow rate, the maximum device temperature was 135°C (see Fig. 13 (b)), which was 3% lower than the single-sided jet impingement cooling and 7% higher than the double-sided jet impingement cooling, but the pressure drop was very low (see Fig. 13 (b)).

Table III shows the comparison among the pin-fin-based heat sink for the BMW I3 module and the single- and double-sided jet impingement cooling. For the same flow rate, 1 L/min, the pin-fin-based heat sink showed a device temperature 3% lower than that of the single-sided jet and 75% higher than that of the double-sided jet. The pressure drop for the pin-fin heat sink was also very low (less than 0.1 psi). But the volume of pin-fin-based heat sink (i.e., the total amount of material) was 4.83 times larger than that of the single jet and 2.39 times larger than that of the double-sided jet. The pin-fin heat sink was 1.33 times longer than the single-sided jets and 1.38 times smaller than the double-sided jets.

TABLE III.   COMPARATIVE STUDY AMONG SINGLE- AND DOUBLE-SIDED JET IMPINGEMENT COOLING AND PIN-FIN-BASED HORIZONTAL COOLING

Type	Device temperature (°C)	Pressure drop (psi)	Volume (L)	Height (mm)
Pin fin	135	0.000817	0.023	13
Single-sided jet	140	1.89	0.00483	9
Double-sided jet	126	0.52	0.0096	18

IV. CONCLUSION

This article presents a comprehensive analysis of jet-impingement-based direct substrate cooling for a power module. A novel design for jet impingement cooling was proposed which satisfied the power density target and reduced the volume and mass of the whole package compared with the BMW I3 module package. Numerical simulations were conducted to study the effects of different design parameters, and the jet diameter and the number of jets were optimized. Double-sided packaging was explored and compared with single-sided packaging. A comparative analysis among the BMW I3 package and the single- and double-sided packages with jet cooling was presented and showed how the proposed design improved the power density. Results showed that for similar thermal performance, volume and corresponding mass decreased by a factor of approximately 4.83 for the single-sided package and by a factor of approximately 2.39 for the double-sided package compared with the BMW I3 package. The only drawback was the pressure drop, which was considerably higher for jet cooling; however, it was within 2 psi. Jet cooling has an inherently higher pressure drop because of the larger flow area change from manifold to nozzle.

ACKNOWLEDGMENT

This material is based upon work supported by the US Department of Energy's (DOE's) Vehicle Technologies Office Electric Drive Technologies Program. The authors thank Susan Rogers of DOE for her support and guidance. The authors also thank Jon Wilkinson for his support in CAD file generation.

REFERENCES

[1] K. O. Armstrong, S. Das, and J. Cresko, "Wide bandgap semiconductor opportunities in power electronics," in *WiPDA 2016 - 4th IEEE Workshop on Wide Bandgap Power Devices and Applications*, 2016, pp. 259–264. https://doi.org/10.1109/WIPDA.2016.7799949.

[2] C. Qian et al., "Thermal management on IGBT power electronic devices and modules," *IEEE Access*, vol. 6, pp. 12868–12884, 2018, doi: 10.1109/ACCESS.2018.2793300.

[3] J. Broughton, V. Smet, R. R. Tummala, and Y. K. Joshi, "Review of thermal packaging technologies for automotive power electronics for traction purposes," *J. Electron. Packag., Trans. ASME*, vol. 140, no. 4, 2018, doi: 10.1115/1.4040828/366154.

[4] Y. P. Zhang, X. L. Yu, Q. K. Feng, and R. T. Zhang, "Thermal performance study of integrated cold plate with power module," *Appl. Thermal Eng.*, vol. 29, no. 17–18, 3568–3573, 2009, doi:

[5] A. J. Robinson, "A thermal-hydraulic comparison of liquid, microchannel and impinging liquid jet array heat sinks for high-power electronics cooling," *IEEE Trans. Compon. and Packag. Technol.*, vol. 32, no. 2, pp. 347–357, 2009, doi: 10.1109/TCAPT.2008.2010408.

[6] B. Kekelia. "Jet impingement cooling of electric machines with driveline fluids." 2021. https://www.allianceforsustainableenergy.org/about.html (accessed Apr. 19, 2022).

[7] "Dielectric jet impingement cooling of electronic chips," in *Proc. 34th ASME Nat. Heat Transfer Conf.*, 2000.

[8] "Jet impingement cooling of electric motor end-windings." Patent US20030102728A1. Available: https://patents.google.com/patent/US20030102728 (accessed Apr. 19, 2022).

[9] M. Fabbri and V. K. Dhir, "Optimized heat transfer for high power electronic cooling using arrays of microjets," *J. Heat Transfer*, vol. 127, no. 7, pp. 760–769, Jul. 2005.

[10] D. Maddox and A. Bar-Cohen, "Thermofluid design of single-phase submerged-jet impingement cooling for electronic components," *J. Electron. Packag.*, vol. 116, no. 3, pp. 237–240, Sept. 1994.

[11] D. Womac, S. Ramadhyani, and F. Incropera, "Correlating equations for impingement cooling of small heat sources with single circular liquid jets," *J. Heat Transfer*, vol. 115, no. 1, pp. 106–115, 1993.

[12] S. Narumanchi, M. Mihalic, G. Moreno, and K. Bennion, "Design of light-weight, single-phase liquid-cooled heat exchanger for automotive power electronics," in *ITherm*, San Diego, CA, USA, 2012, pp. 693–699.

[13] H. Zhang, S. S. Ang, H. A. Mantooth, and S. Krishnamurthy, "A high temperature, double-sided cooling SiC power electronics module," *2013 IEEE Energy Conversion Congr. and Expo.*, pp. 2877–2883, doi: 10.1109/ECCE.2013.6647075.

[14] C. Wang, L. Zheng, L. Han, H. Fang, and J. Xu, "Thermal performance investigation of three-dimensional structure unit in double-sided cooling IGBT module," in *Proc. Electronic Packaging Technology Conf.*, Oct. 2014, pp. 622–625, doi: 10.1109/ICEPT.2014.6922733.

[15] Electrical and Electronics Technical Team Roadmap. 2017. www.uscar.org.

[16] "Copper vs. Aluminum Heatsinks: What You Need to Know." https://www.gabrian.com/copper-vs-aluminum-heatsinks/ (accessed Sept. 24, 2023).

[17] C. Wang, L. Zheng, L. Han, H. Fang, and J. Xu, "Thermal performance investigation of three-dimensional structure unit in double-sided cooling IGBT module," in *Proceedings of the Electronic Packaging Technology Conference*, Oct. 2014, pp. 622–625, doi: 10.1109/ICEPT.2014.6922733.

[18] Y. Xu and D. C. Hopkins, "Misconception of thermal spreading angle and misapplication to IGBT power modules," in *Conf. Proc. - IEEE Applied Power Electronics Conf. and Expo.*, 2014, 545–551. https://doi.org/10.1109/APEC.2014.6803362.

[19] "The 45° Heat Spreading Angle - An Urban Legend?" Electronics Cooling. https://www.electronics-cooling.com/2003/11/the-45-heat-spreading-angle-an-urban-legend/ (accessed Sept. 24, 2023).

[20] "High-Strength Weather-Resistant EPDM Rubber Sheet, 12" x 12", 1/16" Thick." McMaster-Carr. https://www.mcmaster.com/8985K42/ (accessed Sept. 24, 2023).

[21] R. Sahu, E. Gurpinar, and B. Ozpineci, B, "Fourier analysis-based evolutionary multi-objective multiphysics optimization of liquid-cooled heat sinks." *2020 IEEE Energy Conversion Congr. and Expo.*, 2020, 4017–4023. https://doi.org/10.1109/ECCE44975.2020.9235943

An Effective Screening Technique for Early Oxide Failure in SiC Power MOSFETs

Limeng Shi*, Jiashu Qian, Michael Jin, Monikuntala Bhattacharya, Hengyu Yu, Marvin H. White, and Anant K. Agarwal
Dept. of Electrical & Computer Engineering
The Ohio State University
Columbus, OH, USA
Email: shi.1564@osu.edu

Atsushi Shimbori, Zhuxian Xu
Ford Motor Company
Dearborn, Michigan, USA

Abstract—This paper proposes an effective screening technique based on a correlation between initial gate leakage current and oxide failure time at a constant gate voltage for commercial 1.2 kV SiC planar power MOSFETs. During constant-voltage time-dependent dielectric breakdown (TDDB) tests on SiC MOSFETs, it is observed that devices exhibiting higher gate leakage current at the same gate voltage stress are more susceptible to gate oxide breakdown. This observation provides a new perspective for a screening technique, which is to apply a high gate voltage for a short duration (100 ms) to a gate oxide at 150°C in order to find the device with a high gate leakage current. This leaky device can be defined as an unreliable device with high extrinsic defects. Moreover, it is also demonstrated that the screening technique mentioned in this paper does not result in excessive threshold voltage shift or reduce the intrinsic lifetime of gate oxide in good devices. The proposed screening technique can reduce the failures of SiC planar power MOSFETs in electric vehicles by identifying devices with early oxide failures.

Keywords—*SiC planar power MOSFETs, screening, extrinsic failure, gate leakage current, oxide lifetime, threshold voltage*

I. INTRODUCTION

Early oxide breakdown due to high extrinsic defects in the gate oxide is a major issue for commercial SiC power MOSFETs in industrial and automotive applications [1]. To improve the reliability of SiC MOSFETs, manufacturers are required to screen each device after fabrication. The devices with defects in gate oxide would fail during the screening process and be filtered out. However, to minimize the negative impact on threshold voltage and oxide's intrinsic lifetime of SiC MOSFETs caused by screening techniques, the voltage, time, and temperature during the screening process are constrained. This limitation results in the inability to entirely eliminate the long extrinsic failure tail [2], especially in the vicinity of the intrinsic failure. In this case, some devices with relatively long extrinsic failure times may survive during the screening process, posing risks for practical industrial applications such as electric vehicles (EV). It is believed that the present screening techniques are inadequate in terms of effectiveness. According to the local oxide thinning theory [3], the locally enhanced electric field caused by extrinsic defects is the main factor causing early failures. Moreover, the reduction of effective gate oxide thickness leads to higher gate leakage currents for SiC MOSFETs [4]. During the screening process, it is necessary to not only eliminate the failed devices but also remove the devices with gate leakage currents exceeding the critical value under gate voltage stress. The strict screening method can make the oxide lifetime distribution of shipped devices more uniform.

In this work, an innovative screening method is designed based on the relationship between initial gate leakage current and oxide failure time observed during constant-voltage time-dependent dielectric breakdown (TDDB) measurement for commercial SiC planar power MOSFETs. This screening technique uses an instantaneous high gate voltage to reduce the deleterious impact on threshold voltage and oxide's intrinsic lifetime, which saves time and increases efficiency. Thus, removing SiC planar power MOSFETs with high gate leakage current at a specified gate voltage leads to a lower field-failure probability of remaining devices.

II. DEVICE AND TEST METHOD

A. Devices

The commercial 1.2 kV 4H-SiC planar power MOSFETs (packaged in TO-247) are selected for the investigation of screening technology in this work. These devices have been screened by the manufacturer before being shipped. However, some devices with early failures still exist due to the extrinsic oxide failure time being close to the intrinsic lifetime. Information on these commercial 1.2 kV SiC planar power MOSFETs is shown in Table I. The parameters of these devices are obtained with a Semiconductor Parameter Analyzer (B1506A, Keysight, Inc.). The threshold voltages (V_{th}) of the devices under test (DUTs) are extracted at a drain current of 1 mA with drain and gate connected together. The average gate oxide breakdown voltage (V_{BR}) at 150°C is ~ 51 V. The oxide thickness (t_{ox}) is estimated to be ~ 46.4 nm.

TABLE I. GENERAL INFORMATION OF COMMERCIAL SIC PLANAR MOSFETs

Properties	SiC MOSFETs
Mean V_{th} @150°C (V)	2.3
R_{on} @RT (mΩ)	280
Recommended gate use voltage (V)	20
Average oxide V_{BR} @150°C (V)	51
Estimated t_{ox} (nm)	46.4

B. Test Method

Constant-voltage TDDB measurements are performed on commercial 1.2 kV 4H-SiC planar power MOSFETs to obtain the oxide failure lifetime of each DUT at 150°C. During the TDDB test, a constant voltage is applied to the gate electrode of DUT with drain and source grounded. A multichannel digital multimeter (DMM) monitors the gate leakage currents over time until oxide breakdown. The gate oxide lifetimes, with various failure rates under different gate voltage conditions, are extracted for a Weibull distribution. To predict

979-8-3503-3714-3/23 $31.00 © 2023 IEEE

oxide lifetime under typical operating conditions, a linear relationship between oxide failure lifetime and gate voltage is fitted with an E-model [5].

III. RESULTS AND DISCUSSIONS

This section describes a correlation between gate leakage current and oxide failure time for commercial SiC planar power MOSFETs in the TDDB measurements. In addition, the effectiveness of screening devices with gate oxide leakage current is illustrated by measuring the oxide lifetime of devices subjected to this screening treatment. It is also demonstrated that the designed screening technique does not degrade the performance of commercial SiC planar power MOSFETs.

A. TDDB measurements

The gate oxide failure distribution for commercial SiC planar power MOSFETs at a gate voltage of 41 V is depicted in Fig. 1. Based on the Weibull plot of oxide failure times for 40 samples, it can be observed that the intrinsic lifetimes of 38 samples follow a tight distribution with a Weibull slop (β) of 4. The extrinsic failure occurs in two devices marked with the red symbol. This condition indicates that a small number of devices with extrinsic defects still experience early failure even after the manufacturer's screening process. The oxide lifetime with different failure rates can be estimated using (1), where F represents the cumulative failure probability and $t_{F\%}$ means the time when F% of the test population has failed [6]. The $t_{63\%}$, $t_{10\%}$, $t_{1\%}$ and $t_{0.01\%}$ is 65 h, 36.6 h, 20.6 h and 6.5 h respectively for commercial SiC planar power MOSFETs with a gate voltage of 41 V at 150°C.

$$t_{F\%} = t_{63\%} \times \exp\left\{\frac{1}{\beta}(\ln(-\ln(1 - F)))\right\} \quad (1)$$

To estimate the oxide lifetime of commercial SiC planar power MOSFETs at a gate voltage of 20 V, four gate voltages of 39.4 V, 40.4 V, 41 V, and 42.7 V are selected for the TDDB measurement at 150°C. Fig. 2 shows the oxide failure times at different gate voltages with failure rates of 64%, 1% and 0.01%. The extracted intrinsic lifetime of gate oxide scaled to 0.01% at a gate voltage of 20 V is much larger than 20 years. This result indicates the oxide's intrinsic lifetime of commercial SiC planar power MOSFETs is good enough to meet the industrial requirements. However, considering the time-consuming conventional screening techniques due to the need to screen out as many defective devices as possible, it is essential to find a method to quickly pinpoint devices on the long extrinsic failure tail.

B. Correlation of initial gate leakage current and oxide failure time

During the TDDB measurement, the gate leakage currents for DUTs at 41 V are monitored until the gate oxide breakdown at 150°C. As shown in Fig. 3, the gate leakage currents of DUTs are recorded at 0.3 s after the gate voltage of 41 V is applied. The first point of the measured curve is taken as the initial gate leakage current. The corresponding initial gate leakage current and oxide failure time for each DUT are extracted from the TDDB test results, and the obtained correlation is shown in Fig. 4. There is a negative correlation between the oxide lifetime and the initial gate leakage current (the higher initial oxide currents the lower oxide failure times). It can be observed that the majority of

DUTs have an initial gate leakage current in the range of 2.2 μA to 3.2 μA at a gate voltage of 41 V. The two samples previously shown in Fig. 1 that experience early gate oxide breakdown are also shown in Fig. 4 with the red symbol, and the leakage current for these two samples is close to 4 μA. Therefore, it can be concluded that devices with higher gate leakage currents exhibit lower oxide lifetimes in TDDB measurements.

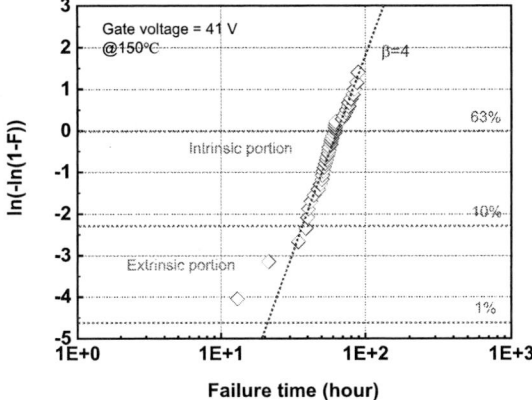

Fig. 1. Weibull distribution of measured oxide lifetimes at a gate voltage of 41 V (estimated E_{ox}= 8.8 MV/cm) for 40 commercial SiC planar MOSFETs.

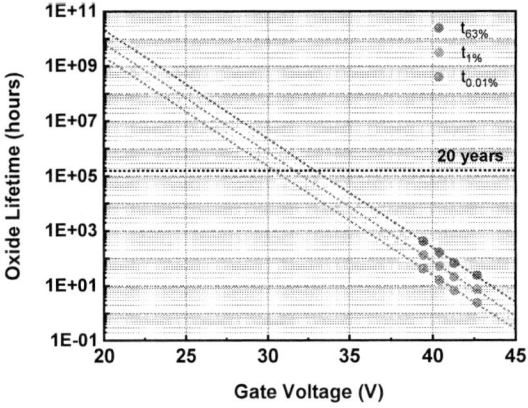

Fig. 2. Gate oxide lifetime at 150°C with the failure rate of 63%, 1% and 0.01% for commercial SiC planar MOSFETs.

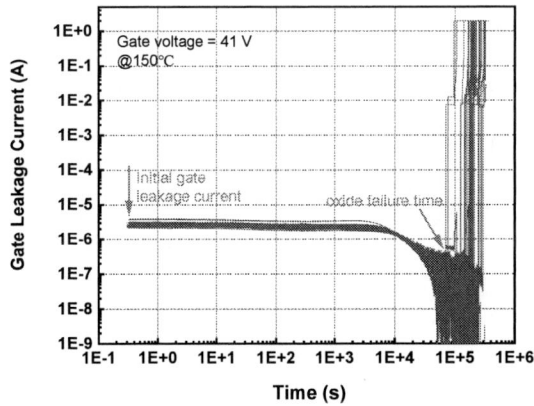

Fig. 3. Gate leakage current versus stress time during TDDB test with a gate voltage of 41 V for commercial SiC planar MOSFETs.

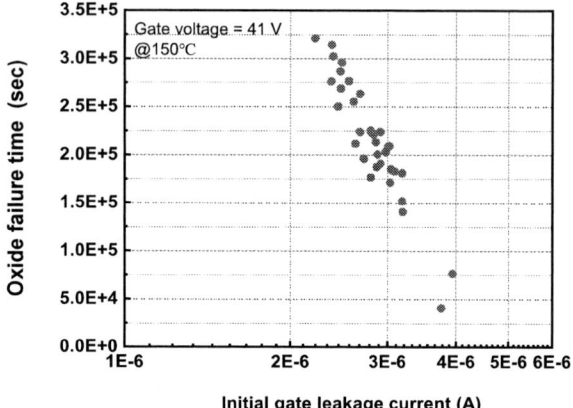

Fig. 4. Correlation of initial gate leakage current and oxide failure time at a gate voltage of 41 V for commercial SiC planar MOSFETs.

C. Investigation of screening technology

To prove the effectiveness of screening devices by the leakage current parameter, a new batch of commercial SiC MOSFETs (50 samples) from the same vendor is selected on the basis of similar threshold voltages for the screening process and TDDB test. For the screening process, a screening time of 100 ms is employed to save the time required for processing. In the previous study [10], it has been found that when the screening time is 100 ms, the maximum screening E_{ox} acceptable for commercial SiC planar power MOSFETs is 9 MV/cm in order to avoid the behavior of negative threshold voltage shift caused by hole trapping. In addition, the high gate voltage and temperature are intended to ensure a high screening efficiency. Therefore, the screening condition used in this work is a gate voltage of 42 V (estimated E_{ox}= 9 MV/cm) for 100 ms at 150°C. The gate leakage current at the gate voltage of 42 V is recorded for each DUT during the screening process. After screening 50 samples, the following TDDB measurements with a gate voltage of 41 V at 150°C are conducted on these 50 samples to obtain the oxide failure time of each DUT.

Fig. 5 shows the correlation between the gate leakage current at 42 V and the oxide lifetime of DUTs at 41 V. It can be observed that five leaky devices with leakage currents exceeding 10 μA at a gate voltage of 42 V are detected. As shown in the plot, the oxide lifetime of the majority of DUTs is in the range of 40 h to 80 h. The oxide failure time of leaky devices is much lower than that of normal devices. This is because the effective oxide thickness is reduced by the extrinsic defects in the gate oxide of devices, which makes the defective devices have higher leakage current and lower oxide lifetime at the same gate voltage. It is believed that extrinsic failures occur in the gate oxide of SiC MOSFETs that display higher leakage current under high gate voltage stress. The results indicate the screening approach using gate leakage current to identify defective devices is feasible. For the commercial SiC planar power MOSFETs used in this paper, when the gate leakage current of the device exceeds 10 μA at a gate voltage of 42 V and a temperature of 150°C, the oxide lifetime of the device is not able to reach the intrinsic lifetime of good devices.

In order to investigate the effect of the screening technique on the threshold voltage of SiC planar power MOSFETs, the threshold voltage shift of DUTs (excluding leaky devices) is measured at 150°C after being subject to a gate voltage of 42 V for 100 ms. The distribution of threshold voltage shift values for 45 samples is presented in Fig. 6. Differences in effective oxide thickness due to defect variation affect the V_{th} shift values of DUTs. Although the same gate voltage is applied to all samples, there is a difference in the effective electric field in the oxide layer, leading to different charging trapping in the oxide during the screening process. Six samples have a negative threshold voltage shift. This phenomenon is due to the higher effective oxide electric field of these devices, which triggers a larger number of holes generated through impact ionization [7]. When the effect of hole trapping exceeds the effect of electron trapping, the threshold voltage exhibits a negative shift behavior [8]. The other samples treated with screening at a gate voltage of 42 V for 100 ms showed a positive shift in the threshold voltage, which is caused by electron trapping in the gate oxide [9, 10]. Overall, the average value of threshold voltage shift is only 0.05 V, so it can be assumed that the screening technique does not have a significant negative impact on the threshold voltage of commercial SiC planar MOSFETs.

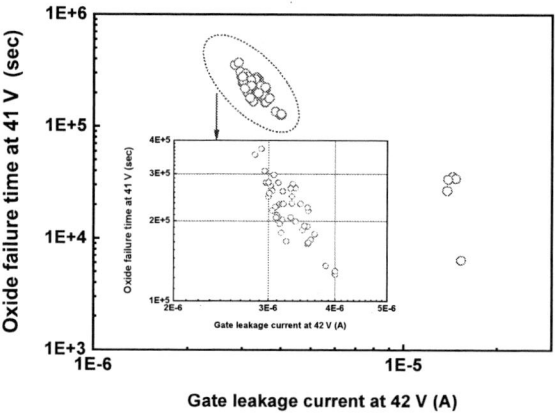

Fig. 5. Correlation between the gate leakage current at 42 V and the oxide lifetime of devices at 41 V for commercial SiC planar MOSFETs.

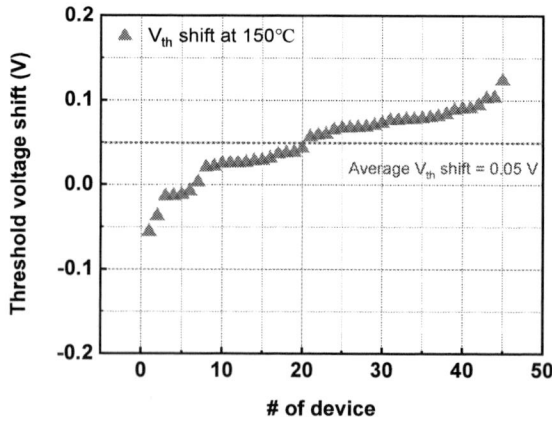

Fig. 6. V_{th} shift at 150°C for commercial SiC planar MOSFETs screened with a gate voltage of 42 V at 150°C for 100 ms.

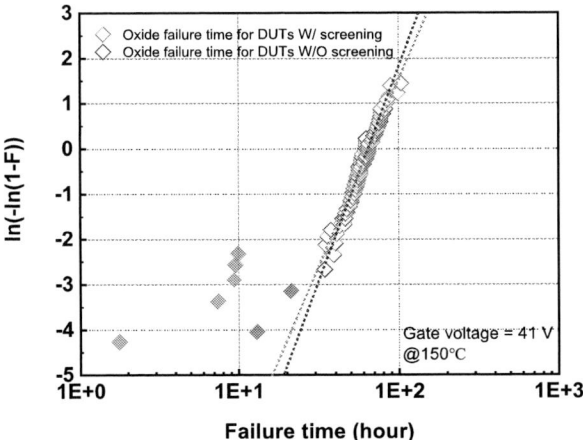

Fig. 7. Weibull distribution of measured oxide lifetimes at a gate voltage of 41 V for commercial SiC planar MOSFETs with and without screening treatment. The screening condition is a gate voltage of 42 V for 100 ms at 150°C.

Fig. 7 depicts the distribution of oxide failure time at 41 V for devices subjected to 42 V for 100 ms screening and devices without screening. It can be found that both batches of samples have a small portion of leaky devices with early oxide breakdown. In addition, the intrinsic portion of gate oxide lifetimes for devices with and without screening has no significant difference. This result indicates that the screening method (gate voltage of 42 V for 100 ms at 150°C) detecting leaky devices does not degrade the intrinsic oxide lifetime of normal commercial SiC planar MOSFETs.

IV. CONCLUSIONS

In this paper, the negative correlation between gate leakage current and oxide failure time is found in constant-voltage TDDB measurements. This leads to the design of an effective screening technique for identifying leaky devices by gate leakage current. This screening technique adopts a high gate voltage of 42 V at 150°C for a duration of 100 ms, which can effectively screen out devices with extrinsic failure while avoiding the negative impact on the threshold voltage and oxide's intrinsic lifetime of normal devices.

ACKNOWLEDGMENT

This work is supported in part by the Ford Motor Company under the Ford Alliance 2019 Project to The Ohio State University and in part by the Block Gift Grant from II-VI Foundation.

REFERENCES

[1] Y. Zheng, R. Potera and T. Witt, "Characterization of Early Breakdown of SiC MOSFET Gate Oxide by Voltage Ramp Tests," 2021 IEEE International Reliability Physics Symposium (IRPS), Monterey, CA, USA, 2021, pp. 1-5.

[2] L. C. Yu, G. T. Dunne, K. S. Matocha, K. P. Cheung, J. S. Suehle and K. Sheng, "Reliability Issues of SiC MOSFETs: A Technology for High-Temperature Environments," in IEEE Transactions on Device and Materials Reliability, vol. 10, no. 4, 2010, pp. 418-426.

[3] T. Aichinger and M. Schmidt, "Gate-oxide reliability and failure-rate reduction of industrial SiC MOSFETs," 2020 IEEE International Reliability Physics Symposium (IRPS), Dallas, TX, USA, 2020, pp. 1-6

[4] C. -T. Yen et al., "Oxide Breakdown Reliability of SiC MOSFET," 2019 IEEE Workshop on Wide Bandgap Power Devices and Applications in Asia (WiPDA Asia), Taipei, Taiwan, 2019, pp. 1-3.

[5] J. W. McPherson and H. C. Mogul, "Underlying physics of the thermochemical E model in describing low-field time-dependent dielectric breakdown in SiO_2 thin films," J. Appl. Phys., vol. 84, no. 3, pp. 1513–1523, 1998.

[6] Weibull, Waloddi. "A Statistical Distribution Function of Wide Applicability." Journal of Applied Mechanics 18 (1951): 293-297.J. Clerk Maxwell, A Treatise on Electricity and Magnetism, 3rd ed., vol. 2. Oxford: Clarendon, 1892, pp.68–73.

[7] P. Samanta and K. C. Mandal, "Hole injection and dielectric breakdown in 6H–SiC and 4H–SiC metal–oxide–semiconductor structures during substrate electron injection via Fowler–Nordheim tunneling," Solid-State Electronics, vol. 114, pp. 60–68, 2015.

[8] L. Shi et al., "Effects of Oxide Electric Field Stress on the Gate Oxide Reliability of Commercial SiC Power MOSFETs," 2022 IEEE 9th Workshop on Wide Bandgap Power Devices & Applications (WiPDA), Redondo Beach, CA, USA, 2022, pp. 45-48.

[9] M. Noguchi, A. Koyama, T. Iwamatsu, H. Amishiro, H. Watanabe and N. Miura, "Gate Oxide Instability and Lifetime in SiC MOSFETs under a Wide Range of Positive Electric Field Stress," 2020 IEEE International Electron Devices Meeting (IEDM), San Francisco, CA, USA, 2020, pp. 23.4.1-23.4.4.

[10] L. Shi et al., "Investigation of different screening methods on threshold voltage and gate oxide lifetime of SiC Power MOSFETs," 2023 IEEE International Reliability Physics Symposium (IRPS), Monterey, CA, USA, 2023, pp. 1-7.

Investigation of the Constant Current Stress for Charge-to-breakdown Extraction in Commercial SiC Power MOSFETs

Jiashu Qian, Limeng Shi, Michael Jin, Monikuntala
Bhattacharya, Hengyu Yu, Marvin H. White, and Anant K.
Agarwal
Dept. of Electrical and Computer Engineering
The Ohio State University
Columbus, OH 43210, USA
qian.539@osu.edu

Atsushi Shimbori, Zhuxian Xu
Ford Motor Co.
Dearborn, MI 48126, USA

Abstract—The constant current stress time-dependent dielectric breakdown (CCS-TDDB) measurement is performed on the commercial 1.2 kV SiC planar power DMOSFETs to determine the charge-to-breakdown (Q_{BD}) of electron through the gate oxide. The CCSs are extracted based on the gate currents in the I_{gss} curve, at which the gate oxide electric fields trigger the hole trapping or not. The analysis of the threshold voltage (V_{th}) and the subthreshold slope (SS) during the stress jointly describe that the stress process in SiC/SiO₂ under the CCS until the gate oxide breakdown is a combination of the interface states generation, the bulk electron trap generation, the tunneling electron trapping, and even the hole trapping. The energy band diagrams imply that the impact ionization/anode hole injection (AHI) caused by the energy band bending in the bulk of the gate oxide can lead to the hole trapping, which accelerates the gate oxide breakdown and makes the critical current stress lower than the gate current corresponding to the critical electric field. The extracted Q_{BD} when the failure rate reaches 63% ($Q_{BD\_63\%}$) at a stable level supports the failure mechanism of the gate oxide as the charge driven breakdown.

Keywords—*SiC, MOSFET, Gate Oxide, CCS, Charge-to-breakdown, Hole Trapping, Charge Driven Breakdown*

I.   INTRODUCTION

Compared to the Si counterparts, commercial SiC power MOSFETs are gradually dominating the market of the power electronics, mainly for the electric vehicle (EV), due to their lower switching loss, higher temperature capability, higher switching frequency, and gradually competitive price because of the reliable body diode as the freewheeling diode to replace the integrated Schottky diode [1, 2]. However, a variety of issues, including the low inversion layer mobility due to the high density of interface states, make the devices not reach their full potential, even negatively influence the gate oxide reliability [3, 4, 5]. Besides, abundant extrinsic defects reflected by the local oxide thinning due to the tiny distortions in the gate oxide have been found to be the cause of the early gate oxide breakdown, which determines the overall gate oxide reliability [6, 7, 8]. These issues challenge the operation of devices at higher oxide electric fields to increase the channel electron density and further improve the device performance [9]. To achieve the improvement, it is necessary to guarantee the gate oxide lifetime. L. Shi et al. have proposed the optimal screening voltage and duration for high gate-voltage pulse screening and long-term burn-in to aggressively screen out the extrinsic defects with acceptable

constraints on the V_{th} shift and the lifetime degradation [10]. What's more, an accurate gate oxide lifetime model is necessary for commercial SiC power MOSFETs.

As a standard industry technique, the time-to-breakdown (t_{BD}) extracted from the constant voltage stress time-dependent dielectric breakdown (CVS-TDDB) measurement is based on the thermochemical E-model [11]. In this model, the gate oxide electric field is assumed to uniformly distribute in the gate oxide. However, P. Moens et al. have found that the t_{BD} extracted from the CVS-TDDB cannot capture the effects of different charge trapping mechanisms, including the hole trapping by the F-N tunneling and the electron trapping by the trap-assisted tunneling (TAT). It is because the local electric field near the cathode or anode interface, which determines the tunneling current, can significantly change, although the average electric field is kept constant during the stress. Instead, P. Moens et al. found that t_{BD} is charge driven rather than field driven. Therefore, the CCS is considered as a better way to stress than the CVS, and a Q_{BD} approach is proposed, clearly showing the transition between the two different charge trapping mechanisms [9].

In this work, different CCSs are applied to stress the commercial 1.2 kV SiC planar power DMOSFETs. The V_{th} and SS under different CCSs are also monitored to analyze the stress process. Furthermore, the critical current stress that starts to trigger the hole trapping is estimated and compared to the critical electric field extracted from the CVS-TDDB results of commercial SiC power MOSFETs. Through the analysis of the Weibull plots under different CCSs, the $Q_{BD\_63\%}$ can be extracted and compared.

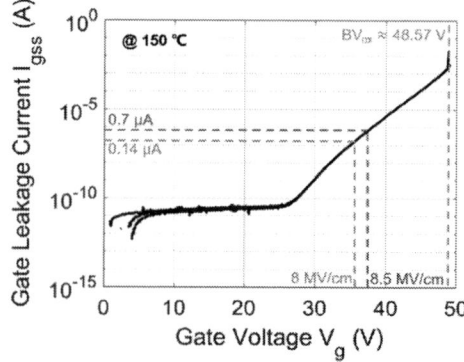

Fig. 1.   The ramped-voltage breakdown (I_{gss}) curves at 150 °C for DUTs from vendor B.

II. EXPERIMENTS

A. Device Under Test (DUT)

The DUTs stressed in this work are commercial 1.2 kV SiC planar power DMOSFETs packaged in TO-247 from Vendor B. The I_{gss} curves of 3 DUTs at 150 °C are shown in Fig. 1. The CCSs applied in this work are 0.14 and 0.7 μA, corresponding to 8 and 8.5 MV/cm, respectively. The thickness of the gate oxide (t_{ox}) is estimated to be around 44.15 nm based on the average gate oxide breakdown voltage (BV_{ox}) with the assumption that the critical oxide breakdown electric field is 11 MV/cm @ 150 °C. The general information of the DUTs is shown in Table I.

TABLE I.   GENERAL INFORMATION OF THE DUTs FROM VENDOR B

Properties	Vendor B
Voltage Rating (V)	1200
Current Rating (A)	11
Structure	Planar
Gate Oxide Breakdown Voltage (V)	48.57
Est. Gate Oxide Thickness (nm)	44.15

B. CCS-TDDB Measurement

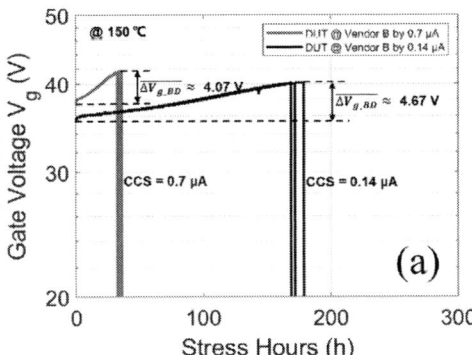

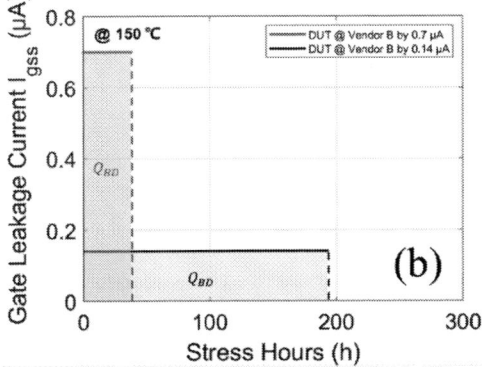

Fig. 2. (a) The gate voltage as a function of the stress hours under the selected CCSs; (b) The schematic diagram of Q_{BD} extraction under the selected CCSs.

A precision source/measure unit (B2901A, Keysight, Inc.) is used to monitor the gate voltage behavior of the DUT under the selected CCSs at 150 °C. During the stress, the drain and source of the DUT are shorted to the ground with a CCS applied on the gate until the gate oxide fails. Fig. 2 (a)

indicates the gate voltage as a function of the stress hours under the selected CCSs. The gate oxide failure time can be extracted when the gate voltage drops rapidly. With the extracted failure time of each DUT, the Q_{BD} can be estimated from the product of the CCS and the failure time as shown in Fig. 2 (b).

III. RESULTS AND DISCUSSION

A. ΔV_{th} and SS under the CCS

In Fig. 3, the gate voltage (V_g), the threshold voltage variation (ΔV_{th}), and the subthreshold slope (SS) as a function of the stress hours are presented. The V_{th} is measured by the linear extrapolation method, and the ΔV_{th} is the variation of the V_{th} at any stress time point compared to the initial V_{th} before the stress. Fig. 3 (a) illustrates the conditions under the CCS of 0.14 μA. The total stress duration until the gate oxide breakdown can be divided into two phases. In Phase I, as the V_g increases, the SS decreases rapidly in the initial stage, followed by a gradual slowdown. The relationship between the SS and the density of the interface states (D_{it}) can be expressed as below:

$$SS \approx \frac{1}{\frac{kT}{q} \cdot ln10 \cdot (1 + \frac{C_D + q \cdot D_{it}}{C_{ox}})} \quad (1)$$

From (1), it indicates that the CCS significantly increases the D_{it} in the beginning. Considering the formula of V_{th} in (2), the significantly increased D_{it} also leads to a significantly increased V_{th}, which makes ΔV_{th} show a rapid increase initially.

$$V_{th} = \Phi_{ms} + 2\Phi_f + \frac{\sqrt{2q\varepsilon_s N_a (2\Phi_f)}}{C_{ox}} + \frac{q \cdot D_{it}}{C_{ox}} - \frac{Q_f}{C_{ox}} \quad (2)$$

However, as the increase of the D_{it} slows down, reflected by the gradual slowdown on the decrease of the SS, the impact of the interface states on the V_{th} and V_g is gradually weakened. On the contrary, the impact of the electron trapping in the bulk of the gate oxide is gradually strengthened and dominates. Since the generation rate of the electron trap and the electron trap capture cross-section under the CCS are constant [12], ΔV_{th} tends to linearly increase as shown in Phase II until the gate oxide breakdown. Therefore, under the CCS of 0.14 μA, the charge trapping mechanism during the stress is electron trapping only.

Fig. 3 (b) illustrates the conditions under the CCS of 0.7 μA. In this case, the total stress duration is divided into three phases. The first two phases are the same as those under the CCS of 0.14 μA. In the extra Phase III, the increase of the ΔV_{th} slows down again until the gate oxide breakdown. Considering the SS is almost saturated in this phase which minimizes the influence of the D_{it}, this phenomenon can be explained by the hole trapping triggered in the bulk of the gate oxide. Although 0.7 μA is extracted at the gate oxide electric field of 8.5 MV/cm below the critical electric field that starts to trigger the hole trapping from the I_{gss} curve at 150 °C, the hole trapping still occurs in this case. Thus, the critical current stress that can distinguish the different charge trapping mechanisms is estimated at an electric field within $8 - 8.5$ MV/cm, lower than $8.5 - 9$ MV/cm of the critical electric field extracted from the CVS-TDDB results of commercial SiC power MOSFETs [13, 14].

979-8-3503-3714-3/23 $31.00 © 2023 IEEE   85

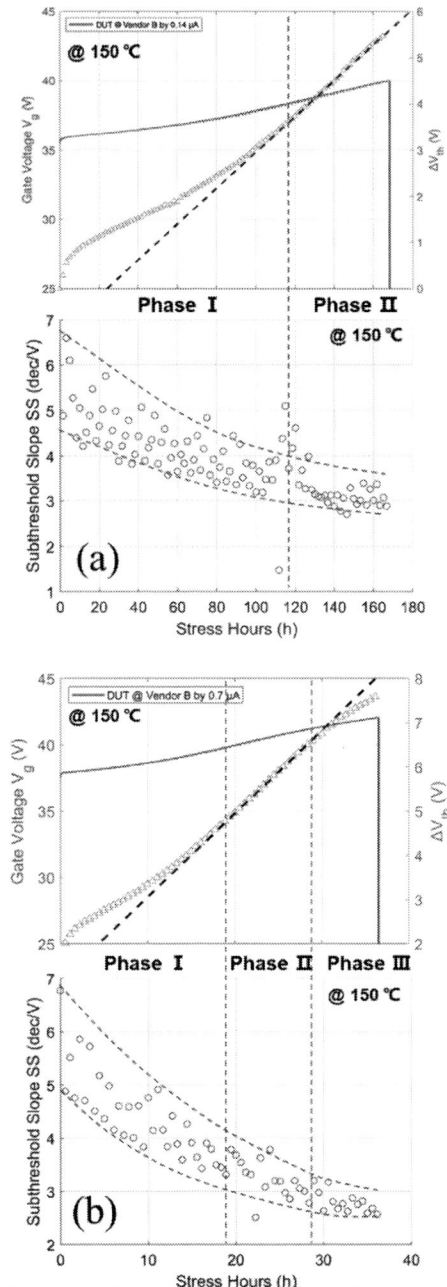

Fig. 3. The gate voltage, the threshold voltage variation, and the subthreshold slope at 150 °C under the CCS of (a) 0.14 μA; (b) 0.7 μA.

B. The Energy Band Diagrams under the CCS

Fig. 4 illustrates the different simplified energy band diagrams under the CCSs of 0.14 and 0.7 μA, respectively, when the gate oxide breakdown just occurs. In Fig. 4 (a), when the CCS is 0.14 μA, the F-N tunneling barrier keeps constant for the CCS with a constant electric field of 8 MV/cm and a constant barrier width of about 3.375 nm. However, the electric field in the remaining gate oxide increases by the energy band bending, which is caused by the increased V_g. In this case, since the gate voltage variation when the gate oxide breakdown occurs (ΔV_{g_BD}) is approximately 4.67 V, the

electric field in the non-F-N tunneling barrier area just reaches 9 MV/cm, which is near the critical electric field to trigger the hole injection from the polysilicon gate. However, since the gate oxide fails right after that 9MV/cm is reached, there is no time for the hole to be trapped and the charge trapping mechanism is the electron trapping only under the CCS of 0.14 μA. In contrast, Fig. 4 (b) shows that the electric field in the non-F-N tunneling barrier area of the gate oxide has reached 9.5 MV/cm when the gate oxide fails, which exceeds the critical electric field. Thus, the hole trapping has already been triggered in this area before the gate oxide breakdown.

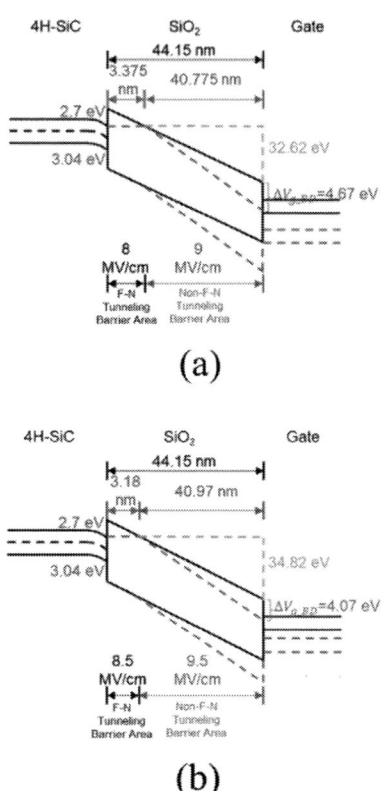

Fig. 4. The simplified energy band diagrams under the CCS of (a) 0.14 μA; (b) 0.7 μA at 150 °C when the gate oxide breakdown just occurs with the ΔV_{g_BD} extracted from the measured data in Fig. 2 (a).

C. The Weibull Plots under the CCS

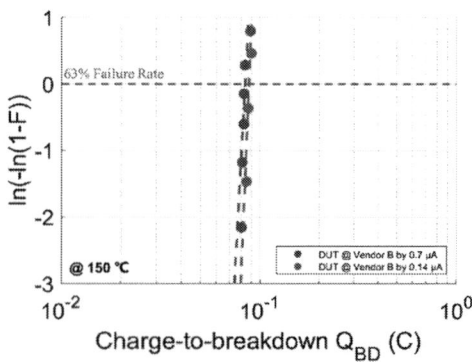

Fig. 5. The gate voltage, the threshold voltage variation, and the subthreshold slope at 150 °C under the CCS of (a) 0.14 μA; (b) 0.7 μA.

In Fig. 5, two Weibull plots under the CCSs of 0.14 and 0.7 μA are shown. It can be observed clearly that the $Q_{BD\_63\%}$ in both cases have no significant variation, concentrated around 0.085 C, although the charge trapping mechanism is different. This helps prove that the failure mechanism of the gate oxide is more likely the charge driven breakdown by electrons through the gate oxide. Once the total electron charge reaches the Q_{BD}, the gate oxide fails.

IV. CONCLUSIONS

In this paper, the CCS-TDDB measurement is performed on 1.2 kV SiC planar power DMOSFETs with the CCSs of 0.14 and 0.7 μA, which correspond to 8 and 8.5 MV/cm in the I_{gss} curves at 150 °C, respectively. Monitoring the ΔV_{th} and the SS, both stress processes indicate a similar procedure with the initial generation of the interface states that dominates, followed by the gradually dominant generation of the electron trap with a constant generation rate in the bulk of the gate oxide. The electron trapping occurs simultaneously during the generation of the electron trap. However, under the CCS of 0.7 μA, an extra step with the hole trapping appears before the gate oxide breakdown. Through the analysis of the energy band diagrams, this phenomenon can be explained by the band bending induced impact ionization/AHI in the bulk of the gate oxide. The Weibull plots under the CCSs of 0.14 and 0.7 μA almost overlap, indicating very close value of the $Q_{BD\_63\%}$ around 0.085 C. Considering that there are different levels of interface states generation in both cases, and even hole trapping due to the variation of the local electric field under the CCS of 0.7 μA, almost no change in the $Q_{BD\_63\%}$ helps prove that the failure mechanism of the gate oxide is more likely the charge driven breakdown. This provides a theoretical basis for establishing a lifetime model of the gate oxide in commercial SiC power MOSFETs based on the Q_{BD}.

ACKNOWLEDGMENT

This research is supported in part by Ford Motor Co. under the Ford-OSU Alliance Project to The Ohio State University.

REFERENCES

[1] Hangseok Choi, "Overview of Silicon Carbide Power Devices", *Fairchild Semiconductor (2016)*.

[2] J. Qian et al., "A Comparison of Ion Implantation at Room Temperature and Heated Ion Implantation on the Body Diode Degradation of Commercial 3.3 kV 4H-SiC Power MOSFETs," *2022 IEEE 9th Workshop on Wide Bandgap Power Devices & Applications (WiPDA)*, Redondo Beach, CA, USA, 2022, pp. 49-53.

[3] D. Okamoto, H. Yano, K. Hirata, T. Hatayama and T. Fuyuki, "Improved Inversion Channel Mobility in 4H-SiC MOSFETs on Si Face Utilizing Phosphorus-Doped Gate Oxide," *IEEE Electron Device Letters*, vol. 31, no. 7, pp. 710-712, July 2010.

[4] Singh, Ranbir, and Allen R. Hefner. "Reliability of SiC MOS devices." Solid-State Electronics, vol. 48, no. 10-11, pp. 1717-1720, Oct. 2004.

[5] Wang, Jun, and Xi Jiang. "Review and analysis of SiC MOSFETs' ruggedness and reliability." *IET Power Electronics*, vol. 13, no. 3, pp. 445-455, Feb. 2020.

[6] Z. Chbili et al., "Modeling Early Breakdown Failures of Gate Oxide in SiC Power MOSFETs," *IEEE Transactions on Electron Devices*, vol. 63, no. 9, pp. 3605-3613, Sept. 2016.

[7] K. P. Cheung, "SiC power MOSFET gate oxide breakdown reliability — Current status," *2018 IEEE International Reliability Physics Symposium (IRPS)*, Burlingame, CA, USA, 2018, pp. 2B.3-1-2B.3-5.

[8] C. -T. Yen et al., "Oxide Breakdown Reliability of SiC MOSFET," *2019 IEEE Workshop on Wide Bandgap Power Devices and Applications in Asia (WiPDA Asia)*, Taipei, Taiwan, 2019, pp. 1-3.

[9] P. Moens, J. Franchi, J. Lettens, L. D. Schepper, M. Domeij and F. Allerstam, "A Charge-to-Breakdown (Q_{BD}) Approach to SiC Gate Oxide Lifetime Extraction and Modeling," *2020 32nd International Symposium on Power Semiconductor Devices and ICs (ISPSD)*, Vienna, Austria, 2020, pp. 78-81.

[10] L. Shi et al., "Investigation of different screening methods on threshold voltage and gate oxide lifetime of SiC Power MOSFETs," *2023 IEEE International Reliability Physics Symposium (IRPS)*, Monterey, CA, USA, 2023, pp. 1-7.

[11] J. W. McPherson, and H. C. Mogul. "Underlying physics of the thermochemical E model in describing low-field time-dependent dielectric breakdown in SiO 2 thin films." *Journal of Applied Physics*, vol. 84, no. 3, pp. 1513-1523, 1998.

[12] Mong-Song Liang and Chenming Hu, "Electron trapping in very thin thermal silicon dioxides," *1981 International Electron Devices Meeting*, Washington, DC, USA, 1981, pp. 396-399.

[13] T. Liu, S. Zhu, M. H. White, A. Salemi, D. Sheridan and A. K. Agarwal, "Time-Dependent Dielectric Breakdown of Commercial 1.2 kV 4H-SiC Power MOSFETs," *IEEE Journal of the Electron Devices Society*, vol. 9, pp. 633-639, 2021.

[14] L. Shi et al., "Effects of Oxide Electric Field Stress on the Gate Oxide Reliability of Commercial SiC Power MOSFETs," *2022 IEEE 9th Workshop on Wide Bandgap Power Devices & Applications (WiPDA)*, Redondo Beach, CA, USA, 2022, pp. 45-48.

Pulse-Voltage Time-Dependent Dielectric Breakdown of Commercial 1.2 kV 4H-SiC Power MOSFETs

Michael Jin*, Limeng Shi, Jiashu Qian, Monikuntala Bhattacharya,
Hengyu Yu, Marvin H. White, and Anant K. Agarwal
Dept. of Electrical & Computer Engineering
The Ohio State University
Columbus, OH, USA
jin.845@osu.edu

Atsushi Shimbori, Zhuxian Xu
Ford Motor Company
Dearborn, Michigan, USA

Abstract—In this paper, the gate oxide lifetime of commercial 4H-SiC planar MOSFETs is evaluated by the pulse-voltage time-dependent dielectric breakdown (PVTDDB) test under various amplitudes of high-frequency gate pulse voltages at 150 °C. The oxide failure time is measured under AC stress to predict the lifetime of the gate oxide in operational conditions. In addition, the behavior of the gate leakage current influenced by hole and electron trapping is monitored to explain the effect of the high frequency signal on charge trapping in the gate oxide. Moreover, a comparison of the conventional constant-voltage TDDB and present PVTDDB test results reveals that there is no significant difference in the measured oxide lifetime between the two methods.

Keywords—oxide lifetime, high frequency, pulse-voltage TDDB, gate leakage current, charge trapping.

I. INTRODUCTION

Silicon Carbide (SiC) power MOSFETs are becoming more attractive to manufacturers in high power applications including electric vehicles and charging stations. Due to safety interests, the gate oxide lifetime of the power devices is a major area of concern. The conventional constant-voltage time-dependent dielectric breakdown (CVTDDB) method is generally employed to estimate the gate oxide lifetime [1-4]. However, during CVTDDB measurements a constant gate stress is applied, which is not analogous to the actual operating conditions of SiC MOSFETs in industrial and automotive applications. The real gate voltage signal in the circuit is a pulse-width-modulated (PWM) signal, consisting of a sequence of high-frequency gate pulses. In this case, an accurate lifetime estimation of the gate oxide becomes a critical reliability issue with regard to test methods. It is desirable to investigate the effects of high-frequency gate pulse voltages on oxide lifetime and evaluate the reliability of commercial SiC power MOSFETs in practical industrial applications.

In this work, a pulse-voltage TDDB (PVTDDB) method is proposed to measure the oxide failure time of commercial 4H-SiC planar MOSFETs under various amplitudes of high-frequency gate pulse voltages at 150°C. PVTDDB involves the application of a pulsed voltage signal to the gate oxide of the devices, which allows for a more accurate prediction of oxide lifetime at normal gate operating conditions [5]. The gate leakage currents during TDDB measurement with DC

and AC gate stress at 150°C are also monitored and analyzed. The charge trapping mechanism is used to investigate the effect of the gate voltage stress on oxide lifetime in both DC and AC cases.

II. DEVICE AND TEST INFORMATION

A. Device Information

The commercial 1.2 kV 4H-SiC planar power MOSFETs (packaged in TO-247-3) are utilized for the PVTDDB and CVTDDB measurement in this work. The details of the planar MOSFET devices under test (DUTs) from Vendor B are shown in Table I. Each device is characterized with a Keysight B1506 curve tracer to obtain the relevant parameters. To estimate the gate oxide thickness (t_{ox}), a ramp-to-breakdown test is performed on three devices. During this measurement, each device is heated to 150°C with the drain and source shorted to ground. The gate voltage is ramped until the gate oxide of the device breaks down. For the planar MOSFETs used in this work, the average breakdown voltage (V_{br}) is ~51 V. By taking the critical oxide electric field (E_{crit}) as 11 MV/cm, t_{ox} is estimated by V_{br}/E_{crit} to be ~ 46.4 nm.

TABLE I. DETAILS OF COMMERCIAL SiC PLANAR MOSFETs FROM VENDOR B

Manufacturer	Vendor B
Structure	Planar
Voltage Rating	1200 V
Current Rating	11 A
Average oxide V_{br} @ 150°C	51 V
Estimated t_{ox}	46.4 nm

B. Evaluation Methodology and Test Setup

The test setup for PVTDDB is shown in Fig. 1. First, ten DUTs are placed onto the high-temperature test board located within an oven at 150°C, with the drain and source of each DUT shorted together. A voltage square wave with the amplitude of the stress voltage (V_{stress}) is applied to the gate. When MOS1 is turned on with MOS2 off, the V_{stress} is dropped across the gate of each DUT. Then, when MOS1 is turned off with MOS2 on, the gate is pulled to ground with MOS2. The frequency of the applied square wave is 10 kHz with a duty

979-8-3503-3714-3/23 $31.00 © 2023 IEEE

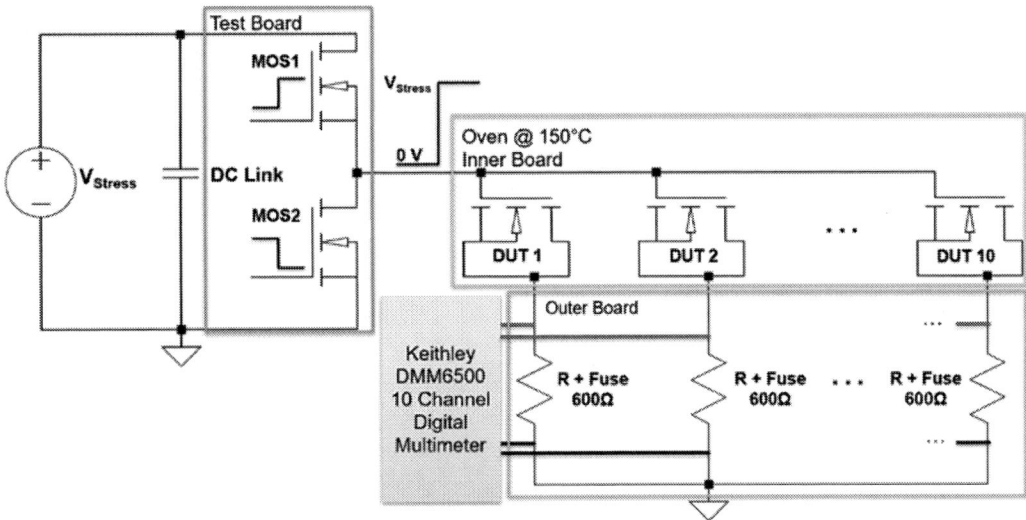

Fig. 1. Test setup with DUTs and auxiliary components.

cycle of 50%. In series with each device is a resistor and fuse (~ 600 Ω), which suppresses voltage spiking on the transient edges of the voltage square wave and protects the circuit in case of an oxide short circuit, respectively. The devices are stressed by high-frequency gate pulse voltages until oxide breakdown occurs in all ten devices, characterized by jumps in the voltages measured across the resistors and fuses.

To observe the effects of different oxide electric fields (E_{ox}) on the gate oxide lifetime, three oxide electric fields are selected as 9.9 MV/cm, 9.3 MV/cm, and 9.0 MV/cm. These correspond to 46 V, 43.2 and 41.8 V, respectively. The stress voltage (V_{stress}) is set by multiplying t_{ox} with a target E_{ox}. The high-frequency gate pulse voltages are delivered simultaneously to all the gate terminals of the ten DUTs. At each E_{ox}, the leakage current through the gate oxide of each DUT is estimated by measuring the voltage across its series resistor and fuse with a ten-channel Keithley digital multimeter (DMM6500), and time-to-failure is recorded. The gate oxide lifetime with a failure rate of 63% at each E_{ox} is extracted as $t_{63\%}$ based on the Weibull distribution. The operating lifetime at the normal gate oxide electric field can be extrapolated using the thermochemical E model [2, 7].

III. RESULTS AND DISCUSSIONS

In this section, the oxide failure times and gate leakage currents of commercial SiC planar MOSFETs in CVTDDB and PVTDDB tests are investigated. In addition, the effects of AC and DC stresses on hole trapping and electron trapping in gate oxide are discussed.

A. Weibull Plots and Lifetime Prediction

The Weibull distributions of the measured oxide lifetimes under DC and AC stress conditions for commercial SiC planar MOSFETs are shown in Fig. 2. The oxide lifetime ($t_{63\%}$) at each E_{ox} is calculated by extracting the time at which the Weibull distribution fit line crosses the line indicating that $\ln(-\ln(1-F))$ equals 0. The variation in device failure time at the same gate voltage stress is attributed to the difference in

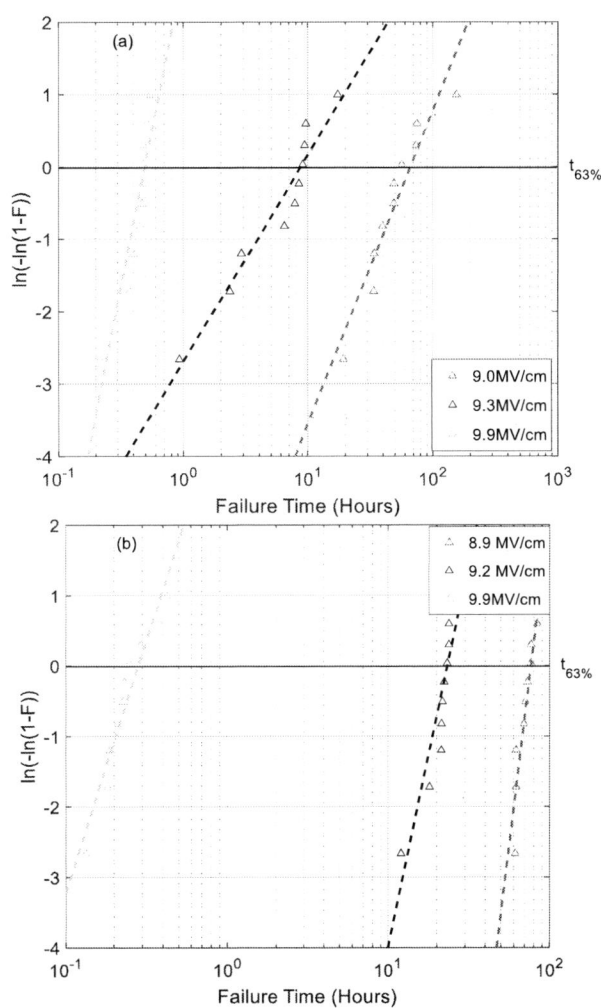

Fig. 2. Weibull distribution of measured oxide lifetimes for (a) PVTDDB measurement with oxide electric field of 9 MV/cm, 9.3 MV/cm and 9.9 MV/cm (b) CVTDDB measurement with oxide electric field of 8.9 MV/cm, 9.2 MV/cm and 9.9 MV/cm.

979-8-3503-3714-3/23 $31.00 © 2023 IEEE

effective gate oxide thickness caused by variations in the device fabrication process, and the random distribution of defects in the gate oxide.

Fig. 3 depicts the comparison of the $t_{63\%}$ extracted from the Weibull plots for the CVTDDB and PVTTDB tests. It can be observed that there is no significant difference in the distribution of oxide lifetimes measured by the two methods. After fitting a straight line based on the E-model, the oxide lifetime of commercial SiC planar MOSFETs at 4 MV/cm predicted by PVTDDB and CVTDDB test methods both exceed the 20-year industrial requirement. However, the estimated lifetime under PVTDDB based on the current data is more optimistic than CVTDDB. Since the gate signal is a pulsed signal rather than a constant voltage under normal operating conditions [5], the estimated lifetime of the gate oxide under PVTDDB should be more representative of the actual oxide lifetime of commercial SiC MOSFETs in application.

The CVTDDB results show an abrupt change in field acceleration factors which is due to the degradation of the oxide lifetime resulting from the initial hole trapping under high electric fields [1-3]. Under lower oxide electric fields, hole trapping never dominates due to the reduced energy of the electrons, and so the lifetime prediction curve has a reduced slope. More details regarding the charge trapping phenomenon are given in the next section. Currently, no transition point is observed in the PVTDDB measurements. Future PVTDDB tests at lower oxide electric fields will be continued to see if this phenomenon also occurs during PVTDDB.

B. Gate Leakage Currents

Fig. 4 (a) and (b) illustrate the gate leakage currents with respect to stress time during the PVTDDB and CVTDDB tests at different oxide electric fields, respectively. It can be seen that the behavior of the leakage current is similar between the two test methods, and can be explained in terms of charge trapping [8].

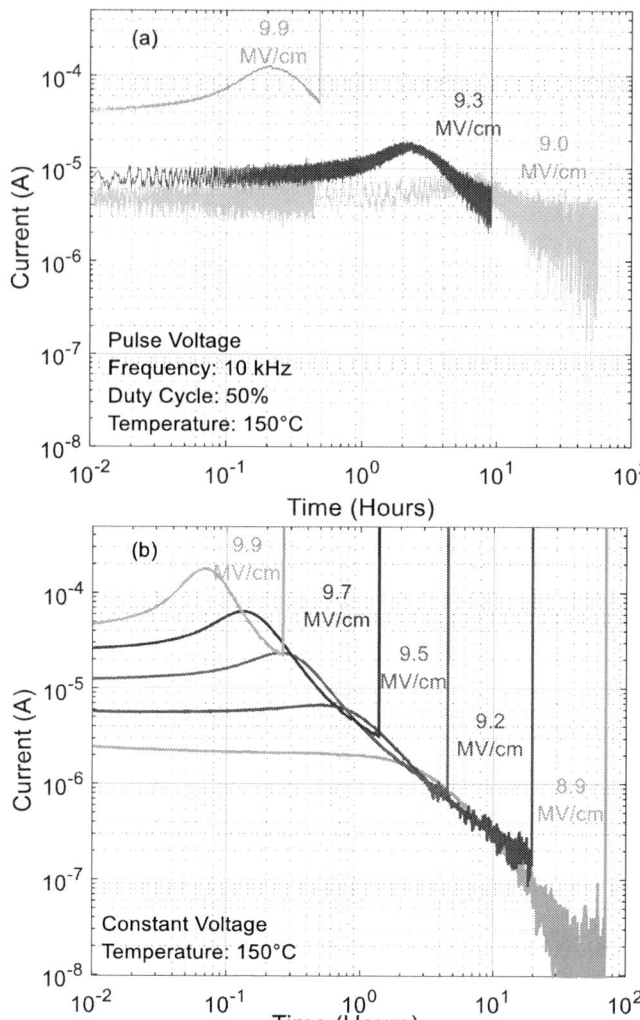

Fig. 4. The gate leakage current under (a) PVTDDB and (b) CVTDDB.

As shown in Fig. 5, at a high electric field (> 9 MV/cm), electrons are injected into and through the gate oxide by the electric field via Fowler-Nordheim tunneling. Some electrons with high energy can reach the poly-Si and cause impact ionization, which can trigger hole generation. The holes can then tunnel into the oxide and get trapped near the SiC/SiO$_2$ interface which would reduce the barrier width. In this case, when the hole trapping dominates, the gate leakage current appears to rise. As stress time increases, electron trapping overtakes hole trapping, which leads to a fall in the gate leakage current, due to the widening of the barrier. From the results, it can be seen that the effects of electron trapping before oxide breakdown in PVTDDB seem to less pronounced when compared to CVTDDB. This phenomenon may be due to the release of trapped electrons during the gate bias period of 0 V in the AC signal which reduces the effect of electron trapping [9].

As the gate oxide field decreases, the effect of the holes from impact ionization reduces, increasing the magnitude of the drop in gate leakage current. Compared to the gate leakage current in the CVTDDB test, the gate leakage current in the

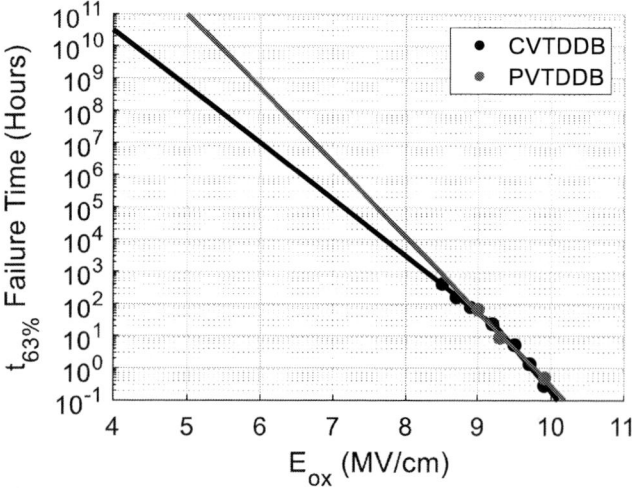

Fig. 3. $t_{63\%}$ as a function of the oxide electric field for CVTDDB and PVTDDB test methods.

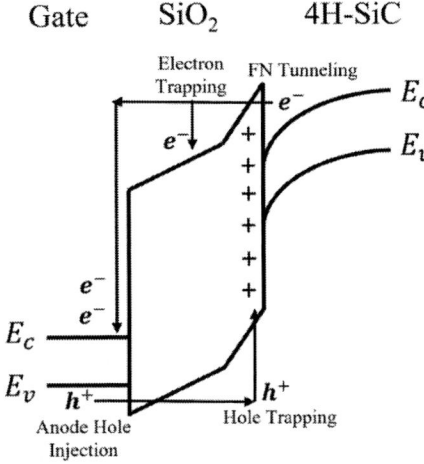

Gate   SiO$_2$ 4H-SiC

Fig. 5. Electron and hole trapping at high oxide electric fields.

PVTDDB test has a lower magnitude of reduction. The oxide failure time measured in the TDDB test is mainly affected by applied positive gate voltage stress time and gate leakage current. On one hand, the effective stress time needs to be long enough to cause the breakdown of gate oxide. On the other hand, a decrease in the leakage current through the oxide causes an expansion of the breakdown time. The PVTDDB proposed in this work adopts a pulsed gate voltage signal with a duty ratio of 50%, which means the device is only stressed for half of the total measurement time, rather than the entire duration. The real positive voltage stress time is half of the total time, which means that the oxide lifetime measured by the PVTDDB method is overstated. However, the decrease in gate leakage current during the PVTDDB test is not large due to reduced electron trapping, and so the AC stress accelerates the breakdown of the oxide when compared to CVTDDB. Overall, the effects of the stress time and leakage current in the PVTDDB test neutralize each other, resulting in a small difference in the lifetime of the oxides measured by PVTDDB and CVTDDB.

IV. CONCLUSIONS

The oxide failure times and gate oxide leakage currents under AC stress conditions for commercial SiC planar MOSFETs are presented in this paper. The lifetime estimation of the gate oxide under pulse-voltage stress provides an optimistic outlook at nominal operating conditions. By analyzing the behavior of the gate leakage currents at high frequency, it can be concluded that the PVTDDB test can

reduce the effect of electron trapping caused by a positive gate voltage. Future tests on other commercial 4H-SiC power MOSFETs are planned to observe the dependence of oxide lifetime and leakage current at different electric fields, higher frequencies, different device structures (planar, double trench, and asymmetric trench), and the effect of a turn-off voltage of -5 V during PVTDDB.

ACKNOWLEDGMENT

This work was supported in part by the Ford Motor Company under the Ford Alliance 2019 Project to The Ohio State University.

REFERENCES

[1] S. Zhu, T. Liu, M. H. White, A. K. Agarwal, A. Salemi and D. Sheridan, "Investigation of Gate Leakage Current Behavior for Commercial 1.2 kV 4H-SiC Power MOSFETs," 2021 IEEE International Reliability Physics Symposium (IRPS), Monterey, CA, USA, 2021, pp. 1-7, doi: 10.1109/IRPS46558.2021.9405230.

[2] T. Liu, S. Zhu, M. H. White, A. Salemi, D. Sheridan and A. K. Agarwal, "Time-Dependent Dielectric Breakdown of Commercial 1.2 kV 4H-SiC Power MOSFETs," in IEEE Journal of the Electron Devices Society, vol. 9, pp. 633-639, 2021, doi: 10.1109/JEDS.2021.3091898.

[3] L. Shi et al., "Effects of Oxide Electric Field Stress on the Gate Oxide Reliability of Commercial SiC Power MOSFETs," 2022 IEEE 9th Workshop on Wide Bandgap Power Devices & Applications (WiPDA), Redondo Beach, CA, USA, 2022, pp. 45-48, doi: 10.1109/WiPDA56483.2022.9955295.

[4] T. Liu et al., "Gate Leakage Current and Time-Dependent Dielectric Breakdown Measurements of Commercial 1.2 kV 4H-SiC Power MOSFETs," 2019 IEEE 7th Workshop on Wide Bandgap Power Devices and Applications (WiPDA), Raleigh, NC, USA, 2019, pp. 195-199, doi: 10.1109/WiPDA46397.2019.8998792.

[5] N. Mohan, T. M. Undeland, and W. P. Robbins, *Power Electronics: Converters, Applications, and Design*, 2nd ed. New York: John Wiley, 1995.

[6] M. Tsuno, M. Suga, M. Tanaka, K. Shibahara, M. Miura-Mattausch and M. Hirose, "Physically-based threshold voltage determination for MOSFETs of all gate lengths," in IEEE Transactions on Electron Devices, vol. 46, no. 7, pp. 1429-1434, July 1999, doi: 10.1109/16.772487.

[7] J. W. McPherson, and H. C. Mogul, "Underlying physics of the thermochemical E model in describing low-field time-dependent dielectric breakdown in SiO2 thin films," J. Appl. Phys., 1 August 1998, pp. 1513–1523. https://doi.org/10.1063/1.368217

[8] Ih-Chin Chen, S. E. Holland and Chenming Hu, "Electrical Breakdown in Thin Gate and Tunneling Oxides," in IEEE Journal of Solid-State Circuits, vol. 20, no. 1, pp. 333-342, Feb. 1985, doi: 10.1109/JSSC.1985.1052311.

[9] L. Shi et al., "Investigation of different screening methods on threshold voltage and gate oxide lifetime of SiC Power MOSFETs," 2023 IEEE International Reliability Physics Symposium (IRPS), Monterey, CA, USA, 2023, pp. 1-7.

979-8-3503-3714-3/23 $31.00 © 2023 IEEE

Common Mistakes in Practical Power Supply Design with Wide Bandgap Devices

Sheng-Yang Yu
Power Design Services
Texas Instruments
Dallas, Texas
seanyu@ti.com

Fei Yang
High Voltage Power – GaN
Texas Instruments
Dallas, Texas
fei.yang@ti.com

Abstract— **Power electronics engineers are motivated to design the next generation power supply with wide bandgap (WBG) devices as they offer better figure-of-merits over Silicon devices. However, several common mistakes have been observed in practical WBG-based power supply designs which do not necessarily cause problems with Silicon devices. Those common mistakes are related to fast switching speed in WBG devices. This paper summarizes some of the mistakes, explains the root causes, and provides solutions.**

Keywords—Gallium Nitride (GaN), Common source inductance (CSI), Silicon Carbide (SiC), Common mode transient immunity (CMTI).

I. INTRODUCTION

Due to better figure-of-merits in wide bandgap (WBG) devices like Gallium Nitride (GaN) high electron mobility transistor (HEMT) and Silicon Carbide (SiC) MOSFETs, the power supply industry has gradually migrated from Silicon FETs to WBG devices in the past decade. Compared to Silicon devices, WBG devices not only provide lower specific resistance but also have other benefits like lower output capacitance (C_{oss}) and low or even no reverse recovery charge [1]-[2] . This enables high-efficiency and high-power-density power converter designs.

In hard switching topologies, the device's turn-on loss consists of the C_{oss} loss, current-voltage overlap loss, and revere recovery loss Q_{rr} depending on the device's structure. As a WBG device with lower capacitance values, GaN and SiC devices can switch faster thus reducing the current-voltage overlap loss. Meanwhile, the enhancement-mode (e-mode) GaN or direct drive depletion-mode GaN devices can eliminate the Q_{rr} loss, which allows more efficient topologies to be adapted in a power supply unit. For example, the totem-pole bridgeless power factor correction circuit (PFC) under continuous conduction mode can only demonstrate its efficiency benefit over the traditional bridge PFC when low or no Q_{rr} FET is used [3].

In soft switching topologies where zero-voltage switching is achieved, there is no turn-on loss as the resonant tank's current is able to circulate the half-bridge device's C_{oss} energy into the resonant tank before the energy is dissipated into the device in a hard-switching condition. The deadtime required to achieve zero voltage switching is related to the half-bridge devices' C_{oss} capacitance as expressed in (1).

$$t_{DT} = \frac{2 \int_0^{V_{DS}} C_{oss}(v_{DS}) v_{DS}}{I}, \qquad (1)$$

where t_{DT} is the minimum deadtime required given a discharge current I and a C_{oss} curve – $C_{oss}(v_{DS})$. A C_{oss} comparison between a GaN FET and a Silicon super junction MOSFET with similar on-resistance ($R_{DS,on}$) is illustrated in Fig. 1. As can be seen, GaN FET has a much lower C_{oss} when FET drain to source voltage (v_{DS}) is low, which helps to reduce t_{DT} to allow longer conduction period and lower RMS current for a high switching frequency operation.

However, the fast-switching WBG devices can also introduce new issues that weren't concerns in power supplies with Silicon devices. GaN and SiC devices are capable of switching at a much faster speed (>100V/nS voltage slew rate) than silicon devices (<50V/nS) during the switching transients. The fast turn-on speed can result in a high common-mode current, leading to system malfunction. Meanwhile, the fast *dv/dt* and *di/dt* during switching transients can induce high voltage ringing that can potentially exceed the device's maximum voltage rating, and cause EMI or other noise-related issues. To mitigate this, the layout and package design need to be considered carefully.

In this paper, the actual mistake examples when designing with WBG devices are provided, the root causes are analyzed, and solutions to these issues are offered focusing on gate driver design, power loop design, design considerations with high *dv/dt*, and noise coupling mitigations.

II. SWITCHING LOSSES HIGHER THAN EXPECTATIONS

There are application cases where the switching loss in the real circuit is higher than the datasheet values validated by the

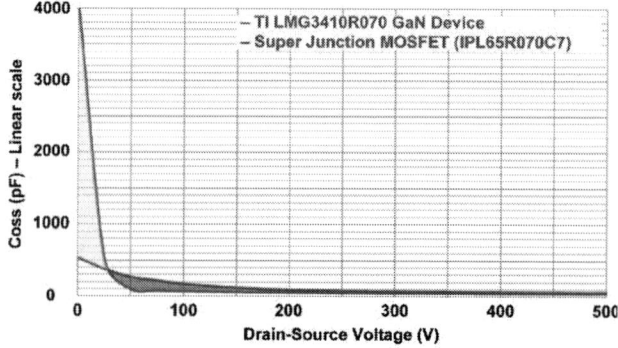

Fig. 1. C_{oss} comparison between GaN and Silicon MOSFET.

industry standard double-pulse test (DPT) [4]. A simplified DPT setup is shown in Fig. 2. The DUT's turn-on and turn-off losses can be calculated through the measured voltage and current across the device. However, a discrepancy can possibly be found between the DPT data and a specific application case as shown in Fig. 3. Depending on the application board design, higher turn-on and turn-off losses can be observed.

Specifically, the parasitic capacitance of the load inductor C_L and the board parasitic capacitance between positive bus and switch-node C_{PCB1} as indicated in Fig. 2 can add additional C_{oss} energy in the turn-on curve. Similarly, the PCB parasitic capacitance between switch-node and negative bus traces C_{PCB2} can also contribute to additional C_{oss} loss. Though this loss is measured during the turn-off process, it is an actual loss that will be dissipated during the hard-switching turn-on process. These parasitic capacitances explain the power loss discrepancies at the zero-current crossing points shown in Fig. 3. In real practice, it is critical to minimize those parasitic capacitances to further reduce the losses. Meanwhile, those parasitic losses need to be accounted for loss estimations.

In addition to the parasitic capacitance, parasitic inductance can also lead to discrepancies in switching loss. Fig. 4 shows the equivalent circuit of a discrete GaN device driven by external gate drivers. First, the common source inductance (L_{cs}) will slow drain-to-source voltage transition and result in higher overlap losses in hard switching topologies, and require longer dead times in soft switching topologies. As shown in Fig. 5 (a), only 5 nH L_{cs} can result in 48% more turn-on loss [5]. On the other hand, the total inductance on the gate driving loop (L_{drv_out} + L_{g_PCB} + L_{g_GaN} + L_{cs} + L_{s_PCB} + L_{drv_gnd} for turn off and L_{drv_vdd} + L_{drv_out} + L_{g_PCB} + L_{g_GaN} + L_{cs} + L_{s_PCB} for turn on) can induce large gate voltages during the fast-switching transient. To mitigate the gate over-voltage stress, the peak gate current will be limited as the gate loop inductance increases thus slowing down gate drive speed and inducing high overlap losses in hard switching applications. Meanwhile, with a large gate loop inductance in the hold-off path, the device's gate voltage will be higher during the cross-talk period where the dv/dt from the active switching device can induce current to the hold-off path through the gate-to-drain capacitance. This increases the Miller shoot-through risk and induces additional cross-conduction losses.

Therefore, in real designs with WBG devices, it is a good practice to minimize the common-source inductance through Kelvin source connections. Also, the gate drive loop inductance

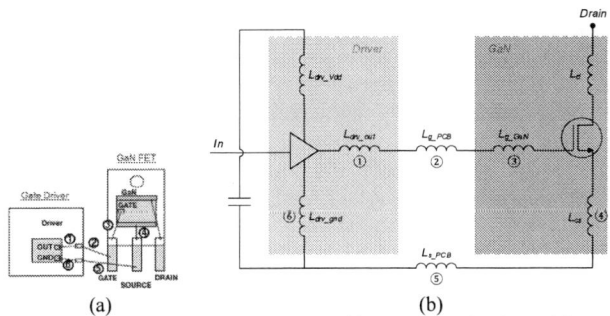

(a)   (b)

Fig. 4. (a) GaN with a discrete gate driver and (b) circuit model.

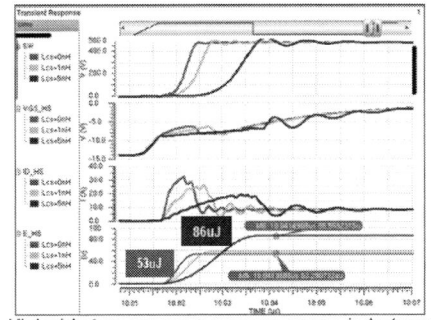

High-side turn on versus common-source inductance:
red = 0 nH, green = 1 nH, blue = 5 nH

(a)

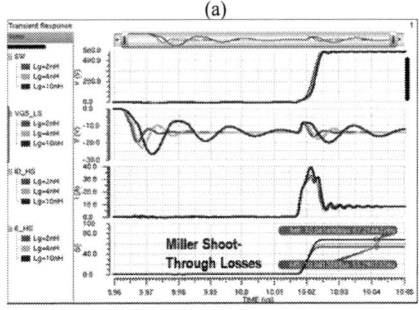

Low-side hold-off versus gate-loop inductance
red = 2 nH, green = 4 nH, blue = 10 nH

(b)

Fig. 5. (a) common source inductance's impact on switching loss (b) gate loop inductance's impact on switching loss.

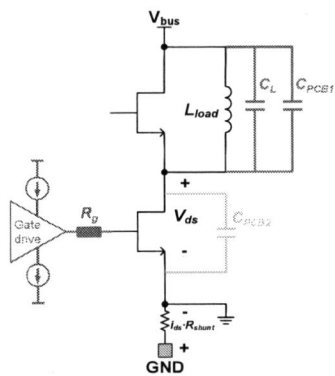

Fig. 2. Double pulse test setup.

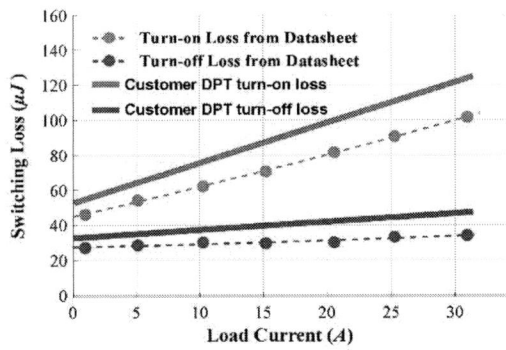

Fig. 3. Switching losses numbers from datasheet and customer side

needs to be minimized as much as possible by placing the bypass bypassing capacitor closer to the driver and optimizing the gate drive layout. To faciliate the appliation use, L_{g_PCB} and L_{s_PCB} can be removed when the driver is integrated with the WBG device.

III. CHALLENGE IN BOOTSTRAP CIRCUIT DESIGN

G aN devices allow a smaller V_{GS} voltage range than Silicon devices. Therefore, it is critical to provide a stable gate voltage to e-mode GaN FETs. In a totem-pole configuration, the bootstrap circuit shown in Fig. 6 is a common method to provide bias power for the high-side device. The high side FET bias voltage V_{dd_H} can be expressed as:

$$V_{dd\_H} \approx V_{dd\_L} - V_{diode} - V_{GaN\_L}, \qquad (2)$$

where V_{dd_L} is the bias supply provided at the low side driver, V_{diode} is the voltage drop across the bootstrap diode, and V_{GaN_L} is low side FET drain-to-source voltage when conducting current either through the channel or third quadrant conduction. As can be seen, V_{dd_H} voltage varies along with the polarity of the load current and how the low-side device conducts. In most WBG devices, the channel conduction voltage is smaller than the third quadrant conduction voltage. Assuming a negative current is going through the low-side device: when the low-side device conducts through the third quadrant, the bootstrap voltage V_{dd_H} is much higher compared to the case where the current conducts through the channel. As a result, the high-side device's gate can be over-charged causing concerns on gate stresses. This can normally happen during the deadtime period when a low-side GaN device has a freewheeling current. On the other hand, assuming another case where the current is positive from drain-to-source in the low-side device, the low-side device's voltage drop V_{GaN_L} is determined by the load current and its on-resistance. At high junction temperatures or large load current conditions, more voltage drop is observed across low-side devices which results in a lower bootstrap voltage on the high-side device. The lower V_{dd_H} voltage can result in higher on-resistance on the high-side device and affect its turn-on switching speed. Combining the different operating conditions in the bootstrap circuit and the narrow gate drive voltage range in GaN, it is essential to provide a steady bias supply for high-side FET. Some GaN vendors provide a regulation stage after the bootstrap circuit that can help to maintain a stable gate drive voltage thus keeping a consistent performance of the device in real applications [6].

IV. THE NECESSARY OF SHORT CIRCUIT PROTECTION

As WBG devices can have the same on-resistance with smaller die size than Silicon devices, a low thermal capacity is expected thus inducing a higher temperature rise given the same short-circuit conditions. Therefore, faster short circuit protection (SCP) is required for WBG devices. For SiC MOSFET with a discrete driver, the desaturation protection method is commonly used [7]. A SiC MOSFET with desaturation SCP design example is shown in Fig. 7. When a short circuit happens with large current flows through the SiC device, higher V_{DS} voltage will be detected through the detection diode at the drain node and cause higher voltage at the non-inverting input of the comparator and trips mid-level voltage generation to make the SiC FET operates under linear region for a short period of time.

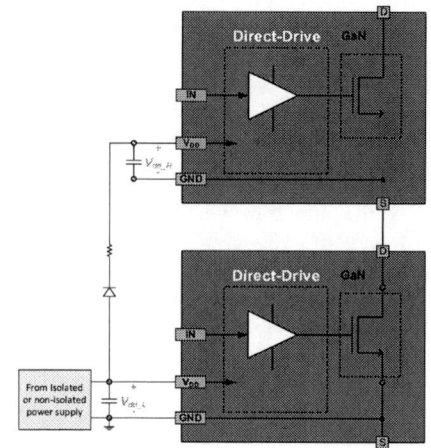

Fig. 6. Totem-pole FETs configuration with Bootstrap circuit.

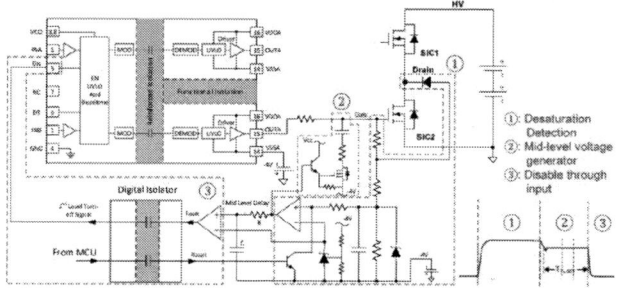

Fig. 7. Desaturation short circuit protection circuit example.

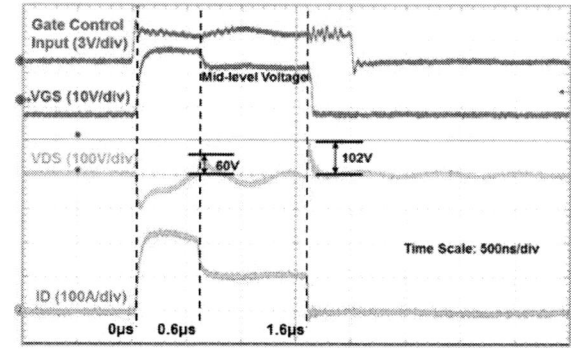

Fig. 8. Desaturation short circuit protection key waveforms.

The comparator output is also used to trip fault protection and force the gate driver to turn off the SiC MOSFET. The waveforms of this 2-level SCP implementation are shown in Fig. 8, which takes 600nS from the de-saturation period to the mid-level voltage generation period and 1.6μS from the mid-level generation period to actually turn off the SiC FET.

An alternative way for SCP is through the integration of driver, short-circuit sensing, and protection together in a WBG device under the IC level. In the example shown from Texas Instruments GaN IC, the SCP could respond much faster than the discrete desaturation SCP mentioned above. As shown in Fig. 9, the drive can quickly detect the short circuit condition when the current hits its protection threshold. Then within 78nS, the device is softly turned off to protect the device and system from further damage.

979-8-3503-3714-3/23 $31.00 © 2023 IEEE   94

V. POWER LOOP LAYOUT

The parasitic inductance in the power loop as shown in Fig. 10 can cause over-voltage spikes, signal integrity issues, and EMI concerns, especially in hard switching applications. With the fast-switching WBG device, these issues are more pronounced.

Power loop placement shown in Fig. 11 is very common in traditional power supply designs with Silicon devices. The decoupling capacitor is placed as close as possible to lower the power loop inductance (L_{s_PCB}), and it is generally acceptable in Silicon-based designs with this kind of lateral layout. However, having a power loop on the same layer won't provide inductance cancellation. The power loop inductance is not minimized in a lateral layout and can cause problems in WBG devices when switching at high dv/dt or di/dt.

An alternative and recommended way is to lay the power loop in a vertical manner as shown in Fig. 12. Current flows out from the decoupling capacitor from one layer and returns from an adjacent layer. In this way, the inductance cancellation is utilized to minimize the power loop inductance. From Ansys Q3D simulation, the power loop inductance for the same devices can be reduced from 12.7 nH to 2.0 nH when the PCB layout is changed from lateral to vertical layout. Fig. 13 shows the circuit simulation results where different values of parasitic inductance are used. When the power loop inductance increases from 2nH to 10nH, more than 100V voltage spike difference can be observed, and the ringing becomes more severe at the 10 nH case causing more EMI and noise concerns.

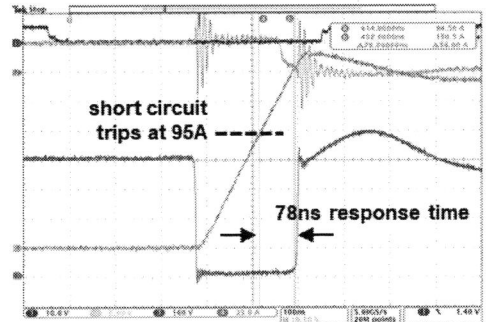

Fig. 9. key waveforms of a short circuit protection on TI's protection integrated GaN device.

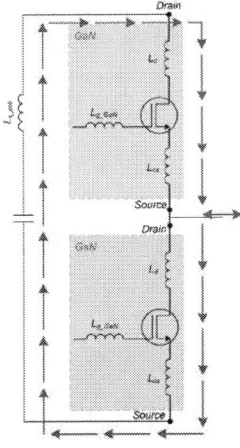

Fig. 10. Power loop inductance in a half-bridge configuration.

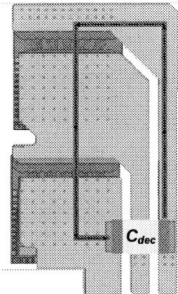

Fig. 11. PCB layout example with power loop mainly on the same layer.

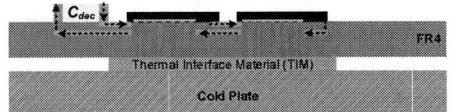

Fig. 12. PCB layout example with power loop in a "vertical loop".

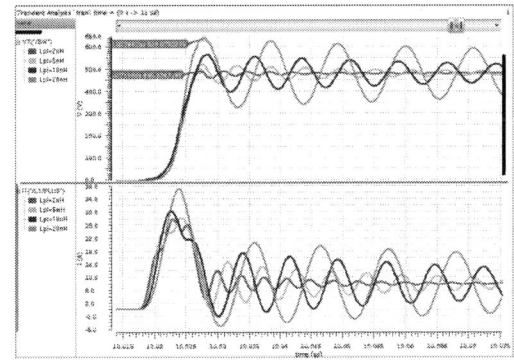

Fig. 13. V_{sw} ringing versus power loop inductance: red = 2nH, green = 5nH, blue = 10nH, orange = 20nH.

VI. COMMON MODE NOISE

The common-mode noise is one of the common mistakes in designing GaN- or SiC-based power converters. Fig. 14 shows the diagram of a half-bridge configuration with the isolator and isolated bias power supply used for high-side devices.

Typically, the coupling capacitances C_{Iso} and C_{Iso_Bias} in Fig. 14 exit in between the primary side and the secondary side of the isolator or isolated bias supply. The parasitic capacitance's value depends on the isolation technology and layout design. For the high-side device, the coupling between the switch node and signal ground can cause issues in practical operation. Specifically, for GaN and SiC devices operating at high slew rates, the high dv/dt can induce a significant current through C_{Iso} and C_{Iso_Bias}. These noise current can affect the signal integrity causing issues like missing pulses, output glitch, excessive propagation delay, output latch, and potentially affect the converter's normal operation.

For instance, Fig. 15 shows the waveform of the GaN half-bridge's switch node operating at ~150V/nS with a bus voltage of 400V. The high-side fault at the isolator's output experiences a 12nS glitch shortly after the high dv/dt transition though the device is not outputting actual fault signals at the isolator's input. To solve this issue, it is recommended to choose an isolator that has a high common mode transient immunity (CMTI) capability that can operate robustly at 150V/nS, 400V bus. Alternatively, additional RC filters can be used at the isolator output to filter out the glitches. Note that the RC filter can induce delays, and need to be carefully used depending on the signals going through the isolator. In some other designs where an isolated gate driver is used, it is critical to make sure it provides enough CMTI to handle the worst-case *dv/dt*. Fig. 16 shows a case where an isolated driver with low CMTI is used to drive SiC devices operating at high *dv/dt*. As can be seen, the output is not following the driver input signals, and the system starts to oscillate.

Similarly, the common mode current also be induced from the isolated bias power supply's parasitic capacitance C_{Iso_Bias}. As a good practice, it is suggested to design the transformer winding with low coupling capacitance for GaN or SiC devices operating at high *dv/dt*. Meanwhile, adding a common-mode choke at the isolated bias supply's output is always helpful to mitigate the noise. Alternatively, when using the integrated bias power supply, the CMTI capability needs to be sufficient to cover the desired *dv/dt* values.

VII. Capacitive Coupling Noises

As can be seen from the previous case, the parasitic capacitance in the isolator or isolated bias supply can induce common mode related noise issues and the CMTI is a key specification to be considered in the design stage. In a practical PCB layout and cooling design, it is also critical to pay attention to the parasitic capacitance. In the following session, two types of issues related to layout parasitic capacitance will be discussed.

A. Capacitive Coupling in PCB

In a practical PCB layout, the capacitive coupling from the switch node to the signal ground needs to be avoided or mitigated as much as possible. Otherwise, the high *dv/dt* can induce noise current to the signal ground causing system malfunctions.

Fig. 17 shows a totem-pole PFC operating waveform during the start-up process. As the bus voltage is starting to build up, it is observed that the positive AC cycle is not regulating. Zooming into the waveform where fault starts to occur, a short-circuit fault is observed from the low-side GaN device's fault indicator when the low-side device turns on. Looking into the details, it is found that the hard-switching turn-on at the low-side GaN at a positive AC cycle induces a high *dv/dt*. This high slew rate then induces a noise to the controller ground through a parasitic capacitance from the PCB layout. Then the high-side PWM signal is affected triggering a false turn-on in the high-side GaN. Consequently, a short circuit is induced and the converter is latched for a half-line cycle. To solve this issue, the layout must be fixed eliminating the coupling capacitance, and the new PCB works normally after the layout fix. As a practical guideline, it is critical to avoid the capacitive coupling from the switch node to the control or signal grounds as indicated in Fig. 18.

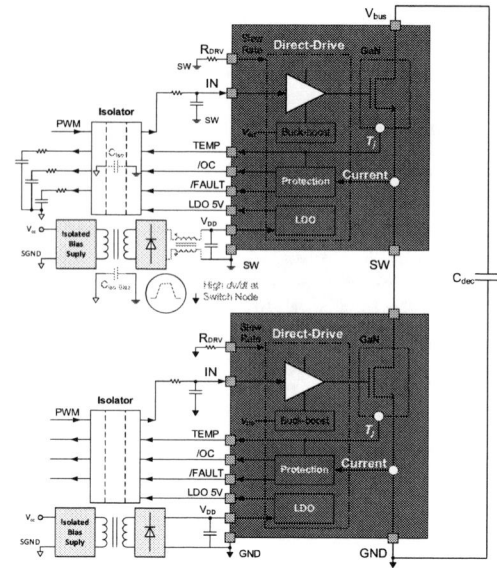

Fig. 14. Circuit diagram for half-bridge with isolated bias power and isolator.

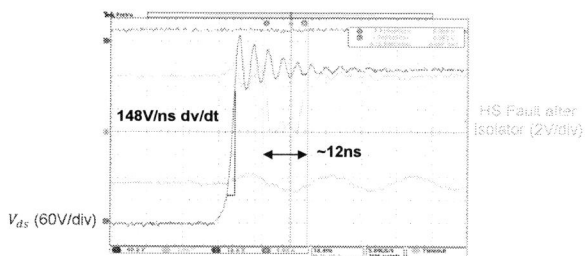

Fig. 15. Isolator glitch due to CMTI issue.

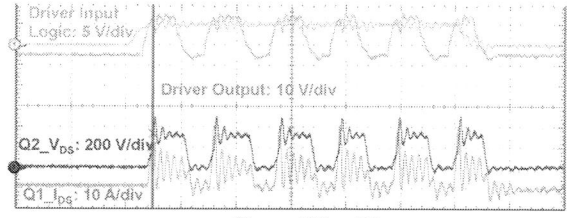

Fig. 16. Example waveform of CMTI failure.

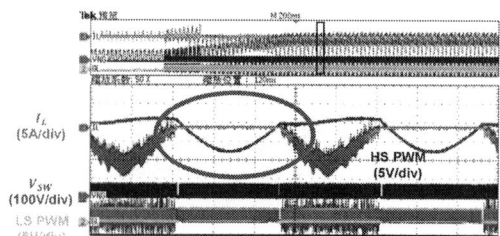

Fig. 17. System malfunction due to capacitive coupling noise from PCB layout.

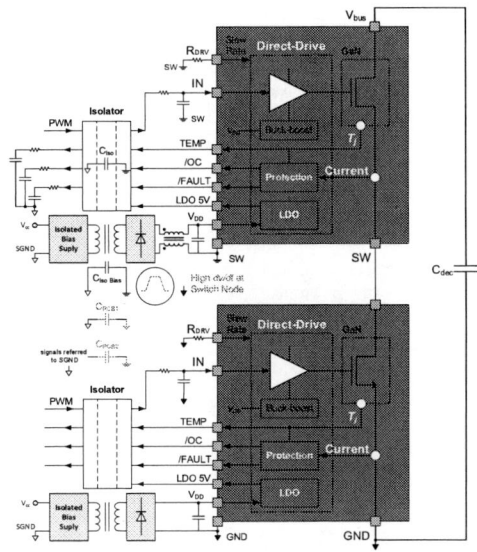

Fig. 18. Eliminating coupling capacitance from switch node to signal side on PCB layout.

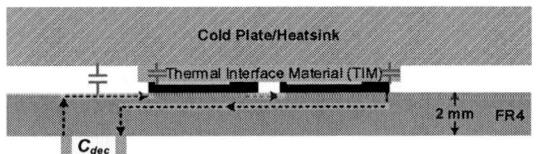

Fig. 19. Capacitive coupling through cooling plane.

B. Capacitive Coupling through Cooling Plane

The capacitance coupling can also occur through the cooling plane, especially for top-side cooling devices shown in Fig. 19. As can be seen, most of the GaN or SiC device's thermal pad is connected to the device's source or drain. In a half-bridge configuration, one of the thermal pads will be connected to the switch-node. Then an insulated thermal interface material (TIM) is inserted in between the device's thermal pad and heatsink to conduct the heat out while providing sufficient insulation. The TIM also forms a capacitive coupling path from the switch node to the cooling plane, and the capacitance value depends on the TIM's thickness and material property. Meanwhile, the cooling plane can also couple to the signal ground through the air or other medium if the cooling plane hangs over the signal ground plane. As a result, a capacitive coupling path is formed in between the switch node and signal ground.

In practical designs, it is suggested to increase the distance from the cooling plane to the PCB's signal plane to reduce the coupling capacitance. Alternatively, it is also a good practice to reduce the top-layer tracings of the PCB where it is directly exposed to the cooling plan.

VIII. SIGNAL SHIELDINGS

In addition to the capacitive coupling, the electromagnetic coupling to signals can also cause system misbehaviors. Fig. 20 shows an example where the PWM signals (at the isolator input) to the GaN half-bridge are distorted during converter operations, and the noises become larger at higher *dv/dt* and *di/dt* operations.

Digging into the layout, it is found that the PWM traces, which travels from the TI C2000 microcontroller to the isolator input of the GaN card, is not well shielded as shown in Fig. 21. Consequently, the PWM signals are affected by the electromagnetic noises when the GaN device starts to operate at high *dv/dt* and high di/dt hard-switching conditions.

To solve the issue, the PWM trace on the PCB is cut, and a dedicated ground-shielded cable is used to deliver the PWM signals from the controller to the isolator input of the GaN devices directly as shown in Fig. 21. With this change, the system can operate normally, and the noises on the PWM signals are eliminated as shown in Fig. 22.

For practical layout design with GaN or SiC devices, it is critical to have the ground shielding to all critical signals including PWM, fault, etc. Fig. 23 shows a good example where the signals to both high-side and low-side GaN devices are well shielded with a dedicated signal ground in the adjacent layer for shielding.

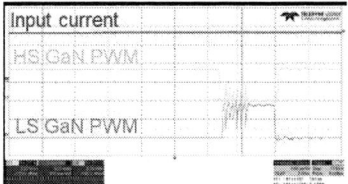

Fig. 20. Capacitive noise coupling issue example.

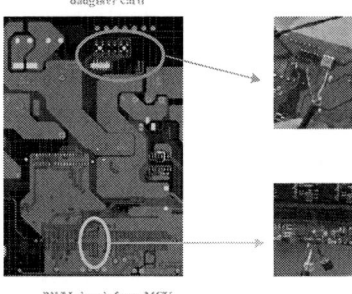

Fig. 21. PCB layout of a design with capacitive noise coupling issue and the solution.

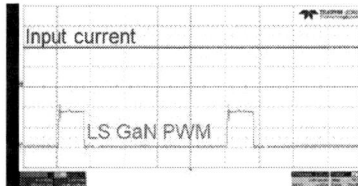

Fig. 22. Shielding PWM helps to mitigate capacitor coupling issue.

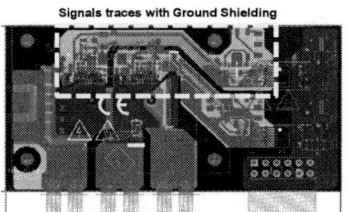

Fig. 23. Ground shielding for all signals.

IX. CONCLUSIONS

This paper summarizes the common mistakes encountered for WBG-based power electronics converter designs. From the lessons learned, designing a power supply with WBG devices needs engineers' extra attention on gate driver design/layout, power loop layout, CMTI rating, and PCB layout.

REFERENCES

[1] D. Reusch and J. Strydom, "Evaluation of Gallium Nitride transistors in high frequency resonant and soft-switching DC–DC converters," in IEEE Transactions on Power Electronics, vol. 30, no. 9, pp. 5151-5158, Sept. 2015.

[2] M. Treu, E. Vecino, M. Pippan, O. Häberlen, G. Curatola, G. Deboy, M. Kutschak, U. Kirchner, "The role of Silicon, Silicon carbide and Gallium Nitride in power electronics," 2012 International Electron Devices Meeting, San Francisco, CA, USA, 2012, pp. 7.1.1-7.1.4.

[3] S. Yu, M. Bhardwaj, G. Wang, and X. Gong, "Designing a high-power bidirectional AC/DC power supply using SiC FETs," Texas Instruments Power Supply Design Seminar SEM2400, 2020. https://www.ti.com/seclit/ml/slup393/slup393.pdf

[4] B. Mondal, R. T. Pogulaguntla and A. K. B, "Double Pulse Test Set-up: Hardware Design and Measurement Guidelines," 2022 IEEE International Conference on Power Electronics, Drives and Energy Systems (PEDES), Jaipur, India, 2022, pp. 1-6.

[5] Xie Y. and Brohlin P., Optimizing GaN performance with an integrated driver, Texas Instruments Whitepaper, 2016.

[6] Brohlin P., Ramadass Y., and Kaya C., Direct-drive configuration for GaN devices, Texas Instruments Whitepaper, 2018.

[7] V. John, Bum-Seok Suh and T. A. Lipo, "Fast-clamped short-circuit protection of IGBT's," in IEEE Transactions on Industry Applications, vol. 35, no. 2, pp. 477-486, March-April 1999

Thermo-Mechanical Analysis of a 650 V/150 A e-GaN HEMT Sandwiched Between a PCB and DBC Substrate

Carl Nicholas[*†], Filip Boshkovski[†‡], Emmanuel Arriola[*†‡], Zichen Zhang[*†] and Guo-Quan Lu[*†‡]

Email: {carl176,bfilip7,erarriola,zichen2013,gqlu}@vt.edu

[*] Bradley Department of Electrical and Computer Engineering, Virginia Tech, Blacksburg, VA 24061, USA

[†] Center for Power Electronics Systems, Virginia Tech, Blacksburg, VA 24061, USA

[‡] Department of Materials Science and Engineering, Virginia Tech, Blacksburg, VA 24061, USA

Abstract—Following successful packaging of a 650 V/150 A e-GaN HEMT between a printed-circuit board and a direct-bond-copper substrate, this study evaluates the thermo-mechanical reliability of the sandwich-style package. Thermo-mechanical fatigue of both soldered and silver-sintered bonds in the package under temperature cycling was studied using ANSYS Mechanical. Deformation properties of eutectic lead-tin solder and sintered silver were modeled after Anand viscoplastic behavior. The inelastic volume-averaged strain energy density accumulated per thermal cycle at each bond was used as a metric of bond fatigue. Simulation results showed that the silver-sintered bond at the gate pad and the solder bond closest to the package center have the highest per-cycle strain energy density. Simulations also evaluated the effect of encapsulation materials on bond fatigue. It was found that using rigid encapsulants, such as one with a coefficient of thermal expansion of 14.9 ppm/°C and an elastic modulus of 17.5 GPa, instead of a commonly used silicone gel reduced the per-cycle bond fatigue by 97%. Findings of this study would help guide the development of high performing and high reliability packaging technology for the 650V/150A GaN HEMT.

Index Terms—Encapsulant, FEA, GaN HEMT, PCB-DBC Hybrid, Simulation, Strain Energy Density, Temperature Cycling, Thermo-mechanical.

I. INTRODUCTION

Gallium nitride (GaN) high-electron-mobility-transistors (HEMTs) are studied for higher power density and higher efficiency inverters and converters in electric drive applications. Device packaging innovations are needed to minimize package parasitic inductances, footprint, and thermal resistance and maximize reliability. Regarding package reliability, temperature is a dominant factor in package failure [1]. The inherent coefficient of thermal expansion (CTE) mismatch of the package components coupled with changes in package temperature causes warpage, which in turn generates mechanical stress in the package [2]. Repeated changes in temperature result in cyclic loading on the bonds that hold the package together, culminating in fatigue failure of those bonds. One method of measuring bond fatigue is with inelastic volume-averaged strain energy density accumulated per thermal cycle, abbreviated as per-cycle fatigue, which measures the average inelastic energy applied to a body over a single temperature cycle. Reference [3] uses per-cycle fatigue to study $Sn_{63}Pb_{Pb37}$ bonds, while [4] studies sintered-silver (Ag) bonds.

A previous study reported a packaging approach for a 150 A, 650 V e-GaN HEMT that offered low parasitic inductances, small footprint, and low junction-to-case thermal resistance [5]. The approach involves Ag-sintering the die on a direct-bond-copper (DBC) substrate and placing a printed-circuit-board (PCB) on top of the device for the external electrical connections. Pins provide the internal electrical connection between the die and the PCB, with the pins being soldered to the PCB and Ag-sintered to the die. The volume separating the DBC and the PCB was filled with polymer encapsulant.

In this study, we focus on evaluating and improving the thermo-mechanical reliability of the package using simulations. We used ANSYS Mechanical, a finite element analysis (FEA) software, to simulate the package deformation under temperature cycling (TC) and find the inelastic volume-averaged strain energy density accumulated per thermal cycle (ΔW_{avg}), or per-cycle bond fatigue, at each bonded interface. As the bond with the greatest ΔW_{avg} within its bond group is predicted to fail first, this method predicts the locations of first failure. Improving the package reliability is focused on reducing the ΔW_{avg} of these first failure bonds. The method of improvement was altering the CTE and elastic modulus of the polymer encapsulant.

II. SIMULATION METHODOLOGY

Fig. 1 describes the methodology for creating and evaluating the simulation. The simulation purpose, CAD model, simulation settings, independent variable, and dependent variables are discussed.

A. Define Type and Intent of Simulation

The intent of this work is to identify which package bonds are likely points of initial failure under TC, and to simulate how the CTE and elastic modulus of the polymer encapsulation

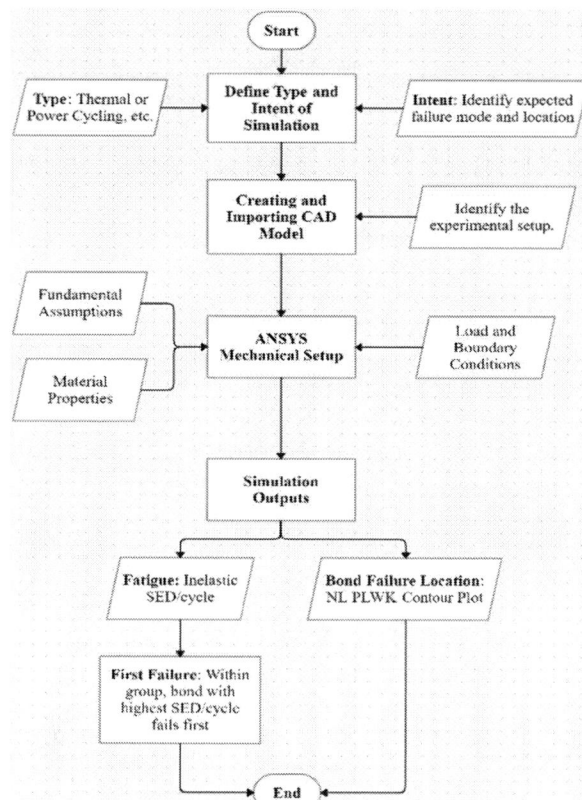

Fig. 1. Flowchart depicting the steps to create and analyze a thermo-mechanical simulation.

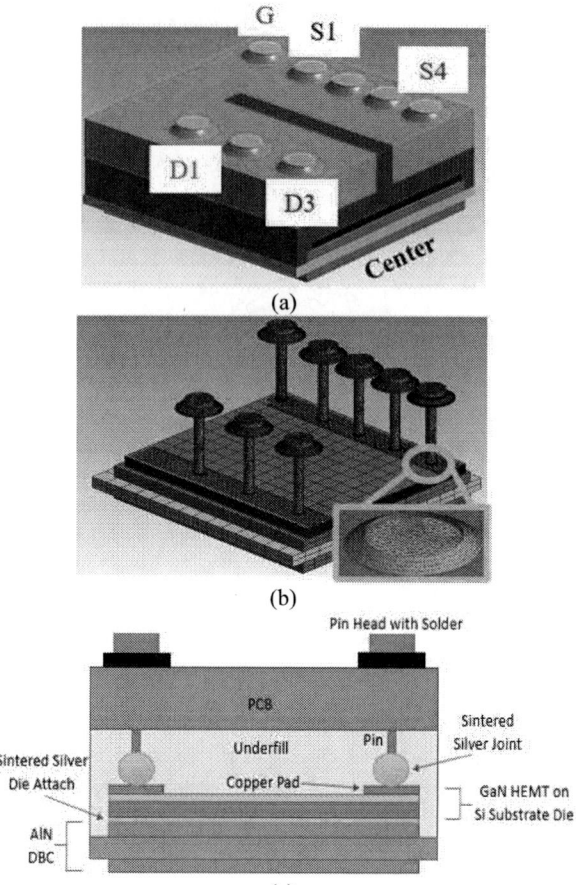

Fig. 2. The package CAD model. (a) Isometric view of package with the gate, source, and drain pins of interest highlighted. (b) Same isometric view with the EMC and PCB hidden. The Source 4 sintered silver bond is also shown (c) A cross-section of the package describing each material layer, along with the black solder bond holding the pin to the PCB, and the grey sintered silver bond holding the pin to the die.

alters those likely failure points. The bonds evaluated are the sintered Ag die attach corners, the sintered Ag bonds that hold the pin interconnects to the die, and the solder bonds holding the pin interconnects to the PCB. The independent variable is the polymer encapsulant, which controls the encapsulant's CTE and elastic modulus. The dependent variables are contour plots of the inelastic work density accumulated in these bonds, and the bond's ΔW_{avg}. Once the initial failure points are identified, which are the bonds possessing the largest ΔW_{avg} of their material group, the effect of the encapsulant material properties can be evaluated against the predicted first failure points.

B. Creating the CAD Model

The CAD model and material layers are displayed in Fig. 2. The pins of interest and their corresponding bonds are highlighted for later analysis. The die passivation, GaN active regions, and metallization are not included. Due to package symmetry, a half model it utilized.

C. Material Models

All materials, except the solder and sintered Ag, are modeled as linearly elastic and isotropic, with material properties in Table I. All materials are temperature dependent except for the encapsulant and FR4 due to a lack of available material data. The solder and sintered Ag are modeled using both the isotropic, linearly elastic model and the Anand viscoplastic constitutive model. Prebuilt into ANSYS, the Anand model describes both the isotropic plastic and creep components of deformation through a set of flow and governing equations. The Anand constants for sintered Ag and Sn60Pb40 are described in [6] and [7]. For this paper, the definition of *inelastic* is identical to *plastic* in ANSYS Mechanical and *viscoplastic* in the Anand constitutive model [8].

D. Load and Boundary Conditions

The thermal load is a Uniform Body Temperature (UBT) condition that controls the temperature of all bodies in the package. Shown in Fig. 3, the UBT simulates the package undergoing four temperature cycles between -40 °C and 125 °C. The thermal cycling parameters are chosen in accordance with JESD22-A104F. Four cycles are modeled to achieve ΔW_{avg} convergence, while the stress-free temperature was chosen to be room temperature. Thermal transients are considered negligible due to the gradual ramp rate intended to remove thermal transients.

TABLE I.
ISOTROPIC, LINEARLY ELASTIC MATERIAL PROPERTIES USED BETWEEN -40 °C AND 125 °C

Material	Elastic Modulus (GPa)	CTE (ppm/°C)	Poisson's Ratio (unitless)
Copper	129 – 119	16.2 – 17.4	0.35
Aluminum Nitride	341 – 338	4.8	0.27
Sintered Silver	9.01 – 2.64	19.6	0.37
Silicon	163 – 162	2.14 – 2.92	0.275
Silicon Nitride	314 – 312	3.88 – 4.09	0.26
Gold	77 – 74	13.5 – 14.2	0.425
Nickel	208 – 200	11.9 – 13.3	0.315
FR4	24.6	15.5	0.136
Sn60Pb40 Solder	30	24	0.4
Silicone Gel	0.0175	80	0.35 (Estimated)
ME-531 Encapsulant	6	21	0.35 (Estimated)
EP-2000 Encapsulant	17.5	14.9	0.35 (Estimated)

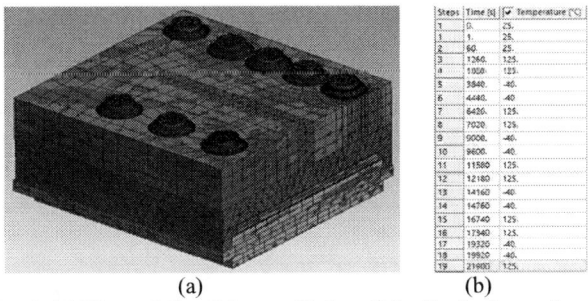

Fig. 3. (a) Thermal Condition applied to all bodies in the package. (b) The temperature profile used.

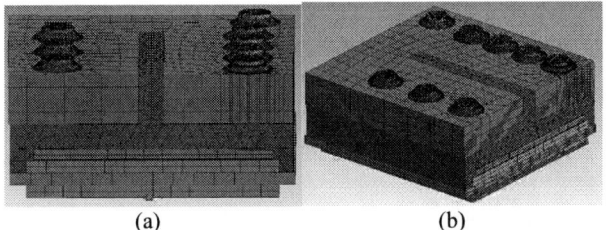

Fig. 4. (a) The fixed support boundary condition applied to the three nodes at the bottom center of the cut plane. (b) The frictionless support applied to the cut plane.

The mechanical boundary conditions are described in Fig. 4. Three fixed nodes on the bottom of the half model cut plane limit the package movement to warpage, while the frictionless support prevents the half model from expanding into the space occupied by the unmodeled half of the package.

E. Fundamental Assumptions

These are remaining assumptions on how the model operated, and thus deviate from experimental work:

- all bodies possess infinite adhesive and cohesive strength;
- all bodies behave uniformly and in accordance with their material data;
- no flaws are modeled or can form;
- PCB orthotropy is not modeled;
- large deflection is enabled.

F. Simulation Independent Variable

The independent variable is the polymer encapsulant that encapsulates the sintered Ag bonds and pins. The three encapsulants chosen for this study are a generic silicone gel, Parker Lord's CoolTherm® ME-531, and Parker Lord's CoolTherm® EP-2000. Their material properties are listed in Table I. Silicone gel has a significant CTE mismatch with the package components, which would normally increase the thermally induced warpage and resulting stress. However, the gel has a low elastic modulus, and therefore exerts minimal influence on the package behavior and bond fatigue. Both ME-531 and EP-2000 are epoxy resins and are therefore stiffer than silicone. Due to their lower CTE mismatch, these rigid encapsulants can instead constrain warpage and reduce the stresses placed on bonds, increasing the cycles to failure [9].

Between the two epoxies, EP-2000 has both the greater elastic modulus and lower CTE mismatch with most package components. Therefore, it is hypothesized that the package bonds will have the lowest ΔW_{avg} when using EP-2000, with ME-531 having a larger but comparable ΔW_{avg}, and silicone gel having a much larger ΔW_{avg}.

G. Simulation Outputs

The two outputs evaluated in this work are the bond's ΔW_{avg}, and the inelastic strain energy density (SED) contour plots.

ΔW_{avg} is a normalized measurement of how much inelastic energy is used to deform the bond per thermal cycle. This value is calculated from the area inside of the stress-strain hysteresis loops at every node within the bond, then averaged across the bond's volume. The post processing code used to generate ΔW_{avg} is from [10]. ΔW_{avg} is the basis of Darveaux's model, which calculates the cycles to crack initiation using ΔW_{avg} and experimentally determined constants. If K_1 and K_2 are positive, experimentally determined constants and N_0 is the cycles until crack initiation, then Darveaux's model is

$$N_0 = K_1 \left(\Delta W_{avg} \right)^{-K_2}. \tag{1}$$

The bond with the largest ΔW_{avg} within its bond group is therefore likely to fail first, while design changes that reduce this maximum ΔW_{avg} will increase the cycles until fatigue cracks form. Only bonds with similar volumes and identical materials can be meaningfully compared, hence why bond groups are used. However, until experimental work determines K_1 and K_2, Darveaux's model cannot predict the cycles to failure of the simulated bonds.

Inelastic SED contour plots identify which bond region accumulates the largest amount of inelastic work. This identifies the cause of loading and the likely location of crack initiation.

III. RESULTS AND DISCUSSION

As the package bonds are comprised of two different materials, there will be a solder and a sintered Ag bond that will fail before all other bonds of the same material. As these bonds can't be directly compared, we are unable to determine which one will fail first. The sintered Ag bonds being studied are the Gate bond, Source 1 bond, Source 4, Drain 1 bond, and Drain 3 bond. These are the bonds that hold the pins to the gate, source, and drain pads of the die. The outermost corner of the sintered Ag die attach (DA) is also evaluated, as it has both the same material and similar volume as the previous mentioned bonds. The solder bonds being studied are also the Gate, Source 1, Source 4, Drain 1, and Drain 3 bonds. These are the bonds that hold the pins to the PCB. All bonds except the DA are highlighted in Fig. 2.

Fig. 5 contains the simulated ΔW_{avg} values for all sintered Ag bonds. The sintered Ag bond expected to fail first in all scenarios is the bond connecting the gate pin to the gate pad. As the most external pin bond, the gate bond is subjected to the greatest shear stresses from warpage, which is also observed in ball grid array (BGA) solder ball studies. As hypothesized, the encapsulant with the lowest maximum ΔW_{avg} is EP-2000, while silicone gel has the highest maximum ΔW_{avg} value. Therefore, EP-2000 is the most optimal encapsulant tested for protecting the sintered Ag Gate bond.

Fig. 6 contains the simulated ΔW_{avg} values for all the solder bonds. The solder bond expected to fail first in all scenarios is the bond connecting the Source 4 pin to the PCB, which is the bond closest to the package center. This is opposite of what is expected in BGA packaging, but whether such comparisons apply to the unique geometry of the solder bonds is uncertain. As hypothesized, the encapsulant with the lowest maximum ΔW_{avg} is EP-2000, while silicone gel has the highest maximum ΔW_{avg} value. Therefore, EP-2000 is also the most optimal encapsulant tested for protecting the Source 4 solder bond.

Fig. 7 displays the inelastic SED contour plots of both the Source 4 solder bond and the Gate sintered Ag bond when using silicone gel. For the solder bond, the highest inelastic work

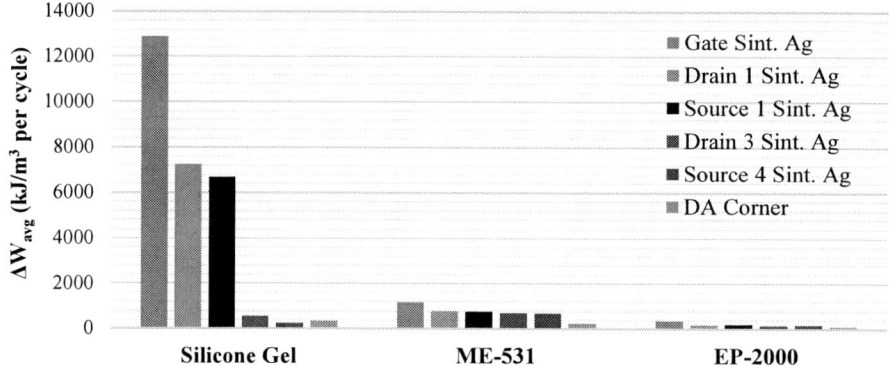

Fig. 5. Simulated ΔW_{avg} of the sintered silver pin bonds and the comparatively sized die attach corner for the three encapsulants. When silicone gel is replaced by EP-2000, the Gate bond's ΔW_{avg} decreases by 97.1 %.

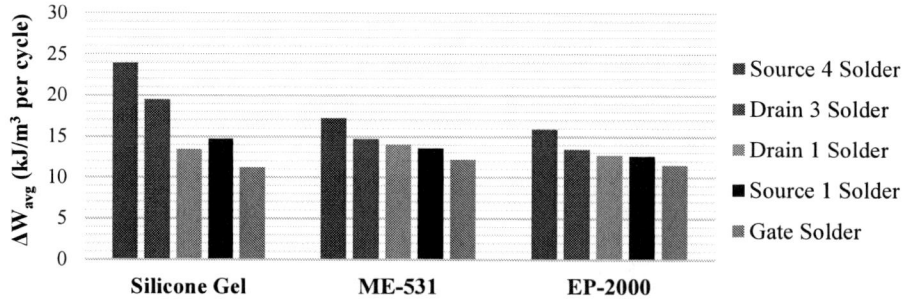

Fig. 6. Simulated ΔW_{avg} of the solder bonds for the three encapsulants. When silicone gel is replaced by EP-2000, the Source 4 bond's ΔW_{avg} decreases by 33.4 %.

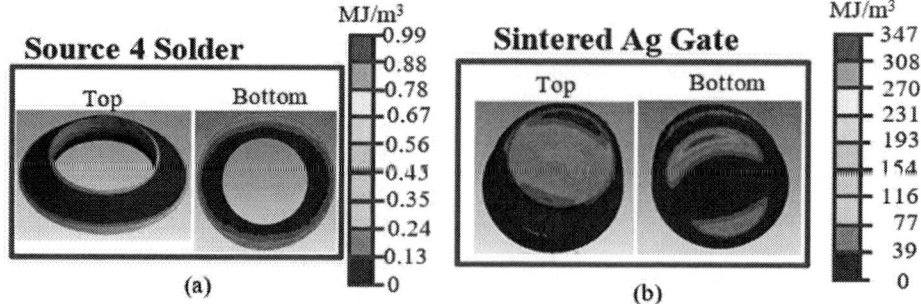

Fig. 7. Inelastic strain energy density contour plots with 30x deformation scale. (a) Source 4 solder bond. (b) Sintered Ag gate bond.

region occurs at the bond edges along the primary axis of the package. During package heating and cooling, the PCB and encapsulant expansion pushes and pulls the Source 4 pin along the long axis of the package during heating and cooling respectively, inelastically deforming the bond with both shear and normal stresses.

For the sintered Ag gate bond, the highest inelastic work region is generated on the base of the bond, angled towards the package corner. During temperature cycling, the PCB and encapsulant push the Gate pin towards the package corner during heating, inelastically deforming the bond through primarily shear stresses.

While it cannot be determined which of the two bonds will fail first by simulation alone, we have established the two likely points of bond fatigue failure in the package, with gate failure leading to complete package failure. Using EP-2000 reduces the per-cycle fatigue placed on both structures, suggesting that a real package with EP-2000 encapsulant would survive a greater number of experimental temperature cycles compared to a package using ME-531 or silicone gel.

V. CONCLUSION

In this work, thermo-mechanical FEA simulations were used to determine the likely points of initial failure in a novel GaN HEMT package. The likely points of failure, defined as the bonds with the highest ΔW_{avg} among their bond group, are the solder bond that connects the Source 4 pin to the PCB and the sintered Ag bond that connects to the gate pin to die. Of these two bonds, the sintered Ag Gate bond is more sensitive to the choice of encapsulant and possesses the lowest ΔW_{avg} when encased in the encapsulant with the highest elastic modulus and lowest CTE of those tested. As the Source 4 solder bond is not in contact with the encapsulant, it is less sensitive to the encapsulant choice, but also possesses the lowest ΔW_{avg} when the lowest CTE, highest elastic modulus encapsulant is used.

The continuation of this work is to perform experimental temperature cycling on these packages to verify the first failure bond and the optimal encapsulant to use in the package. Experimental work will also determine the Darveaux model constants, which can be utilized alongside the verified simulation to predict the cycles to bond failure of any redesigns to the PCB-DBC sandwich package.

ACKNOWLEDGMENT

The authors wish to thank Paul Paret of the National Renewable Energy Laboratory for sharing his knowledge on thermo-mechanical fatigue simulations.

This work was supported by the U.S. Department of Energy Advanced Manufacturing Office through the Wide Bandgap Generation (WBGen) Fellowship at the Center for Power Electronics Systems (CPES) at Virginia Tech, and the CPES High Density Integration Industry Consortium at Virginia Tech.

REFERENCES

[1] Fundamentals of Microsystems Packaging. First ed, ed. R. Tummala. 2001, New York: McGraw-Hill Professional Publishing.

[2] Fundamentals of Thermo-Mechanical Reliability, in Fundamentals of Device and Systems Packaging: Technologies and Applications, Second Edition, R. Tummala, Editor. 2019, McGraw Hill: New York. p. 137-169.

[3] P. P. Paret, D. J. DeVoto and S. Narumanchi, "Reliability of Emerging Bonded Interface Materials for Large-Area Attachments," in *IEEE Transactions on Components, Packaging and Manufacturing Technology*, vol. 6, no. 1, pp. 40-49, Jan. 2016, doi: 10.1109/TCPMT.2015.2499767.

[4] P. Paret, J. Major, D. DeVoto, S. Narumanchi, C. Ding and G. -Q. Lu, "Reliability and Lifetime Prediction Model of Sintered Silver Under High-Temperature Cycling," in *IEEE Journal of Emerging and Selected Topics in Power Electronics*, vol. 10, no. 5, pp. 5181-5191, Oct. 2022, doi: 10.1109/JESTPE.2021.3121195.

[5] S. Lu, T. Zhao, R. Burgos and G. -Q. Lu, "Packaging of (650 V, 150 A) GaN HEMT with Low Parasitics and High Thermal Performance," *2021 International Conference on Electronics Packaging (ICEP)*, Tokyo, Japan, 2021, pp. 39-40, doi: 10.23919/ICEP51988.2021.9451865.

[6] D.-j. Yu, X. Chen, G. Chen, G.-q. Lu, and Z.-q. Wang, "Applying Anand model to low-temperature sintered nanoscale silver paste chip attachment," *Materials & Design*, vol. 30, no. 10, pp. 4574-4579, 2009/12/01/ 2009, doi: https://doi.org/10.1016/j.matdes.2009.04.006.

[7] G. Z. Wang, Z. N. Cheng, K. Becker, and J. Wilde, "Applying Anand Model to Represent the Viscoplastic Deformation Behavior of Solder Alloys," *Journal of Electronic Packaging*, vol. 123, no. 3, pp. 247-253, 1998, doi: 10.1115/1.1371781.

[8] S. B. Brown, K. H. Kim, and L. Anand, "An internal variable constitutive model for hot working of metals," *International Journal of Plasticity*, vol. 5, no. 2, pp. 95-130, 1989/01/01/ 1989, doi: https://doi.org/10.1016/0749-6419(89)90025-9.

[9] W. Wang and T. Nguyen, "A Modeling Study of the Effect of Underfill Materials on Solder Joint Thermal Fatigue of Ball Grid Array Package," presented at the ASME 2014 International Mechanical Engineering Congress and Exposition, 2014. [Online]. Available: https://doi.org/10.1115/IMECE2014-38889.

[10] C. Nicholas, "Thermo-mechanical Analysis of a Custom PCB-DBC Hybrid Package for a (650 V, 150 A) e-GaN HEMT," M.S. thesis, Dept. Elect. Eng., Virginia Tech, Blacksburg, VA, USA, 2023

Avalanche Capability of SiC MOSFET Under High Current

Xuning Zhang, Ehab Tarmoom, Ali Shahabi, Linda Starr and Dennis Meyer
Microchip Inc.
Austin, TX, USA
xuning.zhang@microchip.com

Abstract—This work discusses the avalanche capability of SiC devices under high current conditions. Traditional avalanche capability of SiC MOSFETs is measured with low current and high inductance for avalanche energy. However, this paper focuses on the avalanche capability and failure modes of SiC MOSFETs under high current. An avalanche testing evaluation system was developed with modular implementation. SiC MOSFETs are tested with low inductance and high current for avalanche current capabilities. They are also tested with high inductance and low current for avalanche energy capability. Failure analysis results are presented for two different failure cases. For low inductance high current failures, device fails due to the high energy on the device. For low inductance high current failures, a possible failure mode is proposed, with device turned off with high current through devices, device fails due to high voltage stress and high electrical field since the JFET region of SiC device is saturated and high electric filed is exposed to the gate oxide region.

Keywords—Silicon Carbide, SiC MOSFET, Avalanche, high current.

I. INTRODUCTION

Silicon carbide (SiC) MOSFETs feature lower on-state losses and leakage current, faster switching speeds, smaller size, and higher temperature capability than silicon (Si) power transistors. [1] These features make SiC MOSFETs idea candidates for high voltage high power applications. Recently SiC MOSFETs have received wide adoption on EV traction, on-board charger, off-board charger, solar inverter, and energy storage applications. Moreover, due to the fast development of high voltage DC and AC source converters, the use of SiC in protection applications such as electronic fuse (E-fuse), solid state circuit breaker (SSCB) and solid-state relay (SSR) are also increasing. Many of these applications require high ruggedness performances such as good avalanche energy and long short circuit withstand time. The static and dynamic characteristics of SiC semiconductors are well studied, However, the ruggedness of these devices is still a major research area. A standard approach for assessing the ruggedness of power semiconductors is through an avalanche rating. Previous literature focuses on the avalanche energy analysis where the avalanche capability is tested under a high inductance and low current(HLLI) conditions for total energy capability. [2-8] However, in many applications such as the E-fuse applications, device may enter avalanche condition after switching due to the parasitic inductance in the system, when the current through device can be very high, the failure mode

under low inductance high current (LLHI) can be different from the low current conditions, this work focuses the discussions on the avalanche capability of SiC devices under such high current conditions.

II. AVALANCHE FAILUR MECHNISIUM

Figure 1 shows a structure of a SiC MOSFET with planar gate structure. When SiC MOSFET is in OFF mode, MOSFET channel is completely OFF, if current needs to flow through the device, it has to go through the body diode. During avalanche conditions, devices can fail due to intrinsic thermal limitations. This limit is reached when the number of thermally generated carriers becomes equal to the background doping concentration. For SiC devices, the band gap is wider and the intrinsic carrier concentration is lower than Si devices, therefore it has a much higher intrinsic thermal limit than Si devices, The intrinsic temperature limit for SiC with a doping concentration of 10^{16} cm^{-3} is about 1270 °C and the device will more likely fail due to metallization, gate oxide or package failure before even reaching intrinsic temperature failure. melting of the source metallization which could be attributed to current filamentation due to individual cell failure [7-11]. In this failure mode the ultimate behaviour of the damaged device will show some short connections between drain, source, and gate.

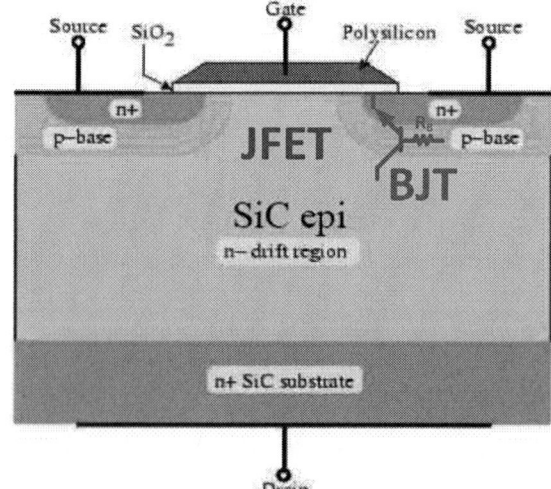

Figure 1. SiC MOSFET structure with planar gate

979-8-3503-3714-3/23 $31.00 © 2023 IEEE

For Si MOSFETs, another common failure mode during avalanche operation is latch-up of the parasitic BJT. During avalanche, the high dv/dt across the drain-to-body capacitance of the MOSFET generates a displacement current. This current flows through the resistance between the body and source of the MOSFET, then produces a voltage drop across the emitter-base junction of the parasitic BJT. If this voltage (determined by the resistance and displacement current) surpasses the built-in potential of the material, it can turn on the BJT. The BJT conduction causes the temperature to increase, which increases the body resistance and reduces the built-in potential. This positive feedback can lead to thermal runaway and ultimately destruction of the semiconductor. For SiC MOSFETs, this type of failure is not as prevalent due to the wider band gap, which results in a greater built-in potential, and the low gain of the parasitic BJT. These features suggest that a larger displacement current would be needed to turn on the BJT. However, it is possible for non-uniformities to exist among the many MOSFET cells, which could result in current crowding and eventually BJT latch-up under high current conditions. Further, SiC MOSFETs switches faster which results in a higher dv/dt and it could produce a current across the body resistance that is sufficient to turn on the BJT[1]. Like the intrinsic thermal failure mode, in this failure mode, the ultimate behaviour of the damaged device will show some short connections between drain, source and gate.

The third failure mode is related to the gate oxide of the SiC MOSFET. This failure mode is a little different between planar and trench gate SiC MOSFETs. In planar gate structures, there is an intrinsic JFET region under the gate of the device as shown in Figure 1. In normal operation, JFET region can shield the high electric field of the devices and help to protect device gate oxide. when device enter avalanche mode when device current is very high such as the protection of E-fuse when system detect over current conditions, the JEFT region can be saturated by the high current and then the gate oxide will be exposed to high temperature and high electric filed which led to gate oxide degradation. In trench gate MOSFETs. the JFET region can be eliminated, or it can be much thinner than planar gate design. In addition, the gate corners will increase electric field. The gate oxide corners are simultaneously exposed to the high electric fields and high temperatures, which degrades the reliability of the gate oxide. As the most vulnerable position inside the trench MOSFET, the gate oxide corner may be firstly destroyed under the avalanche condition, resulting in an avalanche failure. It needs to be note that in this failure mode, when device gate oxide is degraded, the behaviour of the damaged device will first show a gate leakage increase. Before the leaky gate induce secondary thermal damage, if the gate is still turned off with strong negative voltage, the device can still block voltages. No short connections will be measured.

III. AVALANCHE TESTING PLATFORM

In order to measure the avalanche behavior of SiC MOSFETs, a modular avalanche test platform is developed that allows to measure the avalanche energy of a SiC device under both high inductance low current and low inductance

high current conditions. The schematic of the setup is shown in Figure 2 and the actual test hardware is shown in Figure 3.

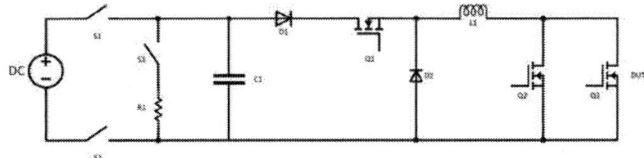

Figure 2. Avalanche test setup schematic

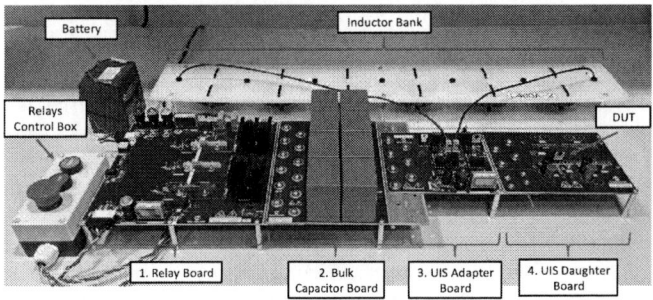

Figure 3. Avalanche test setup hardware

The basic circuit for avalanche testing is the unclamped inductive switching (UIS) test. The testing setup features a modular design and consists of several components. A DC power supply is used to charge capacitor bank C1 through the S1 and S2 charging relays. These relays are closed during charging and then opened to disconnect the setup from the power supply, ensuring safety in case of a fault as well as eliminating ground loops that may affect measurements. Relay S3 is used to discharge C1 when necessary for safety concerns. The energy stored in C1 is then used to charge inductor L1 by activating MOSFETs Q1 and Q2. Q3 is device under test (DUT). Using Q2 instead of Q3 to charge L1 eliminates the possibility of DUT self-heating. When the L1 current or stored energy reaches the target value, MOSFET Q1 is turned off, causing the inductor current to flow through the path comprising Q2 and D2 for a brief period of T2. After this time, MOSFET Q2 is also turned off, interrupting the inductor current and generating a high voltage across Drain-Source of both Q2 and Q3 MOSFETs. In this configuration, Q2 has been selected from a device with a much higher breakdown voltage than Q3, forcing Q3 into the avalanche breakdown voltage without Q2 experiencing an avalanche. During the test, Q3 biased to a negative gate voltage and kept turned off to prevent self-heating, which can affect the avalanche characterization results. This allows for the characterization of the DUT's avalanche energy and capability. This setup has multiple benefits, including: 1) DUT Q3 does not have any self-heating before it enters avalanche conditions allows us to control the starting junction temperature of DUT precisely when needed. 2) Q3 is driven with a negative voltage source (battery powered) with a online monitoring of leakage current, if DUT shows any leaky gate, the setup can immediately monitor the change without taking out of the devices; 3) the total energy during the test is limited to the total energy of the inductor, even if there is a short circuit due to device failure, the energy is controlled and the DUT is protected from a real catastrophic failure which allow the future failure analysis (FA) to give much better results. 4) DUT Q3 is very easy to change to other packages and other devices with all setups kept the same to allow single variable testing for fair comparison.; 5) the load inductor L1 is easily changeable between high inductance and low inductance for different test targets.

With this setup, it ensures a fair and consistent test comparison among different test conditions and different devices.

IV. AVALANCHE TESTING RESULTS

For avalanche testing, since each device is tested to failure, therefore it is not feasible to have the same device tested for all different test conditions to compare the results and find the correlations. When testing different devices, the devices' performance cannot be identical to each other's performance due to the process variation and tolerance. But having all devices in the same lot can help to reduce the variation due to device fabrication. In order to understand the device-to-device variation, five devices from the same lot in the same package is selected for each test to present the trends of impacts of different test conditions. Figure 4 and Figure 5 present the example waveforms of device pass and fail the avalanche test with low inductance and high current. When a device passes the avalanche test, the current decreases from staring current to zero with device voltage rise to the breakdown voltage. the pass energy can be measured by integrating the voltage and current during the whole period All the energy is dissipated on the device. When the device fails the avalanche test, the current through device will see an abrupt increase when device fails. Then the failure energy can also be measured by integrating the voltage and current during the period when device is not damaged.

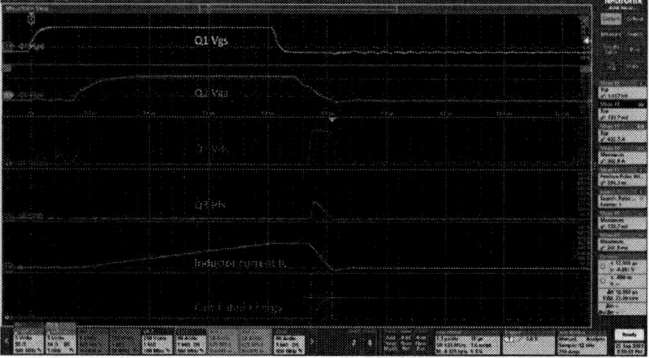

Figure 4. Example waveform of device pass avalanche test with low inductance and high current.

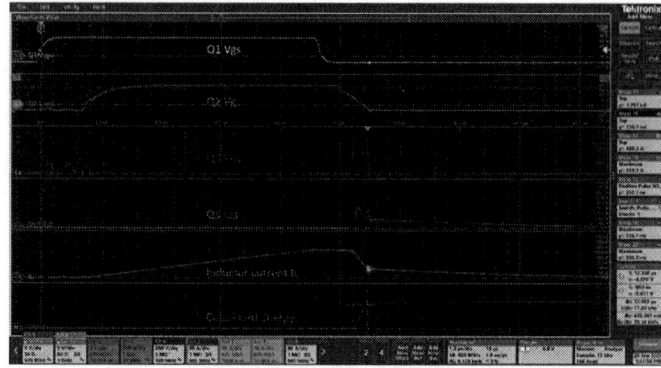

Figure 5. Example waveform of device fails avalanche test with low inductance and high current

Figure 6 presents the test results of single pulse test to failure results with 5.3mH inductor (inductance measured with no load current), the inductor is charged to 40A with 2.2J energy stored in the inductor. The 1200V 80mohm devices (MSC080SMA120B) show a good consistency of avalanche energy with 738mJ. Because the inductor was charged with a relatively high energy, all devices fail before inductor current reaches zero after the 1st test. All devices show a drain-source-gate short connection failure after the test.

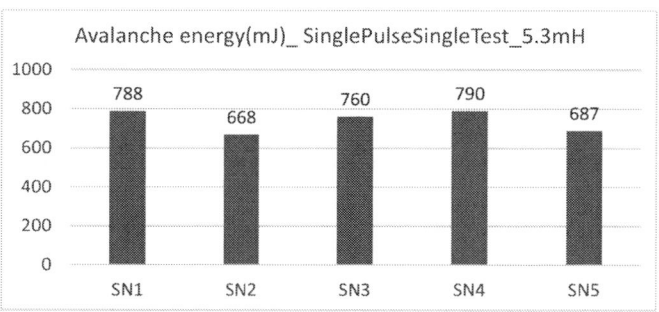

Figure 6. 5.3mH low current single pulse single test avalanche energy.

To mimic the high current low inductance conditions when SiC device need to cut the circuit during an over current condition in E-fuse applications, another five 1200V 80mohm devices (MSC080SMA120B) from the same lot number were tested with a 3.9uH air core inductor. In this test, the device was tested with incremental increase of the inductor current of 20A, all devices were tested with a starting current of 264A and if the device survived the test, then the device will be tested at 284A, if the device survived again, then it will be tested at high current, until the device fails. In the test the highest pass energy/current and highest fail current. Energy is captured. Figures 7 and 8 show that Pass/fail energy and failed peak current.

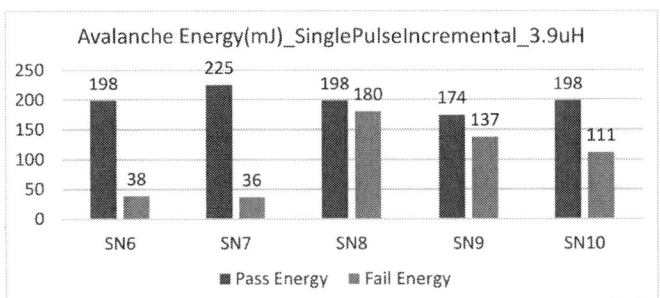

Figure 7. 3.9uH inductance high current single pulse incremental test avalanche energy

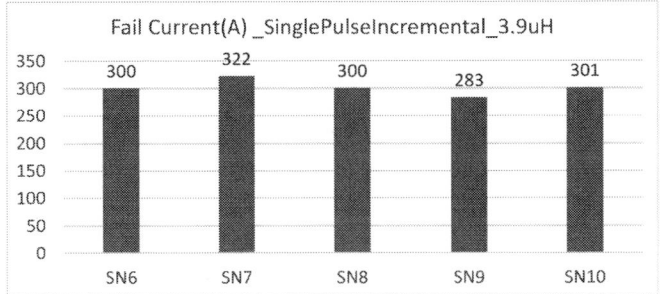

Figure 8. 3.9uH inductance high current single pulse incremental test failure current.

From these results, it can be noted that, when device pass the high current test, device can survive around 200mJ avalanche energy, but when current further increased, the avalanche failure energy reduced significantly. Current has a big impact on the failure energy. Meanwhile, this failure energy is much lower than the high inductance low current failure energy. This is due to the high current during the avalanche starting time along with high breakdown voltage, device will see much higher instant power on the die. Moreover, the high current will also impact and reduce the failure energy significantly. In this test all the devices show a drain-source-gate short connection failure after the failed test and device show no leakage or degradation after the pass test.

To further study the impact of high current on avalanche energy, another five 1200V 80mohm devices (MSC080SMA120B) in from the same lot number are tested with a 1.9uH air core inductor. The devices were tested with incremental increase of the inductor current of 20A, Figure 9 and 10 shows that Pass/fail energy and failed peak current.

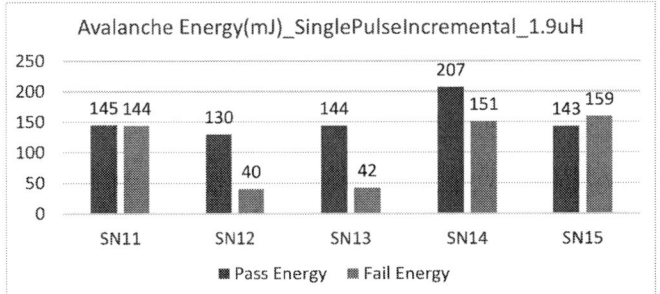

Figure 9. 1.9uH inductance high current single pulse incremental test avalanche energy

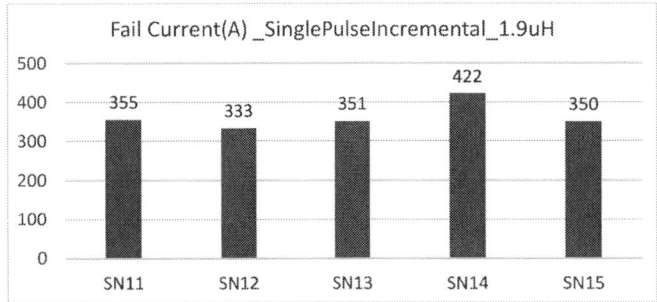

Figure 10. 1.9uH inductance high current single pulse incremental test failure current.

From these results, it can be noted that the failure current is higher than the 3.9uH test results and the pass energy is smaller than the 3.9uH test results. The higher current further reduced the avalanche energy capability of the device. Also in this test, there is one device failure with a leaky gate. When the device passes the previous current avalanche test, the device show no leakage current on the gate with the in circuit gate leakage current monitoring in the setup. But when current increases, after the next test, device shows a much higher gate leakage current, but no short connection among drain, gate, and source.

To further study the gate leakage failure. Another two devices from the same lot number were tested under 100 pulse repetitive testing conditions. 1st device was tested with 1.9uH load inductor starting with 250A current and tested for avalanche with 100 pulses repetitively without any failure, then current was increased to 300A, DUT endured 7 UIS pulses, at the 8th pulse, the gate suddenly became very leaky (read current=7.8uA) but the DS & GS were not shorted yet. The avalanche energy at this point was 121mJ. The 2nd device was tested with 1.9uH load inductor starting with 300A current, DUT endured 6 pulses at pulse 8, the gate showed slight leakage (read current=0.12uA). At pulse 9, the gate became very leaky (read current=12uA). Soon after this test, the readable gate leakage was maxed. The Repetitive testing was continued at the same current level up to 100th UIS pulse. During each test, the device endured 120mJ avalanche energy. The devices drain source and gate were not shorted after 100 UIS tests. Next, the current increased to 350A and repetitive testing was also repeated for this point. DUT with leaky gate passed 10 repetitive pulses. At pulse 11 of this current level, device drain source and gate were finally shorted. The failed test energy was 159mJ.

The test results of the last two devices indicated that under low inductance high current conditions, there is device gate leakage current increase failure that were not observed in the high inductance low current tests. Under certain high current conditions, the device gate becomes very leaky but it can still pass avalanche test at 120mJ. Further increase load current will finally create a drain source gate short failure.

Failure analysis was also performed on certain devices after the avalanche test, since the test setup was designed to control the maximum energy on the die every after a short connection failure the device die remains a good condition for FA analysis. Figures 11 and 12 present the FA images of the die under high inductance low current and low inductance high current conditions.

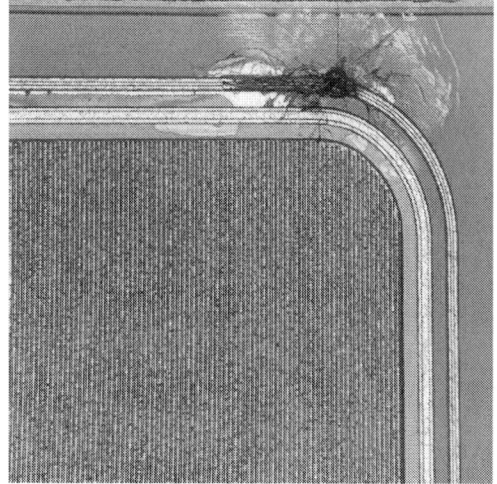

Figure 11. Failure analysis image of high inductance low current avalanche failure.

Figure 12 Failure analysis image of low inductance high current avalanche failure.

CONCLUSIONS

This work discusses the avalanche capability of SiC devices under high current conditions. An modular avalanche testing evaluation system was developed with multiple benefits to ensures a fair and consistent test comparison among different test conditions and different devices . SiC MOSFETs are tested with high inductance low current(HLLI) and low inductance high current(LLHI) unclamped inductive switching (UIS) test for avalanche energy comparison. Testing results indicated a big difference in avalanche energy between low

inductance high current and high inductance low current testing conditions. In LLHI testing conditions, the device shows a high gate leakage current failure which was not observed in HLLI test. For low inductance high current failures, device fails due to the high energy on the device. For low inductance high current failures, a possible failure mode is proposed, with device turned off with high current through devices, device fails due to high voltage stress and high electrical field since the JFET region of SiC device is saturated and high electric filed is exposed to the gate oxide region. Failure analysis results are presented for two different failure cases. Future work is ongoing to analyze the FA results with device design to improve device avalanche capability under high current conditions for E-fuse for solid state circuit breaker applications.

REFERENCES

[1] C. DiMarino and B. Hull, "Characterization and prediction of the avalanche performance of 1.2 kV SiC MOSFETs," 2015 IEEE WiPDA, 2015, pp. 263-267

[2] Jake Choi, Kwangwon Lee, Daiwon Kim, Lieyi Sheng, "Investigations of 900V 4H-SiC Planar Power MOSFET for More Robust Reliability Performance", 2023 IEEE International Symposium on the Physical and Failure Analysis of Integrated Circuits (IPFA), pp.1-6, 2023

[3] Ning Wang, Jianzhong Zhang, Fujin Deng, "Avalanche Dynamics Model of SiC MOSFET Considering Thermal Runaway Phenomenon", IEEE Transactions on Power Electronics, vol.38, no.8, pp.9705-9716, 2023

[4] Junjie An, Shengdong Hu, "Experimental and Theoretical Demonstration of Temperature Limitation for 4H-SiC MOSFET During Unclamped Inductive Switching", IEEE Journal of Emerging and Selected Topics in Power Electronics, vol.8, no.1, pp.206-214, 2020

[5] K. Matocha, I. -H. Ji, X. Zhang and S. Chowdhury, "SiC Power MOSFETs: Designing for Reliability in Wide-Bandgap Semiconductors," 2019 IEEE International Reliability Physics Symposium (IRPS), 2019, pp. 1-8

[6] X. Zhang, G. Sheh, L. Gant and S. Banerjee, "In-Depth Study of Short-Circuit Robustness and Protection of 1200 V SiC MOSFETs," PCIM Europe 2018; International Exhibition and Conference for Power Electronics, Intelligent Motion, Renewable Energy and Energy Management, Nuremberg, Germany, 2018, pp. 1-7

[7] G. Qi, J., Yang, X., Li, X., Chen, W., Long, T., Tian, K., & Wang, X. "Comprehensive Assessment of Avalanche Operating Boundary of SiC Planar/Trench MOSFET in Cryogenic Applications". IEEE Transactions on Power Electronics.

[8] Kelley, M. D., Pushpakaran, B. N., Bilbao, A. V., Schrock, J. A., & Bayne, S. B. "Single-pulse avalanche mode operation of 10-kV/10-A SiC MOSFET". Microelectronics Reliability, 81, 174-180.

[9] Na Ren, Hao Hu, Xiaofeng Lyu, Jiupeng Wu, Hongyi Xu, Ruigang Li, Zheng Zuo, Kang Wang, Kuang Sheng, "Investigation on single pulse avalanche failure of SiC MOSFET and Si IGBT", Solid-State Electronics, Volume 152, 2019,.

[10] Fayyaz, A., Romano, G., Urresti, J., Riccio, M., Castellazzi, A., Irace, A., & Wright, N. " A comprehensive study on the avalanche breakdown robustness of silicon carbide power MOSFETs." Energies, 10(4), 452.

[11] Wang, J., & Jiang, X. "Review and analysis of SiC MOSFETs' ruggedness and reliability". IET Power Electronics, 13(3), 445-455.2019

Scaled Projections of Empirically Verified Hybrid Edge Terminated Vertical GaN Diodes to 20 kV

Tolen Nelson, Prakash Pandey, Daniel G.
Georgiev, Raghav Khanna
EECS Department
University of Toledo
Toledo, OH, USA
Raghav.khanna@utoledo.edu

Michael R. Hontz
Philadelphia Division
Naval Surface Warfare Center
Philadelphia, PA, USA

Alan G. Jacobs, James C. Gallagher, Andrew
D. Koehler, Karl D. Hobart, Travis J.
Anderson
U.S. Naval Research Laboratory
Washington, D.C., USA

Abstract— This work examines how the hybrid edge termination for vertical GaN PiN diodes can be scaled for 20 kV class device epitaxy. Drift layer thickness and doping concentration are designed for an anode layer of 1 x 10^18 cm-3 Mg concentration to enable blocking voltage at 10 kV, 15 kV, and 20 kV. The geometry of the 8-ring hybrid is then optimized for each case and the resulting breakdown voltage and on-resistance examined. These results demonstrate the ability of the hybrid edge termination to be realistically scaled to low on-resistance 20 kV architectures, while also providing motivation for processing improvements needed to grow the low-doped drift layers.

Keywords—Gallium Nitride; electric breakdown; PN diode; edge termination; vertical diodes; modeling; TCAD; avalanche breakdown

I. INTRODUCTION

High voltage power devices (> 10 kV) are needed in modern power electronic applications such as motor drives, renewable energy generation, and pulse arrestor protection circuits [1] - [3]. Gallium Nitride based devices are well suited to meet the requirements of these emerging applications. The high critical electric field and high electron mobility of GaN enable devices with high blocking voltage and low on-resistance. Furthermore, the low minority carrier lifetime of GaN allows fast switching speeds [4].

Low resistance (100.8 m$\Omega \cdot$ cm^2) lateral GaN/AlGaN super heterojunction Schottky barrier diodes have been reported up to 12.5 kV [5]. In lateral devices, the die size must be increased to allow suitable contact spacing, enabling higher blocking voltages. Vertical architectures, however, decouple the breakdown voltage from the die size as the blocking voltage is supported through the thickness of a low doped drift layer. The development of vertical GaN devices have been primarily limited by the availability of suitably thick, low doped drift layers. Proper edge termination of the p-n junction is required to reduce field crowding at the device periphery which must be tolerant to local variations in the epitaxy [6].

Recently, a highest breakdown voltage of 7.86 kV vertical GaN diodes (2.8m$\Omega \cdot$ cm^2) has been reported. An epitaxy

consisting of a 50 µm thick, 1 x 10^15 cm^-3 doped drift layer was achieved to enable this breakdown voltage. The design uses a guard ring and field plate structure over a high-k BTO dielectric for the edge termination [7]. In [8], the authors of this paper have presented the hybrid edge termination which consists of a combination of alternating shallow and deep nitrogen implants to locally compensate charge in the p-layer. The geometry of the rings is designed such that the integrated charge across the termination is fit to a reference bevel edge termination. It was shown this termination offers wide tolerance to wafer variations while remaining simple to fabricate [8].

This paper presents the fabrication of 1350 V hybrid edge terminated vertical diodes, and the validation of a computational physics model. The scaling of the drift layer doping concentration and thickness required to achieve 10, 15, and 20 kV devices is also assessed. The feasibility of the HET design to enable these blocking voltages is also examined. This

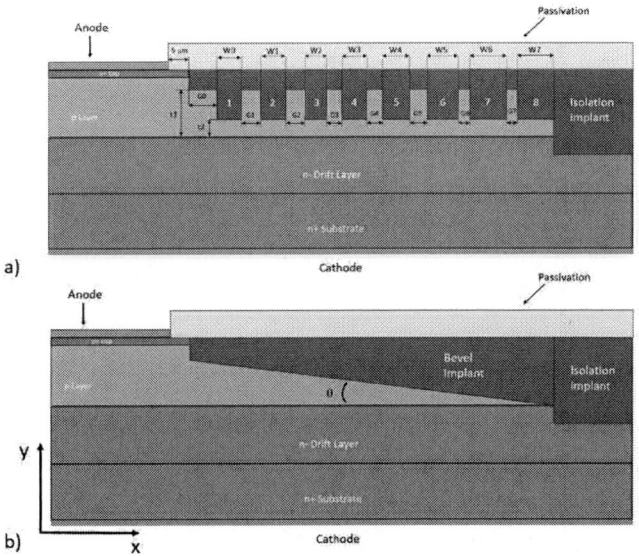

Fig. 1. a) Vertical diode structure with hybrid edge termination b) With bevel edge termination

---

This work was supported in part by U.S. Office of Naval Research under Grant N00014-21-1-2832 and was approved for public release under DCN 543-673-23.

work provides insight into the necessary processing improvements needed to achieve >10 kV vertical GaN diodes.

II. MODEL DEVELOPMENT AND VALIDATION

Fig. 1 compares the proposed HET design (a) with the reference bevel (b) that the HET structure is designed to emulate. A physics based finite element model was developed in the Sentaurus TCAD (Synopsys) platform for the vertical PN diode shown in Fig. 1(a). Doping dependent mobility, incomplete ionization of the Mg p-doping, and anisotropy of the GaN permittivity are considered [9]. The drift layer and p-layers are epitaxially grown via MOCVD on top of a Si doped n^+ substrate. It is important to consider the gradual turn-on of the Mg doping of the p-layer. SIMS measurements were performed on a similarly grown wafer with 2×10^{19} cm^{-3} doped p-layer and then scaled to 1×10^{18} cm^{-3} doping. The gradual Mg profile results in deeper penetration of the depletion region into the p-layer and should be considered when designing the edge termination. The Mg doping profile used in this work is shown in Fig. 2 and included directly via look-up table in the model.

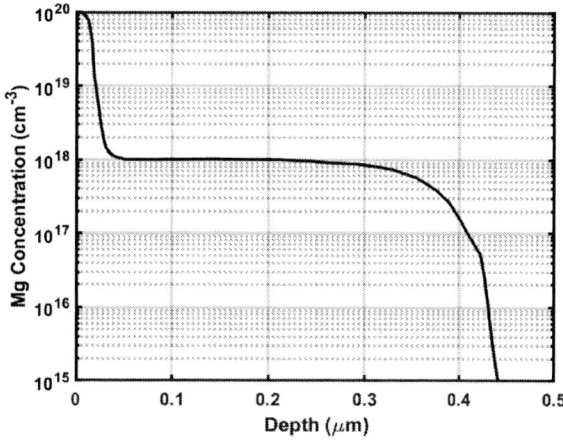

Fig. 2. Mg doping profile scaled from measurements on a similar 2×10^{19} Mg doped wafer.

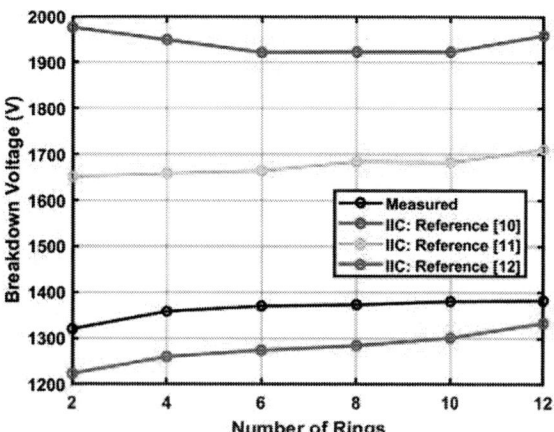

Fig. 3. Measured breakdown voltage on 1.3 kV HET terminated diodes compared with simulated breakdown voltage using various impact ionization coefficients.

In properly designed vertical GaN diodes, the breakdown mechanism is avalanche generation through impact ionization. This process is modeled via Chynoweth's law:

$$\alpha_{n,p} = a_{n,p} \exp\left(-b_{n,p}/E_x\right), \qquad (1)$$

where $\alpha_{n,p}$ is the carrier generation, $a_{n,p}$ and $b_{n,p}$ are impact ionization coefficients, and E_x is the electric field magnitude in the direction of the current. The breakdown voltage is taken at the bias when the impact ionization integral reaches a value of unity. Various measurements of GaN impact ionization coefficients (IIC's) have been presented in literature [10-12]. The differing IIC's leads to a wide range of simulated breakdown voltage.

The HET is designed by fitting its integrated charge profile along the termination to a reference bevel angle as explained further in [8].

$$\rho(x) = \int_0^x \int_0^{y=t(x)} N_A(x,y)\,dy\,dx \qquad (2)$$

Where $\rho(x)$ is the integrated charge along the termination length, N_A is the Mg profile in the p-layer, and $t(x)$ is the thickness of p-layer under the nitrogen implant. HET terminated diodes with number of rings ranging from 2 to 12 were fabricated with an 8 μm, 1.8×10^{16} drift layer by designing the HET geometry to a 0.149 degree reference bevel angle shown in Fig. 1(b). The measured breakdown for each ring number case is ~1375 V. The simulated breakdown voltage for each set of IIC's was examined. A comparison of the measured breakdown voltage with simulated upper and lower bound IIC's is shown in Fig. 3. The different IIC's lead to an 800 V range in simulated breakdown voltage for this epitaxy.

III. SCALED PROJECTIONS TO 20 KV

In vertical p-n diodes, the breakdown voltage increases with the drift layer thickness. For an abrupt junction, the ideal parallel plane breakdown voltage for a non-punch through design can be estimated using [13]:

$$BV = \frac{\varepsilon E_C^2}{2qN_D}, \qquad (3)$$

where E_C is the critical electric field in GaN, N_D is the drift layer doping, q is the electron charge, and ε is the permittivity.

Using the Mg doping profile from Fig. 2, the drift layer epitaxy was scaled to find the maximum doping and necessary corresponding thickness to achieve 10, 15, and 20 kV class devices. The thickness was first increased to be arbitrarily larger than the depletion region at breakdown voltage such that it is never limited due to punch through. Then the doping concentration was decreased to find the maximum allowable doping. Once the doping is selected for each case the drift layer thickness is minimized near the punch through condition to reduce its contribution to the on-resistance. During this process, the lower bound IIC's are examined to ensure the expected breakdown voltage is above this lower bound. To account for the edge termination efficiency the parallel plane breakdown voltage in each case is set accordingly above the target voltage. The HET is designed for a reference bevel of

979-8-3503-3714-3/23 $31.00 © 2023 IEEE

0.0177 degrees shown in Fig. 1(b). As the depletion extends approximately 0.185 µm into the p-layer, this corresponds to an HET width of 600 µm. The same reference bevel angle, and thus HET geometry, is used for each voltage class device. A minimum constraint of 50 nm is considered for the unaltered p-layer thickness under the deep implant to ensure the structure can be realistically fabricated.

A summary of the designed epitaxy for each voltage class device is given in Table 1. The drift layer epitaxy for the 10 kV is 5 x 10^{14} cm^{-3} doping and 130 µm. The ideal parallel plane breakdown voltage ranges from 13.6 kV to 18.16 kV for the lower and upper bound IIC's respectively. The 0.0177 degree angle reference bevel termination for this epitaxy breaks down between 11 kV and 11.98 kV. The optimized HET for this case breaks down between 10.5 and 12.27 kV. This corresponds to 77 percent to 67.5 percent termination efficiency. Achieving higher termination efficiency using this structure would require a smaller reference bevel angle which would increase the width of the termination and resulting total die size. The 20 kV class device has termination efficiency of 79.2 percent to 74.77 percent.

Table 1: Summary of Designed Epitaxy for 10, 15, and 20 kV Class Diodes

Voltage Class	Drift Layer Epitaxy	Lower Bound Planar BV	Upper Bound Planar BV	Lower Bound HET BV	Upper Bound HET BV
10 kV	5x10^{14} cm^{-3}/ 130 µm	13.6 kV	18.16 kV	10.5 kV	12.27 kV
15 kV	3x10^{14} cm^{-3}/ 180 µm	20.02 kV	26.5 kV	15.56 kV	17.66 kV
20 kV	1x10^{14} cm^{-3}/ 200 µm	26.5 kV	32.11 kV	21 kV	24.01 kV

The electric field of the 20kV class HET terminated diode at its breakdown voltage is shown for the lower and upper bound IIC's in Figs. 4 and 5 respectively. At each ring, the electric field peaks with the peak magnitude reducing across the termination. This approximates the reference bevel's smooth electric field profile and shows the HET is properly working even in a 20 kV design. The lower bound IIC's of [12] predict a maximum electric field peak of 2.67 MV/cm while the upper bound IIC's [10] predict a maximum field of 3.94 MV/cm. This results in an increase of 2.1 kV between the two sets of coefficients.

The forward bias characteristics were examined utilizing cylindrical symmetry for each voltage class device and are shown in Fig. 6. Contact resistances of 1 x 10^{-6} Ω * cm^2 were used on both the anode and cathode. The resistance of the drift layer can be estimated as:

$$R = \frac{t}{q\mu N_D A},\qquad(4)$$

where t is the thickness of the drift layer, q is the electron charge, u is the electron mobility, n is the drift layer doping, and A is the device area.

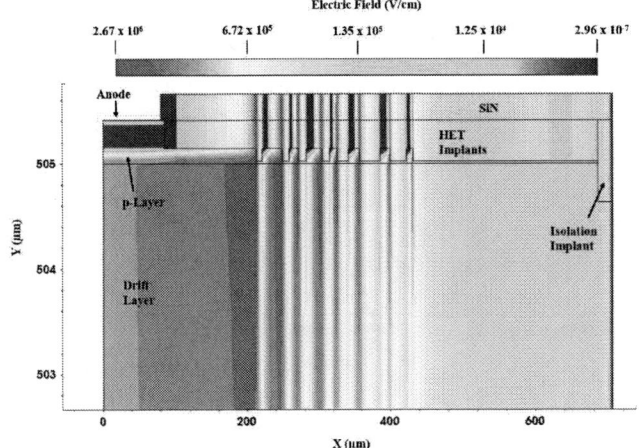

Fig. 4. Electric field for the 20 kV HET device at the simulated breakdown voltage of 21 kV using the lower bound impact ionization coefficients of [12].

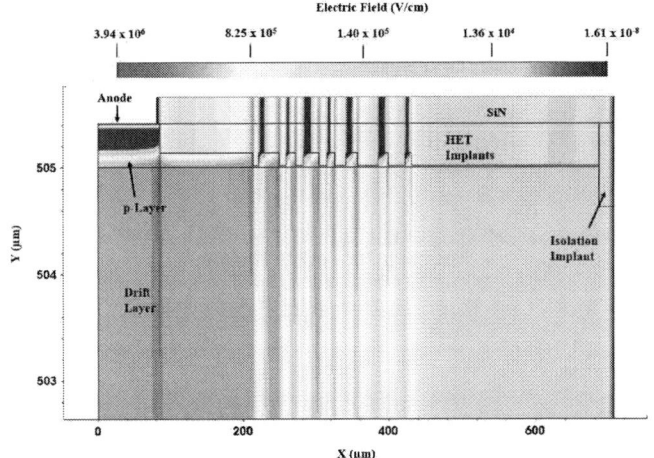

Fig. 5. Electric field for the 20 kV HET device at the simulated breakdown voltage of 24.01 kV using the upper bound impact ionization coefficients of [10].

The specific on-resistance was calculated using the active area of the device and is overlayed for the lower bound IIC's against the 3.75 MV/cm Baliga figure of merit (FOM) theoretical limit as well as various reported high voltage diodes to date in Fig. 7. A significant contribution to the total R$_{ON}$ is due to the inclusion of the Mg growth profile of Fig. 2. The specific on-resistance decreases from ~200 mΩ*cm^2 to ~100 mΩ*cm^2 for the 20 kV class device when an abrupt 1 x 10^{18} cm^{-3} junction is considered. These results show the hybrid edge termination can be realistically scaled to realize high voltage vertical GaN diodes.

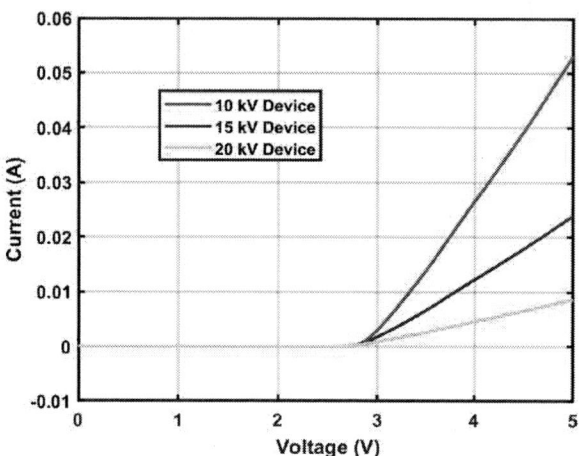

Fig. 6. Simulated forward bias characteristics for the designed 10, 15, and 20 kV class devices.

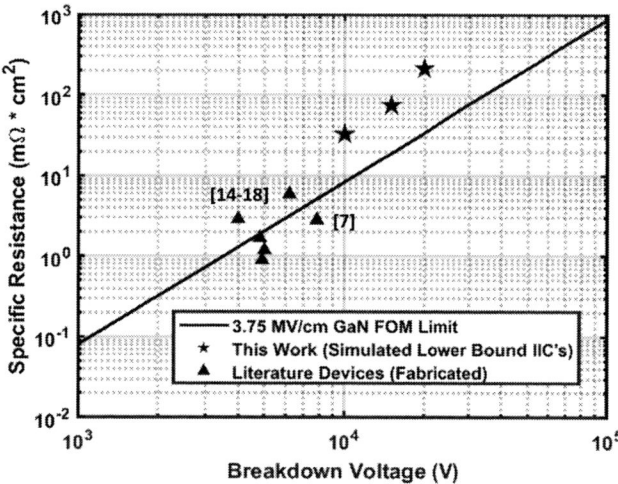

Fig. 7. Specific R_{ON} and breakdown voltage for simulated 10, 15, and 20 kV class devices using the lower bound impact ionization coefficients of [12] and overlayed with various literature reported devices.

IV. CONCLUSION

The scaling of the hybrid edge termination on vertical GaN diodes is assessed using a physics based finite element model. The model's predicted breakdown voltage is validated via fabrication of ~1.3 kV hybrid terminated diodes. Upper and lower bound impact ionization coefficients from literature are evaluated. The drift layer epitaxy is scaled in simulation in the design of 10, 15, and 20 kV class diodes to find the necessary doping and thickness. It is found a 1×10^{14} cm^{-3} / 200 μm drift layer is needed to support ~20 kV operation. The HET provides 75 to 79 percent termination efficiency for the 20 kV device using a reference bevel angle of 0.0177 degrees. This could be further improved by using a smaller reference angle at the cost of increased termination width. These results demonstrate the efficacy of the HET for high voltage devices,

as well as provides targets for the necessary process improvements to realize the high voltage drift layer epitaxy.

V. ACKNOWLEDGEMENT

The authors gratefully acknowledge the support of the U.S. Office of Naval Research (grant number N00014-21-1-2832 and approved under DCN 543-673-23), at the direction of Captain Lynn Petersen, USN(Ret).

REFERENCES

[1] A. K. Morya et al., "Wide bandgap devices in AC electric drives: opportunities and challenges," in IEEE Transactions on Transportation Electrification, vol. 5, no. 1, pp. 3-20, March 2019, doi: 10.1109/TTE.2019.2892807.

[2] Castellazzi, Gurpinar, Wang, S. Hussein and G. Fernandez, "Impact of wide-bandgap technology on renewable energy and smart-grid power conversion applications including storage", Energies, vol. 12, no. 23, pp. 4462, Nov. 2019.

[3] R. Kaplar et al., "20 kV gallium nitride electromagnetic pulse arrestor for grid reliability", Jun. 2019, [online] Available: https://www.osti.gov/servlets/purl/1645441.

[4] B. Baliga, "Semiconductors for high-voltage, vertical channel field-effect transistors," J. Appl. Phys., vol. 53, pp. 1759 – 1764, April 1982.

[5] S. Han et al., "12.5 kV GaN super-heterojunction Schottky barrier diodes," in IEEE Transactions on Electron Devices, vol. 68, no. 11, pp. 5736-5741, Nov. 2021.

[6] H. Fu, K. Fu, S. Chowdhury, T. Palacios and Y. Zhao, "Vertical GaN power devices: device principles and fabrication technologies—Part I," in IEEE Transactions on Electron Devices, vol. 68, no. 7, pp. 3200-3211, July 2021.

[7] Y. Xu et al., "7.86 kV GaN-on-GaN pn power diode with BaTiO3 for electrical field management", *arXiv e-prints*, 2023. doi:10.48550/arXiv.2303.15646.

[8] T. Nelson et al., "Hybrid edge termination in vertical GaN: approximating beveled edge termination via Discrete Implantations," in IEEE Transactions on Electron Devices, vol. 69, no. 12, pp. 6940-6947, Dec. 2022.

[9] P. Pandey et al., "A simple edge termination design for vertical GaN p-n diodes," in IEEE Transactions on Electron Devices, vol. 69, no. 9, pp. 5096-5103, Sept. 2022.

[10] B. J. Baliga, "Gallium nitride devices for power electronic applications," Semicond. Sci. Technol., vol. 28, no. 7, Jul. 2013, Art. no. 074011.

[11] D. Ji, B. Ercan, and S. Chowdhury, "Experimental determination of impact ionization coefficients of electrons and holes in gallium nitride using homojunction structures," Appl. Phys. Lett., vol. 115, no. 7, Aug. 2019, Art. no. 073503.

[12] T. Maeda et al., "Impact ionization coefficients and critical electric field in GaN," J. Appl. Phys., vol. 129, no. 18, May 2021, Art. no. 185702.

[13] I. C. Kizilyalli, A. P. Edwards, H. Nie, D. Disney and D. Bour, "High voltage vertical GaN p-n diodes with avalanche capability," in IEEE Transactions on Electron Devices, vol. 60, no. 10, pp. 3067-3070, Oct. 2013.

[14] A. L. Yates, B.P. Gunning, M. H. Crawford, J. Steinfeldt, M. L. Smith, V. M. Abate, J. R. Dickerson, A. Binder, A. A. Allerman, A. M. Armstrong,and R. J. Kaplar, "Demonstration of 6.0-kV breakdown voltage in large area vertical GaN p-n diodes with step etched junction termination extensions", IEEE Trans. On Electron Devices, Vol 69, No. 4, April, 2022.

[15] V. Telasara, Y. Zhang, V. Vangipuram, H. Zhao, W. Lu, "Vertical GaN-on-GaN pn power diodes with Baliga figure of merit of 27 GW/cm2." Appl. Phys. Lett. Vol.122, no. 12, Mar. 2023.

[16] H. Ohta, N. Kaneda, F. Horikiri, Y. Narita, T. Yoshida, T. Mishima, and T. Nakamura, "Vertical GaN pn junction diodes with high breakdown voltages over 4 kV." IEEE Electr. Device L., vol. 36, pp. 1180-1182, Nov. 2015.

[17] H. Ohta, K. Hayashi, F. Horikiri, M. Yoshino, T. Nakamura, and T. Mishima. "5.0 kV breakdown voltage vertical GaN p–n junction diodes." Jpn. J. Appl. Phys., vol. 57, p. 04FG09, Feb 2018.

[18] I. C. Kizilyalli, T. Prunty, and O. Aktas. "4-kV and 2.8 mΩ-cm2 vertical GaN pn diodes with low leakage currents." IEEE Electr. Device L., vol. 36, p.1073-1075, Oct. 2015.

Investigation of the Impact of Low Thermal Conductivity on Gallium Oxide Power Module Packaging

Mohammad Dehan Rahman, Xiaoqing Song
Department of Electrical Engineering and Computer Science,
University of Arkansas,
Fayetteville, AR, United States
mr117@uark.edu, xsong@uark.edu

Abstract— **The ultra-wide bandgap energy (4.8 eV) of gallium oxide (Ga_2O_3) enables high breakdown electrical strength, low intrinsic carrier concentration and high operation temperatures, making it a promising material for high-temperature, high-density power electronics. The low thermal conductivity of Ga_2O_3 is a limiting factor, which impedes efficient heat removal from the device junction, and increases the junction to case thermal resistance. However, a quantitative study of how much higher thermal resistances and device junction temperature will rise compared to other semiconductor devices like silicon carbide (SiC) are not fully investigated. In this paper, a finite element analysis model of Ga_2O_3 power module is built, and the junction temperature rise of Ga_2O_3 power module is simulated under both bottom sided cooling and double sided cooling packaging. The thermal performance of Ga_2O_3 is compared with SiC power modules under the same heat load and cooling conditions, and the breakdown of total thermal resistance of Ga_2O_3 power module is also calculated and presented to better understand the influence of Ga_2O_3's low thermal conductivity on the power module's overall thermal performances.**

Keywords—gallium oxide, thermal conductivity, power module packaging, thermal modeling

I. INTRODUCTION

Wide bandgap (WBG) power devices are finding increasing applications in transportation electrification, motor drives, energy storage, renewable energy, etc. applications, thanks to their superior performances, like lower conduction resistances, faster switching speed compared to their silicon (Si) counterparts. To further improve the power electronics converter performances (higher efficiency and power density), ultra-wide bandgap (UWBG) power semiconductor devices start to gain increasing interests from researchers owing to the expected even better performances. One of the most promising UWBG semiconductor is gallium oxide (Ga_2O_3), which has superior intrinsic material properties [1-2], e.g., larger bandgap energy [4.4× higher than Si, ~1.5× silicon carbide (SiC) and gallium nitride (GaN)] and lower intrinsic carrier concentration, enabling higher semiconductor device breakdown voltages, smaller conduction resistances and more importantly, higher operation temperature capabilities. Table I summarizes the material properties of some WBG and UWGB semiconductors. Although Ga_2O_3 does not have the largest bandgap compared to other UWBG semiconductors, like aluminum nitride (AlN) and diamond, a modified figure of merit (FOM) in Fig. 1 shows β-Ga_2O_3 promises the best performance among UWBG materials, that is, lowest specific on-resistance (Ron) with the same breakdown voltage (BV), considering the incomplete ionization and background compensation effects due to the availability of shallow donors, low background impurity compensation, and bulk substrates

[3]. In addition, Ga_2O_3 wafer can be fabricated with the same process of Si substrate, making it very cost-effective (>3~5x lower than SiC) [4]. All these prospects make Ga_2O_3 devices a promising candidate for high density, high operation temperature power electronics in automotive and other harsh environment applications.

Table I. Properties of Si and some WBG/UWBG semiconductors [1][2]

Semicoductors	Bandgap (eV)	Critical Electric Field (MV/cm)	Thermal Conductivity (W/m·K)
Si	1.1	0.3	150
SiC	3.3	2.2	490
GaN	3.4	3.9	230
AlN	3.4-6.1	3.0-12	319
Ga_2O_3	4.8	6.7-15	11-27
Diamond	5.5	5-10.1	2290-3450

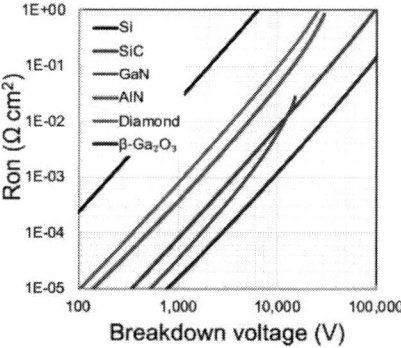

Fig. 1: The specific conduction resistance (Ron) vs. breakdown voltage curves of semiconductor materials considering both incomplete ionization and background compensation effects.[3]

However, the thermal conductivity of Ga_2O_3 is 10× lower than SiC, ~5× lower than Si, which could impede efficient heat removal from the device junction and increases the junction to case thermal resistance of the Ga_2O_3 power module. A quantitative investigation of the influence of the low thermal conductivity on the power module thermal performances are not fully studied [5-6]. This paper will fill up this knowledge gap and investigate the influence of the low thermal conductivity on the power module thermal performances through FEA modeling and thermal analysis, and the suggestions on the power module packaging to address this challenge are provided.

979-8-3503-3714-3/23 $31.00 © 2023 IEEE

II. THERMAL MODELING OF GALLIUM OXIDE POWER MODULES

To study the effects of the low thermal conductivity of Ga_2O_3 on the device junction temperature, the finite element analysis (FEA) model of power modules with different cooling structures are built. Fig. 2 shows three types of cooling methods for the Ga_2O_3 power module, which are bottom side cooling (Fig. 2(a)), top side cooling (Fig. 2(b)) and double side cooling (Fig. 2(c)). Bottom side cooling means the majority heat of the semiconductor chip is dissipated through the substrate bonded to the bottom of the semiconductor chip, which is the most commonly used packaging method for Si power devices. In the double side cooling [8] and top side cooling [9] design, a copper spacer is bonded between the top surface of the device and the top direct bond copper (DBC) substrate. The semiconductor device chip size is 4 mm × 6 mm × 0.18 mm. In this study, the die is attached to the substrate or copper spacer by silver sintering. The default height of the copper spacer is 5 mm, but various copper spacer heights are simulated to investigate its impact on junction temperature and the device total junction

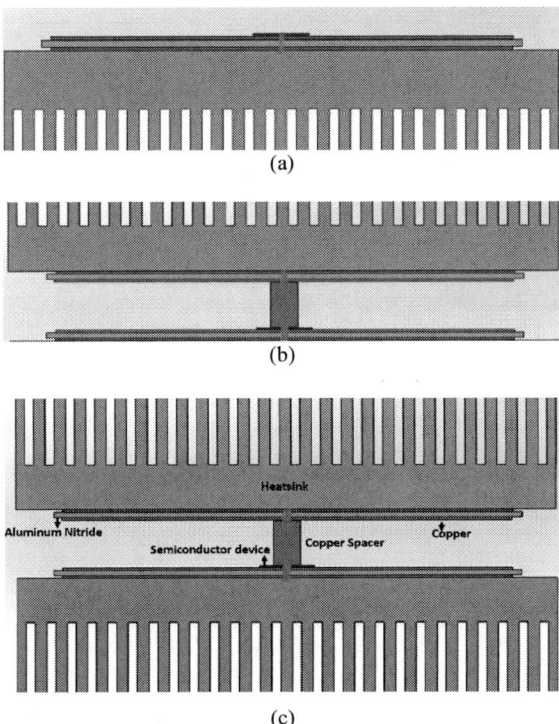

(a)

(b)

(c)

Fig.2: (a) Structure of the bottom side cooling Ga_2O_3 power module. (b) Top side cooling of the architecture. (c) Double side cooling of the architecture

Table II: Material parameters used in the FEA simulation.

Material	Thermal Conductivity W/(m·K)	Heat Capacity (J/(kg·K))	Density (kg/m³)
SiC	490	510	4360
GaN	230	431	6095
Ga_2O_3	27	370	5880
Copper	398	385	8960
Silver Paste	429	235	8600

to case thermal resistance. SiC and GaN [11-12] power modules with the same device structure are also built for comparison. Material properties used to run the FEA analysis are summarized in Table II.

III. THERMAL ANALYSIS BASED ON FEA MODELLING

A. Steady State Maximum Junction Temperature

Based on the built FEA model built in Section II, the steady state device maximum junction temperatures are simulated under different cooling methods. In the FEA thermal simulation, 100 W power losses are assumed to be generated by the Ga_2O_3 chip, and the heat evenly generated by the entire chip.

Fig. 3 shows the steady state maximum junction temperature comparing the bottom side and doubled side cooling of Ga_2O_3 power modules. In the bottom side cooling, a heat transfer coefficient at 10 kW/m²·K, indicating forced liquid cooling is applied at the bottom of the DBC substrate, while in the double side cooling, 10 kW/m²·K heat transfer coefficient is applied to both the top DBC and bottom DBC. It can be seen that the double side cooling Ga_2O_3 power module has ~18 °C (~23%) lower maximum junction temperature rise compared to the bottom side cooling power module, indicating the benefits of adopting double-sided cooling for Ga_2O_3 power modules.

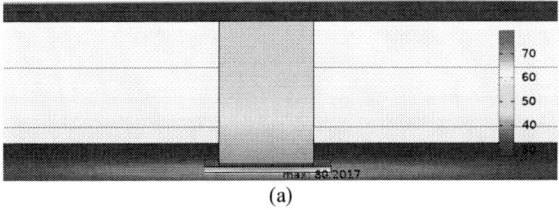

(a)

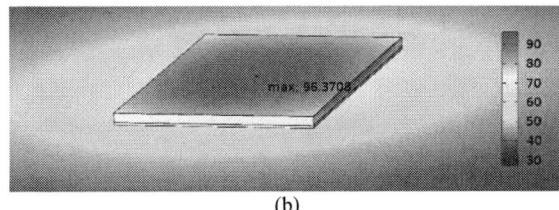

(b)

Fig. 3: Junction temperature comparison between the double side and bottom sided cooling Ga_2O_3 power module (chip size: 6mm × 4mm, power losses: 100 W). (a) double sided cooling Ga_2O_3 power module maximum junction temperature: 80.2 °C. (b) bottom sided cooling Ga_2O_3 power module maximum junction temperature: 97.0 °C.

B. Transient Junction Temperature Rise

The transient junction temperature rise under the 100 W power losses are also simulated. Fig. 4 shows the transient temperature rise of bottom side and double side cooling of Ga_2O_3 power modules (orange lines). In Fig. 4(a), the heat is assumed to be generated by the entire chip, while in Fig. 4(b) the heat is assumed to be generated only by the top surface of the chip, which is applicable to the lateral devices. The junction temperature rise of SiC (blue lines) and GaN (green lines) power modules under the same cooling conditions are also simulated and compared with the Ga_2O_3 power module.

979-8-3503-3714-3/23 $31.00 © 2023 IEEE

From Fig. 4, it can be observed that the junction temperature of the Ga$_2$O$_3$ power module is always higher compared to SiC and GaN power modules owing to the lower thermal conductivity of Ga$_2$O$_3$. By adopting double side cooling for the Ga$_2$O$_3$ power module, it will achieve a similar junction temperature rise as the GaN bottom side cooled power module when the heat is dissipated though the entire chip.

Fig. 4(b) considers another extreme case, where the losses (or heat) is generated mostly from the top surface of the semiconductor chips (which is true for diodes, because of the intrinsic junction voltage drop, more heat is generated at the top surface of the chip). In this case, it can be seen that the

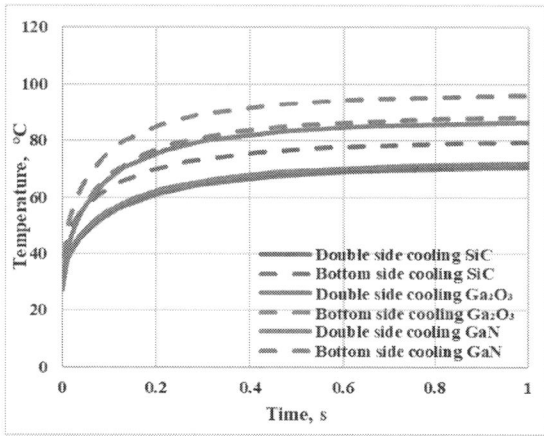

(a)

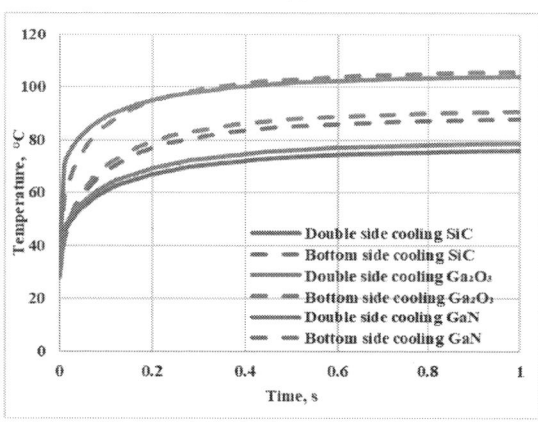

(b)

Fig. 4: SiC, Ga$_2$O$_3$ and GaN power module junction temperature rising curves under 100W power losses.
a) Power losses are generated from the entire chip; (b) Power losses are generated at the top surface of the chip.

double sided cooling method only has slightly lower junction temperature compared to the bottom sided cooling, which means double side cooling method loses its effectiveness. This could be due to the relatively high thermal resistances from the copper space between the top direct bond copper (DBC) and Ga$_2$O$_3$ chip.

In Fig. 5, the top side cooling is also considered and compared with the bottom and double side cooling. Fig. 5(a) shows the device junction temperature rise when the entire

chip is considered as heat source. The maximum junction temperature difference between double side and bottom side cooling is approximately 18 °C. Fig. 5 also indicates that the top side cooling has the maximum junction temperature rise, which is due to the high thermal resistance introduced by the copper spacer. Fig. 5(b), shows the junction temperature rise under the same cooling conditions when only the top surface is considered as heat source. Similarly, the top side cooling has the higher junction temperature, and the steady state junction temperatures under double side cooling and bottom side cooling are also higher than those in Fig. 5(a).

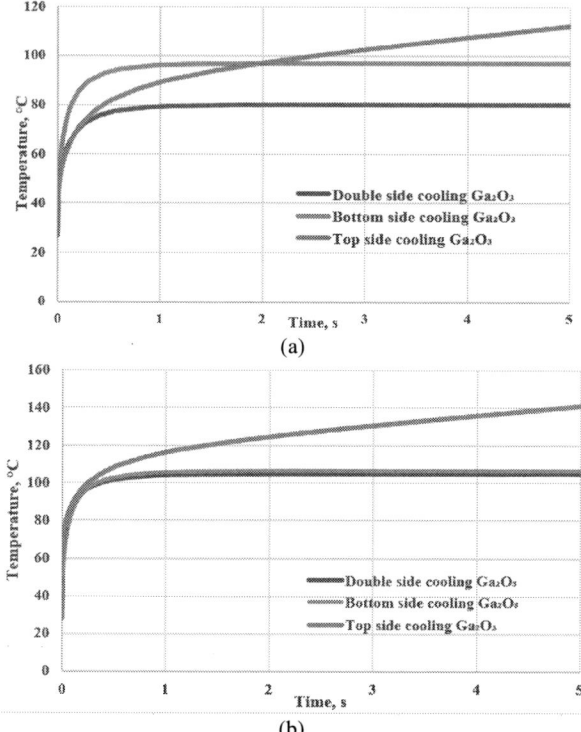

Fig. 5: Junction temperature rise of Ga$_2$O$_3$ power modules with different cooling methods.
a) Power losses are generated from the entire chip; (b) Power losses are generated at the top surface of the chip.

IV. THERMAL RESISTANCE MODELING OF THE GALLIUM OXIDE POWER MODULE

To better understand the effect of the low thermal conductivity of Ga$_2$O$_3$, the thermal resistances of each packaging materials inside the power module is calculated based on the spreading angle thermal resistance model shown in Fig. 6. Conventionally, thermal resistance (R_{Th}) of the bottom side cooling architecture can be calculated according to the following equation.

$$R_{Th} = \frac{T_j - T_C}{P} \qquad (1)$$

However, the equation does not consider the spreading angle of heat transfer in the power module package. In a power module package, heat transfer is not completely linear but at an angle of approximately 45°. This is due to usage of components having different dimensions. Thus, a modified equation needs to be implemented to calculate thermal resistance based on spreading angle of the heat transfer [7]. The concept of the heat transfer is shown in Fig. 6.

Based on Fig. 5, a modified equation can be used [10].

$$R_{Th} = \int_0^{t_s} \frac{1}{k(a+2x.\tan\theta)^2} dz = \frac{1}{2k.\tan\theta} \left(\frac{1}{a} - \frac{1}{a+2t_s.\tan\theta}\right)$$
(2)

Using the equation, thermal resistance of double side and bottom side cooling architectures are calculated.

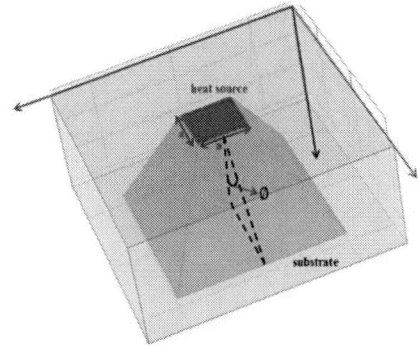

Fig. 6: Thermal resistance model based on the spreading angle [7].

Fig. 7 shows the thermal resistance change with the increase of heat transfer coefficient, which means a more powerful cooling method like liquid cooling, boiling liquid cooling, etc. Thermal resistance of bottom side (blue line) and double side cooling (orange line) of Ga_2O_3 is shown in Fig. 7. It can be observed that overall thermal resistance of double side architecture is lower than bottom side architecture. Another key takeaway is that thermal resistance follows an exponential decay relation with the increase of the heat transfer coefficient. When the heat transfer coefficient is high enough, its effect on the thermal resistance reduction will not be so significant.

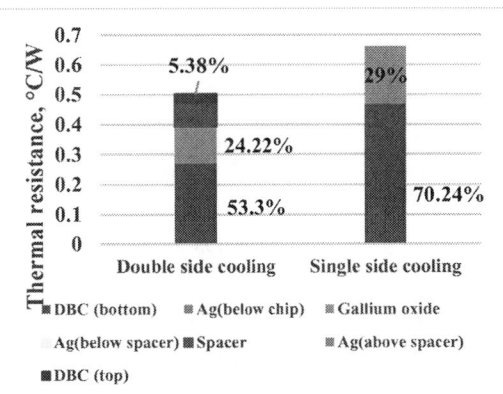

Fig. 8: Thermal resistance comparison between double side and bottom side cooling of Ga_2O_3

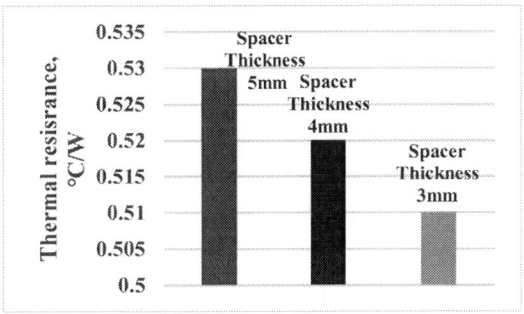

Fig. 9: Thermal resistance comparison with respect to spacer thickness

In double side cooling, another major source of thermal resistance is the copper spacer. It adds about 16.2% of the total thermal resistance. Changing the thickness of the spacer impacts the thermal resistance of the double side cooling as shown in Fig. 9. Change in thermal resistance due to spacer thickness is shown in Fig. 9.

V. CONCLUSIONS

In this paper, the impact of low thermal conductivity on the Ga_2O_3 power module packaging is investigated by the FEA thermal modeling and analysis. A quantitative study of thermal resistances and device junction temperature increase due to the low thermal conductivity of Ga_2O_3 compared to SiC and GaN is provided. It is found that the Ga_2O_3 has 15~30% higher junction temperature rise compared to SiC and GaN power modules under the same cooling conditions. Double side cooling can help reduce the Ga_2O_3 chip's maximum junction temperature by ~20% when the heat is generated by the entre chip. It is also interesting to find that the double side cooling will lose its effect when the majority of the heat is generated at the top surface of the semiconductor chip, and the top side cooling method introduces higher total thermal resistances due to the copper spacer.

REFERENCE
[1] Oleksiy Slobodyanl, Jack Flicker, Jeramy Dickerson, Andrew Binder, Trevor Smith, Robert Kaplat, Mark Hollis, "Analysis of the dependence of critical electric field on semiconductor bandgap",

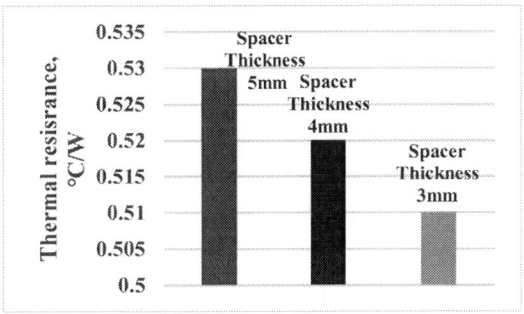

Fig. 7: Thermal resistances of double sided and bottom sided cooling of Ga_2O_3 under different cooling conditions (heat transfer coefficients)

Fig. 8 shows the thermal resistance distribution between different layers of the bottom side cooling and double side cooling. The double side cooling Ga_2O_3 power module has about 30% lower thermal resistance compared to the bottom side cooling Ga_2O_3 module. In the bottom side cooling, the Ga_2O_3 chip takes about 29% of the total junction to case thermal resistance while the DBC substrate contributes to the major thermal resistances due to a much higher thickness compared to the semiconductor chips. In the double side cooling, the Ga_2O_3 chip takes about 24% of total junction to case thermal resistances.

Journal of Materials Research, 2022. https://doi.org/10.1557/s43578-021-00465-2.

[2] Alok Ranjan, Nagarajan Raghavan, Matthew Holwill, Kenji Watanabe, Takashi Taniguchi, Kostya S. Novoselov, Kin Leong Pey, and Sean J. O'Shea. "Dielectric Breakdown in Single-Crystal Hexagonal Boron Nitride", ACS Applied Electronic Materials, 2021. http://doi.org/10.1021/acsaelm.1c00469.

[3] Yuewei Zhang and James S Speck 2020 Semicond. Sci. Technol. 35 125018

[4] M. Baldini, Z. Galazka, and G. Wagner, "Recent progress in the growth of β-Ga2O3 for power electronics applications," Mater. Sci. Semicond. Process., vol. 78, pp. 132–146, May 2018

[5] P. Paret et al., "Thermal and Thermomechanical Modeling to Design a Gallium Oxide Power Electronics Package," 2018 IEEE 6th Workshop on Wide Bandgap Power Devices and Applications (WiPDA), Atlanta, GA, USA, 2018, pp. 287-294, doi: 10.1109/WiPDA.2018.8569139.

[6] B. Chatterjee, K. Zeng, C. D. Nordquist, U. Singisetti and S. Choi, "Device-Level Thermal Management of Gallium Oxide Field-Effect Transistors," in IEEE Transactions on Components, Packaging and Manufacturing Technology, vol. 9, no. 12, pp. 2352-2365, Dec. 2019, doi: 10.1109/TCPMT.2019.2923356.

[7] Y. Xu and D. C. Hopkins, "Misconception of thermal spreading angle and misapplication to IGBT power modules," 2014 IEEE Applied Power Electronics Conference and Exposition - APEC 2014, Fort Worth, TX, USA, 2014, pp. 545-551, doi: 10.1109/APEC.2014.6803362.

[8] H. Chen et al., "Fabrication and Experimental Validation of Low Inductance SiC Power Module with Integrated Microchannel Cooler," 2023 IEEE Applied Power Electronics Conference and Exposition (APEC), Orlando, FL, USA, 2023, pp. 366-371, doi: 10.1109/APEC43580.2023.10131620.

[9] B. Albano, B. Wang, Y. Zhang and C. DiMarino, "Electro-Thermal Device-Package Co-Design for Ultra-Wide Bandgap Gallium Oxide Power Devices," 2022 IEEE Energy Conversion Congress and Exposition (ECCE), Detroit, MI, USA, 2022, pp. 1-7, doi: 10.1109/ECCE50734.2022.9948059.

[10] C. Xu, X. Song and P. Cairoli, "SiC Based Solid State Circuit Breaker: Thermal Design and Analysis," in IEEE Transactions on Industry Applications, doi: 10.1109/TIA.2023.3312055.

[11] Liu, S.Y., Jiang, Y.F., Sung, W.J., Song, X.Q., Baliga, B.J., Sun, W.F., Huang, A.Q., 2017. Understanding High Temperature Static and Dynamic Characteristics of 1.2 kV SiC Power MOSFETs. MSF 897, 501–504. https://doi.org/10.4028/www.scientific.net/msf.897.501

[12] Z. Yang et al., "Overcurrent Capability Evaluation of 600 V GaN GITs under Various Time Durations," 2021 IEEE Applied Power Electronics Conference and Exposition (APEC), Phoenix, AZ, USA, 2021, pp. 376-381, doi: 10.1109/APEC42165.2021.9487155.

A Partial Soft-switching SiC-based ANPC Single-phase Inverter with Low THD for Grid-tied PV Systems

*

1st Wenjie Ma, 2nd Hu Li, 3rd Shan Yin, 4th Xiaohu Pang, 5th Jiayue Fang

Department of Aeronautics and Astronautics
University of Electronic Science and Technology of China
Chengdu, Sichuan
grmn612@163.com, kelly.li@126.com, yinshansamuel@gmail.com, xihu_pang@163.com, jy972160286@stu.xjtu.edu.cn,

Abstract—Single-phase string inverter has been widely applied to grid-tied photovoltaic (PV) rooftop applications for its renewable energy. However, the inherent attribute of intermittency in solar energy may induce unqualified power. To meet the grid-interconnection standards, high demands of efficiency and harmonic distortion need to be imposed on DC/AC inverters. Conventional soft-switching inverters used to sacrificing total harmonics distortion (THD) induced by dead-time effect to exchange for soft switching. Instead, this paper proposes a silicon carbide (SiC) -based active neutral-point-clamped (ANPC) inverter, which makes a balance between efficiency and THD performance. It adopts hybrid PWM modulation to realize partial soft-switching for synchronous switches, while adjusting dead time to achieve lower current harmonics under different power levels. In addition, the mechanism of the dead time effect affected by soft switching is explored. The soft-switching boundaries and design procedure are also analyzed in detail. A 1-kW prototype is demonstrated with the input of 800 V and the rated output of 220 V/ 4.5A. Its efficiency could reach 99.6% with THD=0.95% at full power level.

Index Terms—PV system, Soft switching, Dead time effect.

I. INTRODUCTION

Single-phase string inverter has been widely applied to grid-tied photovoltaic (PV) rooftop applications for its environmental friendliness, small volume and low cost. Its power rating typically ranges from $1 \sim 8$ kW, and relys on two-stage power conversion, which could convert the DC power generated by the PV panel arrays into AC power, thus to fed into a 120 / 220 V single-phase grid connection. However, the inherent attribute of intermittency in solar energy may induce instability or unqualified power. To meet the grid-interconnection standards, high demands of efficiency, harmonic distortion need to be imposed on inverters.

High efficiency, which is one of the most important indexes in PV system of inverters indicates high energy utilization and power density. In order to improve the efficiency, soft

Aircraft Swarm Intelligent Sensing and Cooperative Control Key Laboratory of Sichuan Province

switching is a preferable candidate. In general, those soft-switching topologies can be classified into two categories: DC-side and AC-side soft-switching circuits, according to positions where the resonant tanks are located at [1]. In [2], conventional FB topology is adopted with auxiliary circuit composed of inductors, capacitors and switches added on DC side to supply resonant power. The implementation principle is to adjust the short-circuit pulse of the auxiliary switch to accommodate ZVS for main switches in different power levels. The circuit structure is simple, but full load conducting current will flow through the auxiliary switches, resulting in extra conduction loss. Regarding this point, many literatures choose to shift auxiliary circuits to AC side. In [3], two kinds of resonant tank configurations consisting of discrete inductor and coupled inductor structures are proposed. The AC-side resonant tanks do reduce the unnecessary conduction loss compared with the DC-side. However, the added resonant tanks in AC-side will work synchronously with the high-frequency main switches, so that the high frequency current ripple caused by commutating resonant current will adversely affect the eletromagnetic interference (EMI) performance. Simply put, the implementation of all switches soft switching requires the addition of auxiliary components, which is hard to realize under intermittent input power.

Total harmonics distortion (THD) is the other significant index for grid-tied systems. There are many sources of distortion affecting the current quality, such as complex non-linear devices, PWM modulation, and dead-time effect, among which the dead-time effect is the main source of low-order distortion. In practice, the dead time has to be implemented to prevent the inverter from shoot-through during the switching interval. However, it also introduces potentially harmonic problems. Many literatures have dedicated on the mechanism of the harmonic current injection introduced by dead-time effect [4]. Nevertheless, they mainly focus on the mechanism or compensation method based on hard-switching, and few analysis emphasizes on the effects of dead-time under soft-switching conditions. The difference relys on: The dead time in hard-switching causes long delay times (50~100ns) at the turn-off transition, leading to poor total harmonic distortion (THD) performance. Whereas, this might be much more severe under soft-switching conditions since this dead time interval needs to be further prolonged (100ns~1us) to accommodate the discharging period for the output junction capacitance (C_{oss}) of switches under different power level. Referring to this, it is significant to explore the mechanism of dead time effect affected by soft switching.

In this paper, a partial soft-switching SiC-based ANPC inverter is proposed. Instead of auxiliary circuits, it adopts hybrid PWM modulation to realize synchronous switches partial soft switching, while adjusting dead time to achieve lower current harmonics accommodating fluctuated input power. In addition, a mathematical loss model is established to calculate the specific dead time which could both achieve the peak efficiency and the optimal THD. Furthermore, the mechanism of the dead time effect affected by soft switching is explored. A 1-kW prototype is demonstrated with the input of 800 V and the rated output of 220 V/ 4.5A.

II. THE TOPOLOGY AND OPERATION MODES

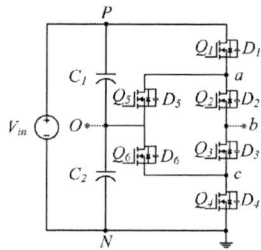

Fig. 1: The ANPC topology.

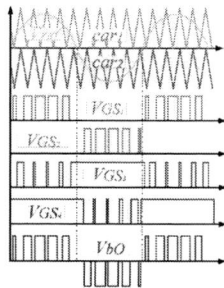

Fig. 2: Hybrid PWM modulation scheme.

The topology of full-SiC ANPC inverter is shown in Fig.1. According to Hybrid PWM modulation scheme, the switching cell of 3L-ANPC inverter can be generated by a positive ANPC (P-ANPC) switching cell and a negative ANPC (N-ANPC) switching cell. Q_1, Q_4 play as the main switches, which operate in positive and negative switching cell, respectively, and are full hard turn-on, full or partial hard turn-off. Q_2, Q_3,

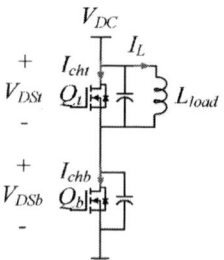

Fig. 3: The phase-leg configuration.

Q_5, Q_6 serve as the synchronous switch to freewheel zero-state current in different switching cell, which could achieve full or partial ZVS and full ZCS in accordance with the power level. The operation scheme and their equivalent circuits are shown in Fig. 2. Due to the symmetry of switching process, only operation modes in P-ANPC will be analyzed here.

III. THE OPTIMAL DEAD TIME ALGORITHM

The range of ZVS for synchronous switches depends on the the zero-state load current $I_o(t)$ and the dead time t_d. In order to achieve the peak efficiency and lower THD, this section proposes a mathematical loss model to help figure out the optimal dead time.

A. The Mechanism of Dead Time

Here takes the phase-leg configuration shown in Fig.3 as an example. In hard switching conditions, due to the nonideal switching behavior of the power devices, it is necessary to add t_d to avoid the shoot through. Conventionally, the optimal t_d depends on the turn-off transition of the main switch (Q_b). The previous works have elaborated the optimal t_d for Q_b to hard turn off [5]. Based on the same mechanism, this section will mainly emphasize on the optimal t_d to achieve less power loss and lower THD. As Q_b is turned off, L_{load} plays as a current source to charge and discharge C_{ossb}, C_{osst} simultaneously. To reduce switching loss and accommodate the ZVS of Q_t, this t_d needs to be dynamically adjusted according to inductance current I_L, thus to ensure the total energy commutation of the

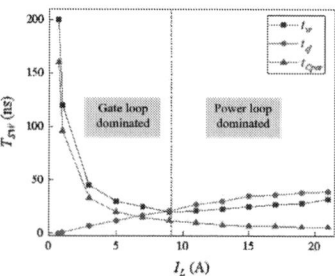

Fig. 4: t_{cf} VS. t_{vr} VS. t_{Cpar} dependence on various operating current.

979-8-3503-3714-3/23 $31.00 © 2023 IEEE

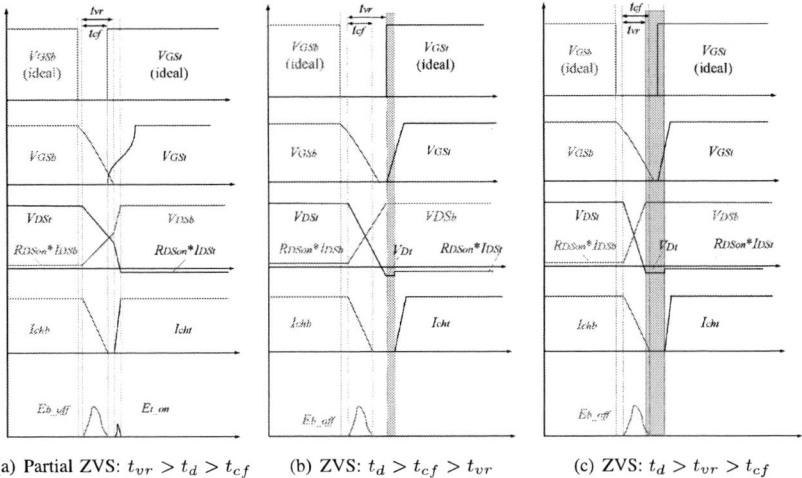

(a) Partial ZVS: $t_{vr} > t_d > t_{cf}$ (b) ZVS: $t_d > t_{cf} > t_{vr}$ (c) ZVS: $t_d > t_{vr} > t_{cf}$

Fig. 5: The schematic diagram of three ZVS modes.

commutaing junction capacitance C_{par} before Q_t turns on. It is noteworthy that the equivalent junction capacitance C_{par} equals to two junction capacitances connected in parallel.

Fig. 4 shows the LTspice simulated turn-off transient for a phase-leg configuration composed of two SiC MOSFETs (ROHM SCT3060AL). The simulation is operated under V_{DC} = 300 V, operating temperature $T_a = 25°C$. It illustrates the relationship among the channel current fall time (t_{cf}), voltage rise time (t_{vr}) of Q_b, and the junction capacitance charging time t_{Cpar} with the change of I_L, where the t_{cf} represents the time for channel current of Q_b to reduce to zero; t_{vr} represents the time for the drain-source voltage V_{DSb} from 0 to rise to V_{DC}; t_{Cpar} represents the time for V_{DSb} from 0 to rise to V_{DC} when all I_L is used to charge C_{par}.

It divides the operation zoom into power loop dominated and gate loop dominated according to the comparison of t_{cf} and t_{vr}. Specifically, in the light power condition ($t_{cf} < t_{vr}$), the majority of I_L is used to charge C_{par}, and only a little is flowing through the channel of Q_b calculated as (2), so that $t_{cf} < t_{Cpar}$. The channel current I_{chb} would reduce to 0 before its drain-source voltage V_{DSb} rises to V_{DC}, thus $t_{cf} < t_{Cpar} < t_{vr}$. In this case, the t_d could be either set as $t_{cf} < t_d < t_{vr}$ as Fig. 5 (a) shows or $t_{cf} < t_{vr} < t_d$ as Fig. 5 (b) shows. The former one has not finish the energy commutation, Q_b, Q_t will be charged and discharged to $V_{DSb0}(V_{DSb0} < V_{DC})$, $V_{DSt0}(V_{DSb0} > 0)$ as (4), (5) shows. Q_b would suffer partial hard turn-off loss, and Q_t would suffer partial hard turn-on loss, for its C_{osst} has not been totally discharged yet. The switching energy can be listed as (6).

$$I_{chb} = g_m(V_{mil} - V_{th}) \quad (1)$$

$$\Delta V_{DS} = \int_0^{t_d} \frac{(I_L - I_{chb})}{C_{par}} dt \quad (2)$$

$$V_{DSt0} = V_{DC} - \Delta V_{DS} \quad (3)$$

$$V_{DSb0} = 0 + \Delta V_{DS} \quad (4)$$

where I_{ch}, V_{mil} indicate the channel current and the Miller voltage, which are stricted by the gate drive loop. g_m, V_{th} indicate the transconductance, and the threshold voltage.

$$E_{sw} = E_{swt-on} + E_{swb-off}$$
$$= \int_0^{t_{vf}} I_{DSt} V_{DSt} dt + \frac{1}{2} C_{par} V_{DSt0}^2$$
$$+ \int_0^{t_d} I_{DSb} V_{DSb} dt - \frac{1}{2} C_{par} V_{DSb0}^2 \quad (5)$$

where E_{sw} indicates the switching energy, the t_{vf} is the voltage fall time of Q_t, whose dV_{DS}/dt is stricted by the gate loop and I_{DS}. The dI_{DS}/dt is related to the gate loop and is assumed to remain the same absolute value with the same gate loop settings.

The latter one could realize ZVS for Q_t, but brings overlong dead time which will induce poor dead time effect and extra dead time loss on D_t.

$$E_{sw} = E_{swt-on} + E_{swb-off}$$
$$= 0 + \int_0^{t_d} I_{DSb} V_{DSb} dt - \frac{1}{2} C_{par} V_{DC}^2 \quad (6)$$

$$E_{Dt} = \int_0^{t_d - t_{vr}} I_{DSt} V_{Dt} dt. \quad (7)$$

At high power condition ($t_{cf} > t_{vr}$), only a portion of I_L is needed to charge C_{ossb}, so that $t_{Cpar} < t_{cf}$. In this case, V_{DSb} will quickly reaches V_{DC} before I_{chb} drops to zero. Thus, t_d could only be set as $t_{vr} < t_{cf} < t_d$ as Fig. 5 (c) shows, so that Q_b would suffer full hard turn-off loss, and the ZVS for Q_t could be ensured anyway. The switching energy equations are the same as (7), (8).

B. The Optimal Dead Time for VSIs

When it comes to voltage source inverter (VSI), I_L flows in the shape of an approximate sine wave. If t_d is set too small as Fig. 5 (a) shows, the most duration of Q_t will suffer partial hard turn-on loss. If t_d is set overlong shown as Fig. 5 (b), (c) shows, poor dead time effect will be induced and the D_t will endure extra dead time loss as well. To make a trade off between efficiency and THD performance, a mathematical loss model is established to help calculate the optimal dead time.

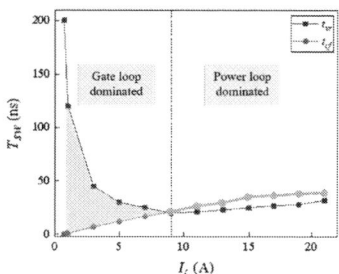

Fig. 6: The configurable range of t_d.

The grey shadow area in Fig. 6 shows the configurable range of t_d. To avoid the shoot-through problem, the lower limit of t_d is t_{cf}. In the gate dominated loop, t_{vr} is the upper limit of t_d, if $t_d > t_{vr}$, Q_t will enjoy ZVS in the whole cycle though, the unnecessary dead time loss of D_t will be induced, thus, the range of t_d is $t_{vr} >= t_d > t_{cf}$. In the power loop dominated area, t_d could only be set as $t_d = t_{cf}$. Since t_d has little effect on the turn-on loss, conduction loss of Q_b, and the turn-off loss, conduction loss of Q_t, the losses here only indicate power losses occurred in turn-off transition of Q_b, including the turn-on loss of Q_t, the turn-off loss of Q_b, the dead time loss of D_t. The design procedure is as Fig. 7 shows.

C. The simulation of Algorithm

According to this method, Fig. 8 presents the 2D variation between t_d and P_{loss} under different power levels. It points out the optimal working point locates at the bottom of each curve. As for ANPC utilized in this paper, the C_{par} equals to $C_{oss1} + C_{oss3} + C_{oss5}$. The design procedure are exactly the same as the half-bridge configuration.

IV. THE EXPERIMENTAL RESULTS

To verify the effectiveness of aforementioned optimal dead time algorithm, a hardware prototype is developed in this section and the inverter performances are evaluated from several aspects, including ZVS ranges, ZVS conditions, THD.

A. The Prototype

As Fig.9 shows, a 1-kW prototype with 800-V DC input and 220-V/ 4.5-A AC output have been built to verfiy the optimal

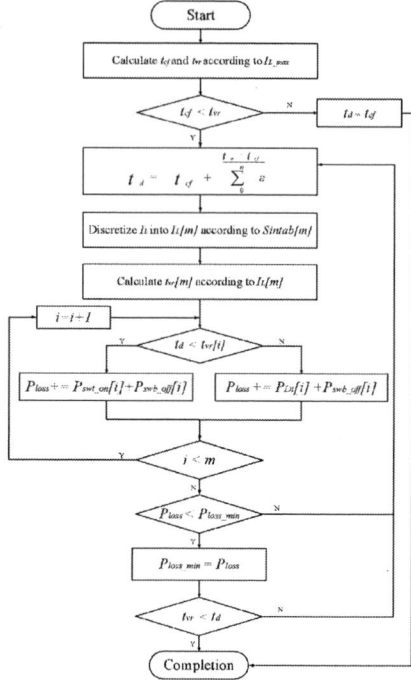

Fig. 7: The flowchart to design the optimal dead time.

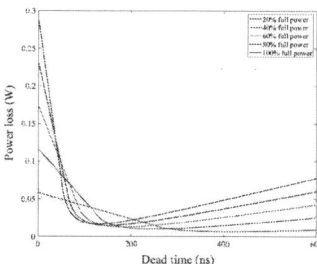

Fig. 8: Power loss VS. dead time under different power conditions.

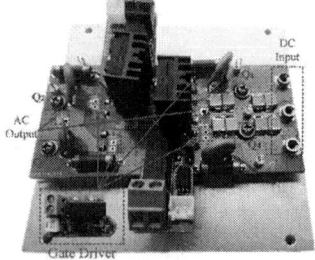

Fig. 9: The ANPC Prototype.

TABLE I: Experimental parameters.					
Input voltage	Rated output	Filter inductance1&2	Filter capacitor	DC-link capacitor	Switching frequency
800 V	220 V/ 4.5 A	3 mH/ 43 uH	6 uF	470 uF	30 kHz

TABLE II: Switch parameters.						
	Voltage rating	Drain current	R_{ds-on} @ 25°C	R_{ds-on} @ 125°C	C_{oss}	Q_g
SCT3060AL	650 V	39 A @ 25°C	60 $m\Omega$ @ 25 °C	79.2 $m\Omega$ @ 125 °C	55 pF @ 500 V	58 nC @ 18 V

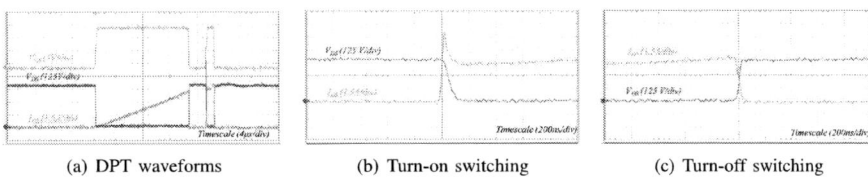

(a) DPT waveforms   (b) Turn-on switching   (c) Turn-off switching

Fig. 10: The experimental results for the DPT on the SiC MOSFET.

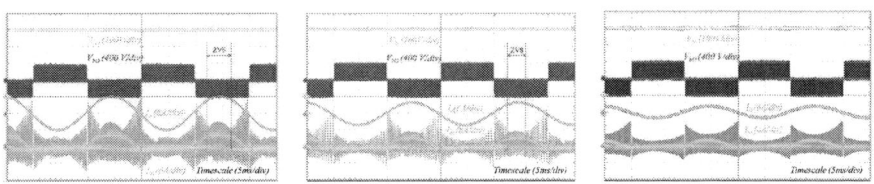

(a) 100% full power level, t_d = 200 ns(b) 50% full power level, t_d = 330 ns(c) 20% full power level, t_d = 400 ns

Fig. 11: Steady state waveforms of $V_{in}, I_{in}, V_{bO}, I_o$ under different power levels and optimal dead time

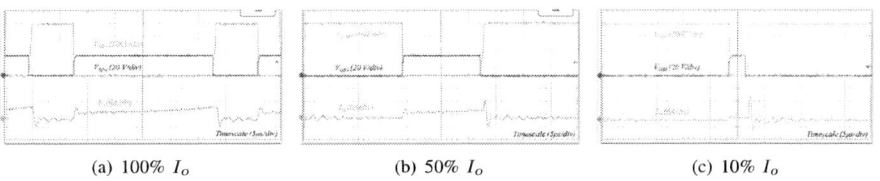

(a) 100% I_o (b) 50% I_o (c) 10% I_o

Fig. 12: Turn-on and turn-off transitions of Q_4 under different I_o at negative cycle in 100% full power level.

dead time algorithm. The experimental parameters are listed in Tab.I. The parameters of SiC MOSFET (ROHM SCT3060AL) are shown in Tab.II. The conduction and switching losses are deduced by the DPT experiment. This DPT is constructed with two SiC MOSFETs (SCT3060AL), and is operated under 400 V/ 4.5A as shown in Fig. 10.

B. The ZVS Ranges

The steady-state switching waveforms of $V_{in}, I_{in}, V_{bO}, I_o$ under different power levels are shown in Fig.11. According to the optimal dead time algorithm mentioned in Section 3, the optimal t_d to achieve higher efficiency and lower THD is 200ns under 100% full power level, 330ns under 50% full power level, and 400ns under 20% full power level. According to Fig.11, I_{in} has large overshoot when it is comparatively low, and as I_{in} increases, this overshoot decreases. Fig.12 shows the switching transitions of Q_4 as the I_o changes in the shape of negative-cell sine wave. It reflects that I_{in} has two times overshoot: The rising edge overshoot and the falling edge overshoot. The former happens when Q_4 is turned on.

This time Q_4 suffers the reverse recovery effect induced by Q_2, Q_6. When I_{in} is low, there is no sufficient energy to discharge C_{oss2} and C_{oss6}, so D_2, D_6 will not conduct within t_d, and reverse recovery effect does not exist. Whereas, as I_{in} rises, D_2 and D_6 start to conduct before Q_4 is turned on, causing severe reverse recovery overshoot at the rising edge of I_{in}. The latter overshoot occurs when Q_4 is turned off, and this overshoot is much larger as I_{in} decreases. This is because when I_{in} is low, C_{oss4} could not be fully charged within t_d. At the time Q_2, Q_6 are turned on, I_{in} will rapaidly increase to charge C_{oss4}. The level of falling edge overshoot depends on the the value of V_{DS4} that already been charged within t_d. The closer V_{DS4} is charged to V_{in}, the less severe the overshoot will be. Based on this mechanism, the ZVS realization range of synchronous switches could be deduced as the area framed by red dashed line in Fig. 11. As the power level turns lighter, the ZVS range becomes narrower. When it comes to the light power conditions, since there is no sufficient energy to discharge the junction capacitance, ZVS will not be realized even to prolong the t_d.

979-8-3503-3714-3/23 $31.00 © 2023 IEEE   123

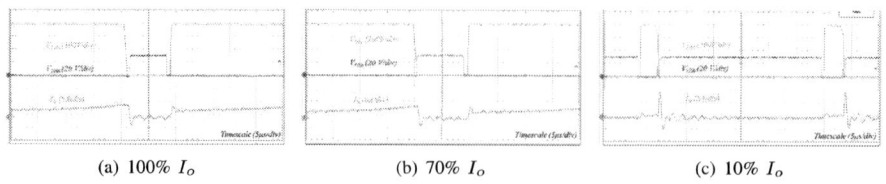

(a) 100% I_o (b) 70% I_o (c) 10% I_o

Fig. 13: Turn-on and turn-off transitions of Q_6 under different I_o at negative cycle in 100% full power level.

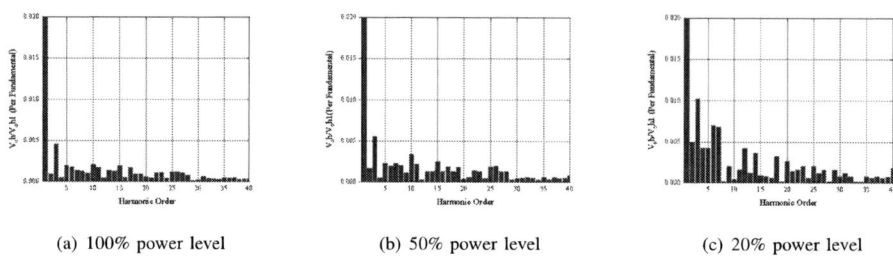

(a) 100% power level (b) 50% power level (c) 20% power level

Fig. 14: The harmonic spectrum of V_{bO}.

C. The ZVS Condition

Fig.12, 13 reflect the switching process of Q_4, Q_6 as I_o changes in the shape of sine wave in negative switching cell under 100% power level. Fig.12 shows the waveforms of V_{GS4}, V_{DS4}, I_{in} when I_o reaches 100%, 50%, 10% peak value, respectively. It can be observed that Q_4 suffers total hard switching on loss regardless of the value of I_o. As for turning-off transitions, a turning point can be observed on the rising edge of V_{DS4} in Fig. 12 (b), (c). This is because as Q_4 is turned off, C_{oss4} could not be fully charged within t_d at low I_o, and when Q_2, Q_6 are turned on, C_{oss4} could be eventually charged by the rapidly increasing I_{in}. Consequently, Q_4 experiences partial hard switch off.

As for Q_6 shown in Fig. 13, when I_o is large and have sufficient energy to discharge C_{oss6} before Q_6 is turned on, Q_6 enjoys ZVS as Fig.13(a) shows, and there will be no falling edge overshoot in I_{in}. However, this will induce extra dead time loss instead, so that the rising edge overshoot will be induced. When I_o reaches 70% peak value as Fig.13(b) shows, Q_6 is close to the boundary of ZVS and partial hard switching-on. This boundary is the optimal operating point which could achieve ZVS and low extra diode loss. Fig.13(c) reflects the waveforms at low I_o. It is clear that the falling edge of V_{DS6} has been cut into two drop rate by a turning point. Before Q_6 is officially turned on, I_o has already discharged part of V_{DS6}, and when V_{GS6} finally reaches the Miller platform, C_{oss6} will immediately release the remaining charge and commute energy with C_{oss4}, so that the falling edge overshoot in I_{in} will be induced. As for its turning-off transitions, Q_6 could achieve ZCS anyway, regardless of the value of I_o.

D. The THD Performance

Fig. 14 shows the Fast Fourier Transform (FFT) results of V_{bO} in Fig. 11, and it shows THD of ANPC under 100% power level reaches 0.95% with dead time =200ns; under 50% reaches 1.11% with dead time =330ns; under 20% reaches 1.91% with dead time =400ns. This t_d is selected to both achieve the higher efficiency and the improved THD. It is observed that the high power conditions possesses less harmonic components in the output, for the shortened t_d has eliminated the spectrum of odd-order harmonics.

V. CONCLUSION

In this paper, an all SiC-based ANPC is proposed to accommodate 1-kW single-phase string PV inverter. The synchronous switches could achieve partial ZVS with the utilization of hybrid PWM modulation scheme. A mathematical loss model is presented to help figure out the optimal dead time. Furthermore, the mechanism of the dead time effect affected by soft switching is explored.

REFERENCES

[1] J. Jana, H. Saha, and K. Das Bhattacharya, "A review of inverter topologies for single-phase grid-connected photovoltaic systems," *Renew. Sust. Energ. Rev.*, p. S1364032116306943, 2016.

[2] Y. Chen, M. Chen, and D. Xu, "A 3kW two-stage transformerless PV inverter with resonant DC link and ZVS-PWM operation," *IEEE Trans. Ind. Electron.*, vol. PP, no. 99, pp. 1–1, 2020.

[3] Y. Xia and R. Ayyanar, "Naturally adaptive, low-loss zero voltage transition circuit for high frequency full bridge inverters with hybrid PWM," *IEEE Trans. Power Electron.*, pp. 1–1, 2017.

[4] Q. Yu, E. Lemmen, K. Wijnands, and B. Vermulst, "Output spectrum modelling of an H-Bridge inverter with dead-time based on switching mode analysis," *IEEE Trans. Power Electron.*, vol. PP, no. 99, pp. 1–1, 2021.

[5] Z. Zhang, H. Lu, D. J. Costinett, F. Wang, L. M. Tolbert, and B. J. Blalock, "Model-based dead time optimization for voltage-source converters utilizing silicon carbide semiconductors," *IEEE Trans. Power Electron.*, 2017.

979-8-3503-3714-3/23 $31.00 © 2023 IEEE

A Physical-Based 3rd-Quadrant Behavioral Model for Power SiC MOSFET

Yuzhi Chen
Department of Electrical Engineering,
Tsinghua University
Beijing, China
chenyz_thu@163.com

Chi Li
Department of Electrical Engineering,
Tsinghua University
Beijing, China
chi.li.2014@ieee.org

Yifan Wu
Department of Electrical Engineering,
Tsinghua University
Beijing, China
wu-yf21@mails.tsinghua.edu.cn

Zedong Zheng
Department of Electrical Engineering,
Tsinghua University
Beijing, China
zzd@tsinghua.edu.cn

Abstract—In this work, a SiC MOSFET third quadrant (3rd-quad) model is built, which is the first physical-based model describing 3rd-quad I-V behavior comprehensively. Physical coupling between channel and body diode is introduced. The impact on channel from body diode conduction is considered. Simple criteria are given for 3 states under different gate bias. Channel current expressions applicable in various situations are proposed. Degradation of 3rd-quad threshold voltage is modeled. Device physical models fitting SiC materials are chosen. Overall model is derived along with the modeling of linear resistance, interface mobility and body diode, which are all physical-based. Validation of the proposed model shows good agreement with the real device measurement at 27 °C and 150 °C and in wide ranges of gate-bias.

Keywords—*SiC MOSFET, third quadrant (3rd-quad) behavior, channel, body diode, body effect, physical coupling.*

I. INTRODUCTION

Wide Bandgap power devices like silicon carbide (SiC) MOSFET operates in higher temperature, faster switching speed and lower loss compared with its silicon counterpart, suitable for power conversion applications. In the past decade, power SiC MOSFET has been gradually implemented in electrical drive system, renewable energy conversion system, DC charger and smart grid for its superior conduction, switching and breakdown performance.

Third quadrant (3rd-quad, voltage and current of the device are both negative, also called reverse conduction) working mode accounts for large proportion in most SiC-based converters. In inductive hard switching topologies like synchronous buck/boost converters and motor drive inverters, SiC MOSFET operates in the 3rd-quad when freewheeling the inductive load current. The 3rd-quad current may distribute in three paths, MOSFET channel, body diode and external anti-parallel Schottky barrier diode (SBD). But in most cases, the external SBD increases the switching loss, and is removed to simplify converter design and to increase power density [1], [2]. To avoid arm shoot-through, alternative driving signals of complementary devices with a dead time is normally adopted as the control strategy. During 3rd-quad freewheeling period, the reverse channel is turned on when positive gate bias is applied and body diode is used at the dead time. But it is more complex than in 1st-quad because of the difference in physics and the coupling of two paths. Hence, it is worthwhile to analyze the 3rd-quad behavior of two paths in SiC MOSFET.

This work was supported by National Natural Science Foundation of China (No. U2106217).

Previous work from Chongqing University indicates that threshold voltage in 3rd-quad is smaller than in 1st-quad because of body effect [3]. But their proposed expression over-estimates the channel current. Researchers from CPES found that the reverse conduction of MOS channel leads a higher turn-on voltage of body diode and a modeling method is given for devices rated at different voltages [4]. However, the impact on channel from the conduction of body diode has not been discussed. In conditions when the gate bias is lower than the threshold voltage, the current calculation method has not been studied yet. Besides, in TCAD numerical tools, the bipolar simulation of SiC is inaccurate and relatively slow. Therefore, a comprehensive model of SiC MOSFET 3rd-quad behavior is meaningful and needs to be conducted.

In this work, a physical-based 3rd-quad behavioral model for power SiC MOSFET is built. Modeling steps adhere to physical natures completely. The rest of this paper is organized as follows. In section II, physical mechanism and physical coupling of channel and body diode are introduced in cellular level. Then the modeling schematic diagram is given. In section III, a channel current calculation method is derived at three different states. Modeling of linear resistance and body diode along with principles of device physics are given. In section IV, the proposed model is performed by MATLAB/SIMULINK and validated by comparing with measured results. Section V summarizes the full paper.

II. THIRD QUADRANT BEHAVIORAL MODEL

A. Channel and body diode 3rd-quad behaviors

Planar SiC MOSFET, commonly used for 1200 V and higher voltage ratings, is investigated in this work. The cell structure of planar SiC MOSFET is shown in Fig. 1. There are two paths for 3rd-quad conduction, MOSFET channel path and body diode path. The channel path is formed when gate voltage exceeds threshold voltage. Unipolar electron current flows via N^+ region, inversion layer below gate oxide, a narrow JFET region, and diffuses to the whole cell pitch in N^- drift region, spreads into substrate and drain finally. The body diode path is opened when source-drain voltage exceeds built-in voltage of SiC PN junction. This path is bipolar, pass through P^+ region initially, and diffuses laterally in the same way with channel path. Conductivity modulation by minority carrier injection occurs when body diode is on, reducing the resistance of drift region. Fig. 1 sketches the current path of MOSFET channel and body diode by orange arrow and blue arrow respectively. Researchers in [4] propose that the body

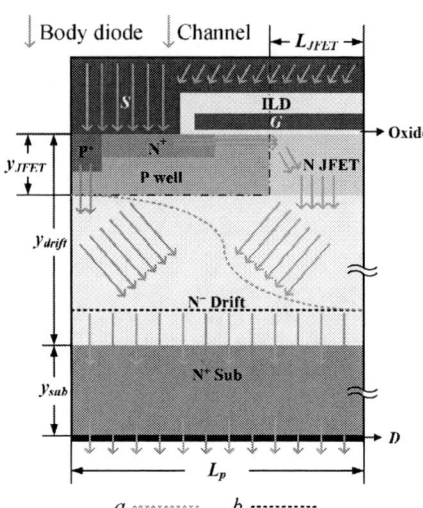

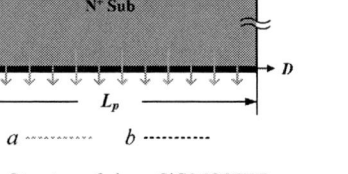

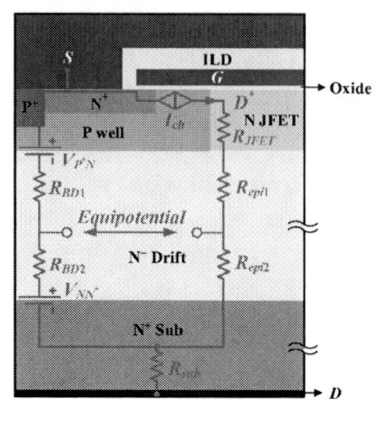

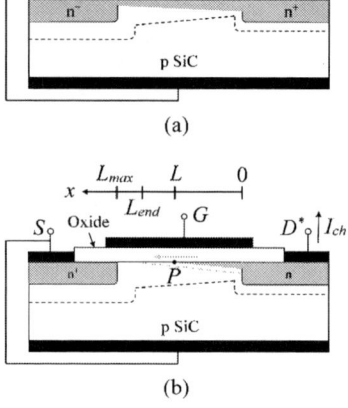

Fig. 1. Structure of planar SiC MOSFET and 3rd-quad current paths.

Fig. 2. Schematic diagram of 3rd-quad behavioral model.

Fig. 3. MOS structure. (a) Channel fully opened. (b) Channel critically opened

diode is shielded when channel conducts. They divide the unipolar path into two parts by an equipotential line penetrating from the PN junction into the drift region below JFET, as sketched by dotted line a. The channel current delays the turn on of body diode by building potential distributions in the cell. But this theory only depicts the influence on body diode turn-on from channel conduction. Actually, when body diode conducts, the voltage across the channel is clamped, which makes the channel current changes quite slowly with drain voltage. In this work, the model is refined from [4] to depict the impact of body diode on channel. The detailed modeling procedure is described as follows.

B. Modeling Procedure and Modeling Diagram

The dotted line b is the refinement in this work, as shown in Fig. 1. It is a boundary of whether the current density is uniformly distributed across the cell and divides the drift region into two parts. This boundary is proved horizontal by TCAD device simulation. The schematic diagram of the proposed model in this work is shown in Fig. 2. Line b is modeled as an equipotential line and divides the bipolar path in the drift region into two bipolar resistance R_{BD1}, R_{BD2}, which presents the effect of carrier injection into the drift. Unipolar resistance is divided as R_{epi1}, R_{epi2} refer to [4]. When body diode is off, line b coincides with line a, the model is not changed. When body diode is on, the clamped channel voltage is exerted by line b.

The components of the model are explained as follows. R_{JFET}, R_{epi1}, R_{epi2}, R_{sub} are linear resistance of JFET region, drift region and substrate. V_{P^+N}, $V_{NN'}$ are built-in voltage of the PN junction and the HL junction. R_{BD1}, R_{BD2} are bipolar resistance of the body PiN diode, varying with injection level. MOSFET channel is modeled by a voltage controlled current source and D^* is the point in the channel end. The resistance of N$^+$ region, P$^+$ region, P well region and source contact is ignored. Channel current I_{ch} is depicted by channel voltage V_{SD^*} and gate bias V_{GS}. The specific modeling steps of these physical structures and quantities is described in the next section.

III. MODELING OF PHYSICAL STRUCTURES AND QUANTITIES

A. MOSFET Channel

The Fermi potential of metal-semiconductor contact ψ_F is defined by

$$\psi_F = \frac{kT}{q}\ln\left(\frac{N_A^-}{n_i}\right) \tag{1}$$

where k is the Boltzmann constant, T is temperature, q is the elementary charge, N_A^- is the ionized acceptor concentration far from the surface and n_i is the intrinsic carrier concentration.

Fig. 3 shows the simplified MOS structure in power SiC MOSFET. An electron inversion layer is formed below the gate oxide when surface potential ψ_S of the p region (P well in Fig. 1) exceeds $2\psi_F$. Voltage potential distribution is built along the channel when a drain voltage is applied. Define x is the positive direction along the channel from D^* to S, and S is grounded ($V_S = 0$). The threshold voltage V_{th} degradation occurs because of body effect [5], and can be written as a function of x, which is given by

$$V_{th} = V_{FB} + 2\psi_F + \frac{\sqrt{2\varepsilon_s qN_A}}{C_{ox}}\sqrt{2\psi_F + V_{(x)}} \tag{2}$$

where $V_{(x)}$ is the electric potential of the channel at point x, ε_s is the silicon carbide dielectric constant, N_A is the acceptor doping concentration far from the surface, C_{ox} is the gate oxide capacitance per unit area, and V_{FB} is the flat band voltage, which is given by

$$V_{FB} = \frac{\Phi_{GS}}{q} - \frac{Q_f}{C_{ox}} - \frac{Q_{it}(\psi_S = 2\psi_F)}{C_{ox}} \tag{3}$$

where Φ_{GS} is the work function difference between metal and semiconductor, Q_f is the fixed-oxide-charge density, and Q_{it} is the interface trap charge density, which is determined by interface parameters and ψ_S.

The electron current $I_{(x)} = -I_{ch}$ is continuous along x axis, which is written as [5]

$$I_{(x)} = -W\mu_{ch}Q_N \frac{dV_{(x)}}{dy} \quad (4)$$

where W is the width of channel in the direction perpendicular to the page, μ_{ch} is the channel interface mobility, and Q_N is the charge per unit area in the inversion layer, which is given by

$$Q_N = -C_{ox}\left[V_{GS} - V_{FB} - 2\psi_F - V_{(x)} \right. $$
$$\left. - \frac{\sqrt{2\varepsilon_s q N_A}}{C_{ox}}\sqrt{2\psi_F + V_{(x)}}\right]. \quad (5)$$

When the device operates in the 3rd-quad, $V_{(x)} < 0$, the threshold voltage V_{th} decreases and electron density of the inversion layer increases along the negative x direction. The expression of 3rd-quad channel current is discussed below at three different states of gate bias.

1) $V_{GS} > V_{th1}$, channel is fully opened.
V_{th1} is the threshold voltage from the datasheet without any body effect, given by

$$V_{th1} = V_{FB} + 2\psi_F + \frac{\sqrt{2\varepsilon_s q N_A}}{C_{ox}}\sqrt{2\psi_F}. \quad (6)$$

At this state, channel is fully turned on, as Fig. 3 (a) shows. The inversion layer is formed when $V_{SD^*} = 0$. Electron current path exists at any positive applied V_{SD^*}. The inversion carrier concentration near D^* increases with V_{SD^*} and the inversion channel length is L_{max} constantly. The expression of channel current I_{ch} is derived by solving (7).

$$\int_{L_{max}}^{0} I_{(x)}dy = -W\mu_{ch}\int_{V_S}^{V_{D^*}} Q_N dV_{(x)} \quad (7)$$

$$I_{ch} = \frac{W\mu_{ch}C_{ox}}{L_{max}}\left\{(V_{GS} - V_{FB} - 2\psi_F)V_{SD^*} + \frac{2}{3}\frac{\sqrt{2\varepsilon_s q N_A}}{C_{ox}}\times\right.$$
$$\left.\left[(-V_{SD^*} + 2\psi_F)^{\frac{3}{2}} - (2\psi_F)^{\frac{3}{2}}\right] + \frac{1}{2}V_{SD^*}^2\right\}. \quad (8)$$

2) $V_{th3} < V_{GS} < V_{th1}$, channel is critically opened.
Define V_{th3} is the threshold voltage in the 3rd-quad. At this state, channel is critically open. There is no inversion layer when $V_{SD^*} = 0$. But with the increase of V_{SD^*}, an inversion layer is formed close to D^* because of body effect before the opening of body diode path. Define P ($\psi_S = 2\psi_F$) is the channel critically open point, as Fig. 3 (b) shows. Body effect does not exist in the source end, so the inversion layer near S will never formed. If the inversion layer near D^* is not formed until body

diode conduction, the channel will never be opened because of the clamping effect of PN junction near D^*. The threshold voltage near D^* is given by

$$V_{th,D^*} = V_{FB} + 2\psi_F + \frac{\sqrt{2\varepsilon_s q N_A}}{C_{ox}}\sqrt{2\psi_F - V_{SD^*}}. \quad (9)$$

The critical condition of whether inversion layer is formed near D^* is judged by

$$V_{GD^*} = V_{GS} + V_{SD^*} = V_{th,D^*}. \quad (10)$$

Define V_f is the turn-on voltage of body diode. When $V_{SD^*} = V_f$, V_{th,D^*} has its minimum value V_{th,D^*min}, and V_{GS} has its critical value V_{th3}, which are given by

$$V_{th,D^*min} = V_{FB} + 2\psi_F + \frac{\sqrt{2\varepsilon_s q N_A}}{C_{ox}}\sqrt{2\psi_F - V_f} \quad (11)$$

$$V_{th3} = V_{th,D^*min} - V_f. \quad (12)$$

In this situation, inversion channel length L varies with V_{SD^*}, and achieves its final value L_{end} when body diode is on, as shown in Fig. 3 (b). To simplify the model, assuming L_{end} is a linear function of V_{GS}, which can be written as

$$L_{end} = L_{max}\frac{V_{GS} - V_{th3}}{V_{th1} - V_{th3}}. \quad (13)$$

For a given V_{GS}, let $V_{SD^*,inv}$, $V_{th,D^*,inv}$ be the critical value of V_{SD^*}, V_{th,D^*} when the electron inversion layer is critically formed near D^*, and exert them into (10), establishing an equation as given by

$$V_{GS} + V_{SD^*} = V_{FB} + 2\psi_F + \frac{\sqrt{2\varepsilon_s q N_A}}{C_{ox}}\sqrt{2\psi_F - V_{SD^*}}. \quad (14)$$

Two solutions of $V_{SD^*,inv}$ are got from (14). The negative root is used for the 3rd-quad. And it is worth noting that the positive root is exactly the solution in the 1st-quad when D^* is critically pinched-off. The critical threshold voltage $V_{th,D^*,inv}$ at D^* for a given V_{GS} is

$$V_{th,D^*,inv} = V_{FB} + 2\psi_F + \frac{\sqrt{2\varepsilon_s q N_A}}{C_{ox}}\sqrt{2\psi_F - V_{SD^*,inv}}. \quad (15)$$

Similarly, assuming channel length L ($0 < L \le L_{end}$) is a linear function of V_{SD^*}, which can be written as

$$L = L_{end}\frac{-V_{SD^*,inv} + V_{SD^*}}{-V_{SD^*,inv} + \left[I_{ch,BDon}\times(R_{epi1} + R_{jfet}) + V_f\right]}$$
$$V_{SD^*} \in \left(V_{SD^*,inv}, \ I_{ch,BDon}\times(R_{epi1} + R_{jfet}) + V_f\right] \quad (16)$$

where $I_{ch,BDon}$ is the channel current when body diode is critically turned on.

979-8-3503-3714-3/23 \$31.00 © 2023 IEEE

TABLE I.	OTHER 4H-SiC MATERIAL PARAMETERS			
Parameter	Value	Unit	Ref.
$\Delta E_{D,0}$	201.3	meV	[6]
$\Delta E_{A,0}$	61.4	meV	[6]
α_D	4	10^{-8} eV cm	[7]
α_A	1.9	10^{-8} eV cm	[7]
$\mu_{n,p}^{min}$	1140, 125	cm^2 V^{-1} s^{-1}	[8]
$\mu_{n,p}^{max}$	40, 15.9	cm^2 V^{-1} s^{-1}	[8]
$\alpha_{n,p}$	-0.5		[9]
$\beta_{n,p}$	$-2.35, -2.15$		[7], [8]
$\gamma_{n,p} = -\delta_{n,p}$	0.61, 0.34		[10]
$N_{n,p}^{crit}$	1.94×10^{17}, 1.76×10^{19}	cm^{-3}	[10]
$N_{n,p}^{ref}$	5×10^{16}	cm^{-3}	[11]
η	1.84		[6]

TABLE II.	INTERFACE MOBILITY CALCULATION PARAMETERS		
Parameter	Unit	Value	Ref.
Γ_C	eV^{-1} cm^{-2}	1.5×10^{11}	[7]
n_C	cm^{-3}	1.2×10^{12}	[13]
ζ_C		0.8	[13]
A	cm/s	7.82×10^7	[14]
B	(V cm^{-1})$^{-2/3}$ K cm s^{-1}	9.92×10^6	[14]
Γ_{SR}	V/s	0.4×10^{13}	

For $V_{SD^\bullet} < V_{SD^\bullet,inv}$, P coincides with D^*. For any $V_{SD^\bullet} > V_{SD^\bullet,inv}$, point P is constantly at the critical state that the inversion layer is being formed. The electric potential of P is given by

$$V_{PS} = \begin{cases} -V_{SD^\bullet} & , \quad 0 < V_{SD^\bullet} \leq V_{SD^\bullet,inv} \\ -V_{SD^\bullet,inv} & , \quad V_{SD^\bullet} > V_{SD^\bullet,inv} \end{cases} \quad (17)$$

The expression of channel current I_{ch} is derived as

$$\int_{L_{max}}^{0} I_{(x)}dy = -W\mu_{ch}\int_{V_P}^{V_{D^\bullet}} Q_N dV_{(x)} \quad (18)$$

$$I_{ch} = \frac{W\mu_{ch}C_{ox}}{L}\left\{(V_{GS}-V_{FB}-2\psi_F)V_{PD^\bullet} + \frac{2}{3}\frac{\sqrt{2\varepsilon_s qN_A}}{C_{ox}}\times \right.$$

$$\left.\left[(-V_{SD^\bullet}+2\psi_F)^{\frac{3}{2}} - (V_{PS}+2\psi_F)^{\frac{3}{2}}\right] + \frac{1}{2}V_{SD^\bullet}^2 - \frac{1}{2}V_{PS}^2\right\}. \quad (19)$$

When $L = L_{end}$, $I_{ch,BDon}$ can be solved by (19) and Ohm's law, which is expressed by

$$I_{ch,BDon} = \frac{V_f - V_{SD^\bullet}}{R_{epi1} + R_{jfet}}. \quad (20)$$

3) $V_{GS} < V_{th3}$, channel is closed.
If the leakage current is ignored, there will be no current flowing through the channel at this state in whole $V_{SD^\bullet}$ range, the 3rd-quad behavior is dominated by body diode.

B. Models of Device Physics

1) Mobility

a) Bulk mobility: Classic Caughey-Thomas mobility model is selected in this work [12]. The bulk mobility of electron and hole is given by

$$\mu_{n,p}^b = \mu_{n,p}^{min}\left(\frac{T}{300}\right)^{\alpha_{n,p}} + \frac{\mu_{n,p}^{max}\left(\frac{T}{300}\right)^{\beta_{n,p}} - \mu_{n,p}^{min}\left(\frac{T}{300}\right)^{\alpha_{n,p}}}{1 + \left(\frac{T}{300}\right)^{\delta_{n,p}}\left(\frac{N_{tot}}{N_{n,p}^{crit}}\right)^{\gamma_{n,p}}} \quad (21)$$

where N_{tot} is the total dopant concentration, $\mu_{n,p}^{min}$, $\mu_{n,p}^{max}$ are the minimum and maximum mobility for electron and hole at 300K, $N_{n,p}^{crit}$, α, β, γ, δ are temperature and doping dependent parameters, which are listed in Table I.

b) Interface mobility: The channel interface mobility μ_{ch} is modeled by [7]

$$\frac{1}{\mu_{ch}} = \frac{1}{\mu_b} + \frac{1}{\mu_C} + \frac{1}{\mu_{SR}} + \frac{1}{\mu_{SP}} \quad (22)$$

where μ_b is the channel bulk mobility calculated by (21), μ_C is the mobility due to Coulomb scattering, μ_{SR} is the mobility due to surface roughness scattering, μ_{SP} is the mobility due to surface phonon scattering. μ_C, μ_{SR}, μ_{SP} are given by

$$\mu_C = \frac{\Gamma_C}{N_f + N_{it}}T\left(1 + \frac{n_s}{n_C}\right)^{\zeta_C} \quad (23)$$

$$\mu_{SP} = \frac{A}{E_{EFF}} + \frac{B}{TE_{EFF}^{1/3}} \quad (24)$$

$$\mu_{SR} = \frac{\Gamma_{SR}}{E_{EFF}^2} \quad (25)$$

where n_S is the electron density per unit volume at the surface, E_{EFF} is the interface effective normal field, N_f is the total density of fixed charges and takes the value 2×10^{12} cm^{-3} in this work, N_{it} is the total density of charged interface states. Γ_C, n_C, ζ_C, A, B, Γ_{SR} are fitted parameters listed in Table II. In this work, n_s, E_{EFF}, N_{it} are calculated by the charge sheet model proposed in [15], [16], as a function of V_{GS} and T. The interface state distribution parameter for N_{it} calculation in midgap region and band edge take the value 4.45×10^{11} eV^{-1} cm^{-2} and 1.7×10^{13} eV^{-1} cm^{-2} reported in [16].

2) Incomplete ionization

The ionized donor concentration, acceptor concentration N_D^+, N_A^- can be written as

$$N_D^+ = \frac{N_D}{1 + G_D \exp\left(\dfrac{E_F - E_C}{kT}\right)} \quad , \quad G_D = g_D \exp\left(\frac{\Delta E_D}{kT}\right) \quad (26)$$

$$N_A^- = \frac{N_A}{1 + G_A \exp\left(\dfrac{E_V - E_F}{kT}\right)} \quad , \quad G_A = g_A \exp\left(\frac{\Delta E_A}{kT}\right) \quad (27)$$

where N_D, N_A are the doping concentration of donor and acceptor, N_D^+, N_A^- are the ionized concentration of donor and acceptor, g_D, g_A are the degeneracy factor for donor and acceptor, which equal to 2 and 4 respectively, and ΔE_D, ΔE_A are the ionized energy for donor and acceptor, given by

$$\Delta E_D = \Delta E_{D,0} - \alpha_D N_{tot}^{1/3} \qquad \Delta E_A = \Delta E_{A,0} - \alpha_A N_{tot}^{1/3} \quad (28)$$

where the value of $\Delta E_{D,0}$, $\Delta E_{A,0}$, α_D, α_A for nitrogen and aluminum are listed in Table I.

3) Bandgap narrowing

Bandgap narrowing exert considerable impact on physical properties especially at highly doped regions. Persson's model for 4H-SiC is proved accurate by [6] and is used in this work [17], where the decrease in bandgap energy for conduction and valence band edges is given by

$$\Delta E_{g,n} = A_{n,c}\left(N_D^+/10^{18}\right)^{1/3} + B_{n,c}\left(N_D^+/10^{18}\right)^{1/2}$$
$$+ A_{n,v}\left(N_D^+/10^{18}\right)^{1/3} + B_{n,v}\left(N_D^+/10^{18}\right)^{1/2} \quad (29)$$

$$\Delta E_{g,p} = A_{p,c}\left(N_A^-/10^{18}\right)^{1/4} + B_{p,c}\left(N_A^-/10^{18}\right)^{1/2} + A_{p,v}\left(N_A^-/10^{18}\right)^{1/3}$$
$$+ B_{p,v}\left(N_A^-/10^{18}\right)^{1/2} + C_{p,v}\left(N_A^-/10^{18}\right)^{1/4} \quad (30)$$

where $A_{n/p,c/v}$, $B_{n/p,c/v}$, $C_{p,v}$ are listed in Table III.

C. Linear Resistance

The resistance modeled in this part is unipolar linear resistance and only dependent on temperature. The resistivity of n-SiC is expressed by

$$\rho = \frac{1}{nq\mu_n} \quad (31)$$

where μ_n is the electron mobility in bulk material derived from (21), and n is the electron concentration. Assuming the electron current diffusing angle from JFET to drift is 45°, The resistance of drift region is given by

$$R_{epi} = \frac{1}{n_{drift}q\mu_{n,drift}}\left[\frac{\ln\left(L_p/L_{JFET}\right)}{W}\right.$$
$$\left. + \frac{y_{drift} - y_{JFET} + L_{JFET} - L_p}{WL_p}\right] \quad (32)$$

where L_p, L_{JFET}, y_{drift}, y_{JFET}, y_{sub} are geometry parameters, as shown in Fig. 1. $R_{epi1} = 0.4R_{epi}$ and $R_{epi2} = 0.6R_{epi}$ are taken from [4]. The resistance of JFET and substrate are modeled by

TABLE III. BANDGAP NARROWING CALCULATION PARAMETERS

p-SiC	Value	n-SiC	Value
$A_{p,c}$	16.15	$A_{n,c}$	17.91
$B_{p,c}$	1.07	$B_{n,c}$	2.20
$A_{p,v}$	35.07	$A_{n,v}$	-28.23
$B_{p,v}$	-6.74	$B_{n,v}$	-6.24
$C_{p,v}$	56.96		

$$R_{JFET} = \frac{1}{n_{JFET}q\mu_{n_{JFET}}} \cdot \frac{y_{JFET}}{WL_{JFET}} \quad (33)$$

$$R_{sub} = \frac{1}{n_{sub}q\mu_{n,sub}} \cdot \frac{y_{sub}}{WL_p} \quad (34)$$

D. Body Diode

In this work, the physical-based modeling method for body diode is modified from [11]. Body diode is separated as V_{P^+N}, V_{NN^+}, R_{BD1}, R_{BD2}, modeled as

$$V_{BD} = V_{P^+N} + V_{NN^+} + \int_{0^+}^{y_{drift} - y_{JFET}} E(y)dy + R_{sub}J_{BD}$$
$$R_{BD1} = 0.4 \times \int_{0^+}^{y_{drift} - y_{JFET}} E(y)dy$$
$$R_{BD2} = 0.6 \times \int_{0^+}^{y_{drift} - y_{JFET}} E(y)dy \quad (35)$$

where J_{BD} is the body diode current density, $E(y)$ is the electric field along y direction in the drift region. The bipolar carrier lifetime is modeled as [18]

$$\tau_{n,p} = \frac{\tau_{0,n,p}\left(T/300\right)^\eta}{1 + \left(N_{tot}/N_{n,p}^{ref}\right)} \quad (36)$$

where η, $N_{n,p}^{ref}$ are listed in Table I.

IV. MODEL VERIFICATION

The model is realized by MATLAB/SIMULINK. Sequences to obtain the 3rd-quad behavior are sketched in the flowchart in Fig. 4. For a given temperature, the charge-sheet parameters (for mobility calculation), linear resistance and body diode I-V characteristics are calculated offline. Then current of channel and body diode is calculated online for different V_{GS} and V_{SD}. SiC MOSFET C2M0080120D from Cree, rated at 1200 V 36 A, is chosen to verify the model.

It is noted that the channel current in 1st-quad can also be calculated by expression (8). Hence the 1st-quad operation is first performed to prove the validity and generality of the proposed model. The modeled results of output and transfer characteristics at 300 K are shown in Fig. 5. And the 3rd-quad model is performed at 300 K and 423 K, the results are shown in Fig. 6. When body diode path opens, as the knee points of those curves show, the slope of the 3rd-quad current is changed because the channel current is clamped. The modeled results show good matchings with measured results in wide ranges of V_{GS} and including all three different states analyzed above.

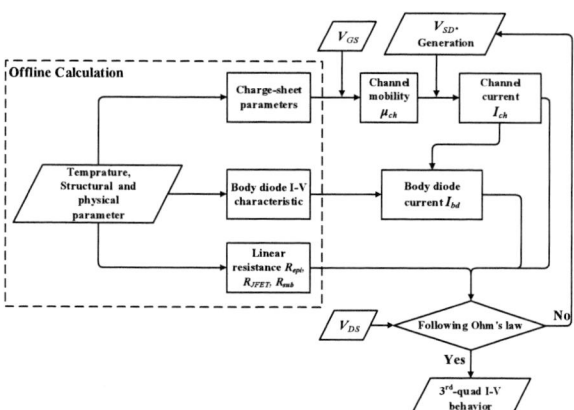

Fig. 4. Modeling flowchart of SiC MOSFET 3rd-quad behavior

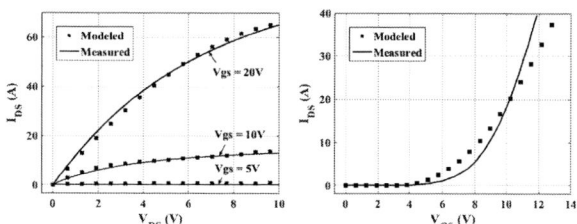

Fig. 5. Modeled and measured results of output (left) and transfer (right) (V_{DS} = 20 V) characteristics for C2M0080120D at 300 K.

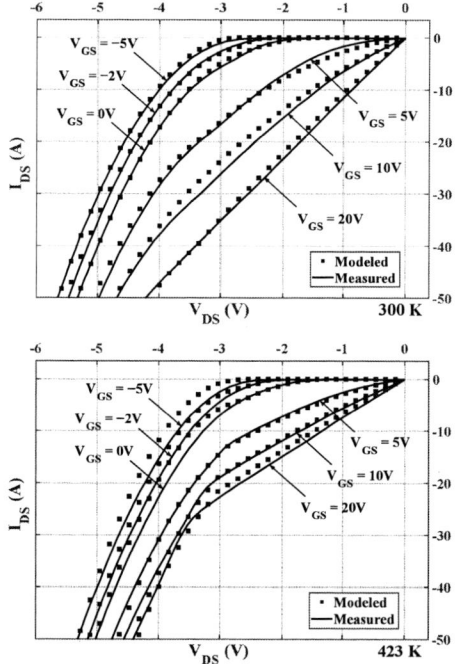

Fig. 6. Modeled and measured results of 3rd-quad I-V characteristics for C2M0080120D at 300 K (above) and 423 K (below).

V. CONCLUSION

SiC MOFET has two paths in 3rd-quad working mode, body diode path and channel path. In this paper, the physical coupling between two paths are analyzed. It is found out that the body diode conduction clamps the channel current. After analyzing the current paths inside the cell, a comprehensive 3rd-quad behavioral model is built. Three states (fully opened,

critically opened and closed) for channel and their judgement criteria are given. The 3rd-quad channel length variation theory is proposed and precise current expressions are derived based on device physics. The distribution of current in two paths can be obtained in different situations after performing this model. This model explains the physical mechanism of SiC MOSFET 3rd-quad behavior in cellular level. It also reveals and reflects the relationships between device structural parameters, material parameters and external electrical behaviors, which provides theoretical guidelines for cell design, device selection and control strategy in SiC MOSFET applications.

ACKNOWLEDGMENT

This present work was supported by national natural science foundation of China No. U2106217.

REFERENCES

[1] S. Yin *et al.*, "Comparison of SiC voltage source inverters using synchronous rectification and freewheeling diode," *IEEE Trans. Ind. Electron.*, vol. 65, no. 2, pp. 1051–1061, Feb. 2018.

[2] K. Yamaguchi, K. Katsura, T. Yamada, and Y. Sato, "Criteria for using antiparallel SiC SBDs with SiC mosfets for SiC-Based inverters," *IEEE Trans. Power Electron.*, vol. 35, no. 1, pp. 619–629, Jan. 2020.

[3] L. Tang *et al.*, "Investigation into the third quadrant characteristics of silicon carbide MOSFET," *IEEE Trans. Power Electron.*, vol. 38, no. 1, pp. 1155–1165, Jan. 2023.

[4] R. Zhang *et al.*, "Third quadrant conduction loss of 1.2-10 kV SiC MOSFETs: impact of gate bias control," *IEEE Trans. Power Electron.*, vol. 36, no. 2, pp. 2033–2043, Feb. 2021.

[5] S. M. Sze, *Physics of Semiconductor Devices.* NJ USA: Wiley, 1981.

[6] K. Tian *et al.*, "Modelling the static on-state current voltage characteristics for a 10 kV 4H-SiC PiN diode," *Materials Science in Semiconductor Processing*, vol. 115, p. 105097, Aug. 2020.

[7] T. Kimoto and J. A. Cooper, *Fundamentals of Silicon Carbide Technology: Growth, Characterization, Devices and Applications.* NJ USA: Wiley, 2014.

[8] T. Ayalew, "SiC Semiconductor Devices, Technology, Modeling, and Simulation," *Ph.D.thesis, Technology Univ.*, Vienna, Austria, 2004.

[9] M. Roschke and F. Schwierz, "Electron mobility models for 4 H, 6 H, and 3 C SiC," *IEEE Trans. Electron Devices*, vol. 48, no. 7, pp. 1442–1447, Jul. 2001.

[10] W. J. Schaffer *et al.*, "Conductivity anisotropy in epitaxial 6H and 4H-SiC," in *Proc. Mat. Res. Soc. Symp., Diamond, SiC Nitride Wide Bandgap Semicond.*, vol. 339, pp. 595–600, San Francisco, CA, 1994.

[11] S. Bellone, F. G. Della Corte, L. F. Albanese, and F. Pezzimenti, "An analytical model of the forward I-V characteristics of 4H-SiC p-i-n diodes valid for a wide range of temperature and current," *IEEE Trans. Power Electron.*, vol. 26, no. 10, pp. 2835–2843, Oct. 2011.

[12] D. M. Caughey and R. E. Thomas, "Carrier mobilities in silicon empirically related to doping and field," *Proc. IEEE*, vol. 55, no. 12, pp. 2192–2193, 1967.

[13] S. Dhar *et al.*, "Inversion layer carrier concentration and mobility in 4H-SiC metal-oxide-semiconductor field-effect transistors," *J. Appl. Phys.*, vol. 108, no. 5, p. 054509, Sep. 2010.

[14] S. Potbhare, N. Goldsman, G. Pennington, A. Lelis and J. McGarrity, "Numerical and experimental characterization of 4H-silicon carbide lateral metal-oxide-semiconductor field-effect transistor," *J. Appl. Phys.*, vol. 100, no. 4, p. 044515, Aug. 2006.

[15] E. Arnold, "Charge-sheet model for silicon carbide inversion layers," *IEEE Trans. Electron Devices*, vol. 46, no. 3, pp. 497–503, Mar. 1999.

[16] A. Pérez-Tomás *et al.*, "Field-effect mobility temperature modeling of 4H-SiC metal-oxide-semiconductor transistors," *J. Appl. Phys.*, vol. 100, no. 11, p. 114508, Dec. 2006.

[17] C. Persson, U. Lindefelt and B. E. Sernelius, "Band gap narrowing in n-type and p-type 3C-, 2H-, 4H-, 6H-SiC, and Si," *J. Appl. Phys.*, vol. 86, no. 8, pp. 4419–4427, 1999.

[18] Synopsys, *Sentaurus Device User Guide*, Synopsys, Inc., Mountain View, CA, Jun. 2018.

979-8-3503-3714-3/23 $31.00 © 2023 IEEE

Performance Comparison of GaN-Based Multilevel Converters for Electric Vehicle Powertrain Application

Seyed Iman Hosseini Sabzevari, Armin Ebrahimian, Nathan Weise

Department of Electrical and Computer Engineering, Marquette University, Milwaukee, WI, USA
iman.hosseini@marquette.edu, armin.ebrahimian@marquette.edu, nathan.weise@marquette.edu

Abstract—Recently, the advantages of utilizing 800V batteries in electric vehicles have been investigated in the literature. However, adopting the DC link voltage of 800V in conventional topologies prevents the utilization of GaN semiconductors. The reason is that the majority of GaN devices are commercially available up to 650V. Consequently, the application of GaN devices will be limited to low-voltage systems. In this paper, the application of 650V GaN-based multilevel converters for electric vehicle powertrains with 800V batteries is studied. Five GaN-based and hybrid converters are simulated in PLECS and their electrical and thermal performance are compared. At first, the overall performance of the selected converters is investigated in the open-loop control mode by connecting them to an RL load. Then, the converters efficiency and power loss are analyzed while driving an electric motor. Simulation results show that the performance of the GaN-based FCMC converter is highly affected by the additional losses in the booster filter. In addition, the efficiency of the hybrid ANPC converter is slightly higher than the GaN-based ANPC converter while the hybrid T-Type converter has the highest peak efficiency. Also, the hybrid NPC converter has the most consistent performance over the entire operation range.

Index Terms—Electric vehicle, gallium nitride, multilevel converter, powertrain, wide bandgap devices.

I. INTRODUCTION

Gallium Nitride (GaN) semiconductors are promising solutions for the next generation of power converters due to their superior material properties such as high mobility 2-D electron gas (2DEG) channel and high-electron saturation velocity. In general, GaN devices have smaller on-resistance, lower switching loss, and higher power density compared to conventional devices [1], [2]. Despite the preferable properties of GaN devices, utilization of GaN devices is limited to low voltage applications. This is because, at the moment, the blocking voltage of the majority of the commercially available GaN devices is less than or equal to 650V.

Conventionally, 400V batteries are used in Electric Vehicles (EVs) powertrains. However, recently, the utilization of an 800V DC bus/battery in EVs has been studied. Implementation of 800V batteries can provide desirable features such as reduction of current, decreasing battery charging time, and improving system efficiency by reducing power loss [3]–[5].

This material is based upon work supported by PowerAmerica and its members. The results and use of any data are governed by the PowerAmerica bylaws.

However, depending on the topology of the power converters, components such as capacitors and semiconductors with high voltage capability are required. For example, a standard two-level six-switch converter is widely used in the EV automotive market. In this topology, the blocking voltage of the switches is equal to the DC bus voltage. Therefore, the implementation of an 800 V battery requires power switches with a blocking voltage of higher than 800V. As a result, the commercially available 650V GaN semiconductors cannot be used in this configuration.

Multilevel converters are gaining an increasing interest in industry and academia and many researchers reported implementation of multilevel converters in EV applications [6]–[13]. In [6] a novel bidirectional T-Type converter is presented. The presented topology generates three-level phase voltage while the inverse voltage across the switches is half of the bus voltage. The performance of multilevel buck-boost converter for EV applications is investigated in [7]–[9]. However, the performance of this topology is analysed while a 400V battery was used as the input voltage of the converter. In [10] the advantages of the modular battery pack with multilevel NPC inverter topology are analyzed. Patel et al. [11] investigated the performance of multilevel converters with different voltage levels (up to 9 voltage levels) in EV applications. In [12] the modular multilevel converters are studied. The strengths and weaknesses of the selected topologies are analyzed with respect to various indicators such as but not limited to switches voltage stress, power density, and fault tolerant capability. As shown in these studies, in multilevel converters, only a fraction of the DC bus voltage will be applied to the switches. Consequently, the 650V GaN devices can be used in an 800V powertrain system. This characteristic of multilevel converters is specifically highlighted in [14], [15]. In [14] the cascaded H-bridge inverter is used as the base module for an n-submodule design. In [15] performance of a three-level neutral point clamped (NPC) inverter is compared with a two-level six-switch inverter. Comprehensive power loss analysis and converters performance for a standard drive cycle are also presented.

In order to leverage commercially available 650V GaN devices and meet the requirement of an 800V DC bus for automotive applications, in this paper, performance of five 650V GaN-based three-level converters for 800V EV powertrain is

979-8-3503-3714-3/23 $31.00 © 2023 IEEE

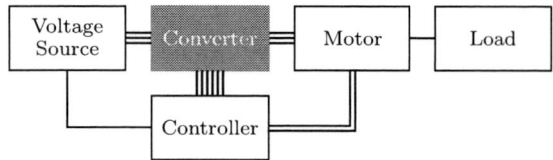

Fig. 1: Simulation platform block diagram.

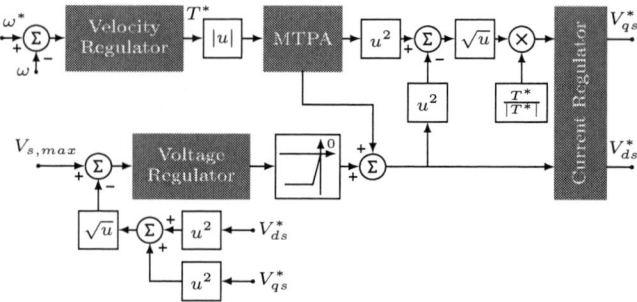

Fig. 2: Implemented speed controller for IPM machines.

studied. The goal of this paper is to analyze and compare the selected converters with respect to efficiency and thermal performance. The converters are simulated in PLECS software, in which their performance is analyzed in two modes: open-loop and motor speed control mode. This paper is organized as follows: in section II the simulation model is presented. In section III the simulation results are discussed. Lastly, the paper is concluded in section IV.

II. SIMULATION PLATFORM

Fig. 1 illustrates the simulation platform which has 5 main subsystems. The Voltage Source block provides the input DC voltage of the Converter subsystem. It includes two series DC voltage sources in parallel with two capacitors with the midpoints connected to each other. The Motor subsystem has two modes of operation, i.e. a series RL load for open-loop test and the model of an IPM motor based on the 2012 Nissan Leaf motor (EM61) for speed-control simulation. The Load subsystem represents the mechanical load torque of the motor. The Controller subsystem senses the required signals for controlling the system and generates proper gate signals for the switches.

A. Converter Subsystem

In this paper, five different converters based on four multi-level topologies are studied: GaN-based Flying capacitor multilevel converter (FCMC), GaN-based Active neutral-point-clamped (ANPC) converter, hybrid NPC converter, hybrid T-Type converter, and hybrid ANPC converter. Fig. 4 shows the single-phase schematic of these topologies. The three-phase configuration is created by properly connecting three single-phase converters. As can be seen in Fig. 4(a), the FCMC converter has a series RLC booster filter in its output shown in red color. In FCMC topology, generating multiple voltage levels is directly related to the voltage balance of the flying capacitors. In single-phase configuration adopting the phase-shift PWM (PSPWM) technique results in natural voltage balancing of the capacitors. However, in three-phase FCMC topology, the natural voltage balance is not preserved due to the imbalance harmonics of phase voltage. Consequently, an auxiliary circuit is required [16], [17]. Fig. 4(b) shows the ANPC topology. In this topology, the switches that are shown in red color are switching at low frequency. This results in almost zero switching loss in these devices. Hence, in the hybrid ANPC converter, these devices are replaced by SiC semiconductors. In the NPC converter, Fig. 4(c), high voltage SiC Schottky diodes are used for minimizing the power loss in the converter. There are four switches in T-Type topology, in

which two of them (shown in red color in Fig. 4(d)) operating at a voltage equal to the input DC voltage. Therefore, in hybrid T-Type converter high voltage SiC switches are used.

B. Test Scenarios

There are two sets of simulations, i.e. open-loop and speed-control tests. In the open-loop test, the Motor subsystem contains a three-phase series RL connected in star configuration and the Controller block works in open-loop control mode. The reference voltages is a set of balanced three-phase sinusoidal waveforms with the desired frequency. The amplitude of the reference signals can be controlled with the value of the modulation index. The Load block has no contribution in this test scenario.

In the speed-control test, the Motor subsystem has the dq model of an IPM motor which is based on the 2012 Nissan Leaf motor. In addition, the load torque on the motor shaft is applied by the Load subsystems. In this scenario, the Controller subsystems uses the measured phase current, and rotor position to generate proper gate signals for operating at the desired speed. Fig. 2 illustrates the simulated speed controller. The controller is capable of field-weakening operation mode and it can track reference speed while utilizing the current and voltage limiters. The MTPA block contains a maximum torque per ampere algorithm for IPM machine that minimizes the stator current magnitude at the desired torque value. The outputs of the controller are the reference voltages that will be fed to the modulation unit for generating gate signals for switches. The implemented modulation techniques for the studied topologies are shown in Fig. 3. For three-level FCMC

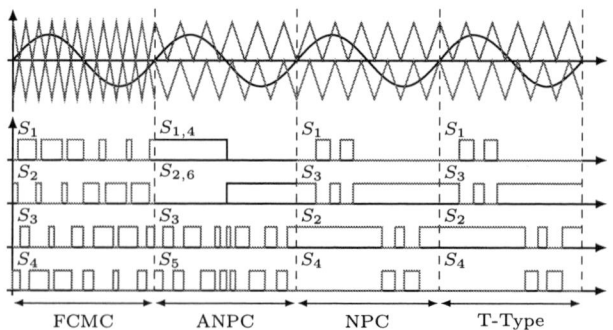

Fig. 3: Implemented modulation techniques.

979-8-3503-3714-3/23 $31.00 © 2023 IEEE

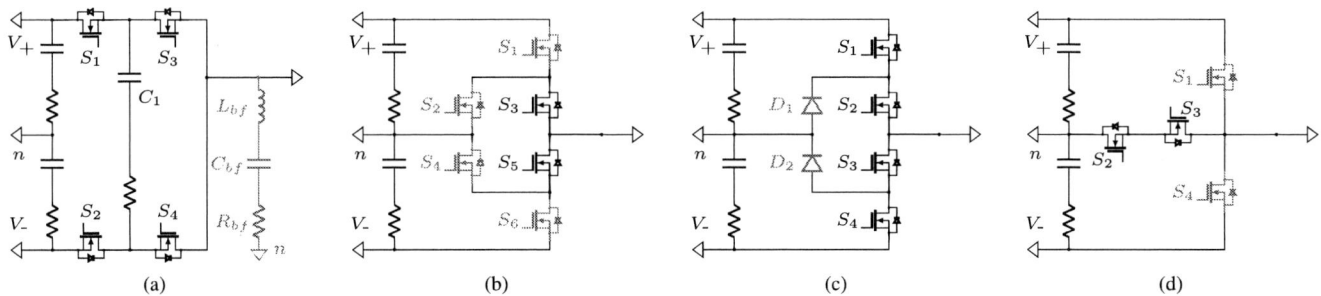

Fig. 4: Single-phase configuration of a) FCMC, b) ANPC, c) NPC, and d) T-Type topologies.

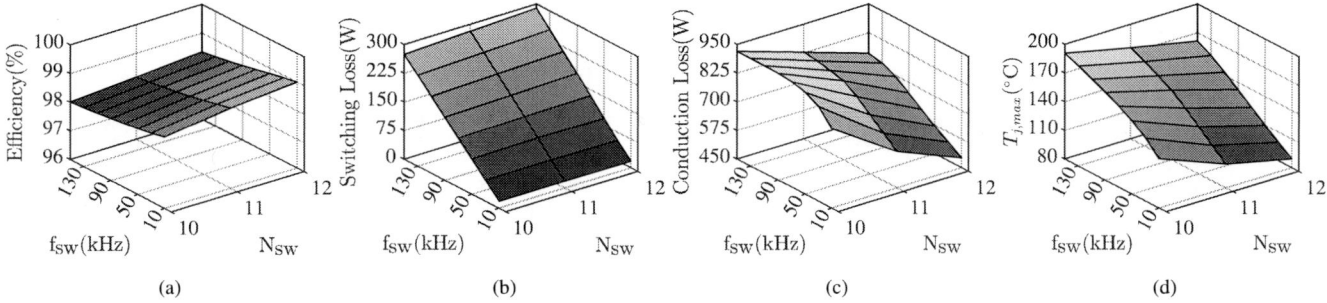

Fig. 5: GaN-based ANPC open-loop simulation result with modulation index of m=0.95. a) Efficiency. b) Total switching loss. c) Total conduction loss. d) Maximum junction temperature.

topology, the PSPWM technique is utilized by using two triangular carrier signals that are phase-shifted by 90 degree. For three-level ANPC, NPC, and T-Type topologies, the phase disposition PWM (PDPWM) technique is implemented. The gate signals can be generated by using two triangular carrier signals above and below zero that are in phase. However, in ANPC topology, four of the switches are working at low frequency which is equal to the reference signal frequency.

III. SIMULATION RESULTS

In this section, the simulation results are discussed. The converters performance is compared in open-loop and speed-control modes. In the open-loop test, the converter switching frequency and number of paralleled devices are swept, and based on the results a feasible operating point is selected for the speed-control test.

A. Open-loop Test

Table I listed the values of parameters and variables in the open-loop test. The converter input DC voltage is equal to 800V. The value of the load resistor and inductance are selected based on the motor characteristics in section III-C. It is also assumed that the power factor of the load is equal to 0.8 (lagging). In addition, all capacitors are precharged to their final value. In these tests, all system parameters and variables are kept the same and only the converter topology and the modulation technique are changed. Fig. 5 shows the simulation result for GaN-based ANPC in open-loop mode. As can be seen, the efficiency increases as the switching frequency is reduced and the number of paralleled devices is increased. This can be explained by looking at the converter

power loss results. Fig. 5(b) and Fig. 5(c) show the switching power loss and conduction loss of the switches, respectively. It is evident that the switching loss decreased significantly by decreasing the switching frequency and the conduction loss decreased by increasing the number of paralleled devices. This is because the conduction loss is exponentially proportional (with the power of two) to the value of current and by increasing the number of paralleled devices, the current that goes through each single switch will decrease. Fig. 5(d) shows the resultant maximum junction temperature of the switches. It should be noted that the typical value for the absolute maximum junction temperature of the selected semiconductors

TABLE I: Open-loop simulation parameters.

Parameters	Value	Description
R_{load}	842.1 mΩ	load resistance
L_{load}	555.3 uH	load inductance
f_{load}	181 Hz	load frequency
$V_{dc1,2}$	400 V	input DC voltage source amplitude
$C_{1,2}$	100 uF	input capacitor value
T_{dt}	50 ns	switching dead time value
T_{init}	25 °C	initial temperature
R_{tim}	1 K/W	thermal interface material resistance
R_{hs}	0.1 K/W	heatsink thermal resistance
C_{hs}	1 uJ/K	heatsink thermal capacitance
f_{sw}	[10:20:150] kHz	switching frequency
N_{sw}	{10,11,12}	switching frequency
S_n	GS-065-060-5-TA	650V 60A GaN switch
S_n	C2M0025120D	1200V 63A SiC switch
D_n	C4D40120D	1200V 54A SiC schottky diode
I_{rated}	420A	rated load current

979-8-3503-3714-3/23 $31.00 © 2023 IEEE   133

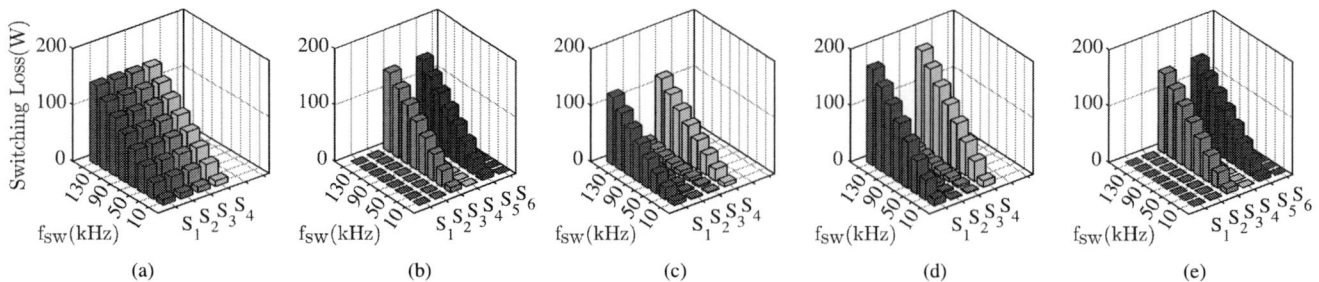

Fig. 6: Switching loss distribution among semiconductors in a) GaN-based FCMC, b) GaN-based ANPC, c) hybrid NPC, d) hybrid T-Type, and e) hybrid ANPC converters ($N_{sw} = 11$).

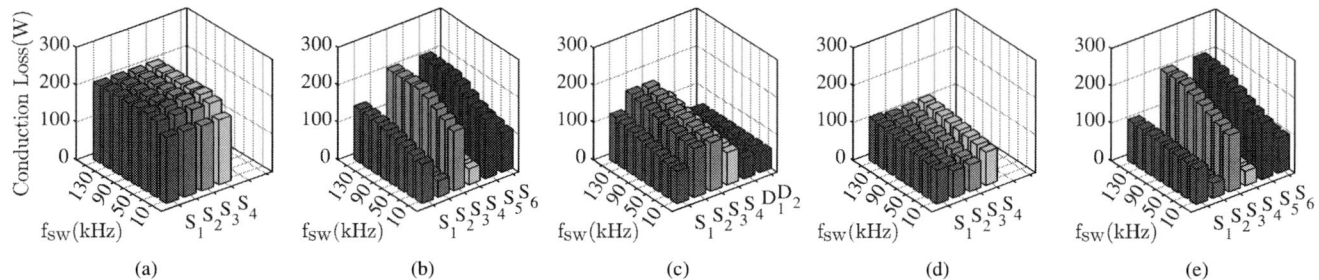

Fig. 7: Conduction loss distribution among semiconductors in a) GaN-based FCMC, b) GaN-based ANPC, c) hybrid NPC, d) hybrid T-Type, and e) hybrid ANPC converters ($N_{sw} = 11$).

is 150°C. However, in practice, 125°C would be considered as the safe operation limit for the maximum junction temperature of switches. Therefore, based on Fig. 5(d), working at high switching frequencies with a low number of paralleled devices is not permissible.

As an example, fig. 6 and Fig. 7 show the value of switching and conduction losses in each switch position for all converters with the modulation index of m=0.95 and 11 parallel devices. The effect of the selected modulation technique on power

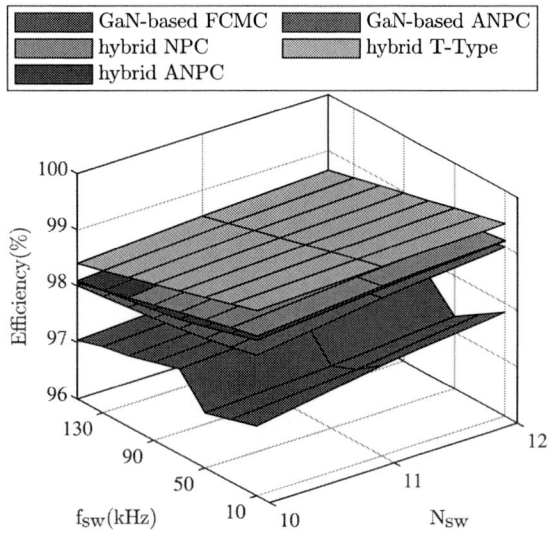

Fig. 8: Efficiency comparison of converters in open-loop with modulation index m=0.95.

loss is evident in these figures. In the FCMC converter with the PSPWM technique, the switching and conduction losses are equal in all four switches showing evenly distributed power loss among switches. However, in the GaN-based and hybrid ANPC converters with the PDPWM method, four of the switches are switching at the load frequency with almost zero switching loss. Similarly, two of the four switches in NPC and T-Type converters have small switching power loss. Fig. 8 compares the efficiency of all converters with respect to the switching frequency and number of paralleled devices at the modulation index of m=0.95. It can be seen that the hybrid T-Type converter has the highest efficiency over the entire range while the GaN-based FCMC has the lowest. Hybrid ANPC converter has slightly higher efficiency than GaN-based ANPC converter, however, both converters have lower efficiency than hybrid NPC converter except at high frequencies. The efficiency of the GaN-based FCMC converter is mainly affected because of the additional loss in the booster filter damping resistor.

B. GaN-based ANPC vs Hybrid ANPC

Comparing the performance of the GaN-based ANPC and the hybrid ANPC converters highlights one of the disadvantages of GaN semiconductors. As mentioned before, in the the ANPC topology, four switches are switching at load frequency which results in negligible switching loss. This is evident in the switching loss of the GaN-based ANPC converter shown in Fig. 6(b). Therefore, these four switches are replaced with SiC switches in the hybrid ANPC converter. Based on the components datasheet, the typical value of the selected SiC switch on-resistance is the same as the selected GaN device,

979-8-3503-3714-3/23 $31.00 © 2023 IEEE

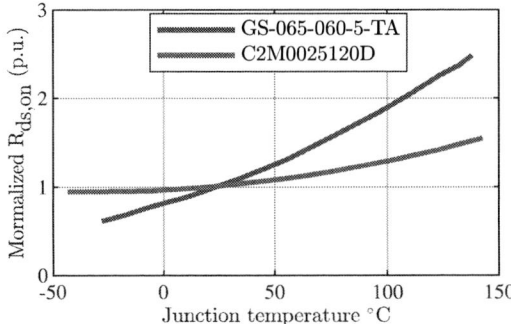

Fig. 9: Normalized on-resistance of GS-065-060-5-TA and C2M0025120D with respect to junction temperature.

equal to 30mΩ. However, simulation results show that value of the conduction loss in the GaN-based ANPC converter is higher than the hybrid ANPC converter. Take switch position 1 as an example where SiC devices are used in the hybrid ANPC converter. Fig. 7(b) and Fig. 7(e) show that the conduction loss in this switch position in the hybrid ANPC converter is smaller than the GaN-based ANPC converter. This can be explained by Fig. 9 which shows the normalized on-resistance of selected devices with respect to junction temperature. Although both devices have the same typical value of on-resistance at 25°C, the rate of the increase in the value of the on-resistance with respect to the junction temperature in the GaN device is higher than the SiC device. In the open-loop test with 11 parallel devices and f_{sw}=10kHz, the junction temperature of the switch position 1 at steady-state is 7.25°C higher in the GaN-based ANPC converter (90.75°C) than the hybrid ANPC converter (83.5°C). Therefore, the actual value of the switch position 1 on-resistance in the GaN-based ANPC converter is higher than the hybrid ANPC converter results in a higher conduction loss.

C. Speed-Control Test

For this test, in order to avoid excessive complications and be able to compare converters performance, the converter variables are selected based on the open-loop results, and other factors such as the effect of switching dv/dt on the efficiency of the converter are ignored. Therefore, 11 parallel devices and a switching frequency of 10kHz are selected for the speed-

TABLE II: Simulated motor parameters based on 2012 Nissan Leaf motor.

Parameters	Value	Description
R_s	11.06 mΩ	stator resistance
L_d	150.4 uH	stator d-axis inductance
L_q	406.2 uH	stator q-axis inductance
Phi	88.8873 mVs	magnet induced flux
P	8	number of poles
J	0.028729 kgm²	motor inertia
P_m	80 kW	rated power
ω_b	2728.37 rpm	base speed
T_{max}	280 Nm	maximum mechanical torque

control test. In this test, a motor model based on the 2012 Nissan Leaf motor (EM61) is used. The Ansys Motor-CAD software is used to extract the motor model in the dq reference frame. Table II listed the motor parameters used in the speed-control test.

By using the designed speed controller, all converters are simulated with respect to the permissible torque-speed range of the motor. The speed is increased from 500rpm to 9000rpm and the motor load torque is changed from 25Nm to 280Nm. The simulation time was set high enough until the thermal steady state was reached before saving the value of the parameter. The loss in the converter is calculated by adding

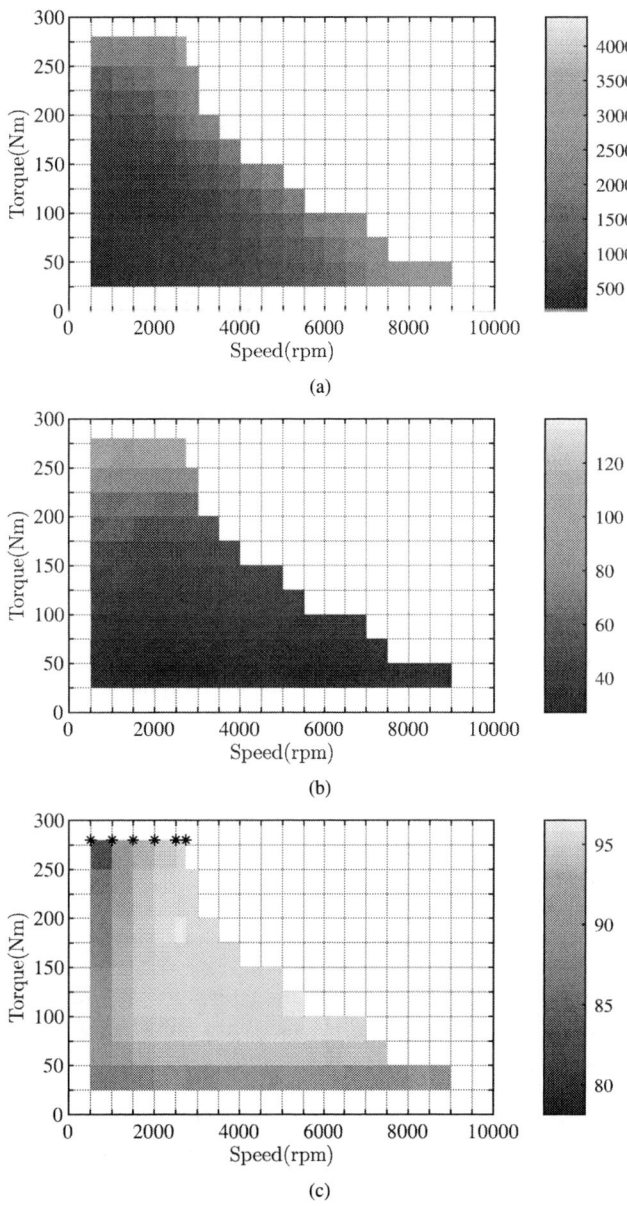

Fig. 10: GaN-based FCMC speed-control simulation result. a) Total loss. b) Maximum junction temperature. c) Efficiency.

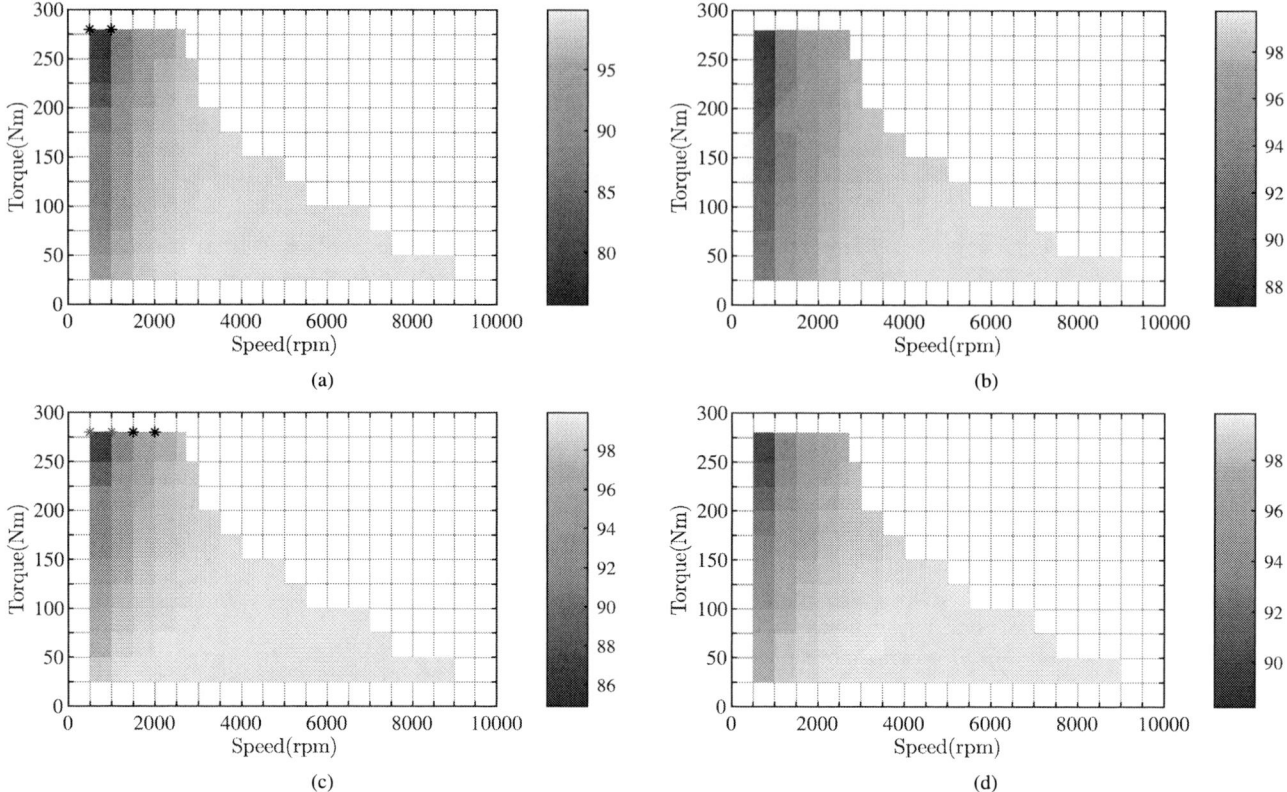

Fig. 11: Speed-control efficiency simulation result in a) GaN-based ANPC, b) hybrid NPC, c) hybrid T-Type, and d) hybrid ANPC converters.

the value of all semiconductor conduction loss, switching loss, and loss in parasitic components in the Converter and Source subsystems.

Fig. 10 shows the speed-control simulation results for the GaN-based FCMC. It can be seen that the value of total loss in the converter increased as the value of load torque on the motor and the motor speed increased. As the load torque on the motor increases, the motor needs to generate more mechanical torque or power to maintain its speed. Therefore, the phase current of the motor increases which results in higher conduction loss in the converter. In addition, as the speed increases, the value of the loss in the booster filter increases as well. A similar trend can be seen in the maximum junction temperature of semiconductors. As a result, the efficiency of the GaN-based FCMC decreases as the load torque increases. In addition, the converter efficiency drops at the low load torque values. As mentioned earlier, the safe operation limit for the maximum junction temperature of the selected semiconductors is 125°C. In fig. 10(c) the operating points where the maximum junction temperature exceeds this value are marked with black color. It can be seen that, at the rated value of load torque, the junction temperature is higher than the limit. It is noteworthy that it is unlikely to operate at the rated value of load torque except during the vehicle acceleration.

Fig. 11 shows the efficiency of other converters in the speed-control test. As can be seen, similar to the GaN-

based FCMC, the efficiency of the converters decreases as the motor load torque increases. However, unlike in the GaN-based FCMC, the efficiency of other converters stays high at low values of load torque. It should be noted that the black and red markers indicate operation points where the maximum junction temperature was higher than 125°C and 150°C, respectively. Consequently, based on the maximum junction temperature results, a better thermal solution system or a higher number of parallel devices are required for proper operation over the entire torque-speed range of the motor in the GaN-based ANPC and T-hybrid T-Type converters. Table III shows the simulation results of the speed-control test at a few important operating points for all converters. It can be seen that the GaN-based FCMC has the lowest peak efficiency throughout the entire operation range while the hybrid T-Type converter has the highest value. This is mainly because of the additional losses in the boaster filter of the GaN-based FCMC. In addition, the GaN-based FCMC has the lowest efficiency at the base speed and the maximum load torque. The simulation results indicate that the GaN-based ANPC converter has the lowest value of efficiency among the studied converters. In addition, based on Table III, the GaN-based ANPC converter has the highest value of efficiency changes ($\Delta\eta = \eta_{max} - \eta_{min}$=24.08%) while the hybrid NPC converter has the most consistent performance ($\Delta\eta$=12.57%) in the entire operation range with respect to the efficiency value.

TABLE III: Speed-control simulation results comparison.

Converter	Maximum Efficiency					Minimum Efficiency (ω=500rpm & T=280Nm)			Base Speed & Maximum Torque		
	η	ω	T	P_{loss}	$T_{j,max}$	η	P_{loss}	$T_{j,max}$	η	P_{loss}	$T_{j,max}$
GaN-based FCMC	96.49%	3500rpm	125Nm	1657.3W	43.2°C	78.16%	3031.1W	136.3 °C	94.52%	4418.3W	136.4°C
GaN-based ANPC	99.83%	9000rpm	25Nm	39.19W	26.7°C	75.74%	3219.4W	138.5°C	96.39%	2829W	120.7°C
Hybrid NPC	99.73%	9000rpm	25Nm	63.54W	27.4°C	87.16%	2233.8W	110.4°C	97.51%	2056.6W	95.6°C
Hybrid T-Type	99.89%	9000rpm	25Nm	25.02W	26°C	84.89%	2629.5W	159.9°C	97.73%	1873.9W	113.4°C
Hybrid ANPC	99.82%	9000rpm	25Nm	41.49W	26.8°C	88.21%	2051.8W	113.1°C	97.56%	2016.3W	109.7°C

IV. CONCLUSION

In this paper, the performance of 650V GaN-based multi-level converters for 800V EV powertrain was investigated. Five three-phase converters based on four well-known multilevel topologies were simulated in PLECS software. The converter performance and its power loss were analyzed in two test scenarios. At first, the converters were connected to an RL load with an open-loop control method, in which the switching frequency and number of paralleled devices swept within predefined sets. The test results were used to select a switching frequency and a number of paralleled devices for the speed-control test. Then, by using a motor model based on the 2012 Nissan Leaf motor and a speed vector controller in the d-q reference frame, the performance of the converter was analyzed for the entire permissible operation range of the motor.

The simulation results indicated that the performance of FCMC topology is highly related to the booster filter performance. The additional loss in the booster filter decreased the converter efficiency resulting in the lowest efficiency among other converters. In addition, the hybrid ANPC converter showed better performance than the GaN-based converter with respect to the efficiency and maximum junction temperature of semiconductors. The hybrid T-Type converter had the highest overall efficiency in open-loop tests and the highest peak efficiency in speed-control tests. In addition, the hybrid NPC converter had the most consistent efficiency over the entire range of operations.

REFERENCES

[1] K. J. Chen et al., "GaN-on-Si Power Technology: Devices and Applications," in IEEE Transactions on Electron Devices, vol. 64, no. 3, pp. 779-795, March 2017, doi: 10.1109/TED.2017.2657579.

[2] E. A. Jones, F. Wang and B. Ozpineci, "Application-based review of GaN HFETs," 2014 IEEE Workshop on Wide Bandgap Power Devices and Applications, Knoxville, TN, USA, 2014, pp. 24-29, doi: 10.1109/WiPDA.2014.6964617.

[3] A. Allca-Pekarovic, P. J. Kollmeyer, P. Mahvelatishamsabadi, T. Mirfakhrai, P. Naghshtabrizi and A. Emadi, "Comparison of IGBT and SiC Inverter Loss for 400V and 800V DC Bus Electric Vehicle Drivetrains," 2020 IEEE Energy Conversion Congress and Exposition (ECCE), Detroit, MI, USA, 2020, pp. 6338-6344, doi: 10.1109/ECCE44975.2020.9236202.

[4] D. Atkar, P. Chaturvedi, H. M. Suryawanshi, P. Nachankar, D. Yadeo and S. Krishna, "Solid State Transformer for Electric Vehicle Charging Infrastructure," 2020 IEEE International Conference on Power Electronics, Smart Grid and Renewable Energy (PESGRE2020), Cochin, India, 2020, pp. 1-6, doi: 10.1109/PESGRE45664.2020.9070447.

[5] I. Aghabali, J. Bauman and A. Emadi, "Analysis of Auxiliary Power Unit and Charging for an 800V Electric Vehicle," 2019 IEEE Transportation Electrification Conference and Expo (ITEC), Detroit, MI, USA, 2019, pp. 1-6, doi: 10.1109/ITEC.2019.8790562.

[6] A. Sheir, M. Z. Youssef and M. Orabi, "A Novel Bidirectional T-Type Multilevel Inverter for Electric Vehicle Applications," in IEEE Transactions on Power Electronics, vol. 34, no. 7, pp. 6648-6658, July 2019, doi: 10.1109/TPEL.2018.2871624.

[7] R. González, C. A. Rojas and L. Callegaro, "Three-level DC-DC GaN-based Converter with Active Thermal Control for Powertrain applications in Electric Vehicles," 2021 22nd IEEE International Conference on Industrial Technology (ICIT), Valencia, Spain, 2021, pp. 502-507, doi: 10.1109/ICIT46573.2021.9453595.

[8] H. Moradisizkoohi, N. Elsayad, A. Berzoy, C. R. Lashway and O. A. Mohammed, "A multi-level bi-directional buck-boost converter using GaN devices for electric vehicle applications," 2017 IEEE Transportation Electrification Conference and Expo (ITEC), Chicago, IL, USA, 2017, pp. 742-746, doi: 10.1109/ITEC.2017.7993362.

[9] C. A. Rojas, R. Gonzalez, L. Callegaro and H. Young, "Mission Profile-Oriented Active Thermal Control of a Bidirectional Three-Level Buck-Boost GaN-Based DC-DC Converter for Electric Vehicles Powertrains," IECON 2021 – 47th Annual Conference of the IEEE Industrial Electronics Society, Toronto, ON, Canada, 2021, pp. 1-6, doi: 10.1109/IECON48115.2021.9589692.

[10] Busquets-Monge, S., S. Alepuz, G. García-Rojas, and J. Bordonau. " Electric Vehicle Powertrains with Modular Battery Banks Tied to Multilevel NPC Inverters," Electronics 2023, 12, 266, doi: 10.3390/electronics12020266

[11] V. Patel, M. Tinari, C. Buccella and C. Cecati, "Analysis on Multilevel Inverter Powertrains for E-transportation," 2019 IEEE 13th International Conference on Compatibility, Power Electronics and Power Engineering (CPE-POWERENG), Sonderborg, Denmark, 2019, pp. 1-6, doi: 10.1109/CPE.2019.8862373.

[12] R. Hariri, F. Sebaaly and H. Y. Kanaan, "A Review on Modular Multilevel Converters in Electric Vehicles," IECON 2020 The 46th Annual Conference of the IEEE Industrial Electronics Society, Singapore, 2020, pp. 4987-1993, doi: 10.1109/IECON43393.2020.9255037.

[13] A. Poorfakhraei, M. Narimani and A. Emadi, "A Review of Multilevel Inverter Topologies in Electric Vehicles: Current Status and Future Trends," in IEEE Open Journal of Power Electronics, vol. 2, pp. 155-170, 2021, doi: 10.1109/OJPEL.2021.3063550.

[14] F. Chang, O. Ilina, M. Lienkamp and L. Voss, "Improving the Overall Efficiency of Automotive Inverters Using a Multilevel Converter Composed of Low Voltage Si mosfets," in IEEE Transactions on Power Electronics, vol. 34, no. 4, pp. 3586-3602, April 2019, doi: 10.1109/TPEL.2018.2854756.

[15] S. Satpathy, S. Bhattacharya and V. Veliadis, "Comprehensive Loss Analysis of Two-level and Three-Level Inverter for Electric Vehicle Using Drive Cycle Models," IECON 2020 The 46th Annual Conference of the IEEE Industrial Electronics Society, Singapore, 2020, pp. 2017-2024, doi: 10.1109/IECON43393.2020.9254520.

[16] R. H. Wilkinson, T. A. Meynard and H. du Toit Mouton, "Natural Balance of Multicell Converters: The General Case," in IEEE Transactions on Power Electronics, vol. 21, no. 6, pp. 1658-1666, Nov. 2006, doi: 10.1109/TPEL.2006.882951.

[17] B. P. McGrath and D. G. Holmes, "Natural Capacitor Voltage Balancing for a Flying Capacitor Converter Induction Motor Drive," in IEEE Transactions on Power Electronics, vol. 24, no. 6, pp. 1554-1561, June 2009, doi: 10.1109/TPEL.2009.2016567..

Improved non-destructive mutual inductance estimation method for multi-chip power modules

Arthur Boutry, Sergio Jimenez, Andrew Lemmon
Department of Electrical and Computer Engineering
The University of Alabama
Tuscaloosa, USA
ajboutry@ua.edu

Abstract—In multi-chip power modules, the extraction of mutual inductance is an enduring challenge. Recently, it has been demonstrated with an empirical, non-invasive, and non-destructive method the ability to measure the mutual inductance between the positive and negative terminals. This method relies on measuring the mutual inductance of a power module in the same fashion as a transformer. A problem of this method is the low Signal-to-Noise Ratio (SNR) due to the weak coupling between the module terminals. In this contribution, we will increase the effectiveness of this method by improving the hardware setup. The results will also be cross-validated with FEA model predictions for a module geometry.

Index Terms—Mutual inductance, power module, characterization, modeling.

I. INTRODUCTION

In power electronics applications needing high voltage and high current modules with multiple chips (MCPM - multi-chip power module), such as industry applications, traction, or renewable energy, the module terminals geometry evolves to improve the module's overall performance. As such, the flux cancellation phenomenon is used to reduce the resulting inductance at play in the module during its functioning. Reducing the stray inductance allows for the reduction of losses and overvoltages and, therefore, improves the overall behavior of the converter.

The way to achieve flux cancellation in an MCPM half-bridge type module is to put in parallel and close to each other the Bus+ terminal (or drain) and the Bus- (or source). The evolution between modules is depicted in Fig. 1. The different inductances in play, including the mutual inductance, are displayed in Fig. 2. Accurately measuring mutual inductance in power modules is essential to assess the performances of a module and the validity of its geometry, as well as modeling the power module in circuit simulation software such as LTspice. Indeed, maximizing the mutual inductance is a goal of manufacturers, so bigger mutual inductances are expected with time, and mutual inductances of at least a few nH are expected in MCPM nowadays.

Known methods for estimating mutual inductance involve destroying the module [2] or requiring the full CAD geometry for simulation in FEA software like ANSYS Q3D [3]–[6]. An improved method was published recently [1], which did not need to destroy the module or to have the CAD geometry. This method allows measuring the mutual inductance of the MCPM

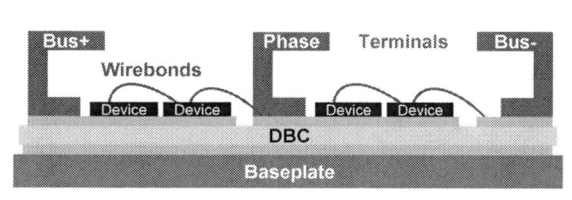

(a) Module without flux cancellation.

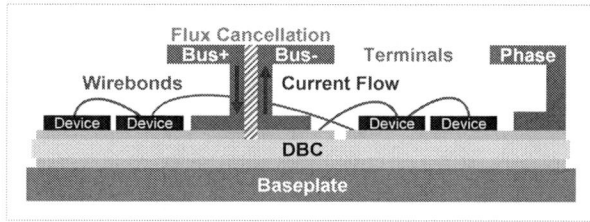

(b) Module with flux cancellation.

Fig. 1: Flux cancellation in power modules [1].

in the same way a transformer would be characterized. The principle of the measure is illustrated in Fig. 3. This method is promising but was not validated against many module geometries and had a bad Signal-to-Noise Ratio (SNR). In order to improve the SNR issue, an amplifier was added to the circuit, on the stimulus side. The second section of this article will describe the validation of the use of the amplifier. Then, the third section will confront the experimental method and FEA simulation on a typical module geometry. Lastly, conclusions on the method and perspectives will be discussed.

II. EXPERIMENTAL SETUP AND VALIDATION

A. General Experimental setup

The improved setup proposed in this paper is described in Fig. 4, showing the module and its parasitic inductances (L_m being the targeted measure), as well as the stimulus part (in red, signal generator + amplifier) and the measure points (voltage measurement A between Bus- and K2, voltage measurement B between Bus- and phase). Voltage measurement A is recommended when possible, as it should show only L_m.

979-8-3503-3714-3/23 $31.00 © 2023 IEEE

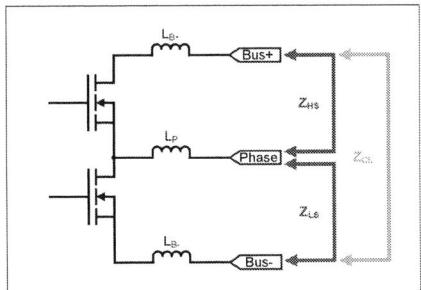

(a) Schematic of a half-bridge module without flux cancellation.

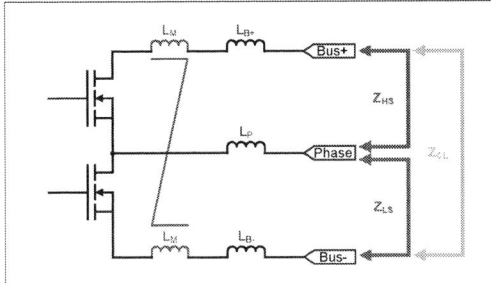

(b) Schematic of a half-bridge module with flux cancellation.

Fig. 2: Half-bridge MCPM with and without flux cancellation [1].

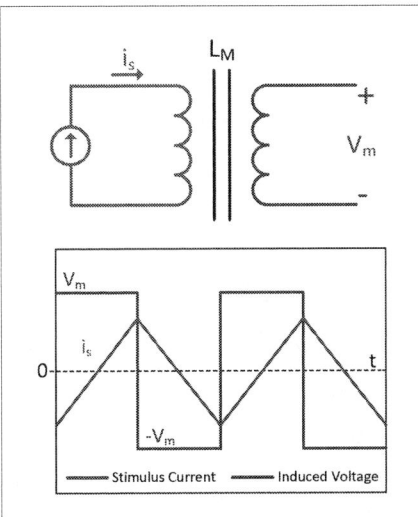

Fig. 3: Principle of the measure, illustrated with a triangle stimulus [1].

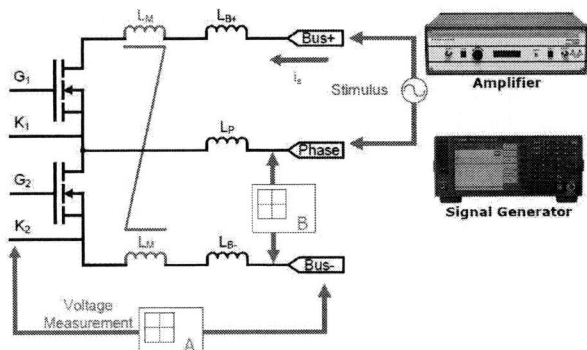

Fig. 4: New test setup that includes an amplifier.

B. Validation of the improved test setup

To validate the use of the amplifier in the circuit, a test has been run on the same simplified mutual inductance boards (see Fig. 7) described in [1]. This test subject comprises two boards: one stimulus board - where current is injected - and one pickup board - where the output voltage is measured. Each board consists of a PCB where a rectangular conductive path is buried. The two boards can rotate against each other, allowing the testing of different values of L_m for different angles.

In Fig. 5, the results with the setup that comprises the amplifier are compared with the results of the setup without amplifier and COMSOL FEA mutual inductance simulation. It can be seen that for different gains and different angles, the results with the amplifier correspond to previous results without the amplifier and FEA simulation. This validates the use of the amplifier by comparing it with a previous study. Furthermore, the interest in using an amplifier is illustrated in Fig. 6. With the amplifier turned on and with a noticeable gain (20%), the voltage output on the pickup board is increased to another order of magnitude, indicating that the parasitic noise is lowered and SNR is increased. The results demonstrate that the presence of the amplifier allows us to have more precise measures.

III. STUDY CASE: ECONODUAL-TYPE TEST SUBJECT

In order to validate the method with more results than the previous study [1], a typical geometry in power modules is also used in this article, the Econodual. It is important to give the precision that the obtained mutual inductances with this geometry in this article do not correspond to the mutual inductance of a real module. This adapted test subject was designed within a broader study [7]. The real module mutual inductance, if the module has similar terminals, could be more or it could be less. The test subject, as tested, is visible in Fig. 8a. To have a simplified version of the modules and avoid the influence of the semiconductors, only wire bonds and metal brackets are used to connect the different copper surfaces. No semiconductor chip has been used. The whole test setup is shown in Fig. 8b. Triple shielded cables are used to improve the precision and reduce parasitics. To connect the terminals

979-8-3503-3714-3/23 $31.00 © 2023 IEEE

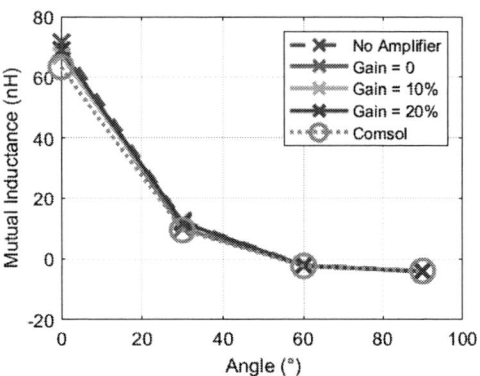

Fig. 5: Results for mutual inductance using the rotating boards, without and with the amplifier, and with the COMSOL model results.

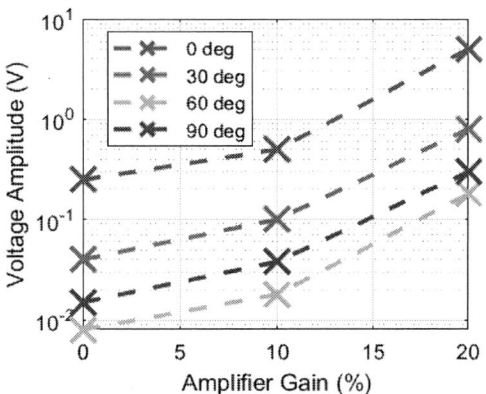

Fig. 6: Voltage level at the output for the rotating boards circuit, for different gains of the amplifier.

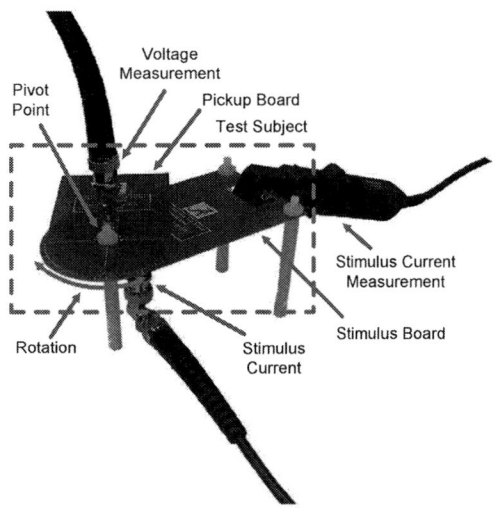

Fig. 7: Rotating boards test subject [1].

on the DUT, the cables have to be cut, and the outer shield is used as a wire. Unlike what is shown in Fig. 3, sinusoidal stimulus current is used, in order to avoid high-frequency noise due to the triangular stimulus. The resulting waveforms with sinusoidal stimulus are displayed in Fig. 9.

The mutual inductance results for the Econodual test subject are displayed in Fig. 10, where the mutual inductance varies from 4nH to 4.5nH between 2 to 5 MHz and is steady at 5nH from 6 to 9MHz. In ANSYS Q3D, the obtained mutual inductance with the full module is 6nH at all these frequencies. A few points have to be noted about the experimental and simulation results. In [6], due to parasitic effect at higher frequencies (less effective chokes, capacitive effects), mutual inductance measurement values for frequencies higher than 5MHz are expected to be higher, and less precise. For the simulation results, slight changes in the definition of sinks and sources can lead to differences up to 1nH. With

(a) Econodual Test Subject.

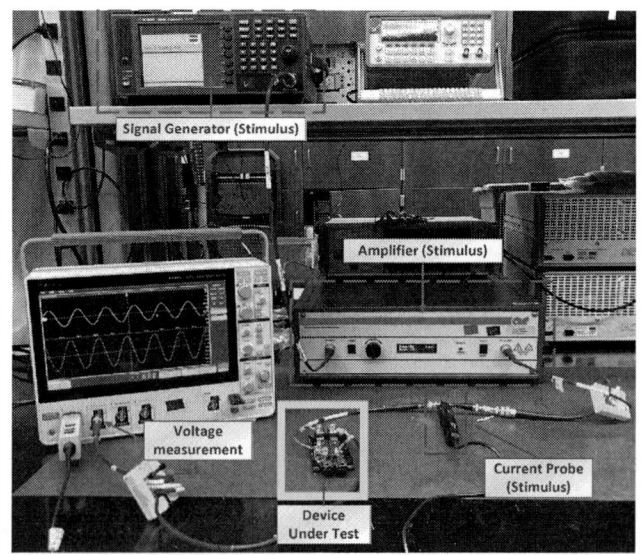

(b) Test setup with the Econodual as the DUT.

Fig. 8: Econodual test subject: experimental setup.

979-8-3503-3714-3/23 $31.00 © 2023 IEEE

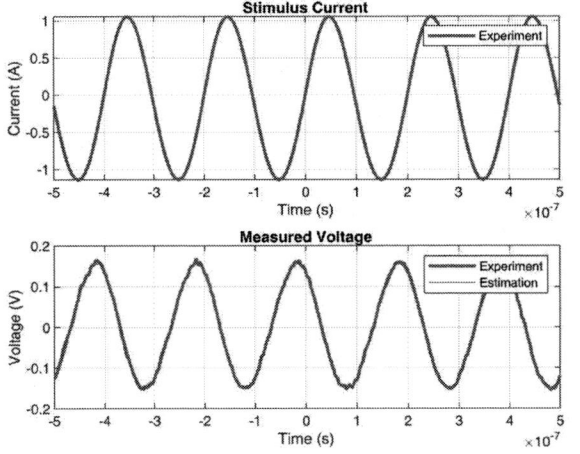

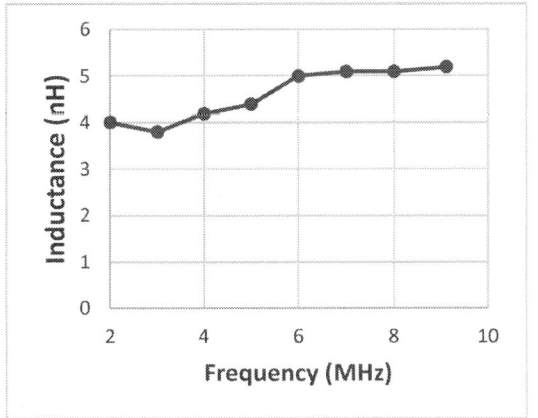

Fig. 9: Sinusoidal stimulus and result, with the Econodual Test Subject, at 5MHz.

Fig. 10: Results for the Econodual type test subject.

IV. CONCLUSION AND RECOMMENDATIONS

A. Results of this study

This article presented an improved version of a mutual inductance measurement technique. The addition of an amplifier allowed a better Signal-to-Noise Ratio and has been validated against a previous study. In addition to this validation, the method has been used with a new module geometry, and results in accordance with the simulation have been found. In a nutshell, this method allows an estimation of the mutual inductance of a module, but not an exact measure. Though the advantages of this method are important: it is non-destructive, easy and cheap to put in place, and does not require a full 3D model of the module.

B. Recommendations

As described in [6], results should be looked at between 2 to 5MHz. The cables should be shielded, and chokes should be used. As described in [6] again, this method should not be used when the mutual inductance is expected to be low.

C. Perspectives

This method has been improved compared to our previous study, but further investigations should be led to fully validate this method. First, more module geometries experimental results should be compared with their simulation results. Then compensation techniques should be looked into in order to improve the precision of the measure.

REFERENCES

[1] A. Shahabi, A. Lemmon, B. DeBoi, T. Beechner, and R. Mayo, "Empirical procedure for estimating mutual coupling in high-performance power modules," in *2021 IEEE Electric Ship Technologies Symposium (ESTS)*, pp. 1–7, 2021.

[2] B. Nelson, A. Lemmon, B. DeBoi, M. Olimmah, and K. Olejniczak, "Measurement-based modeling of power module parasitics with increased accuracy," in *2020 IEEE Applied Power Electronics Conference and Exposition (APEC)*, pp. 1430–1437, 2020.

[3] Z. Chen, R. Burgos, D. Boroyevich, F. Wang, and S. Leslie, "Modeling and simulation of 2 kv 50 a sic mosfet/jbs power modules," in *2009 IEEE Electric Ship Technologies Symposium*, pp. 393–399, 2009.

[4] D.-P. Sadik, K. Kostov, J. Colmenares, F. Giezendanner, P. Ranstad, and H.-P. Nee, "Analysis of parasitic elements of sic power modules with special emphasis on reliability issues," *IEEE Journal of Emerging and Selected Topics in Power Electronics*, vol. 4, no. 3, pp. 988–995, 2016.

[5] E. Falck, M. Stoisiek, and G. Wachutka, "Modeling of parasitic inductive effects in power modules," in *Proceedings of 9th International Symposium on Power Semiconductor Devices and IC's*, pp. 129–132, 1997.

[6] B. DeBoi, *Characterization and Modeling of Sic Multi-Chip Power Modules*. PhD thesis, University of Alabama, 2022.

[7] B. T. DeBoi, A. N. Lemmon, B. McPherson, and B. Passmore, "Improved methodology for parasitic analysis of high-performance silicon carbide power modules," *IEEE Transactions on Power Electronics*, vol. 37, no. 10, pp. 12415–12425, 2022.

SiC MOSFET Device for Radio Frequency Power Conversion

Amaury Gendron
Analog Power Conversion
Bend, USA
agendron@apowerc.com

Dumitru Sdrulla
Analog Power Conversion
Bend, USA
dsdrulla@apowerc.com

Nathaniel Barr
Analog Power Conversion
Bend, USA
nbarr@apowerc.com

Tetsuya Takata
Kyosan Electric Mfg. Co., Ltd
Yokohama, Japan
takata-t@kyosan.co.jp

Dick Frey
Analog Power Conversion
Bend, USA
dfrey@apowerc.com

Su-Wen Chen
Analog Power Conversion
Bend, USA
schen@apowerc.com

Albert Gu
Analog Power Conversion
Bend, USA
agu@apowerc.com

Abstract—A new SiC MOSFET specifically designed for power conversion above 30 MHz is presented. This 600 V / 400 mΩ SiC MOSFET includes features (metallized gate, raised gate oxide, two metal layers) to minimize the switching losses. The layout consists of multiple independent MOSFET cells placed on the same SiC substrate, which strongly enhances the die's power dissipation. RF tests on a 54.24 MHz class C power amplifier give a RF gain of 30.7 dB at 250 W of output power and 24.8 dB at 700 W.

Keywords—*SiC MOSFET Device, RF Power Amplifier, Thermal Management, Switching Power Losses*

I. INTRODUCTION

In 1989, B. Jayant Baliga predicted that power devices made on wide bandgap materials should perform significantly better than silicon-based radio frequency (RF) devices [1]. Thanks to the high critical field and thermal conductivity of silicon carbide (SiC), a SiC RF device would have 73% lower power loss and a take up only 4% of the die area of a silicon RF transistor. The vertical-double-implanted SiC MOSFET (VDMOS) is a promising device to "translate" that prediction into a successful product. The SiC MOSFET technology has matured during the last few years [2], and a VDMOS could deliver very high amounts of power [3], [4], leveraging high-voltage and high-current capabilities.

RF power amplifiers are widely adopted in diverse industrial (semiconductor plasma processing [5]-[7], wireless power transfer [8]-[10]) and medical (implantable devices) applications, with operating frequencies in the lower Industrial, Scientific, and Medical (ISM) bands at 6.78, 13.56, 27.18, and 40.68 MHz. Wide-bandgap (silicon carbide and gallium nitride) power devices unlock opportunities for smaller and faster RF amplifiers [11], ultimately improving the efficiency and power density of high-power RF systems. However, to the authors' knowledge, no commercial SiC MOSFETs have been designed for optimal RF performance. In addition, reports on electrical tests for SiC MOSFETs in RF packages are very scarce in the literature.

In this paper, a 600V / 400 mΩ SiC MOSFET designed for power conversion at very high frequency (above 30 MHz) is reported for the first time. Its performances are showcased in a 54.24 MHz class C amplifier. In section II, the designs are described for both the SiC MOSFET and its custom RF package. DC and low-frequency dynamic data are presented in section III, thermal simulation results in section IV, and design and test results of the class C power amplifier in section V.

II. RF POWER DEVICE AND PACKAGE DESIGNS

A. RF SiC MOSFET

High-power conversion in the RF domain requires systems with both a high maximum power dissipation rating and low switching losses. To fulfill these requirements, our approach consists of designing a SiC device comprising multiple independent MOSFET cells. These cells are placed on the same large die in a pattern optimized for low thermal resistance, hence improving the thermal performance (Fig. 1a). The region between the cells covers most of the die and is inactive, meaning it does not perform any electrical function. The active area is contained within the MOSFET cells, which allows us to keep its footprint small. This design can provide very low MOSFET capacitances, enabling efficient switching at very high frequency.

Such a design with a large inactive region is called a "low packing density" design and provides trade-offs between the main parameters of a MOSFET. For sake of comparison, we drew the layout of a "high packing density" design (Fig. 1b). Both designs have the same active area footprint and the same active cell layout (same gate series resistance), which results in similar switching performance at very high frequency. However, due to its smaller die size, the "high packing density" design has a significantly higher thermal resistance. For applications in high power conversion, the "low packing density" design offers enhanced thermal management and is a much more attractive option.

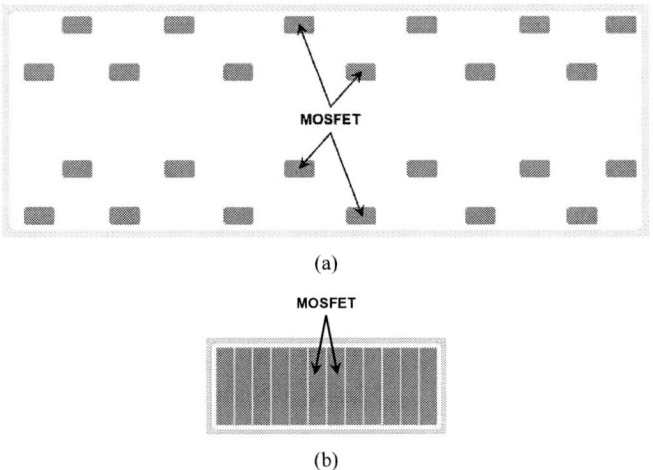

(a)

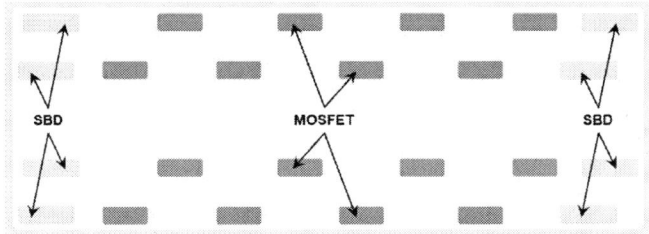

(b)

Fig. 1: Simplified layouts for a "low packing density" design with 24 MOSFET cells (a), and a "high packing density" design with 12 MOSFET cells (b). The two layouts are drawn to scale.

This cell design is very well suited for integrating a Schottky Barrier Diode (SBD) by converting a few MOSFET cells (Fig. 2). For optimal operation, the SBD cells should be placed in the coolest regions of the die, near the periphery. This device provides a low forward voltage and virtually zero-recovery free-wheeling diode for the MOSFET in hard-switching topologies, like in class D power amplifiers [12].

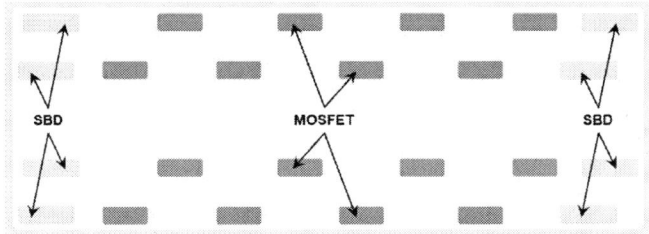

Fig. 2: Simplified layout for 16 MOSFET cells integrated with 8 SBD cells.

Our SiC MOSFET includes specific features to enhance the RF performance (Fig. 3). A custom epitaxy was defined to minimize the RF figure of merit $R_{DS,ON}$ x Q_{OSS} (the on-state resistance multiplied by the output charge of the MOSFET). Compared to a typical 600 V epitaxy for low-frequency switching, a thicker and lower-doped epitaxy was selected. A low gate series resistance (ESR) is mandatory for a MOSFET to operate at very high frequency. To reduce its value, a comb layout with very short fingers (less than 200 µm) was drawn and a metalized gate process (1.0 µm-thick aluminum cap on top of the polysilicon gate) was implemented. We also focused on reducing the MOSFET capacitances. A 0.8 µm-thick raised oxide is placed over the JFET region of the MOSFET to minimize the reverse capacitance, and a second metal layer allows us to "fold" the source and gate pads over the MOSFET active area, hence drastically reducing the source-to-drain and gate-to-drain parasitic capacitances.

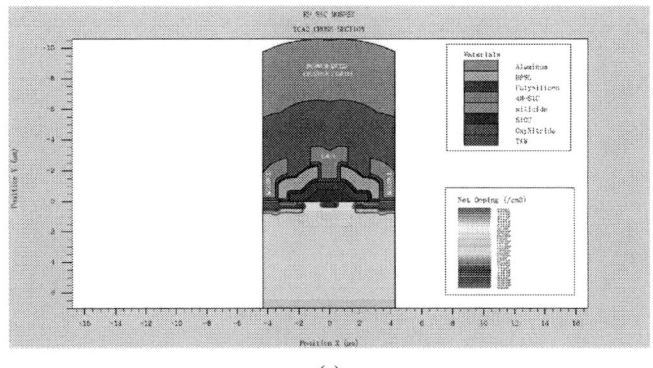

(a)

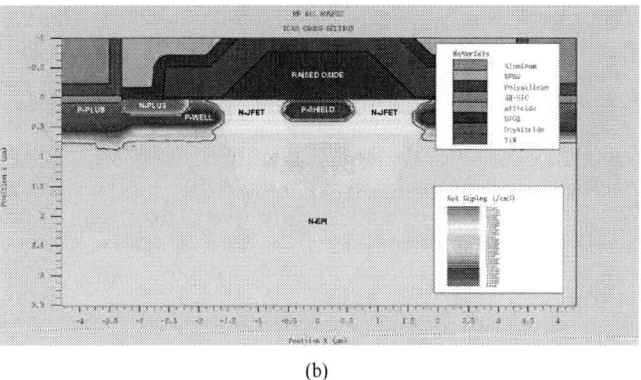

(b)

Fig. 3: Cross-sections of the MOSFET active cell over the whole epitaxy thickness and BEOL material stack (a), and near the implanted regions (b) from TCAD simulation with Silvaco's Victory Process tool.

It has been reported in the literature that the high-voltage termination has a strong effect on the switching losses [13], including the losses from charging and discharging the output capacitance [14], [15]. These results are of particular importance for our cell design, where each cell is enclosed inside its own high-voltage termination. The total termination length is the sum over all the cells and reaches typically several centimeters.

Simulations with Silvaco TCAD tools (Victory Process and Victory Device) were performed to compare different termination designs. Our study covers junction termination extension (JTE), floating field rings (FFR), and deep trench terminations, with the results reported in TABLE I. A narrow JTE was designed, shrunk to equal the thickness of the epitaxy. However, it still has the highest junction capacitance (C_J) and stored capacitive charge (Q_C) of any termination examined. The JTE lightly doped region is tied to the source termination and provides a large area for the source-to-drain capacitance. For this reason, any design variations based on a JTE would be detrimental to an RF design. The FFR design is a much more efficient option with C_J and Q_C approximately half that of JTE. This high-voltage termination was selected for our current SiC MOSFET. From the simulation results, a deep trench termination is the best option and its effect on the switching losses should be negligible. The development of a deep trench filled with dielectric material is on our technology roadmap and the plan is to implement this design in our second generation of RF SiC MOSFETs.

979-8-3503-3714-3/23 $31.00 © 2023 IEEE   143

TABLE I.   High-Voltage Terminations Electrical Parameters

	BV	C_J	Q_C
	V	$pF.cm^{-1}$	$nC.cm^{-1}$
		V_R=300 V	
JTE	876	1.36	0.77
FFR	1069	0.75	0.29
TRENCH	772	0.31	0.10

B. Discrete RF Package

Standard packages for power devices (e.g. TO-220, TO-247, and TO-267) are poorly suited for RF power conversion. These packages' general configuration is to have the drain connected to a metal plate on the backside. Such a configuration could be problematic in RF systems where the RF output signal comes from the drain. Instead, we designed a custom package tailored for RF applications (Fig. 4). The source, gate, and drain terminals are connected through wire bonds to pads on the topside of the package. The substrate consists of a 40 mil-thick beryllium oxide (BeO) and 40 mil-thick copper-tungsten stack. The BeO provides electrical isolation to the backside without strongly degrading the thermal resistance, thanks to its high thermal conductivity (247 $W \cdot m^{-1} \cdot K^{-1}$). The backside copper-tungsten plate could be tied to ground and a cold plate directly attached to it for efficient heat removal. Another important feature is the symmetry of this package design, which ensures uniform operation between the MOSFET cells.

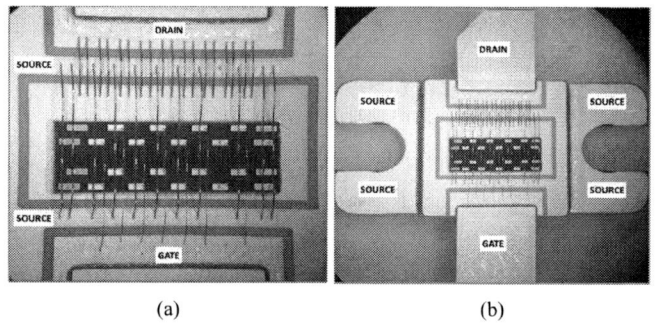

(a)   (b)

Fig. 4: SiC MOSFET with an integrated SBD in the custom RF package (a) and full package overview (b).

III. Electrical Characterizarion

A. Static Characteristics

An initial electrical characterization was performed on the parts in RF packages before inserting them in the class C amplifier. From the output characteristics (Fig. 5), the low-current on-state resistance at 20 V gate bias is 400 mΩ, which corresponds to approximately 5.0 mΩ.cm². And thanks to the excellent thermal performance of our "low packing density" design, our SiC MOSFET can operate at high DC drain current, close to the current saturation limit.

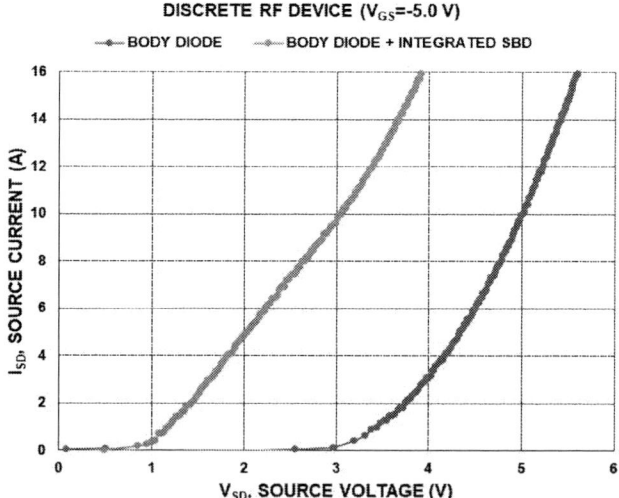

Fig. 5: Output characteristics and on-state resistance at 20 V gate bias (inset) for the SiC MOSFET of Fig. 1a.

Diode forward characteristics were acquired for the SiC MOSFETs both with and without an integrated SBD (Fig. 6). The body diode of the MOSFET turns on around 3 V, which is typical for SiC technology, and then provides a low-resistivity current path. The integrated SBD turns on around 1 V and is the sole conduction path up to 10A, at which point the body diode also turns on. This test confirms that the integrated SBD is perfectly suited to operate as a free-wheeling diode.

DISCRETE RF DEVICE (V_{GS}=-5.0 V)
BODY DIODE   BODY DIODE + INTEGRATED SBD

Fig. 6: Reverse diode characteristic for the SiC MOSFET of Fig 1a and the SiC MOSFET with an integrated SBD of Fig. 2.

B. Dynamic Characteristics

Low MOSFET capacitances are mandatory for applications at very high frequency. The input, output, and reverse capacitances (C_{ISS}, C_{OSS}, C_{RSS}) were measured up to a drain voltage of 500 V (Fig. 7). These capacitances are extremely low with 203, 26, and 2.7 pF at 300 V for C_{ISS}, C_{OSS}, and C_{RSS}, respectively. The output capacitance for the MOSFET with an integrated SBD was also measured (Fig. 7) and the value at

979-8-3503-3714-3/23 $31.00 © 2023 IEEE

300 V is 42 pF. The 16 pF increase corresponds to the contribution from the junction capacitance of the SBD.

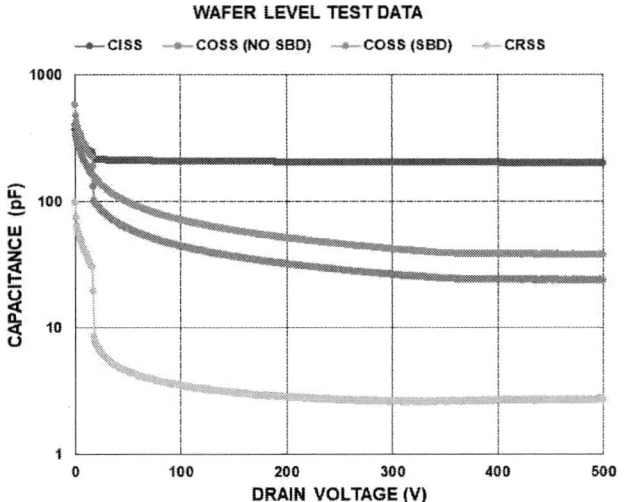

Fig. 7: MOSFET capacitances for the SiC MOSFET of Fig 1a and output capacitance for the SiC MOSFET with an integrated SBD of Fig. 2.

IV. THERMAL MANAGEMENT

We relied on thermal simulations from Ansys Icepak tool to study the thermal performance. Our simulation model is a simplified representation of the custom RF package. It includes the BeO substrate and the copper-tungsten base plate. The SiC die is placed on top of the BeO and covered with die coat. The power dissipation is modeled by 2D power sources placed on the surface of each MOSFET cell with the total power equally divided between them. In each source, the power is generated uniformly over the whole surface. The temperature of the copper-tungsten baseplate's backside is kept constant at 20 °C, to model an ideal heat sink.

To illustrate the improvement in thermal performance from the "low packing density" design, the maximum temperature variation vs dissipated power was simulated for both the high and low packing density designs (Fig. 8). The ratio in power dissipation is approximately 4. For a 200 °C temperature limit, the maximum power dissipations are 67 and 259 W for the high and low packing density designs, respectively.

The benefits of the low packing density designs are particularly strong for SiC technology, due to the high thermal conductivity of SiC (333 $W \cdot m^{-1} \cdot K^{-1}$ in the direction perpendicular to the surface of the die and 500 $W \cdot m^{-1} \cdot K^{-1}$ in the plane of the die). The heat spreading in the die is excellent, as illustrated by the temperature maps for the top and bottom surfaces of the SiC die at 258 W (Fig. 9). On the top surface, the temperature is the highest in the MOSFET cells where the heat is generated. As the heat spreads in the die, the temperature distribution becomes very uniform. Therefore, the entire backside of the die is leveraged for efficient heat removal.

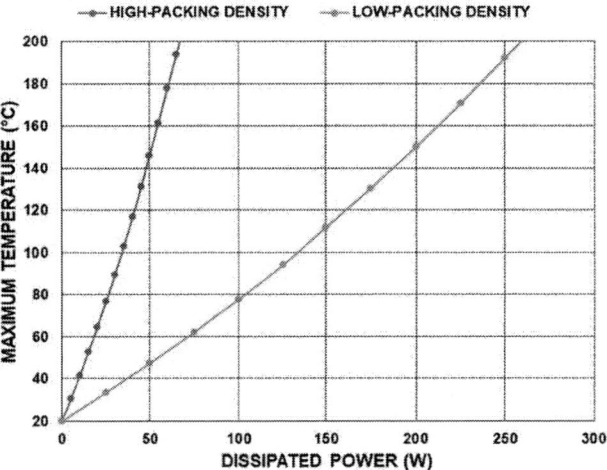

Fig. 8: Dependence of the maximum die temperature with the total power dissipated in the MOSFET cells for the "low and high packing density" designs (Fig. 1) in the custom RF package.

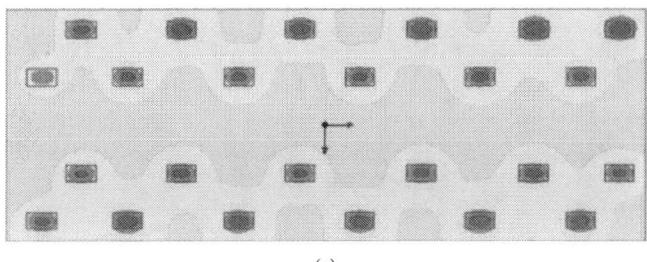

(a)

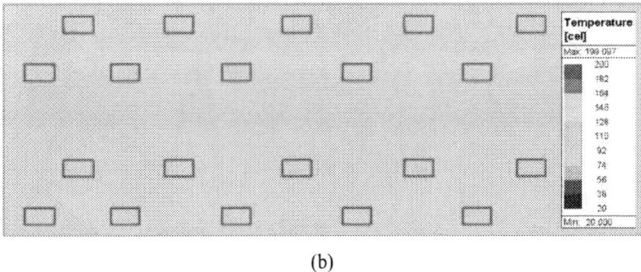

(b)

Fig. 9: Temperature maps on the top (a) and bottom (b) surfaces of the SiC MOSFET die (Fig. 1a) at 258 W total power dissipation.

Thermal simulations were also performed for the MOSFET with an integrated SBD. In these devices, most of the power is dissipated in the MOSFET cells while the power dissipated in the SBD cells is very small, negligible for our simulations. The locations of the SBD cells, near both ends of the die, are much cooler than the rest of die, as shown on the temperature map for the top surface at 212 W (Fig. 10). The maximum temperature reaches 198 °C in the MOSFET cells and only 91 °C in the SBD cells. These results confirm that our design could effectively mitigate the forward voltage increase of the SBD with temperature and enable us to integrate the MOSFET and SBD on the same SiC die.

979-8-3503-3714-3/23 $31.00 © 2023 IEEE   145

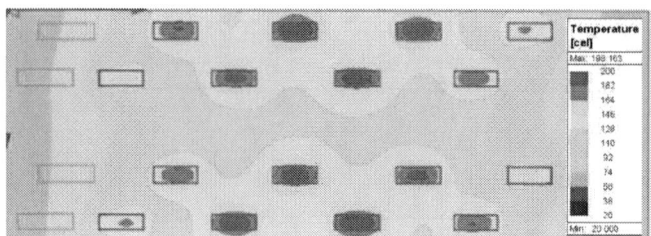

Fig. 10: Temperature map on the top surface of the SiC MOSFET die with an integrated SBD (Fig. 2) at 212 W total power dissipation.

V. RF CHARACTERIZATION

A. Class C Amplifier Design

A test board with adjustable loads for class C operation at 54.24 MHz was designed to characterize the RF performances of our SiC MOSFET (Fig. 11). The tests were performed with a 150 V drain power supply (V_{DD}) and a carefully adjusted gate voltage to a value slightly higher than the threshold voltage. The drain power supply was selected to ensure the drain voltage remains below the rating of the part (600 V) during resonant operation of the class C amplifier. The gate voltage brought the SiC MOSFET into weak conduction, at a drain current of about 10mA. During the RF tests, a 54.24MHz input signal (sine wave) was applied to the input terminal. Most of the information on the tuning of the board to achieve resonant operation is presented in Fig. 11b.

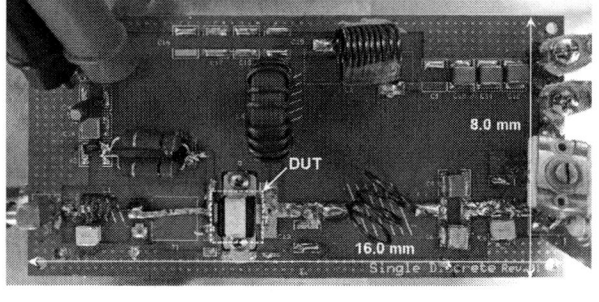

(a)

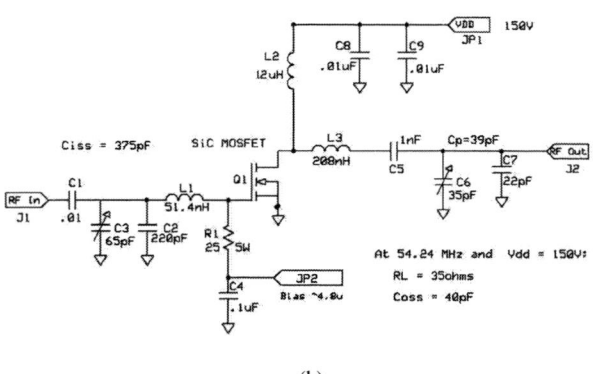

(b)

Fig. 11: Picture (a) and schematic (b) of the class C amplifier for RF tests at 54.24 MHz.

To ensure efficient heat removal, the substrates of the SiC MOSFET packages were brazed to cooling flanges that allowed us to attach the parts to a heatsink with circulating coolant (water) at 16 °C. The temperature of the SiC MOSFET was monitored with an infra-red camera (Fluke Ti480 PRO with FLK-LENS/MAC2 macro lens).

B. Test Results

The RF gain, efficiency, and output power were recorded as the input power was varied (Fig. 12). The RF gain is outstanding, reaching 30.7 dB for 250 W output power. And the amplifier can deliver a large amount of power with a single SiC die, up to at least 500 W with 90 % efficiency. The recorded temperature at 500 W output power was only 84 °C. A second set of tests with a 175 V drain power supply gives 700 W of output power with 93 % efficiency and 24.8 dB RF gain. The temperature at 700 W is 125 °C, which leaves room to further increase the power density.

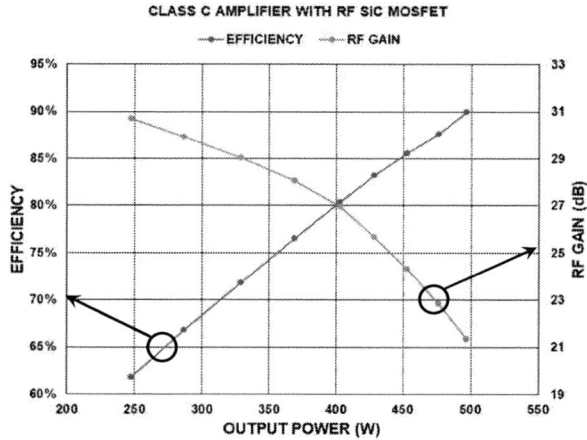

Fig. 12: Class C amplifier test results at 54.24 MHz for efficiency, RF gain and output power. The tests were performed with a 150 V drain power supply.

The drain and gate voltage waveforms at 500 W (Fig. 13) confirm the operation at 54.24 MHz. The gate voltage varies above and below the threshold voltage, turning the MOSFET on and off, and the drain voltage reaches 400 V, safely below the 600V rating.

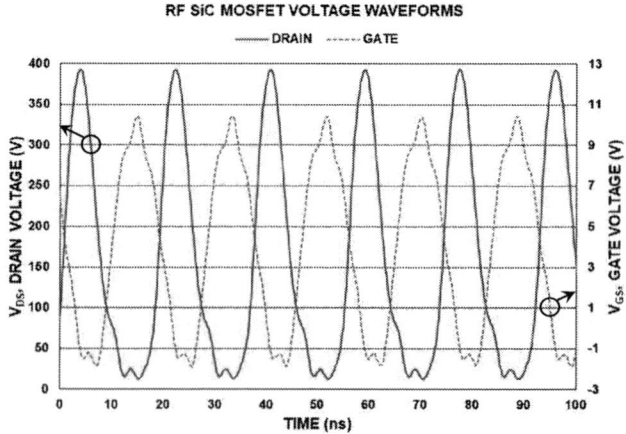

Fig. 13: Drain and gate voltages waveforms for the SiC MOSFET at 500 W output power.

VI. CONCLUSION

SiC VDMOSs hold great promise for power management at very high frequency (above 30 MHz), and thanks to the maturity of SiC MOSFET technology, new commercial products are ready to be introduced in the market. In this paper, we present a 600 V SiC MOSFET developed for industrial RF applications. Our "low packing density" design allows us to combine very low capacitances (203, 26, and 2.7 pF for C_{ISS}, C_{OSS}, and C_{RSS}) with exceptional thermal performance. Our device could dissipate 4 times more power than a conventional "high packing density" SiC MOSFET with similar capacitances. The RF performance is demonstrated for a 54.24 MHz class C power amplifier. The tests were performed up to an output power of 700 W, which resulted in a maximum temperature of 125 °C. The efficiency and RF gain at 700 W are 93 % and 24.8 dB, respectively. The maximum RF gain is 30.7 dB, reached at 250 W output power. To further characterize the RF performance of our SiC MOSFET, we plan to build a single ended module (SEM) with the gate driver and MOSFET combined into the same package. For class E operation, this module should deliver unmatched efficiency and output power.

ACKNOWLEDGMENT

The authors thank Ed Maxwell, Brody McMann, and the whole team at SiCamore Semi for manufacturing the RF SiC MOSFETs.

REFERENCES

[1] B. J. Baliga, "Power Semiconductor Device Figure of Merit for High-Frequency Applications," IEEE Electron Device Lett., vol. 10, no. 10, pp. 455–457, October 1989.

[2] A. Gendron-Hansen, C. Hong, Y. Yang, J. May, D. Meyer, D. Sdrulla, et al., "High-Performance 700 V SiC MOSFETs for the Industrial Market," in IEEE WiPDA Proc., pp. 410–415, October 2019.

[3] J. Xu, Z. Tong, and J. M. Rivas-Davila, "1 kW MHz Wideband Class E Power Amplifier," IEEE Open Journal of Power Electronics., vol 3, pp. 84–92, January 2022.

[4] G. Zulauf, Z. Tong, J. D. Plummer, and J. M. Rivas-Davila, "Active Power Device Selection in High- and Very-High-Frequency Power Converters," IEEE Trans. Power Electron., vol 34, no. 7, pp. 6818–6833, July 2019.

[5] T. Yoneyama, T. Hosoyamada, S. Kobayashi, , and I. Yuzurihara, "RF Band PWM Generator with High Efficiency and Wide-Band Control," in IEEE ECCE Proc., pp. 1889–1894, October 2021.

[6] K. Palanisamy, K. Sawant, and J. Choi, "Paralleling Devices in a 13.56 MHz Class Φ_2 Inverter to Achieve Current Splitting and Improve Device Thermal Performance," in IEEE WiPDA Proc., pp. 152–157, November 2021.

[7] S. E. Cabrera, T. Gehring, F. Denk, Q. Jin, J. Dycke, M. Renscheler, et al., "SiC-Based Resonant Converters With ZVS Operated in MHz Range Driving Rapidly Variable Loads: Inductively Coupled Plasmas as a Case of Study," IEEE Trans. Power Electron., vol 37, no. 7, pp. 7775–7788, July 2022.

[8] B. Regensburger, A. Kumar, S. Sinha, and K. Afridi, "High-Performance 13.56-MHz Large Air-Gap Capacitive Wireless Power Transfer System for Electric Vehicle Charging," in IEEE COMPEL Proc., pp. 1–4, June 2018.

[9] L. Gu, G. Zulauf, A. Stein, P. A. Kyaw, T. Chen, and J. M. Rivas-Davila., "6.78-MHz Wireless Power Transfer With Self-Resonant Coils at 95% DC–DC Efficiency," IEEE Trans. Power Electron., vol 36, no. 3, pp. 2456–2460, March 2021.

[10] Y. Wang, R. Kheirollah, F. Lu, and H. Zhang, "Exploring Switching Limit of SiC Inverter for MultikW Multi-MHz Wireless Power Transfer System," in IEEE APEC Proc., pp. 2952–2957, March 2023.

[11] J. Perreault, J. Hu, J. M. Rivas-Davila, Y. Han, O. Leitermann, R. C. N. Pilawa-Podgurski, et al., "Opportunities and Challenges in Very High Frequency Power Conversion," in IEEE APEC Proc., pp. 1–14, February 2009.

[12] F. Denk, K. Haehre, C. Simon, S. Eizaguirre, M. Heidinger, R. Kling, et al., "25 kW high power resonant inverter operating at 2.5 MHz based on SiC SMD phase-leg modules," in IEEE PCIM Europe Proc., pp. 1–7, June 2018.

[13] X. Li, B. Tan, A. Q. Huang, B. Zhang, Y. Zhang, X. Deng, et al., "Impact of Termination Region on Switching Loss for SiC MOSFET," IEEE Trans. Electron Devices., vol 66, no. 2, pp. 1026–1031, February 2019.

[14] Z. Tong, G. Zulauf, J. Xu, , J. D. Plummer, and J. M. Rivas-Davila., "Output Capacitance Loss Characterization of Silicon Carbide Schottky Diodes," IEEE J. Emerg. Sel. Top. Power Electron., vol 7, no. 2, pp. 865–878, June 2019.

[15] Z. Tong, J. Roig-Guitart, J. D. Plummer, and J. M. Rivas-Davila., "Origins of Soft-Switching Coss Losses in SiC Power MOSFETs Diodes for Resonant Converter Applications," IEEE J. Emerg. Sel. Top. Power Electron., vol. 9, no. 4, pp. 4082-4095, August 2021.

Edge Termination Design Considerations for 1.2kV 4H-SiC MOSFETs While Utilizing Room Temperature Ion Implantations

Stephen A Mancini[a], Seung Yup Jang[a], Zeyu Chen[b], Dongyoung Kim[a], Balaji Raghothamachar[b], Michael Dudley[b], and Woongje Sung[a]

[a] University at Albany, State University of New York, College of Nanoscale Science and Engineering, Albany NY, 12203, USA

[b] Department of Material Science and Chemical Engineering, Stoney Brook University, Stoney Brook NY, 11794, USA

(e-mail: smancini@albany.edu)

Abstract- **Several 1.2kV-rated 4H-SiC MOSFETs were fabricated with various edge termination (ET) structures to investigate the effects of both Junction Termination Extension (JTE) based and P+ ring-based ETs, which were implanted at room temperature (RT). The study revealed that JTE-based ETs exhibited reduced generation of Basal Plane Dislocations (BPD) and consequently suppressed leakage currents when the P+ dose exceeded the previously established RT critical aluminum dose of $1x10^{15}$ cm^{-2}. In contrast, the P+ ring-based termination structures led to elevated leakage currents, even in the absence of BPDs within RT-implanted devices. Consequently, when producing devices using RT ion implantation, JTE-based ETs proved to be the ideal choice.**

Index Terms- **4H-Silicon Carbide (SiC), MOSFET, Forward Blocking, Leakage Current, Breakdown Voltage, Room Temperature Implantation, Edge Termination, Basal Plane Dislocations (BPDs), X-Ray Topography.**

I. Introduction

The wide bandgap of 4H-SiC offers significant advantages over silicon as the material of choice in high-voltage power devices and applications. This large bandgap, combined with the material's ability to withstand high electric fields allows for the creation of a thin and heavily doped epitaxial drift layer to block the required voltage. As a result, devices made utilizing 4H-SiC exhibit lower resistances when compared to silicon at voltage ratings greater than or equal to 600V [1]. While the material properties of 4H-SiC make it the preferred choice for power device applications, it also presents unique challenges not encountered in traditional silicon processing. In silicon processing, ion implantation can be conducted at room temperature (RT) allowing for the photoresist to be used as a blocking mask [2]. However, in the case of 4H-SiC the ion implantation process is typically carried out at elevated temperatures to mitigate the generation of Basal Plane Dislocations (BPDs), which can lead to device degradation under high bipolar current stress [3,4]. The high temperature (HT) implantation process necessitates the use of an oxide blocking layer rather than relying on a photoresist layer,

introducing additional processing steps. Consequently, this increases the overall processing complexity, cost, and time required for fabrication while potentially hindering overall device performance [5].

Due to the aforementioned drawbacks associated with all HT ion implantation, there have been studies aimed at integrating full-scale RT ion implantation without generating a significant amount of BPDs. Previously, it was reported that a critical aluminum ion dose of $1x10^{15}cm^{-2}$ could effectively prevent this significant BPD generation while implementing RT ion implantation [6]. Subsequent developments in this field have demonstrated that precise control over both the implantation energy and dose simultaneously allows for devices with an implant dose nearly 10x greater than the previous critical dose while suppressing BPD generation, thereby enhancing device performance and reliability [4]. However, in that study, it was observed that a high BPD concentration was originating within the edge termination regions of these devices [4]. Given the crucial role of the edge termination structure in ensuring the necessary blocking voltage is achieved, further investigation is warranted. In this study, we examined the effectiveness of Junction Termination Extension (JTE) based edge terminations and Floating Field Ring (FFR) based edge terminations, fabricated using both elevated and RT ion implantation techniques.

II. Active Area and Edge Termination Designs

A. Active Area Design

To accommodate the 1.2kV blocking voltage rating of thedevices, an n-epitaxial drift layer was designed. This layer has a thickness of 10μm and a background nitrogen doping concentration of $8x10^{15}$ cm^{-3}. This design allowed for the achievement of a parallel plane blocking voltage of approximately 1600V, while also minimizing the $R_{on,sp}$ of the MOSFET devices. Once the breakdown voltage was successfully achieved with the edge termination, 2-D forward blocking electrical simulations were conducted. These simulations aimed to identify the optimal dimensions within the active area of the MOSFETs. For this, the doping of the JFET region within these devices was enhanced from the low

---

979-8-3503-3714-3/23 $31.00 © 2023 IEEE

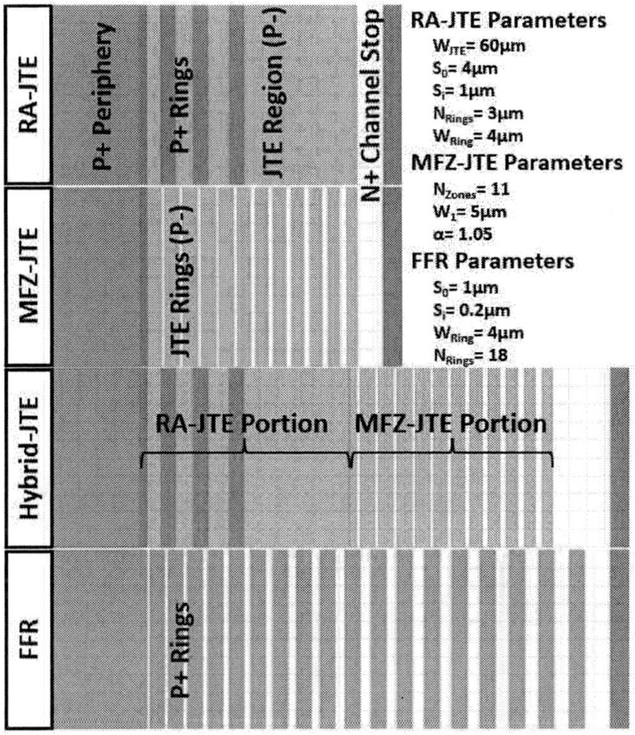

Figure 1. Top layout views of the various edge terminations utilized within this study. Key components and design parameters for each edge termination are labeled above. The Hybrid edge termination optimized parameters are the same as the standalone RA-JTE and MFZ-JTE structures. Each parameter was optimized through the use of 2D electrical breakdown simulations.

background doping of $8x10^{15}$ cm^{-3} to $5x10^{16}$ cm^{-3}. The results of these optimizations led to the determination of the ideal dimensions for the JFET width and channel length. These dimensions, 1.2µm and 0.5µm respectively, were found to result in a high breakdown voltage, minimal leakage current, and low Specific on-resistance ($R_{on,sp}$).

B. Edge Termination Designs

In this study, four different edge terminations were optimized using 2-D blocking electrical simulations. These examined edge terminations utilized two distinct methods: the Junction Termination Extension (JTE) based methods and the P+ ring-based method. The JTE-based edge terminations included the Ring Assisted JTE (RA-JTE), Multiple Floating Zone JTE (MFZ-JTE), and the Hybrid JTE structures. The final edge termination using the P+ ring-based edge termination employed the non-equally spaced Floating Field Ring (FFR) structure. Each structure along with its optimized dimensions can be observed in Fig. 1.

The RA-JTE edge termination (ET) was optimized with specific parameters: spacing between the first P+ ring and the main junction (S_o), incremental spacings between rings (S_i), the number of rings (n), ring width (W), all within the JTE region with a width (W_{jte}). The final dimensions can be determined using the equation $S_n=S_o+S_i(n-1)$, where Sn represents the spacing between each P+ ring relative to the ring number. The RA-JTE's optimized parameters for S_o, S_i, n, W, and W_{jte} are 4 µm, 1 µm, 3, 4 µm, and 60 µm, respectively.

The MFZ-JTE was optimized to feature multiple JTE regions with gradually decreasing widths, defined by the equation $W_n=W1/\alpha^{(n-1)}$, where n corresponds to the zone number, W_n represents the width of the nth region, and α is a decreasing parameter. The MFZ-JTE's optimized parameters for n, W1, and α are 11, 5 µm, and 1.05, respectively.

The Hybrid JTE structure combines both the RA-JTE and a MFZ-JTE into a single edge termination structure. In this arrangement, the RA-JTE portion is situated closer to the main junction of the device, followed by the MFZ-JTE portion. This approach enables a wide range of JTE doses during device fabrication while simultaneously ensuring high breakdown voltages [7].

The non-equally spaced FFR ET utilizes concentric P+ rings which consist of several key parameters: the spacing between the first P+ ring and the main junction (S_o), incremental spacings between rings (S_i), the number of rings (n), and ring width (W_{ring}). Unlike previous ET structures, the JTE region(s) was not used in the creation of this structure. Since these P+ rings can be formed alongside the main P+ junction, the overall processing time and complexity can be reduced due to the omission of an implantation step. The optimized FFR parameters of S_o, S_i, W_{rings}, and N_{rings} are 1 µm, 0.2 µm, 4 µm, and 18, respectively.

C. Implantation Profiles and Conditions

Fig. 2 shows the three distinct simulated P+ aluminum implantation profiles along with the JTE implantation profile employed to create the various edge termination structures analyzed in this study. These profiles are varied with high (H), medium (M), and low (L) doping concentration within the surface (S) and body (B) of the implanted junction. The 'LSLB (meaning Low Surface Low Body)' (simplified to '1x'), 'MSHB' ('5x'), and 'HSMB' ('9x') has a total Al dose of 1×10^{15} cm^{-2}, 5×10^{15} cm^{-2}, and 9×10^{15} cm^{-2}, respectively [4]. Due to this profile shift towards the surface, the total BPD generation can be reduced despite having a greater aluminum dose. All implantations were performed at RT; however, the '5x' implant condition was also fabricated at an elevated temperature of 600°C for comparison purposes.

III. Device Fabrication

Several 1.2kV Vertical MOSFETs with various edge termination structures have been successfully fabricated for this study. This voltage rating was achieved by initially growing a heavily doped N+ buffer layer, followed by a 10µm thick, $8x10^{15}$cm-3 n-doped epitaxial layer on top of the 4H-SiC substrates with a 4° offset. Ion implantation was utilized to create the various doping regions throughout the MOSFETs. Aluminum ions were used to form the P+/P-Well/JTE regions, while Nitrogen ions were used for the formation of the JFET/N+ regions. All implantations were conducted at room temperature, 25°C, except for the P+ implant for the '5xHT' wafer. The P+ regions for these '5xHT'

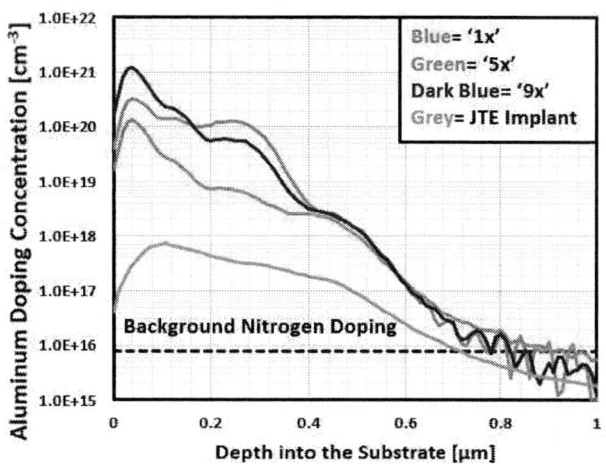

Figure 2. Simulated aluminum profiles of the various P+ implantations along with the P- JTE implant profile. The dose of the '1x', '5x', and '9x' profiles are 1x10^{15}, 5x10^{15}, and 9x10^{15}cm^{-2}, respectively. The '9x' profile peak was shifted towards the surface resulting in less BPD generation when compared to the '5x' condition [4].

samples were implanted at an elevated temperature of 600°C to mitigate potential ion implantation damage, such as BPDs. Following the ion implantation processes, a carbon capping layer was deposited, and the wafers underwent an activation annealing process for 10 minutes at 1650°C.

The remaining fabrication steps and designs were consistent across all wafer conditions. To create the MOSFET gates, a 50nm-thick gate oxide layer was formed through thermal oxidation, followed by the deposition of a polysilicon layer. Subsequently, both the oxide and polysilicon layers underwent etching to shape the MOSFET gate. After the gate formation, an interlayer dielectric layer was uniformly deposited on the wafer's surface and then patterned to facilitate the creation of an ohmic contact within the source regions. Nickel was subsequently deposited, and a pre-annealing process was conducted at 750°C for 2 minutes using a rapid thermal anneal (RTA) process. Any residual nickel that did not form silicide was removed. On the backside of the wafer, additional nickel was deposited. Subsequently, all nickel metal, both on the front and backside of the wafer, underwent an annealing process at 1000°C for 2 minutes. This annealing process facilitated the formation of both the front and backside ohmic contacts for the MOSFET devices. Following the ohmic contact formation, a thin layer of titanium nitride was applied, followed by the deposition and patterning of a 4μm thick aluminum top metal layer to establish the final source and gate contacts. As the concluding steps of the process flow, a nitride layer and a thick polyimide layer were deposited and patterned to passivate the front side of the wafer.

IV. Results

A. Physical Analysis of Fabricated Devices

To analyze the BPD density generated from each fabricated edge termination, X-Ray topography was employed. These X-Ray topography images were obtained from 4H-SiC wafers and captured on high resolution X-Ray films at 1-BM, Advanced Photon Source at Argonne National Laboratory. X-Ray topography images of the RA-JTE, MFZ-JTE, Hybrid-JTE, and FFR edge termination structures fabricated under each P+ implantation condition can be observed in Fig. 3-6 respectively.

From this analysis, it was found that negligible BPD generation occurred within all the '1xRT' and '5xHT' implanted edge termination structures. However, the same cannot be observed for the edge terminations fabricated using the '9xRT' and '5xRT' P+ implantation conditions. Within these structures, BPD nucleation was detected in the P+ periphery of the MFZ-JTE, the rings and periphery of the RA-JTE and Hybrid-JTE, and throughout the entirety of the

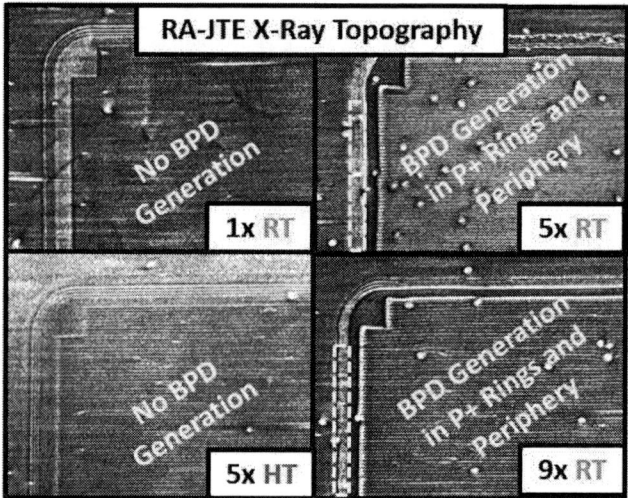

Figure 3. X-Ray topography images of the MOSFETs utilizing the RA-JTE ET. Both the '1x RT' and '5x HT' structures resulted in negligible BPD generation; however, BPD generation is observed originating in the P+ periphery and rings in the '9x RT' and to a greater extent in the '5x RT' structures.

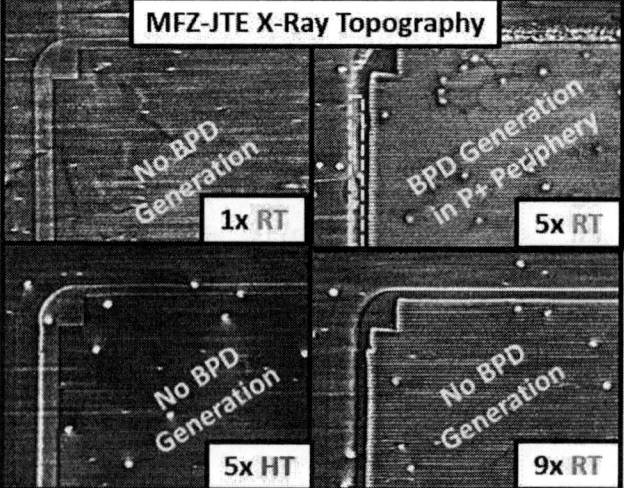

Figure 4. X-Ray topography images of the MOSFETs utilizing the MFZ-JTE ET. BPD generation is observed originating in the P+ Periphery of the '5x RT' and '9xRT structures.

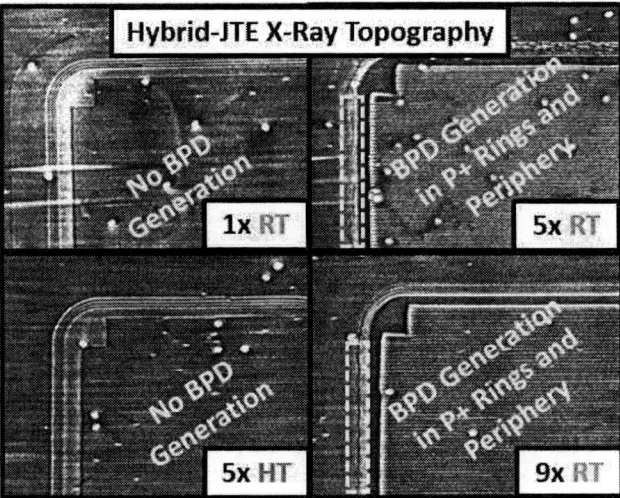

Figure 5. X-Ray topography images of the MOSFETs utilizing the Hybrid-JTE ET. BPD generation is observed originating in the P+ Periphery and P+ rings of the '5x RT' and '9xRT structures.

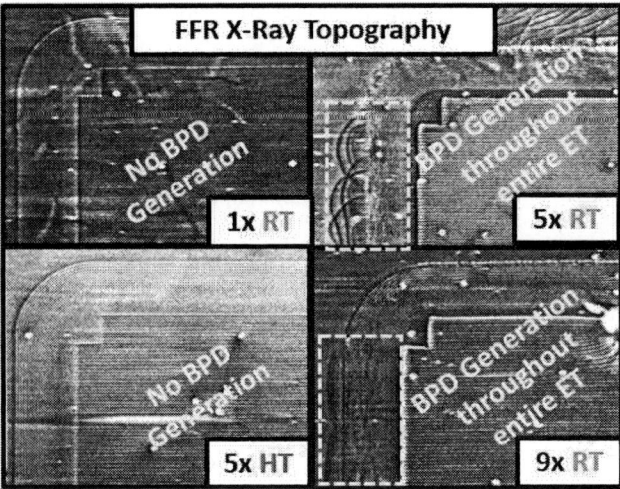

Figure 6. X-Ray topography images of the MOSFETs utilizing the non-equally spaced FFR ET. BPD generation is observed originating throughout the entirety of the '5x RT' and '9xRT edge terminations and periphery.

Table I. BPD densities of each edge termination structure fabricated using the four different P+ implantation profiles. Negligible BPD density were calculated for all the '1xRT' and '5xHT' edge termination structures. BPD density calculations were based on the BPD nucleation from device edges and assuming the entire device volume.

Implant Condition	RA-JTE [cm⁻²]	MFZ-JTE [cm⁻²]	Hybrid-JTE [cm⁻²]	FFR [cm⁻²]
1x RT	~0	~0	~0	~0
5x HT	~0	~0	~0	~0
5x RT	0.5×10^4	0.3×10^4	0.5×10^4	2.1×10^4
9x RT	0.2×10^4	~0	0.2×10^4	0.5×10^4

FFR ET. Since this BPD generation can be clearly defined along P+ implanted regions, this damage generation is process related as opposed to being native to the 4H-SiC wafer. This is a result of the activation annealing process, in which regions between the P+ and P- region (JTE implant)

and between P+ and non-implanted regions encounter a large lattice mismatch strain due to a large difference in implanted Aluminum concentration [8, 9]. Due to the profile shift towards the surface, the total BPD generation was suppressed in the '9xRT' implanted edge terminations despite having a greater aluminum dose when compared with the '5xRT' P+ implantation condition.

The total BPD densities of each edge termination fabricated under the various implantation conditions used can be observed in Table I. The '5xRT' JTE-based edge terminations resulted in approximately $0.3 \times 10^4 \text{cm}^{-2}$ more BPDs when compared to their '9xRT' implanted counterparts. This effect was amplified even further for the ring-based edge termination, with the '5xRT' implantation generating approximately $1.6 \times 10^4 \text{cm}^{-2}$ more BPDs when compared to the '9xRT' P+ implant condition. In the case of the '5xRT' FFR ET structure, nearly a 4x greater BPD density generation can be observed when compared to its RA-JTE and Hybrid-JTE counterpart. This can be attributed to the 4x larger P+ implanted ring area within the FFR structure, which is necessary to effectively block the required voltage. Consequently, the FFR ET is not ideal for mitigating BPD generation while utilizing RT ion implantation when compared to a JTE-Based structure.

It is also important to note that negligible BPD generation was observed withing the active area of the MOSFETs as the P+ source regions are fully enclosed by the P-well. Similar to the JTE in the edge termination, the P-well implanted regions act as a P- implanted area therefore suppressing the lattice mismatch strain subsequently mitigating BPD generation within the active region for RT implanted structures.

B. Forward Blocking Characteristics

Fig 7. shows the typical forward blocking, and leakage characteristics for each ET structure fabricated using the various P+ implantation conditions. In the case of JTE-based ET structures, the '5xRT' implantation resulted in elevated leakage currents; however, the '1xRT' and '9xRT' structures exhibited leakage current results comparable to the '5xHT' implanted structure used as a reference. Despite differences in leakage current for the '5xRT' P+ implanted devices, all variations in implant conditions showed a breakdown voltage of approximately 1610V, 1360V and 1620V for the RA-JTE, MFZ-JTE and Hybrid JTE structures respectively. Conversely, the P+ ring-based ET resulted in elevated leakage currents for all RT implanted conditions, even in cases where lattice damage such as BPD generation was not observed, as seen in the '1xRT' P+ implanted structure. This elevated leakage current resulted in a soft breakdown of the '9xRT' and '5xRT' P+ implanted devices as they both reached a drain-source current compliance of 1mA prior to avalanche behavior, which was not observed in any of the JTE-based edge terminations. The resulting breakdown voltage for the '5xHT', '1xRT', '9xRT' and '5xRT' P+ implanted FFR ETs are approximately 1500V, 1400V, 1290V and 1210V respectively.

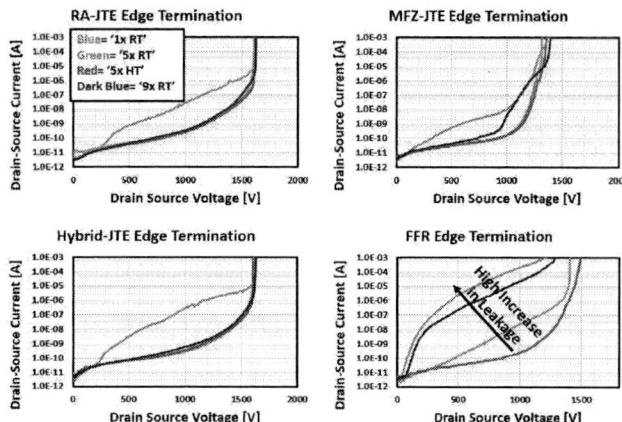

Figure 7. Typical blocking behaviors of the MOSFETs utilizing the various edge termination structures and P+ implantations. All JTE based ETs resulted in a lower leakage current when compared to that of the P+ ring-based ETs for RT implanted devices. A gate bias of Vgs= -5V was applied to each measurement to minimize any potential leakage through the channel.

	Device Active Area	Device Periphery	Edge Termination Region
Hybrid PiN	Top Metal / P+ / Epitaxial Drift Layer		Oxide/ Passivation / P- JTE Region
Hybrid MOSFET	Epitaxial Drift Layer	P+	Oxide/Passivation / P- JTE Region
FFR MOSFET	Epitaxial Drift Layer	P+	Oxide/Passivation / P+ P+ P+

Figure 8. Simplified cross-sectional schematics of the Hybrid PiN Diode, Hybrid MOSFET, and FFR MOSFET structures. The active area, device periphery, and the edge terminations regions for each structure are indicated and divided by the red dashed lines. The only differences between the Hybrid PiN and Hybrid MOSFET structures are found within the device active area, while the distinctions between the Hybrid MOSFET and FFR MOSFET lie in the edge termination region.

To determine the origins of the observed leakage currents, we conducted an analysis of both the Hybrid MOSFET and FFR MOSFET alongside a PiN diode fabricated using the same implant conditions and Hybrid-JTE edge termination for the purpose of comparison. Fig 8. highlights these three different structures, each comprising a device active area, device periphery, and edge termination as their main components. For the Hybrid PiN Diode and Hybrid MOSFET devices, the only difference lies within the active region of the devices, as both share identical peripheries and edge terminations. However, differences for the Hybrid MOSFET and FFR MOSFET can be seen in the edge termination regions with both devices having identical active area and periphery regions.

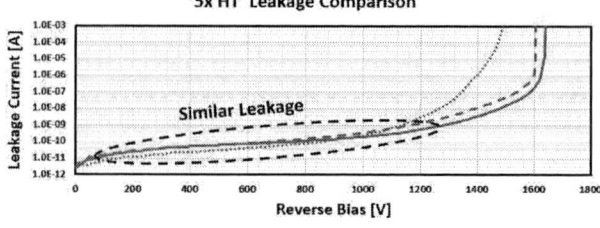

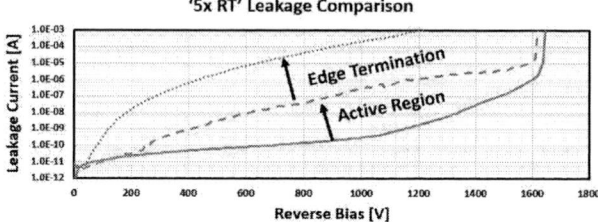

Solid= Hybrid PiN Diode, Dashed= Hybrid MOSFET, Dotted= FFR MOSFET

Figure 9. Typical forward blocking characteristics for the Hybrid PiN Diode, Hybrid MOSFET, and FFR MOSFET fabricated using either the '5xHT' (top) or '5xRT' (bottom) P+ implantation condition. All '5xHT' devices all showed comparable blocking characteristics. However, a large increase in leakage current can be seen originating within the '5xRT' implanted structures. This leakage, either originating from the MOSFET active region or the FFR edge termination is labeled above.

The typical leakage for the Hybrid PiN, Hybrid MOSFET and FFR MOSFET devices fabricated with either the '5xHT' or '5xRT' P+ implantation conditions can be observed in Fig 9. The '5xHT' implanted devices, used as a baseline, all showed comparable leakage levels at low drain biases, indicating the ideal blocking capabilities of each device without the interference of BPDs. However, for the '5xRT' devices, a significant increase in leakage current can be observed between the PiN and MOSFET structures. Both devices employ the same Hybrid edge termination and device periphery, meaning this 100x increase in leakage current originates from the active region of the MOSFET. Another substantial increase in leakage current is evident between the FFR MOSFET and Hybrid MOSFET devices where all components except the edge termination were identical. This '5xRT' FFR ET resulted in an additional 1000x increase in leakage current when compared to the Hybrid MOSFET structure. Since it has been confirmed that these structures exhibit near-ideal blocking and leakage when fabricated with the '5xHT' P+ condition, this additional leakage can be attributed to the significantly higher BPD density of the '5xRT' FFR MOSFET.

Due to the JTE-Based edge terminations having either near-ideal or substantially suppressed leakage currents when compared to their P+ ring-based counterparts, it is preferable to employ this edge termination method when fabricating devices using RT ion implantation.

V. Conclusions

Several 1.2kV MOSFETs with various edge terminations were successfully fabricated using different P+ implantation conditions. It was confirmed that using a high dose and energy Al ion implantation at RT generated a significant amount of BPDs. However, the generation of BPDs was reduced in edge terminations structures in which the P+ regions were fully enclosed by a lighter doped P-region (JTE implant) when compared to the non-enclosed P+ rings in the FFR edge termination structure. This reduction in BPD generation subsequently lead to a reduction in leakage current similar to the edge terminations formed through HT ion implantation. Therefore, the use of JTE-Based edge terminations is ideal over its FFR-based counterparts, when fabricating devices employing RT ion implantation due to the mitigated BPD generation and suppressed leakage currents.

Acknowledgement

This work was supported by the National Renewable Energy Laboratory ("NREL"), U.S. Department of Energy, the Advanced Manufacturing Office, DE-AC36-08GO28308. This research used resources of the Advanced Photon Source, a U.S. Department of Energy (DOE) Office of Science User Facility operated for the DOE Office of Science by Argonne National Laboratory under Contract No. DE-AC02-06CH11357. The Joint Photon Sciences Institute at SBU provided partial support for travel and subsistence for access to Advanced Photon Source.

References

[1] B Jayant Baliga, *Fundamentals of power semiconductor devices.* New York, NY: Springer, 2008, Chap.3, pp. 91-155.

[2] S.Wolf and R.N. Tauber, *Silicon Processing for the VLSI Era Volume 1:Process Technology.* Sunset Beach, U.S.A. : Lattice Press, 1986. pp. 280-283.

[3] A. Agarwal, H. Fatima, S.Haney, S.-H. Ryu, "A New Degradation Mechanism in High-Voltage SiC Power Mosfets". *IEEE Electron Device Letters vol. 28, no. 7, pp.587-589, 2007.* doi.: 10.1109/LED.2007.897861.

[4] S. A. Mancini, S. Y. Jang, D. Kim and W. Sung, "Static Performance and Reliability of 4H-SiC Diodes with P+ Regions Formed by Various Profiles and Temperatures," 2022 IEEE International Reliability Physics Symposium (IRPS), 2022, pp. P62-1-P62-6, doi: 10.1109/IRPS48227.2022.9764538.

[5] D. Kim, N. Yun and W. Sung, "Advancing Static Performance and Ruggedness of 600 V SiC MOSFETs: Experimental Analysis and Simulation Study," 2021 IEEE International Reliability Physics Symposium (IRPS), 2021, pp. 1-4, doi: 10.1109/IRPS46558.2021.9405109.

[6] T. Kimoto, J. A. Cooper, *"Fundamentals of Silicon Carbide Technology: Growth, Characterization, Devices and Applications".* John Wiley & Sons Singapore Pte. Ltd. 2014. pp. 197-200.

[7] W. Sung and B. J. Baliga, "A Near Ideal Edge Termination Technique for 4500V 4H-SiC Devices: The Hybrid Junction Termination Extension," *IEEE Electron Device Letters*, vol. 37, no. 12, pp. 1609–1612, Dec. 2016, doi: 10.1109/led.2016.2623423.

[8] K. Konishi, R. Fujita, Y. Mori, and A. Shima, "Inducing defects in 3.3 kV SiC MOSFETs by annealing after ion implantation and evaluating their effect on bipolar degradation of the MOSFETs," *Semiconductor Science and Technology*, vol. 33, no. 12, p. 125014, Nov. 2018, doi: 10.1088/1361-6641/aae814.

[9] Z. Chen et al, "Analysis of Basal Plane Dislocation Motion Induced by P+ Ion Implantation Using Synchrotron X-ray Topography" *2022 19th International Conference on Silicon Carbide and Related Materials (ICSCRM)*, September 2022, doi: 10.4028/p-6dx2v3 .

Scalable Test System for Long Term Reliability Assessment of SiC MOSFET Stability in Extreme dV/dt Stress Conditions

Lisi Zhu[*], David C. Sheridan[†], Kiran Chatty[†], Zhan Liu[*], Arash Salemi[†], Jin Zhang[*], and Madhur Bobde[†]

Email: {david.sheridan, kiran.chatty, arash.salemi, madhur.bobde}@us.aosmd.com {lisi.zhu, zhan.liu, jin.zhang}@sh.aosmd.com

[†]Alpha and Omega Semiconductor, Sunnyvale, CA 94085, U.S.A.

[*]Alpha and Omega Semiconductor, Shanghai, 200070, China

Abstract—A scalable test system was developed to evaluate the stability of SiC MOSFETs operating in a continuous mode at extreme dV/dt conditions. The half-bridge test system can test multiple devices simultaneously with minimal power consumption and thus easily expandable to quickly generate the required device characterization statistics to ensure accurate conclusions. Several SiC MOSFETs with different rated voltages and oxide thickness were evaluated for potential degradation when subjected to both reverse-recovery and forward hard switched stress of 10V/ns and 150V/ns over 1500+ hours at elevated temperature with less than 4% parameter shifts.

Keywords—SiC MOFSET, reliability, hard switching, dv/dt, burn-in system, DHTOL

I. INTRODUCTION

In the last few years, the power semiconductor market has seen an accelerated use of SiC MOSFETs for use in many applications including automotive, solar inverters, and a wide variety of industrial power supplies. One of the primary factors driving the rapid adoption of SiC MOSFETs is due to the unipolar physical nature enabling fast switching performance that typically results in increased system efficiency. In the drive for higher system performance, increasing the switching frequency and switching speed of the SiC MOSFETs is often a key parametric trade-off evaluated during system design.

For semiconductor designers, however, understanding the reliability impacts of technology and process design on the complicated system-level dynamic stresses on the device are difficult to predict. Standard qualification and reliability tests outlined by JEDEC [1] or AEC-Q101 [2] such as HTRB and HTGB only stress the steady state operation of the device and are used to accelerate a response to a very specific physical mechanism or area of the device. Under typical application conditions, the device is subject to intermittent and dynamic stresses combining high current, high temperature, fast transients, on both gate and drain nodes. This can result in new degradation mechanisms or failure modes.

For SiC MOSFETs, the issues with gate oxide quality have been well published for many years. The inherent SiC/SiO$_2$ interface has a high population of traps compared to Si

MOSFETs, and that can lead to V_{TH} drift, hysteresis, or even failure when operating even in steady state environment [3]. Under dynamic operation of fast and repetitive transients, results have shown more complicated degradation phenomenon of both the oxide interface and intrinsic body diode [4-6].

As previously mentioned, circuit designers are pushing power electronics to higher frequency to increase efficiency. This often results in higher dV/dt. Early SiC diodes were shown to have failure modes when pushed to extreme dV/dt [7], but few results have been reported on the long-term stability of SiC MOSFETs under repetitive dV/dt. In order to ensure reliability over a wide range of dV/dt conditions, a suitable reliability test system should have the ability to stress both forward operating mode, the MOSFET body diode reverse recovery, easily adjust dV/dt, maintain a constant Tj, and scalable for large sample size testing for generating statistics.

II. SYSTEM OVERVIEW

As outlined in [8], evaluating the switching reliability can be approached in two different methods: application level test systems or test-vehicle circuits. In this work, we have selected a scalable test-vehicle that is able to evaluate SiC MOSFET under different working conditions, including temperature, voltage, current, switching speed and switching frequency. This system design is similar to [9] but differs in the focus on scalability, flexibility in switching speed, and variable test current levels. Fig. 1 shows a picture of the H-bridge stress test system used in this work. One stress test system includes drivers and controls for 6 subsystem boards to operate simultaneously.

Fig. 1. H-Bridge stress test system.

Figs 2 shows the diagram of the H-bridge stress test system. Q1 and Q2 works as device under test (DUT) which is compatible with TO247-3L or TO-247-4L SiC MOSFETs in the system. Each subsystem integrates powerful cooling system, over-current and over-temperature protection independently.

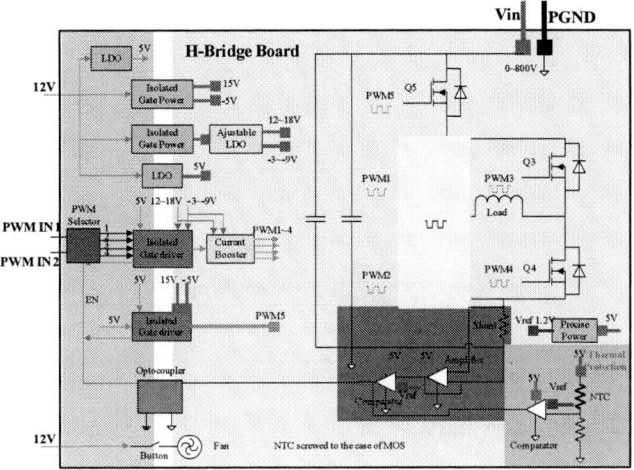

Fig. 2. H-Bridge test system diagram.

A. Gate Drive Circuit

Each MOSFET has its own gate driver circuit, including isolated driver and isolated power supplies. To stress SiC MOSFET under extreme dv/dt, gate driver needs to have a high Common Mode Transient Immunity (CMTI) and drive current. Fig. 3 shows the gate drive circuit of the SiC MOSFET; a non-isolated driver with a drive current of 30A was connected between isolated driver and MOSFET.

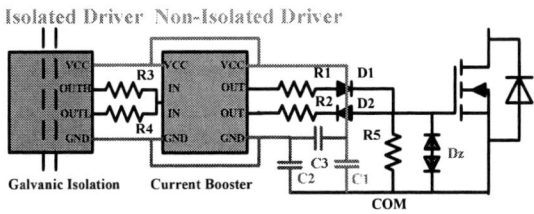

Fig. 3. SiC MOSFET Gate Drive Circuit

B. System working theory

To reduce power consumption of the test system, full bridge topology with inductor which can recycle the energy from the load is used. Fig. 4 shows the generalized switching sequence of the H-bridge stress test system. Q1 will be operated in hard switching mode and Q2 will be operated in free-wheeling mode.

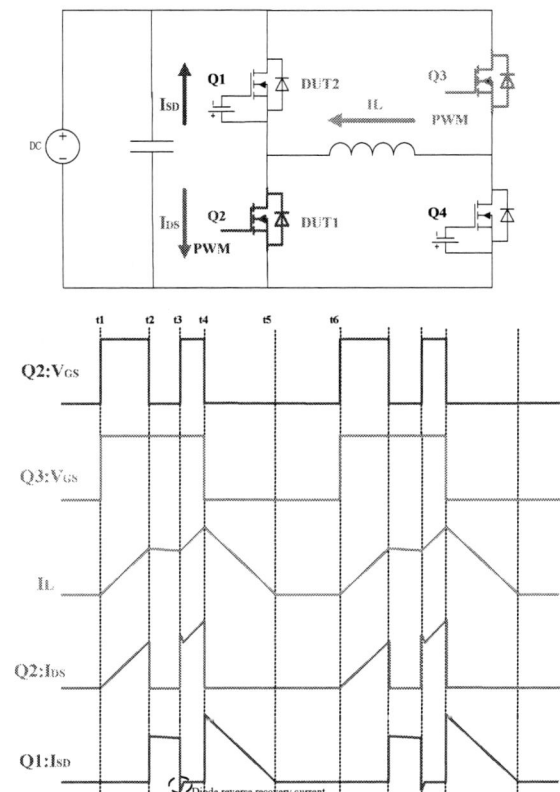

Fig. 4. Generalized switching sequence of the SiC MOSFETs.

In the basic test mode, the gate of SiC MOSFET Q2 is pulsed to allow the current in the inductor to ramp to the desired test current level and then turned off to allow MOSFET Q1 to conduct the current in reverse conduction mode through the body diode and through Q3 in the forward conduction mode. At this time, Q2 is turned back on causing Q1 to commutate the body diode, resulting in dV/dt and dI_{rr}/dt stress on Q1. Since the inductor current is ramped down to 0A, no re-circulating energy has to be dissipated. To ensure the inductor current has enough time back to 0A, the duty cycle should be equal to or less than 50%.

Fig. 5 shows a picture of the scalable test setup running both 10V/ns and 150V/ns stress over several devices simultaneously. As the circuit is relatively simple, tuning of the current can be easily achieved to stress corner use conditions such as extreme negative switching spikes on the drain or gate. Fig. 7 shows a test of -15V repetitive gate stress during switching transients which shows the ability to tune the stress on the gate as well as dV/dt. Fig. 8 shows the switching waveforms of Q2 under 150V/ns switching stress.

To reduce the dimension of subsystem board and simplify the design of cooling system, all the SiC MOSFETs are attached on a single wind tunnel heatsink with a single high-

979-8-3503-3714-3/23 $31.00 © 2023 IEEE   155

Fig. 5. Systems operating multiple DUTs simultaneously at 10V/ns and 150V/ns.

speed fan. The cooling system is efficient allowing the SiC MOSFETs to work at high frequency and large currents. It is not possible to test Q1 and Q2 under same junction temperature due to the differences in the power loss. To get the desired junction temperature, different material or thickness of isolation sheets with different thermal resistances are implemented to adjust the total thermal resistance from junction to heatsink. Since it is difficult to measure the junction temperature directly and the junction temperature is estimated from the case temperature above the center of die. Fig. 5 shows the temperature measurements on a DUT using a thermal camera; the case temperature above the center of die is approximately 85°C and the junction temperature is estimated to be 100°C.

Fig. 6. Thermal measurements of the DUT by thermal camera.

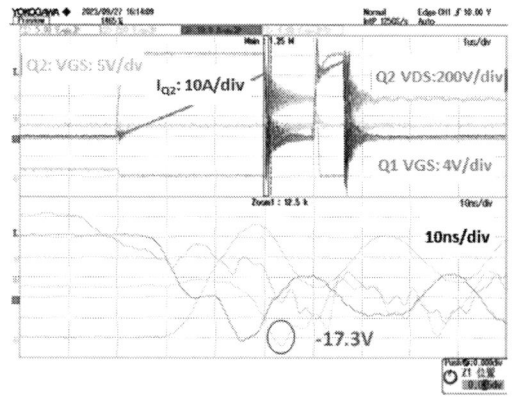

Fig. 7. Switching waveforms of Q1 and Q2.

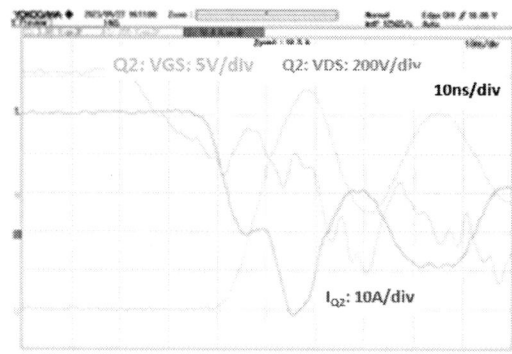

Fig. 8. Switching waveforms of Q2 under 150V/ns switching stress.

III. TEST DEVICES AND RESULTS

All SiC MOSFETs used in this work were planar MOSFETs designed by the authors [10]. The MOSFETs have a nominal threshold voltage (V_{TH}) of 2.8V, and typical $R_{DS,ON}$ ranges from 15 mΩ to 65 mΩ, with voltage ratings of either 750V or 1200V. All devices belong to an AEC-Q101 qualified process family.

The initial test device was a 1200V/33mΩ SiC MOSFET packaged in a TO-247-4L package. Devices were stressed in the test system for over 1500 hours at a frequency of 20kHz, estimated junction temperature (T_j) of 100°C, V_{DC}=800V, and I_D=40A. Fig. 9-11 show the read point measurements of the critical parameters of both high-side and low-side MOFSETs with negligible shift in the V_{TH}, $R_{DS,ON}$, I_{GSS} or body diode V_F. Production devices with rated breakdown voltages of 750V as well as experimental devices with 10% thinner gate oxide were also tested with similar positive results.

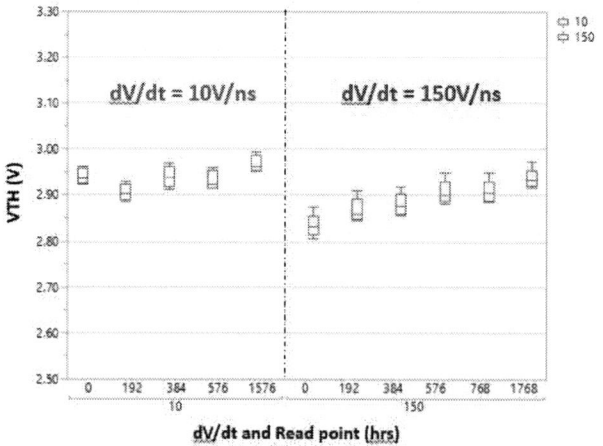

Fig. 9. V_{TH} read point measurements up to 1768 hours for both 10V/ns and 150V/ns dV/dt on hard switched Q2.

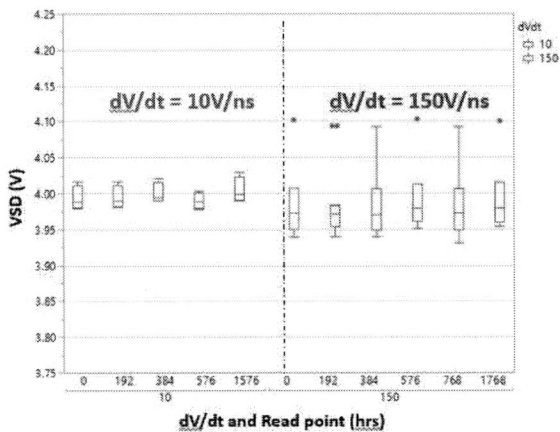

Fig. 9. Body diode forward voltage measurement up to 1768 hours for both 10V/ns and 150V/ns dV/dt on body diode stressed Q1.

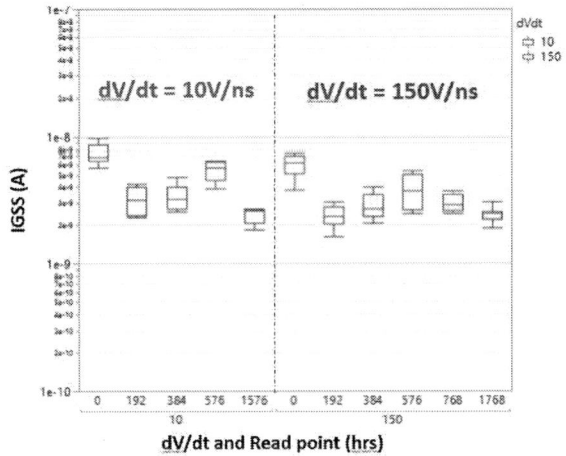

Fig. 10. Gate leakage measurements to 1768 hours for both 10V/ns and 150V/ns dV/dt on body diode stressed Q1 and Gate waveform of Q1 during body diode reverse recovery of hard switched Q2.

A. Test Failures

While the test system was designed to test devices with large dV/dt, this also resulted in a high dI_{rr}/dt of the Q1 switch. Under normal testing, Q1 had a measured dI_{rr}/dt approximately 3000A/us. Large test devices with corresponding larger capacitances than the standard DUTs were also tested with a $V_{DS} = 800V$ and much higher di/dt (4500A/us) for body diode reverse recovery.

The gate leakage current I_{GSS} of the large Q1 which operated in free-wheeling mode was over specification after being stressed for more than 500 hours. No failures were found when same devices were tested under $V_{DS} = 800V$ and a reduced dI_{rr}/dt of 3000A/us. High di/dt is a potential root cause for serious electrical overstress of Q1 gate at die level during Q1 body diode reverse recovery as previously reported in [5,11].

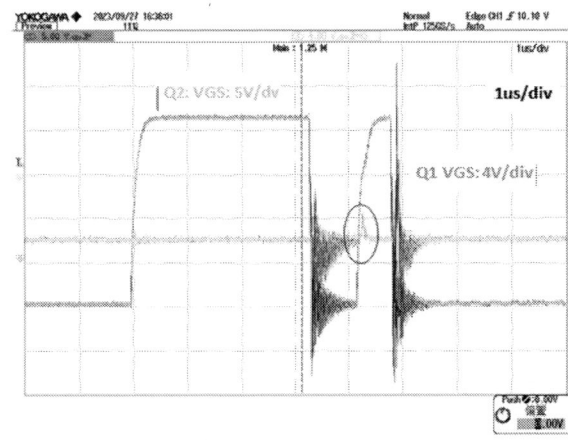

Fig. 11. Gate waveform of Q1 during body diode reverse recovery

Fig. 12 shows the gate waveform of Q1 during body diode reverse recovery. While the measured gate voltage is in a safe range, the real gate waveform can't be measured from the gate pin of Q1 outside. The oscillation and electrical overstress will occur in the gate at die level. Fig. 12 shows the Q1 body diode reverse recovery waveform under different dI_{rr}/dt. The reverse peak recovery current I_{rm} in the second current recovery phase (tb) is much higher (> 7000A/us) when it turned back to 0A. The drain and gate voltage overshoot are proportional to this recovery di/dt of I_{RR} and will result in severe overstress during each switching cycle.

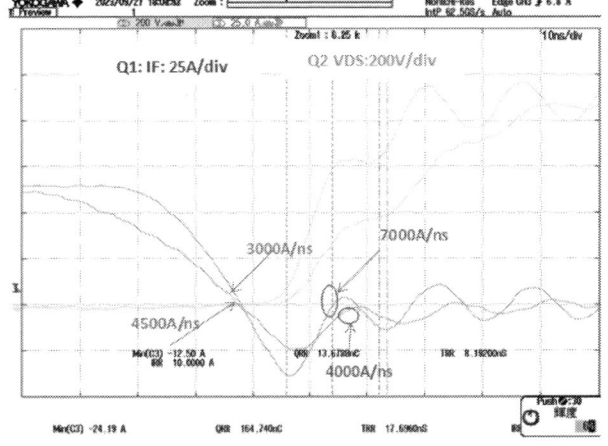

Fig. 12. Q1 body diode reverse recovery waveform under different di/dt

IV. SUMMARY

A scalable test system was developed to evaluate both the forward and reverse long-term switching reliability of SiC MOSFETs under extreme dV/dt and dI/dt conditions. The commercial SiC MOSFETs tested all passed > 1500hours of 150V/ns and 3000A/us stress with minimal shift in device parameters. Larger devices when tested with a higher 4500A/us reverse recovery dIrr/dt showed increased gate leakage due to the induced high internal V_{GS} and V_{DS} overstress. These results highlight the need for designers targeting fast switching and high dV/dt operation of the SiC MOSFETs to optimize the layout of gate and power loop to reduce the parasitic impedance, select suitable gate resistors, and to carefully monitor the device overstress in order to ensure reliable device operation.

REFERENCES

[1] JEDEC; https://www.jedec.org/

[2] AEC - Q101 - Rev - E March 1, 2021 [Online]. Available: https://www.aecouncil.com/

[3] K. Puschkarsky, T. Grasser, T. Aichinger, W. Gustin, and H. Reisinger,. "Review on SiC MOSFETs High-Voltage Device Reliability Focusing on Threshold Voltage Instability",. IEEE Transactions on Electron Devices ,. 66, 2019, S. pp. 4604 – 4616

[4] S. Nakata and S. Tanaka, "Temperature Dependence of dV/dt impact on the SiC MOSFET", *Mat. Sci Forum*, Vol. 963, pp 596-599, 2019.

[5] S. Palanisamy, T. Basler, X. Liu, C. Herrmann, R. Elpelt and P. Sochor, "Overcurrent turn-off robustness and stability of the switching behavior of SiC MOSFET body diodes," *2022 IEEE 34th International Symposium on Power Semiconductor Devices and ICs (ISPSD)*, Vancouver, BC, Canada, 2022, pp. 257-260, doi: 10.1109/ISPSD49238.2022.9813611.

[6] R. Ouaida *et al.*, "Gate Oxide Degradation of SiC MOSFET in Switching Conditions," in *IEEE Electron Device Letters*, vol. 35, no. 12, pp. 1284-1286, Dec. 2014, doi: 10.1109/LED.2014.2361674.

[7] G. Wang, E. Van Brunt, T. Barbieri, B. Hull, J. Richmond and J. Palmour, "On Developing a dV/dt Rating for Commercial 650 V- and 1200 V-Rated SiC Schottky Diodes," *PCIM Europe 2017; International Exhibition and Conference for Power Electronics, Intelligent Motion, Renewable Energy and Energy Management*, Nuremberg, Germany, 2017, pp. 1-6.

[8] S. R. Bahl, F. Baltazar and Y. Xie, "A Generalized Approach to Determine the Switching Lifetime of a GaN FET," *2020 IEEE International Reliability Physics Symposium (IRPS)*, Dallas, TX, USA, 2020, pp. 1-6, doi: 10.1109/IRPS45951.2020.9129631.

[9] G. Sheh, "Large Scale Test Bed for In-Circuit Reliability Testing of Silicon Carbide Diodes and MOSFETs Emulating Real Life Voltage and Current Stress:, *2017 IEEE APEC*, USA, pp. 2282-2289, 2017.

[10] A. Salemi, B. Zhu, P. Bui-Quang, Y. Ding, K. Chatty, A. Liu, and D. Sheridan, "Optimized 750V SiC MOSFETs for Electric Vehicle Inverter Operation", Trans Tech Publications Ltd, Vol. 945, pp 67-70, 2023, DOI: https://doi.org/10.4028/p-6yitf5

[11] M. Pulvirenti, A. G. Sciacca, L. Salvo, M. Nania, G. Scelba and G. Scarcella, "Body Diode Reverse Recovery Effects on SiC MOSFET Half-Bridge Converters," *2020 IEEE Energy Conversion Congress and Exposition (ECCE)*, Detroit, MI, USA, 2020, pp. 2871-2877, doi: 10.1109/ECCE44975.2020.9236330.

Enhanced Conduction and Switching Performance of 1.2 kV 4H-SiC MOSFETs through High JFET Doping Concentration

Dongyoung Kim, Skylar DeBoer, Seung Yup Jang, Adam J. Morgan, and Woongje Sung

College of Nanoscale Science and Engineering, State University of New York Polytechnic Institute, Albany, NY, 12203, USA, kimd1@sunypoly.edu

Abstract— **This paper aims to provide a comprehensive analysis of the impact of JFET doping concentration on the static and dynamic characteristics of 1.2 kV 4H-SiC MOSFETs. Different channel and JFET designs were implemented by varying the JFET doping concentrations. The utilization of high JFET doping concentrations resulted in an 18% reduction in specific on-resistance ($R_{on,sp}$) compared to low JFET doping concentrations. This improvement was achieved by enhancing channel mobility and reducing JFET resistance. Additionally, narrow JFET widths were achieved. However, the high channel doping concentrations led to increased leakage currents, resulting in lower breakdown voltages. Nevertheless, under negative gate voltages, the issues associated with leakage currents were effectively resolved. The high JFET doping concentration demonstrated a 14% reduction in total switching loss due to the low switching loss during the turn-on transient, although it exhibited high switching loss during the turn-off transient owing to the high C_{gd}.**

Keywords— **4H-SiC, MOSFETs, Output characteristics, JFET implant, Transfer characteristics, Capacitance, Blocking characteristics, Switching characteristics, 2D Simulation**

I. INTRODUCTION

Silicon Carbide (SiC) MOSFETs offer numerous advantages for high voltage and high-frequency applications, including power inverters and fast chargers for electric vehicles [1]. The primary driving force behind the commercialization of SiC MOSFETs in most applications is their low on-resistance coupled with superior switching performance, outperforming conventional Silicon IGBTs. Extensive research endeavors have been undertaken to minimize the specific on-resistance ($R_{on,sp}$) while maintaining a specified breakdown voltage (BV). Particularly, the design of the channel and JFET plays a crucial role in achieving optimum performance for 1.2 kV 4H-SiC MOSFETs [1], [2]. The channel resistance remains a significant factor due to low channel mobility, whereas the JFET strongly influences specific on-resistance, leakage currents, reliability, and ruggedness. However, there is a lack of research focusing on channel and JFET designs with

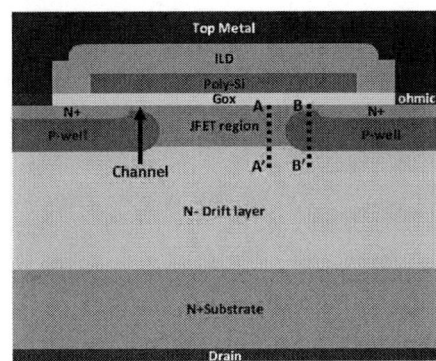

Fig. 1. The cross-sectional view of 1.2 kV 4H-SiC MOSFETs. The designed half JFET width of 0.6 µm, channel length of 0.5 µm, and cell pitch of 5.4 µm were utilized.

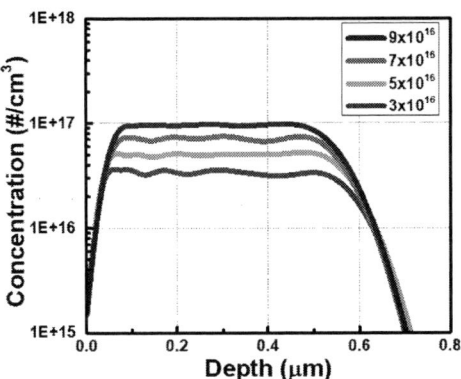

Fig. 2. Designed implant profiles for JFET region (A-A' in Fig. 1).

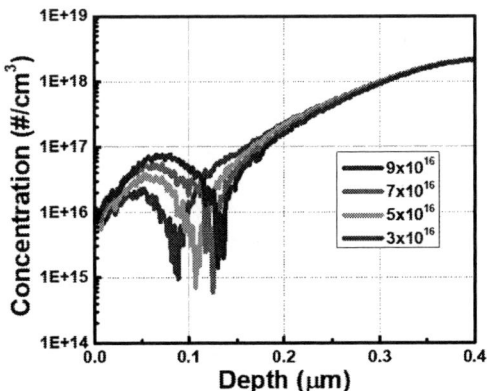

Fig. 3. Designed implant profiles for channel and P-well (B-B' in Fig. 1).

The information, data, or work presented herein was funded in part by the Office of Energy Efficiency and Renewable Energy (EERE), U.S. Department of Energy, the Vehicle Technologies Program Office Award Number DE-EE0008710.

979-8-3503-3714-3/23 $31.00 © 2023 IEEE   159

varying JFET doping concentrations.

In this study, we explore the impact of JFET doping concentration by achieving different accumulation channel and JFET doping concentrations. The measured field effect channel mobility extracted from the FATFET structure, the measured static characteristics, and the simulated switching characteristics will be discussed in relation to different JFET doping concentrations.

II. DEVICE DESIGN

Fig. 1 illustrates the cross-sectional views of the 1.2 kV 4H-SiC MOSFETs, incorporating a designed half JFET width of 0.6 μm, channel length of 0.5 μm, and cell pitch of 5.4 μm. In order to investigate the impact of high JFET doping concentration, Varying JFET doping concentration of 3×10^{16} cm^{-3}, 5×10^{16} cm^{-3}, 7×10^{16} cm^{-3}, and 9×10^{16} cm^{-3} were utilized, as depicted in Fig. 2. The designed implant profiles for channel and P-well (B-B' in Fig. 1) are presented in Fig. 3. To achieve the high channel mobility, an accumulation mode channel was designed, and a high channel doping concentration was implemented in the high JFET doping concentration.

III. FABRICATION TECHNOLOGY

The devices were fabricated at Analog Devices, Inc. (ADI) fabrication facility in Hillview, San Jose, CA, utilizing the same base process line [1]. For the fabrication of the 1.2 kV 4H-SiC MOSFETs, a 10 μm thick drift layer with N-type doping concentration of 8×10^{15} cm^{-3} on a 6-inch, N+ 4H-SiC substrate was employed. The formation of P-well/P+ source/JTE, and JFET/N+ source was achieved through Aluminum and Nitrogen ion implants, respectively, with all implants carried out at 500 ℃. After the implantation steps, a 1650 °C, 10-minute activation anneal was conducted using a carbon cap. The gate oxide, with a thickness of 50 nm, was formed through ultrathin (2 nm) thermal oxide followed by 48 nm of deposited oxide, and then subjected to a post-oxidation anneal (POA) in N$_2$O ambient. N-type polysilicon was deposited and patterned to form the gate. Subsequently, Borophosphosilicate glass (BPSG) was deposited as the interlayer dielectric (ILD), patterned, and etched to establish ohmic contact regions. Nickel (Ni) was deposited on the front side, and rapid thermal annealing (RTA) was performed for the silicidation process. Unsilicided Ni metals were then removed and subjected to RTA at 965 °C for 2 minutes. On the backside, a Ni deposition was carried out, followed by the same RTA process. For the source and gate metal, a 4 μm thick Ti/TiN/Al stack was deposited. Silicon nitride and polyimide were utilized for passivation. Finally, a solderable metal stack was deposited on the backside.

IV. RESULTS AND DISCUSSIONS

Fig.4 shows the measured transfer characteristics and the extracted field effect channel mobility of the fabricated FATFETs, which is the lateral MOSFETs with a channel

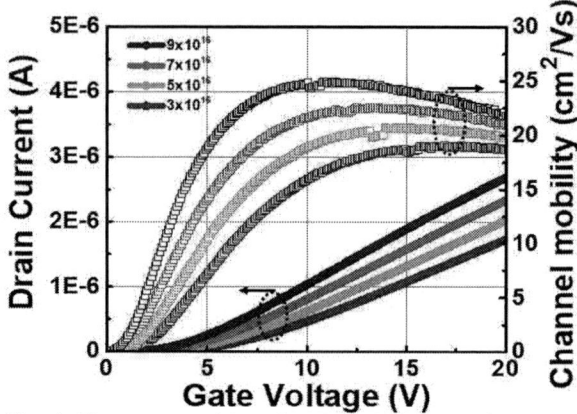

Fig. 4. The measured transfer characteristics and extracted filed effect channel mobility of the fabricated 4H-SiC FATFETs.

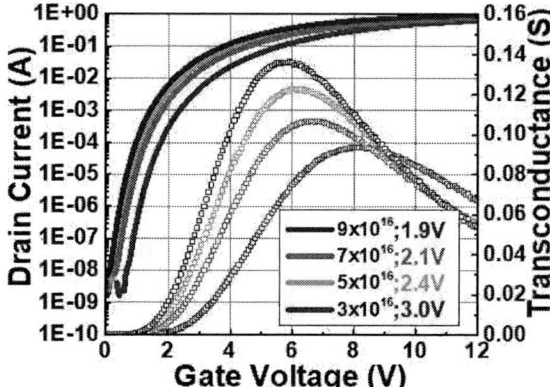

Fig. 5. The measured transfer characteristics of the fabricated 4H-SiC MOSFETs.

length of 200 μm. The transfer characteristics were measured at V$_{ds}$ of 0.1 V. High channel doping results in high channel mobility due to the low effective normal field [3]. Moreover, a higher nitrogen doping concentration in the channel can lead to lower interface traps, resulting in improved mobility caused by Coulomb scattering [4].

Fig. 5 presents the measured transfer characteristics and the extracted transconductance of the fabricated MOSFETs. The high channel mobility, resulting from the increased channel doping, leads to a higher maximum transconductance.

Fig. 6 shows the measured output characteristics of the fabricated MOSFETs. The combination of high channel mobility and reduced JFET resistance, achieved through high JFET doping, improves the conduction behaviors. Additionally, as shown in Fig. 7, the high JFET doping allows for a narrow JFET width, which enhances the reliability and ruggedness [5]. This is attributed to the wider effective JFET width formed by the depletion boundary during conduction behaviors, as shown in Fig. 8. MOSFETs with high JFET doping concentration exhibit wider current paths formed by the depletion boundary, even without JFET implantation in the region.

The measured blocking characteristics of the fabricated MOSFETs under different gate voltages are demonstrated in Fig. 9. High JFET doping concentrations result in high

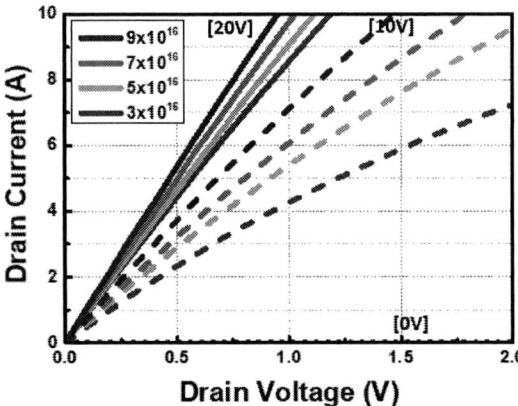

Fig. 6. The measured output characteristics of the fabricated 4H-SiC MOSFETs. $R_{on,sp}$ of 5.01, 4.80, 4.51, and 4.25 $m\Omega\text{-}cm^2$ were achieved for 3×10^{16}, 5×10^{16}, 7×10^{16}, and 9×10^{16}, respectively.

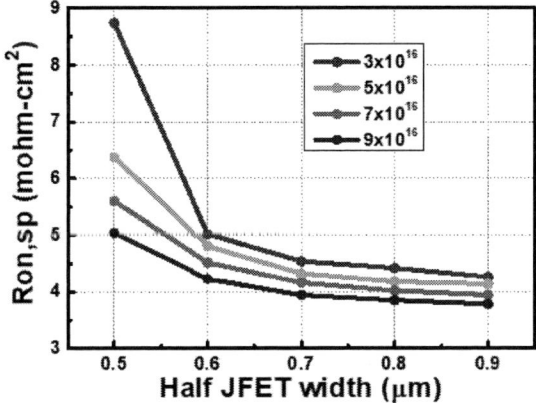

Fig. 7. Summary of the measured $R_{on,sp}$ of different JFET widths.

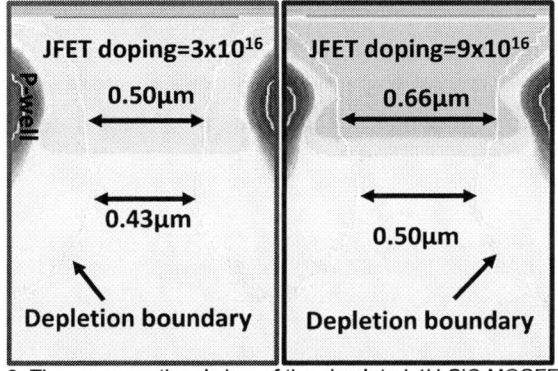

Fig. 8. The cross-sectional view of the simulated 4H-SiC MOSFETs.

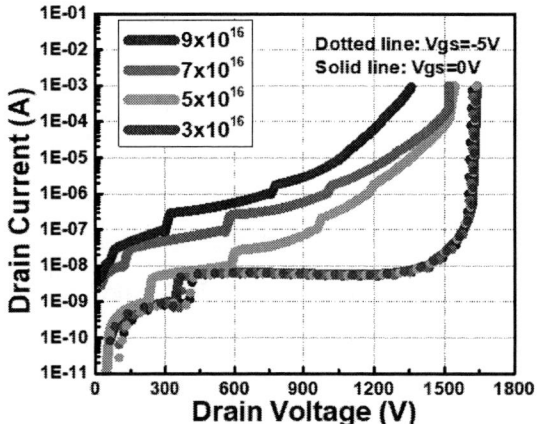

Fig. 9. The measured blocking characteristics of the fabricated 4H-SiC MOSFETs under different gate voltages.

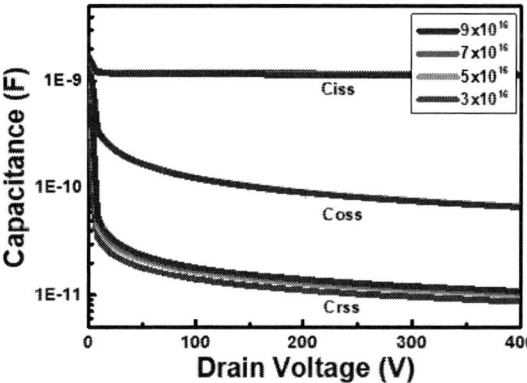

Fig. 10. The measured capacitance of the fabricated 4H-SiC MOSFETs.

leakage current from the channel, leading to lower breakdown voltages. This is due to the low channel potential barrier in the channel region of MOSFETs with high JFET doping concentrations. However, under negative gate voltages, identical breakdown voltages were achieved as the channel is closed, suppressing the leakage current from the channel. Utilizing a negative gate voltage during switching characteristics effectively reduces energy loss, making blocking characteristics with negative gate voltages commonly employed [6], [7]. As a result, the large leakage current associated with high JFET doping can be mitigated by

utilizing negative gate voltages.

Fig. 10 shows the measured capacitance of the fabricated MOSFETs. The input capacitance (C_{iss}) and output capacitance (C_{ds}) remain identical regardless of the JFET doping concentrations because C_{iss} is determined by the gate oxide thickness, and C_{oss} is governed by drift specifications. However, reverse transfer capacitance (C_{rss}) increases with an increase in JFET doping.

Fig. 11 and 12 depict the simulated switching characteristics during the turn-on transient and turn-off transient, respectively. The off/on gate voltages with R_g of 20 Ω are -4/20 V, respectively. A DC supply voltage of 800 V and an on-state current of 18 A were used. The switching energies for turn-on and turn-off were calculated at 10% V_{gs} to 10% V_{ds} and 90% V_{gs} to 90% V_{ds}, respectively. High JFET doping provides low switching loss during the turn-on transient despite the high C_{rss} due to the high transconductance [8]. However, MOSFETs with high JFET doping concentrations exhibit high switching loss during the turn-off transient, which is attributed to the high C_{rss}.

Table I summarizes the experimental and simulated results of MOSFETs with different JFET doping concentrations. High JFET doping concentrations enhance the conduction behaviors due to the high channel mobility and low JFET resistance. Although C_{rss} increased with the high JFET doping

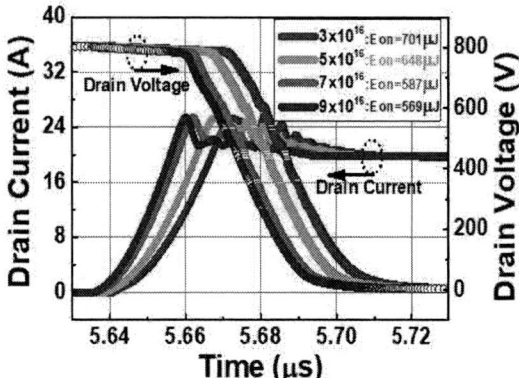

Fig. 11. The simulated switching on characteristics of the 4H-SiC MOSFETs.

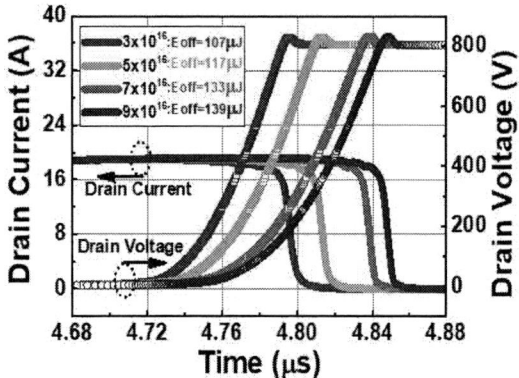

Fig. 12. The simulated switching off characteristics of the 4H-SiC MOSFETs.

Table I. Summary of experimental and simulated results.

	3×10^{16}	5×10^{16}	7×10^{16}	9×10^{16}
$R_{on,sp}$ [mΩ-cm²]	5.01	4.80	4.51	4.25
V_{th} [V]	3.0	2.4	2.1	1.9
BV [V]	1637	1635	1623	1618
C_{iss} [nF]	1.14	1.15	1.15	1.17
C_{oss} [nF]	67.4	67.6	67.0	66.9
C_{rss} [pF]	8.7	10	10.4	10.9
Simulated E_{on} [μJ]	701	648	587	569
Simulated E_{off} [μJ]	107	117	133	139
Simulated E_{total} [μJ]	808	765	720	708

*Measured $R_{on,sp}$ was extracted at V_{gs} of 20V and V_{ds} of 0.1V.
Measured V_{th} was extracted at V_{ds} of 0.1V and I_{ds} of 5mA.
Measured BV was extracted at V_{gs} of -5V and I_{ds} of 1mA.
Measured capacitance was extracted at V_{ds} of 400V.
Simulated E_{on} was extracted at 10% V_{gs} to 10% V_{ds}.
Simulated E_{off} was extracted at 90% V_{gs} to 90% V_{ds}.

concentrations, the switching loss during the turn-on transient significantly decreased due to the high transconductance attributed to high channel mobility. As a result, an overall improvement in total switching loss was achieved with the high JFET doping concentrations.

In summary, this study provides valuable insights into the performance trade-offs associated with different JFET doping concentrations in 1.2 kV 4H-SiC MOSFETs. The findings shed light on the potential for optimizing device characteristics for specific application requirements, balancing factors such as on-resistance, switching losses, and breakdown voltages. The outcomes of this research contribute to advancing the understanding and design of high-performance SiC MOSFETs for a range of high-power applications.

V. CONCLUSION

This paper provides a comprehensive analysis of the impact of JFET doping concentration on the static and dynamic characteristics of 1.2 kV 4H-SiC MOSFETs. By increasing the JFET doping concentration, significant improvements were observed in both the static and dynamic characteristics. Furthermore, the utilization of high JFET doping concentrations allowed for a narrow JFET width, enhancing the reliability and ruggedness. As a result, the 1.2 kV 4H-SiC MOSFETs with high JFET doping concentrations hold promise for future power electronics applications.

ACKNOWLEDGMENT

The authors would like to thank Department of Energy (DOE) for supporting this project. The authors acknowledge the fabrication of the devices by Analog Devices, Inc. (ADI) fabrication facility in Hillview, San Jose, CA.

REFERENCES

[1] D. Kim, N. Yun, S. Y. Jang, A. J. Morgan, and W. Sung, "An Inclusive Structural Analysis on the Design of 1.2kV 4H-SiC Planar MOSFETs," *IEEE Journal of the Electron Devices Society*, vol. 9, pp. 804–812, 2021, doi: 10.1109/JEDS.2021.3109605.

[2] D. Kim, N. Yun, A. J. Morgan, and W. Sung, "The Effect of Deep JFET and P-Well Implant of 1.2kV 4H-SiC MOSFETs," *IEEE Journal of the Electron Devices Society*, vol. 10, pp. 989–995, 2022, doi: 10.1109/JEDS.2022.3218689.

[3] K. Matocha, "Challenges in SiC power MOSFET design," *Solid-State Electron.*, vol. 52, no. 10, pp. 1631–1635, Oct. 2008, doi: 10.1016/j.sse.2008.06.034.

[4] F. Ciobanu, T. Frank, G. Pensl, V. V. Afanas'ev, S. Shamuilia, A. Schöner, and T. Kimoto, "Nitrogen Implantation - An Alternative Technique to Reduce Traps at SiC/SiO2-Interfaces," *Mater. Sci. Forum*, vol. 527–529, pp. 991–994, 2006, doi: 10.4028/www.scientific.net/MSF.527-529.991.

[5] D. Kim, A. J. Morgan, N. Yun, W. Sung, A. Agarwal, and R. Kaplar, "Non-Isothermal Simulations to Optimize SiC MOSFETs for Enhanced Short-circuit Ruggedness," in *2020 IEEE International Reliability Physics Symposium (IRPS)*, April 28 – May 30, 2020. doi: 10.1109/IRPS45951.2020.9128324

[6] F. Yukawa, T. Takaku, and K. Yano, "Effect of Negative Gate Voltage on the Turn-off Performance of Si-IGBT Device," *IEEJ Journal of Industry Applications*, vol. 9, no. 5, pp. 557–562, 2020. doi: 10.1541/ieejjia.9.557.

[7] L. Abbatelli, C. Brusca, and G. Catalisano, "How to fine tune your SiC MOSFET gate driver to minimize losses," *AN4671 Application note*, April 2015. https://www.st.com

[8] K. Han, A. Kanale, B. J. Baliga and S. Bhattacharya, "Static, Dynamic, and Short-Circuit Performance of 1.2 kV 4H-SiC MOSFETs with Various Channel Lengths," *2019 IEEE 7th Workshop on Wide Bandgap Power Devices and Applications (WiPDA)*, Raleigh, NC, USA, 2019, pp. 47-52, doi: 10.1109/WiPDA46397.2019.8998803.

979-8-3503-3714-3/23 $31.00 © 2023 IEEE

Co-Optimization Design and Analysis of WBG and UWBG Power Diodes with Operational Regimes

Lee Gill[1,2], Jonah Shoemaker[3], Jack Flicker[1], Stephen Goodnick[3], Robert Kaplar[1], and Alan Michaels[2]

[1]Sandia National Laboratories, Albuquerque, NM, USA
[2]Virginia Tech, Blacksburg, VA, USA
[3]Arizona State University, Tempe, AZ, USA

Abstract—**Ultra-wide-bandgap (UWBG) materials are recognized for their potential to address performance limitations inherent in wide-bandgap (WBG) devices. This paper presents a comprehensive optimization design methodology for power diodes, targeting minimized power dissipation across specified system operational regime-based reverse voltage, forward current density, frequency, duty cycle, and temperature for diverse device types and materials. Juxtaposed with traditional WBG devices, such as SiC and vertical GaN diodes, UWBG materials like diamond, Ga2O3, and AlGaN have been evaluated and optimized. The derived optimized device loss metrics, encompassing both conduction and switching losses, are used in circuit simulations that assess UWBG device efficacy within a single-phase three-level boost power factor correction (PFC) converter topology. This serves as a tangible application benchmark, contrasting WBG and UWBG material performances. The established framework introduced in this work underscores a holistic co-design and optimization approach, considering distinctive device attributes with converter behavioral insights and comparing different material systems and device categories within a practical power conversion application context.**

Index Terms—**Wide-bandgap (WBG), ultra-wide-bandgap (UWBG), power diodes, boost PFC, co-design, optimization**

I. INTRODUCTION

Ultra-Wide-Bandgap (UWBG) semiconductors represent the fourth generation of semiconductor science and technology, standing at the forefront of contemporary semiconductor research [1]. This category of materials encompasses semiconductors with a bandgap exceeding that of GaN (E_g=3.4 eV), including notable examples like Diamond, AlN, BN, and Ga2O3, as well as their alloy derivatives. The predominant merit of these materials in the realm of power electronics applications lies in their inherent capacity to endure elevated electric fields without being subject to avalanche breakdown. This characteristic is important for device geometry scaling and facilitating the development of higher-voltage devices, which can enhance converter switching efficiency, and increase power density. Realizing these advantages, however, requires a comprehensive understanding of how the intrinsic properties of UWBG semiconductors influence the performance metrics of power switching devices and converter operation derived from them. Such comprehension is realized from a holistic

Sandia National Laboratories is a multimission laboratory managed and operated by National Technology & Engineering Solutions of Sandia, LLC, a wholly owned subsidiary of Honeywell International Inc., for the U.S. Department of Energy's National Nuclear Security Administration under contract DE-NA0003525.

approach, termed "co-design", where the interplay between critical material properties — including high-field transport (i.e., breakdown), low-field transport, and thermal attributes — is carefully evaluated. An integral aspect of this approach is the assessment of the "Figures of Merit" (FOM) for power devices.

Yet, the decision to insert devices into end-use power electronic frameworks should not be fixated on the FOM metrics alone. Instead, the overarching system-level advantages offered by a device under specific operational regimes, such as switching frequency, hold-off voltage, and current-handling capability have to be carefully considered. Given their elevated ionization thresholds and substantial critical fields, UWBG devices need lower doping levels and thinner drift regions compared to their WBG counterparts to achieve similar breakdown voltages. Therefore, a simple evaluation of device IV curves across these material classes may not provide an insightful or pragmatic comparison in power conversion applications.

This paper demonstrates an advanced diode optimization process and converter integration analysis, in which the optimization tool takes system operational metrics, selects device drift region dimensions and doping concentrations to optimize system power dissipation, based on models suited for diverse material systems and device architectures. PiN and Schottky diodes have been evaluated, attributed to their inherent simplicity. The optimization algorithm has factored important effects, like the Mott-Gurney conduction and incomplete dopant ionization phenomena. Emerging from the device optimization process are distinct loss characteristics for diverse materials and diode configurations. These characteristics, once extracted, are compared in their performance within the context of a single-phase 3-level boost power factor correction (PFC) topology. This paper delves into the details of the device design optimization in Section II whereas Section III expands on the findings from the optimization development, and Section IV integrates the device attributes into converter configurations, detailing both the integration strategy and simulation outcomes.

II. DEVICE CO-OPTIMIZATION DESIGN PROCESS

A. Device Optimization Constituent Equations

During system operation, the power dissipation in a diode is composed of three parts: static power dissipation due to conduction loss or reverse leakage current, dynamic power dissipation due to switching, and power loss due to displacement currents as expressed in (1):

$$P_{\text{loss}} = P_{\text{static}} + P_{\text{dynamic}} + P_{\text{displacement}} \qquad (1)$$

The static power dissipation, P_{static}, is composed of power dissipated in the blocking state and in the conduction state along with the reverse power dissipation, which is the product of the required system-level blocking voltage (V_{reverse}), the leakage current density (J_{reverse}), and fractional amount of time in the off-state, $1 - D$. This expression is shown in (2).

$$P_{\text{static}} = J_{\text{reverse}} \cdot V_{\text{reverse}} \cdot (1 - D) + P_{\text{forward}} \qquad (2)$$

The reverse bias, V_{reverse}, is constrained by the doping and drift thickness-dependent breakdown voltage of the device and the breakdown voltage is computed according to whether the diode is in non-punch-through configuration (the depletion width x_d is less than the device thickness t_{drift}) or punch-through configuration in (3):

$$V_{\text{breakdown}} = \begin{cases} \frac{\epsilon_s \cdot E_{\text{crit}}^2}{2qN} & \text{for } x_d \leq t_{\text{drift}} \\ E_{\text{crit}} \cdot t_{\text{drift}} - \frac{qN t_{\text{drift}}^2}{2\epsilon_s} & \text{for } x_d > t_{\text{drift}} \end{cases} \qquad (3)$$

The critical field is calculated according to the method described in [2], [3]. By setting the carrier multiplication integral equal to unity, the fraction of ionized carriers diverges.

To calculate power dissipation in the conduction mode, it is determined if the given material, device thickness, and doping concentration would demonstrate space-charge limited (SCL) current according to the method described in [4]. SCL effects are assumed to take place if the drift thickness of the device falls within the interval defined by (4):

$$0.1 \ \mu m \leq t_{\text{drift}} \leq \sqrt{\frac{\varepsilon_s(V_{\text{SCL}} + V_{\text{turn-on}})}{q \cdot n}} \qquad (4)$$

where V_{SCL} is the bias drop across the device due to the space-charge limited current, calculated self-consistently with (4) and (5):

$$V_{\text{SCL}} = \sqrt{\frac{8}{9} \frac{J_{\text{forward}} \cdot t_{\text{drift}}^3}{\mu \cdot \varepsilon_s}} \qquad (5)$$

μ is the mobility for the given doping concentration and electric field. n is the density of free carriers (electrons for n-type devices and holes for p-type), computed by self-consistently solving for the Fermi level and the free charge density that simultaneously satisfy (6)-(7):

$$n = \frac{N}{1 + g_D \exp\left(\frac{q(E_F - E_D)}{kT}\right)} \qquad (6)$$

$$n = N_c \int_0^\infty \frac{(E - E_c)^{0.5} \, dE}{1 + \exp\left(\frac{E - E_F}{kT}\right)} \qquad (7)$$

The on-state resistance is primarily a function of the device thickness (t_{drift}), free carrier concentration (n), carrier mobilities (μ_n, μ_p), and the minority carrier concentration (holes for all materials except Diamond). The average minority-carrier concentration is equal to the ratio of the stored charge in the device drift region [5]. Since the stored charge, found by integrating the distribution of injected carriers, is equal to the product of the forward current density J_{forward} and the blocking layer ambipolar lifetime τ, which reduces to (8).

$$R_{\text{on}} = \frac{t_{\text{drift}}}{q(\mu_n \cdot n) + \frac{q(\mu_n + \mu_p)J_{\text{forward}} \cdot \tau}{t_{\text{drift}}}} \qquad (8)$$

The dynamic power dissipation is equal to the frequency of switching, f, multiplied by the sum of energies dissipated in the on (E_{on}) and off (E_{off}) transitions. The stored charge per unit area is the product of the forward current density and the ambipolar lifetime. This reduces the dynamic power dissipation to (9).

$$P_{\text{dynamic}} \approx V_{\text{reverse}} \cdot J_{\text{forward}} \cdot \tau \qquad (9)$$

Finally, each switching event induces a displacement current proportional to the capacitance of the diode and the rate of change of the bias in the transition, dV/dt. By integrating the product of the voltage and current over the voltage range V_{forward} to $-V_{\text{reverse}}$, the power dissipation per switch event due to this effect can be determined. This must be multiplied by twice the switching frequency to get the total power displacement. By assuming $V_{forward} \approx V_{on}$, and inserting the classical equation for PN-junction capacitance, the displacement power can be described by (10):

$$P_{\text{displacement}} = \frac{f}{3}\sqrt{\frac{qN_d\varepsilon_s}{2}}(V_{\text{on}} + V_{\text{reverse}})^{\frac{3}{2}} \qquad (10)$$

B. Diode Device Types

There are several device types that are considered in this study. This includes the operation of PiN, Schottky, and MPS diodes. The simplest type of diode is a simple PiN diode. It is composed of an n-doped region in contact with a p-doped region, both uniformly doped. In the forward conduction regime, PiN diodes exhibit high turn-ON voltage due to the built-in voltage of the PiN junction. For the purposes of the optimization tool, it is assumed that the built-in voltage is equal to the bandgap of the material comprising the junction. In the reverse blocking regime, PiN diodes are characterized by low leakage currents, as the thermal generation/recombination process is the only source of free carriers. For simplicity, the optimization tool calculates the reverse leakage current density of PiN diodes to be 1 pA/V.

A Schottky barrier diode (SBD) is characterized by lower turn-on voltage, but higher reverse current leakage compared to the PiN diode. The turn-on voltage is determined by the Schottky barrier height between the semiconductor and the metal contact (φ), subject to field-induced barrier lowering from the initial barrier height (φ_0) due to image charges. This

979-8-3503-3714-3/23 $31.00 © 2023 IEEE

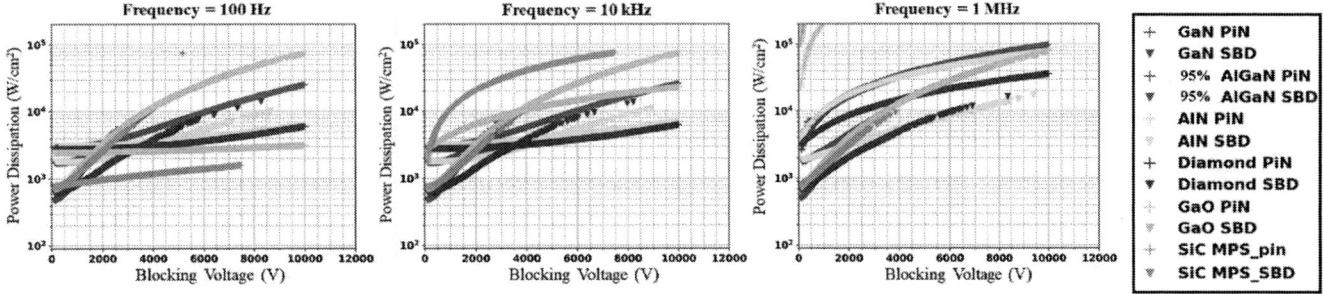

Fig. 1. Minimum power dissipation results from the optimization across blocking voltage with different device materials and types. Figures are generated at different switching frequencies (100 Hz, 10 kHz, and 1 MHz), all at room temperature.

effect is modeled with (11) where E_{max} is the maximum value of the electric field across the semiconductor/metal interface and is described by (12)

$$\varphi = \varphi_0 - \sqrt{\frac{qE_{\max}}{4\pi\varepsilon_s}} \qquad (11)$$

$$E_{\max} = \sqrt{\frac{2qN(\varphi_0 - V_{\text{reverse}})}{\varepsilon_s}} \qquad (12)$$

In the reverse blocking regime, leakage current is driven by thermionic emission (J_{TE}) over the Schottky barrier (dominant at high temperatures), tunneling of carriers through the barrier via field emission (J_{FE}, dominant at very low temperatures), and thermal-assisted thermionic field emission of carriers (J_{TFE}, dominant at mid-range temperatures and higher doping levels). Unlike a PiN diode, the SBD is a majority-carrier device. Therefore SBDs do not exhibit reverse recovery from turn-off and effectively have a negligible carrier lifetime. While this is advantageous in high-frequency switching since dynamic power dissipation is negligible, it comes at the cost of eliminating conductivity modulation in forward conduction, increasing R_{on} compared to PiN diodes.

Lastly, the MPS diode consists of interdigitated regions containing Schottky barriers and PiN junctions. In blocking mode, the MPS diode acts much like a PiN diode, with very low reverse current leakage. However, in forward conducting mode, either the Schottky barrier (JBS mode) or the PiN junction (MPS mode) may be the primary conduction pathway depending on the magnitude of the applied voltage. JBS mode will have lower dynamic power dissipation due to the lack of minority carriers, while MPS mode will exhibit conductivity modulation due to the presence of minority carriers. Optimizing a device with two possible conduction modes will yield instabilities in the numerical computation, as the optimization will tend to oscillate between metastable local minima. In order to avoid this oscillation, the MPS device was optimized as two separate devices (one in JBS mode and one in MPS mode) and the minimum power dissipation between each mode was chosen.

C. Optimization Design Algorithm

The optimization tool developed for UWBG devices is an advancement of the work described in [2], [5]. The process flow of this optimization tool is described in Algorithm 1. For a given device type and material, a mobility model is used to determine carrier mobilities at a given temperature, doping level, and electric field. For some materials, this mobility model is computed from various scattering contributions, while for the rest of the materials empirical models available in the literature are used instead. The operational parameters of the diode in the system are the reverse operating bias, forward current density, switching frequency, duty cycle, and device operational temperature. The critical fields are tabulated for various values of doping concentration and device thickness using the method described before. The optimization program then minimizes power dissipation in the system by altering device drift region thickness and doping concentration subject to the constraint that device breakdown is greater than or equal to twice the required blocking voltage. By iterating across all available system operating parameters, it is possible to develop a map in system operating space and determine which devices/materials produce the lowest power loss for any given system operational point based on materials properties.

Algorithm 1 Optimization for Minimizing Power Dissipation

Input: Device type, Material, V_R, J_F, f_s, D, Temperature (K)

Output: Optimal t_{drift} and N_d

   Initialisation :

1: Set initial values for diode thickness (t_{drift}) and doping concentration (N_d)

   LOOP Process:

2: **for** each iteration until convergence or satisfaction **do**

3:    Adjust t_{drift} and N_d to minimize power dissipation

4:    Ensure constraints:

5:       1. $V_R \leq \frac{1}{2}V_{BR}$

6:       2. $0.1\ \mu m \leq t_{\text{drift}} \leq 100\ \mu m$

7:       3. $10^{15}\text{cm}^{-3} \leq N_d \leq 10^{17}\text{cm}^{-3}$

8: **end for**

9: **return** Optimal t_{drift} and N_d

979-8-3503-3714-3/23 $31.00 © 2023 IEEE

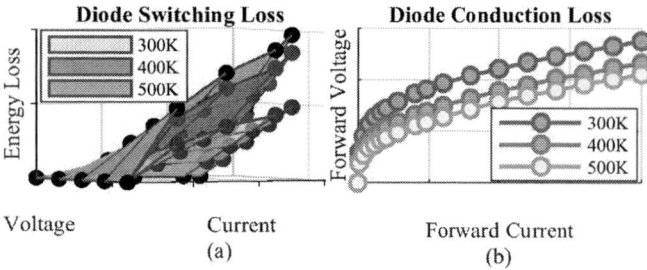

Fig. 2. The power diode loss map delineates: (a) switching losses, encompassing contributions from P_{static}, $P_{dynamic}$, and $P_{displacement}$, when relevant, and (b) the forward bias voltage derived from the forward current characteristics.

III. UWBG AND WBG DEVICE OPTIMIZATION RESULTS

Fig. 1 illustrates the outcomes of the optimization program for diverse device materials and types at switching frequencies of 100 Hz, 10 kHz, and 1 MHz. The program assesses the power dissipation per unit area in relation to the blocking voltage, specifically for SiC MPS diodes, while PiN and Schottky diodes are evaluated for all other materials.

The simulation results show the optimal material and device type that yields the minimal power dissipation for each switching frequency and reverse bias condition. For instance, at 10 kHz frequency, GaN SBDs are the optimal choice for voltages below 500 V. Diamond SBDs are favored in the 500 V to 3.1 kV range, AlN SBDs from 3.1 kV to 4.1 kV, and diamond PiN diodes are preferred between 4.1 kV and 10 kV. It's pertinent to highlight that the optimization tool encounters challenges in discerning solutions at extreme reverse bias levels and switching frequencies, shown from the absence of SiC PiN-mode MPS data points exceeding 7.5 kV at 100 Hz. The tool does not make extrapolations beyond the highest logged solution for a particular material and device category, leading to the selection of β-Ga2O3 PiN diodes as the optimal choice above 7.5 kV in this instance.

The optimization tool is also suitable for considering several other metrics. For example, the switching and conduction loss traits are depicted in Fig. 2. Heat maps that showcase optimized doping concentrations, drift thicknesses, and the resultant unipolar figure of merit, denoted as $VB^2/R_{on,sp}$, can be also produced.

IV. DEVICE TO CIRCUIT INTEGRATION AND CONVERTER PERFORMANCE

A. Device to Converter Integration

To evaluate the optimized WBG and UWBG device loss characteristics, the single-phase 3-level boost power factor correction (PFC) topology [6] was used in various current and frequency operation regimes. Such a topology is selected due to its usage of a full-bridge diode configuration in the front end along with the boost conversion stage to address the challenges of power quality and efficiency in single-phase power systems. The topology produces three voltage levels from a single-phase input, thereby enhancing the power factor and reducing total harmonic distortion (THD), especially at high power levels.

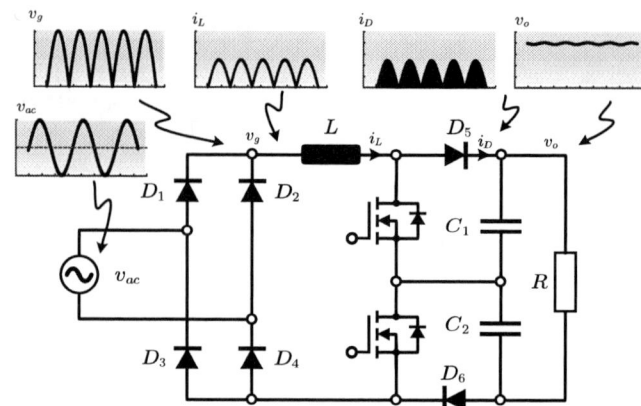

Fig. 3. Single-phase 3-level boost PFC circuit schematic and illustration of important current and voltage waveforms at different nodes in the converter.

The intermediate voltage level allows for smoother transitions between voltage states, resulting in reduced voltage stress on components and minimized switching losses. Consequently, the circuit can operate at higher frequencies, leading to a better assessment and comparison of the switching frequency impacts on the WBG and UWBG power diodes.

Fig. 3 depicts the single-phase 3-level boost PFC circuit topology, with associated input and output waveforms. The input waveform, originally sinusoidal, becomes full-wave rectified post the diode bridge. This transformation produces a pulsating DC voltage, with the PFC-induced input current waveform closely mirroring the voltage waveform for a high power factor. Ideally, when waveforms are in phase, it indicates a near-unity power factor.

Through the diode bridge ($D_1 - D_4$), the inductor current waveform can vary based on the control strategy. Freewheeling diodes ($D_5 - D_6$) conduct during the primary switch's off periods. With switches deactivating via a phase shift, the inductor current is routed through the freewheeling diode, producing a pulsed waveform synchronized with the rectified input voltage at v_g.

The output voltage of the 3-level boost PFC is a DC voltage with a value higher than the input voltage and through the output capacitors the output voltage pulsations are smoothened out to a DC characteristic.

The rectifying diode bridge ($D_1 - D_4$) and the freewheeling diodes ($D_5 - D_6$) experience distinct voltage and current stress profiles in the PFC converter due to their respective roles. Fig. 4 showcases the voltage and current behaviors across these diodes. Rectifying diodes, synchronized by the PFC to align current and voltage phase, experience only conduction losses during their ON-state, eliminating switching losses. Conversely, the freewheeling diodes, influenced by the voltage and current pulsations from the boost converter switches, are subject to both switching and conduction losses. These dynamics, while unique, are pivotal in assessing total losses and differentiating between switching and conduction contributions.

979-8-3503-3714-3/23 $31.00 © 2023 IEEE

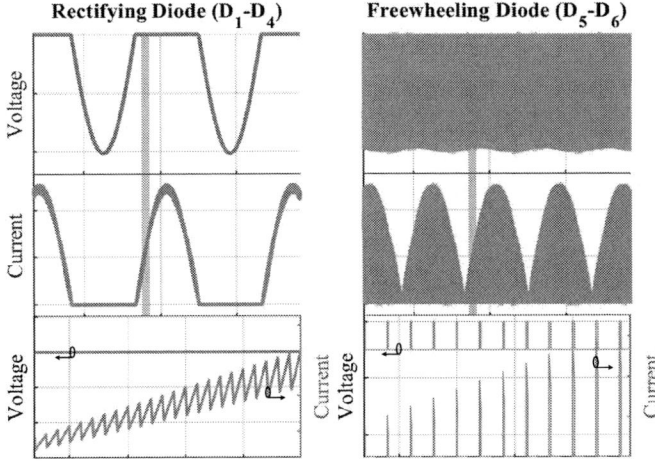

Rectifying Diode (D₁-D₄)

Fig. 4. The voltage and current stress waveforms differ between the rectifying and freewheeling diodes. Rectifying diodes primarily incur conduction loss as they conduct the input current during zero voltage. In contrast, freewheeling diodes experience both conduction and switching losses, the latter arising from the hard-switching dynamics imposed by the converter switches. The bottom panels provide a magnified view of the voltage and current waveforms, focusing on the areas highlighted by the lightly shaded regions.

B. Converter Performance with Device Loss Characteristics

The loss characteristics of both UWBG and WBG devices, as depicted in Fig. 2, were integrated into the diode models with proper loss calculation. Following this integration, the single-phase 3-level boost PFC converter was used to evaluate the loss performance over different device materials and types. The PFC simulation circuit maintains a power factor of 5% for all test cases in order to comply with the international power quality standards. Moreover, by ensuring this low power factor, the converter operation reduces potential electronic interference, prompting smoother interactions with other electronic systems and significantly diminishing the strain on the power grid infrastructure.

Fig. 5 shows the output of the simulation model, highlighting the input voltage and current waveforms, boosted output voltage, and the power loss characteristics from both the conduction and switching of the diodes. Note that the conduction loss closely follows the rectified positive half-cycle input current waveform as the rectifying diodes account most of the conduction loss whereas the switching loss is induced from the pulsating voltage and current through the freewheeling diodes of the boost PFC.

Utilizing the simulation model with distinct device loss characteristics, a detailed analysis of the converter's performance was analyzed over various current densities and switching frequencies. Fig. 6 provides a comprehensive parametric analysis, mapping out the interplay between different current densities and switching frequencies across a range of materials and device types.

Observations from Fig. 6(a) illustrate the efficiency density against varied current densities. Furthermore, it contrasts the contributions of conduction and switching losses relative to

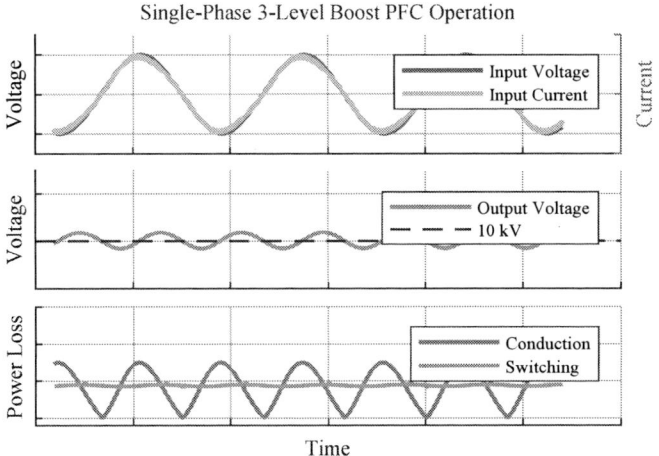

Fig. 5. Single-phase 3-level boost PFC simulation results, illustrating ≤5% THD, 10 kV desired output voltage with small ripple, and power loss comparison between the conduction and switching loss of the power diodes.

the converter's total loss at a benchmark of 1000A/cm². A notable observation is the superiority of diamond in this analysis, which exhibits minimal conduction loss contributions, especially at elevated current densities. In parallel, both GaN and GaO present a noteworthy and similar trend, demonstrating analogous efficiency density performances across current densities.

The parametric assessment of PIN diodes, as depicted in Fig. 6(b), shows comparable high efficiency density performance between AlGaN, Diamond, and GaN materials whereas GaO and SiC show increasing impacts from the switching loss characteristics. Furthermore, Fig. 6(c) and (d) show the switching frequency analysis between different device materials for SBD and PIN diodes, respectively. With increasing switching frequency, the efficiency density is reduced with the impact of insignificant switching loss contributions, though for the SBD the switching losses still account for ≤1% for each device material in Fig. 6(c). However, the switching losses become the majority of the loss contributions the switching frequency increases as illustrated in Fig. 6(d). Note that the loss contributions of SiC is only partially shown in order to show proper scaling for the rest of the devices.

In summary, the presented analyses provide operational dynamics of various materials and device types under different conditions, underscoring the intricate balance between efficiency, conduction, and switching losses between device materials and types.

V. CONCLUSION AND FUTURE WORK

This paper introduces an optimization algorithm for WBG and UWBG power diodes. The results reveal optimal diode materials and designs across varied switching frequencies and blocking voltages at a temperature of 300 K and a forward current density of 1000 A/cm². The simulation studies assess the devices' performance when employed as rectifying diodes in a full-wave rectifier and as freewheeling diodes in a

979-8-3503-3714-3/23 $31.00 © 2023 IEEE

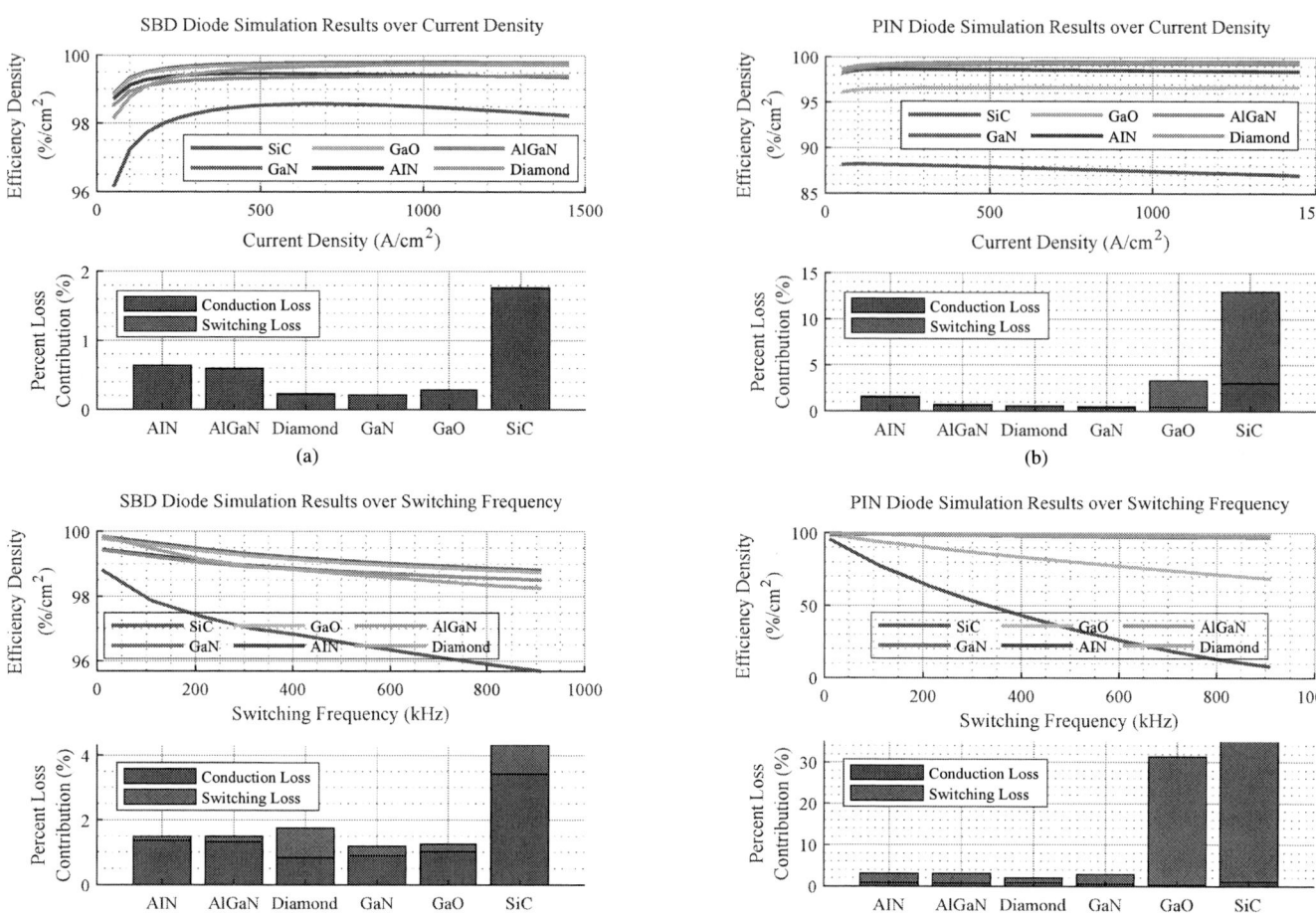

Fig. 6. Analysis of efficiency density and power loss contribution comparison between different UWBG and WBG device materials and types. (a) depicts the efficiency density (%/cm²) over current density (A/cm²) between UWBG and WBG materials for the SBD diode construction along with the percentage loss contributions in regards to the total powre loss whereas (b) shows the results for the PIN diodes. (c) and (d) illustrate the results across different switching frequencies for SBD and PIN, respectively.

boost converter, collectively configured for PFC operation. The outcomes highlight the synergy achieved through a co-design approach, optimizing devices for specific parameters while enhancing converter performance across broader operational domains.

Future endeavors should encompass thermal interactions between the die and package, development of accurate thermal models for circuit simulations, and exploration of diverse diode combinations to ensure optimal performance in desired circuit configurations while evaluating the materials' FOM [7], [8].

REFERENCES

[1] J. Y. Tsao, S. Chowdhury, M. A. Hollis, D. Jena, N. M. Johnson, K. A. Jones, R. J. Kaplar, S. Rajan, C. G. Van de Walle, E. Bellotti, C. L. Chua, R. Collazo, M. E. Coltrin, J. A. Cooper, K. R. Evans, S. Graham, T. A. Grotjohn, E. R. Heller, M. Higashiwaki, M. S. Islam, P. W. Juodawlkis, M. A. Khan, A. D. Koehler, J. H. Leach, U. K. Mishra, R. J. Nemanich, R. C. N. Pilawa-Podgurski, J. B. Shealy, Z. Sitar, M. J. Tadjer, A. F. Witulski, M. Wraback, and J. A. Simmons, "Ultrawide-bandgap semiconductors: Research opportunities and challenges," *Advanced Electronic Materials*, vol. 4, no. 1, p. 1600501, 2018.

[2] J. A. Cooper and D. T. Morisette, "Performance limits of vertical unipolar power devices in gan and 4h-sic," *IEEE Electron Device Letters*, vol. 41, no. 6, pp. 892–895, 2020.

[3] D. Morisette and J. Cooper, "Theoretical comparison of sic pin and schottky diodes based on power dissipation considerations," *IEEE Transactions on Electron Devices*, vol. 49, no. 9, pp. 1657–1664, 2002.

[4] H. Surdi, F. A. M. Koeck, M. F. Ahmad, T. J. Thornton, R. J. Nemanich, and S. M. Goodnick, "Demonstration and analysis of ultrahigh forward current density diamond diodes," *IEEE Transactions on Electron Devices*, vol. 69, no. 1, pp. 254–261, 2022.

[5] J. Flicker and R. Kaplar, "Design optimization of gan vertical power diodes and comparison to si and sic," in *2017 IEEE 5th Workshop on Wide Bandgap Power Devices and Applications (WiPDA)*, 2017, pp. 31–38.

[6] A. D. B. Lange, T. B. Soeiro, M. S. Ortmann, and M. L. Heldwein, "Three-level single-phase bridgeless pfc rectifiers," *IEEE Transactions on Power Electronics*, vol. 30, no. 6, pp. 2935–2949, 2015.

[7] D. Cittanti, E. Vico, and I. R. Bojoi, "New fom-based performance evaluation of 600/650 v sic and gan semiconductors for next-generation ev drives," *IEEE Access*, vol. 10, pp. 51 693–51 707, 2022.

[8] K. Shenai, "The figure of merit of a semiconductor power electronics switch," *IEEE Transactions on Electron Devices*, vol. 65, no. 10, pp. 4216–4224, 2018.

Monitoring Current of a GaN HEMT at Ultra-High Magnetic Fields

Brett Setera[1,2], Aristos Christou[2], and Natalia Gudino[1]

Email: bsetera@umd.edu   christou@umd.edu   natalia.gudino@nih.gov

[1]MRI Engineering Team, Laboratory of Functional and Molecular Imaging, National Institutes of Health, Bethesda, MD

[2]Department of Materials Science & Engineering, University of Maryland, College Park, MD

Abstract— **The performance of commercially available GaN HEMT and LDMOS devices were investigated inside ultra-high static magnetic fields. Device performance was assessed by measuring output current for its use in innovative on-coil current-source, switch mode power amplifiers in ultra-high field MRI.**

Keywords— **GaN, MRI, magnetic field, robustness**

I. Introduction

Magnetic Resonance Imaging (MRI) utilizes strong magnetic fields combined with radiofrequency (RF) pulse transmission to detect hydrogen nuclei in the body. Increasing the strength of the magnetic field (B_0) up to ultrahigh-field strengths (≥ 7 T) improves image resolution, sensitivity, and contrast, allowing for unprecedented neurological research opportunities [1]. Increasing B_0 to 7 T requires increasing the transmitted RF signal to 300 MHz. The shorter excitation wavelength interacting with the dielectric tissue leads to image artifacts caused by the inhomogeneity of the excitation field (B_1) [2]. B_1 homogeneity can be improved by replacing the single channel volume transmitter by an array of independently controlled transmit coils. This approach is known as parallel transmission (pTx) [3].

Power RF amplifiers are typically built with laterally diffused metal-oxide semiconductor (LDMOS) devices. In MRI, these amplifiers are remotely located from the MRI coils so power is transferred through long coaxial cables. In a pTx implementation, this setup results in cable coupling, high cable losses and inefficient power monitoring because of the number of coaxial connections required. Optically controlled, on-coil switch-mode current-source amplifiers built with GaN HEMTs eliminates cable losses and coupling and allows direct control and monitoring of the coil current [4]. Thus, this technology is promising for the implementation of pTx systems with high number of channels.

Commercially available GaN HEMTs have been successfully implemented in on-coil switch-mode pTx technology due to their higher power density, ultra-small package, and their switching performance within the MRI bandwidth [5]. This configuration, however, places the HEMT inside the strong magnetic field. The performance of other III-V semiconductors, specifically GaAs low-noise amplifiers used for receivers that operate inside the MRI bore, have been found more sensitive to the B_0 field than Si devices [6], [7]. This increased sensitivity can be attributed to the higher electron mobility and typically longer channels in GaAs, causing carriers in GaAs FETs to be more susceptible to Lorentz forces [6],[7]. These forces are orientation dependent, wherein a perpendicular orientation between the current carrying channel and B_0 produces a stronger force on electrons than does a parallel orientation. Lagore et al. [6] reported more than 50% decrease in current draw for a commercially available GaAs pseudomorphic HEMT switching at 128 MHz oriented perpendicularly to a 9.4 T field, whereas an n-channel, dual-gate Si MOSFET experienced no change under identical conditions. Their research indicates that device material, geometry, and orientation should be considered for successful implementation of FETs inside a strong magnetic field.

The work presented in this paper builds on the understanding of III-V device performance inside strong magnetic fields by monitoring the impact of the field on the output current of an RF switching block implemented with a commercially available GaN HEMT device. A commercially available LDMOS device is also tested under identical conditions. Preliminary experimental results for a GaN HEMT operating at cryogenic temperatures inside the strong magnetic field is included as well. These results will assist in development of optimized GaN-based devices for transmit coil arrays used in next-generation ultrahigh-field MRI technologies.

II. Theory

The motion of electrons inside transverse magnetic and electric fields is determined by the Lorentz force [8]:

$$F = qE + q(v \times B) \qquad (1)$$

Where q is elementary charge, E is the electric field, B is the magnetic field, and v is the electron velocity. This force induces cyclotron motion on electrons moving normal to the magnetic field, and the radius of the cyclotron is proportional to the electron velocity [9]. In semiconductor devices, electrons experience scattering events that interrupt the cyclotron motion. Although GaN has a higher theoretical electron saturation velocity than Si, the various scattering mechanisms that occur in the AlGaN/GaN heterojunction impart a higher scattering frequency than in Si devices. The mean free time between scattering events can be estimated using the mobility equation [8],

$$\mu_n = q\tau/m^* \qquad (2)$$

979-8-3503-3714-3/23 $31.00 © 2023 IEEE

where m^* is the effective mass of an electron and τ is mean free time between scattering events. Maximum mobility and the effective electron mass, used to calculate τ, are listed in **Table 1** for Si, AlGaN/GaN, and AlGaAs/GaAs.

Table 1. Semiconductor material parameters and calculated mean free time.

	Units	Si	AlGaN/GaN	AlGaAs/GaAs	Ref.
Electron Mobility	cm^2/V-s	1450	2000	8000	[8], [9]
Electron effective mass	m^*	$0.98m_0$	$0.22m_0$	$0.067m_0$	[8], [9]
Mean free time, τ	ps	0.8	0.2	0.3	Calculated

The mean free time between scattering events is on the order of picoseconds. As shown in **Table 1**, Si presents a τ four times greater than GaN and nearly three times that of GaAs. Therefore, based on these material properties it is expected that electrons in AlGaN/GaN and AlGaAs/GaAs are less affected by the magnetic field than those in Si because cyclotron motion is interrupted more frequently. Besides material, device geometry and channel length must be considered for this analysis.

Inside the device channel, the average trajectory of drift electrons subject to transverse magnetic and electric fields is tilted by an amount given by the Hall angle, and a longer mean free time produces a larger Hall angle [9]. The disparity between the zero-field trajectory and the Hall trajectory increases proportionally to channel length, resulting in a larger magnetoresistance effect in longer channels. This means that to determine the impact of magnetic fields on different devices and their performance, both device material and channel geometry should be considered. For these reasons, analysis of commercial devices in this paper produces only device-specific conclusions that focus on the EPC8009 GaN HEMT and MRFE6VS252 LDMOS tested herein. These devices present similar high-frequency performance and have non-magnetic packaging.

In *Section IV B*, preliminary results for an experiment where the HEMT DUT is held at cryogenic temperatures inside the strong magnetic field are provided. At liquid nitrogen temperatures (~77K), electron drift velocity in GaN HEMTs can increase by more than 25% compared to that at room temperature, assuming device operation under identical gate and drain voltages [10]. As temperature decreases from room temperature towards cryogenic temperatures, electron mobility and drift velocity in GaN increase due to a reduction of scattering events [11]–[13]. With fewer scattering events and the resulting increased drift velocity, it is expected that the Lorentz Force will act more strongly on the drift electrons according to (1) and that the cyclotron motion will be interrupted less frequently. Therefore, it is expected that a device kept at cryogenic temperatures will present a more significant change in device output current with magnetic field strength than a device at room temperature. Due to the proprietary nature of the materials composition and layout inside of the commercial devices we tested, it is difficult to estimate the possible impacts on device performance for its operations at cryogenic temperatures and strong magnetic fields. Furthermore, cryogenic temperatures are far outside the operating temperature range provided in the device datasheet. Nonetheless, the HEMT was tested under this extreme condition in search of a stronger effect on device performance by the high magnetic field.

III. METHODS

A. Orientation Dependence Experiment

To ensure that experimental data is sensitive to magnetic field effects on the DUT and not other possible field effects on peripheral components, circuit simplification was prioritized. Therefore, a preamplification stage was not implemented, instead the DUTs were DC biased to the threshold gate voltage of each device and the RF signal was superimposed. To minimize cable cross talking, the RF input and output signals were transmitted through an optical fiber and a coaxial cable, respectively. The output stage of the switching block was connected through an LC filter to a surface coil tuned to 297.2 MHz (^{1}H frequency at 7 T) and loaded with a solution that has similar electrical conductivity and permittivity of average brain tissue, referred to as a 'brain phantom' [14]. Separate test circuits with identical input/output stages were made for the EPC8009 GaN HEMT and MRFE6VS252 LDMOS.

The test circuit block diagram in **Fig 1(a)** shows both the test circuit and remote connections for signal generation and acquisition. Each test circuit comprised of two printed circuit boards joined by copper foil connections, with one section containing the input and output stages and the other containing two transversely oriented devices, labelled DUT1 and DUT2. Each device can be connected to the circuit by a jumper and monitored individually, allowing for the analysis of two coplanar orientations between the device and B_0. The gate DC bias was set at threshold voltage and pulsed with 2% duty cycle (33600A Series Trueform Waveform Generator; Keysight) to 2.3V for the GaN HEMT and 2.8V for the LDMOS. The 297.2 MHz RF signal provided from the signal generator (SML01, 9 kHz to 1.1 GHz; Rohde & Schwarz, Munich, German) was synced with the DC pulse and transmitted to the test circuit via optical fiber. The drain was biased to 10 V and supplied by a separate power supply (GPD-4303S; GW Instek). The output stage surface loop, loaded with the brain phantom, was coupled to a 4-turn pick-up coil connected coaxially to the oscilloscope (DSOX6004A, 1GHz; Keysight). The output signal is recorded as a peak-to-peak voltage from the pick-up loop and is assumed to be directly proportional to the drain current of the DUT.

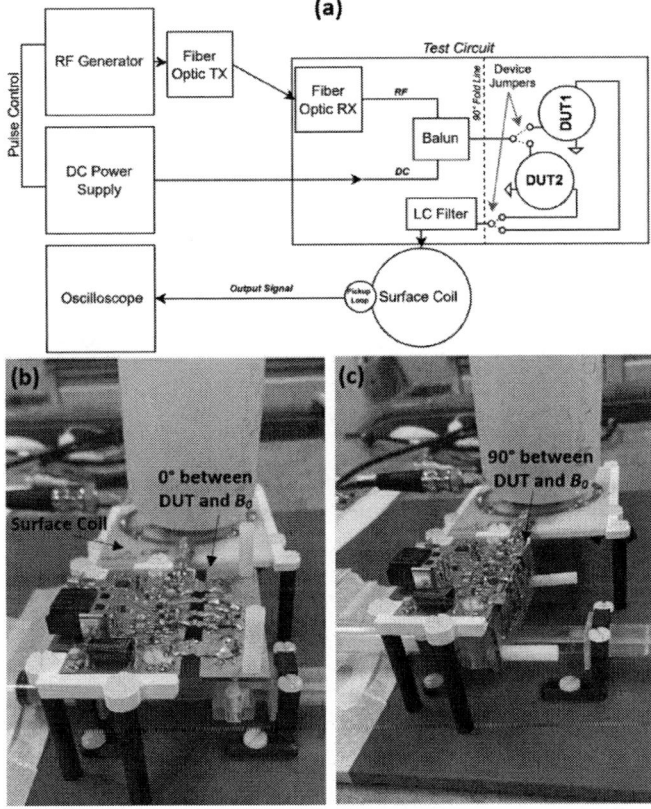

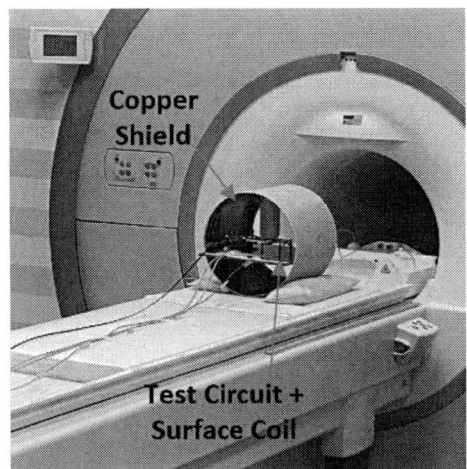

Fig. 1. (a) Test circuit block diagram. (b) Test circuit including surface coil with brain phantom, with the angle between DUT and B_0 set to 0° and (c) set to 90°.

To determine orientation dependence between the DUT and B_0, the test circuit was mounted on a mechanical structure capable of rotating the test circuit section containing the DUT 90° as shown in **Fig 1 (b)** and **(c)**. This mechanical setup combined with the two different coplanar device orientations on each test circuit allows the monitoring of four distinct device orientations. To determine the dependence of the device output signal on magnetic field strength, the experimental setup was placed on the moveable patient table of a 7 T MRI Scanner (Siemens, Erlangen) and data points were acquired while locating the table/DUT at selected static field contours, as mapped by the magnet's manufacturer. Three data points were taken at each position to account for oscilloscope error at specific locations within the magnetic field (0.2T, 1T, 3T, 5T, 7T), and the table/DUT was swept through the same locations three times to account for any error due to shifting of the output coaxial cable. To minimize interference and possible coupling between the surface coil and conductive elements inside the MRI scanner, the entire setup was placed inside a cylindrical copper shield, shown in **Fig 2**. To analyze the relative change of the signal, the output signal recorded at each magnetic field strength was normalized to the output signal recorded at 0.2 T.

Fig. 2. Experimental setup on patient table of 7 T MRI Scanner.

B. Low Temperature Perpendicular Orientation Experiment

Electron velocity and resulting Lorentz force (1) are higher at lower temperatures. Therefore, to increase the effect of the magnetic field on the device performance, the experiment described in Section II-A was repeated at cryogenic temperatures (~77 K, liquid nitrogen). The DUT was held at perpendicular orientation to B_0 and submerged in liquid nitrogen inside a double-layered, air insulated Styrofoam container. Temperature of the DUT was monitored by a type-T thermocouple connected to a thermocouple monitor (SR630, Stanford Research Systems). In this setup, the DUT was operating outside the temperature range specified on the manufacturer's datasheet.

IV. RESULTS

A. Orientation and Field Strength Dependence

Preliminary experiments determined that devices oriented transversely but coplanar showed no differences in performance when switching device inside the strong magnetic field. Thus, results will only be discussed in terms of parallel and perpendicular orientations between the DUT and B_0.

The change in output signal for the HEMT and LDMOS devices when moving the test board from 0.2 T to 7 T is shown in **Fig 3** for both orientations of the board relative to B_0. The change in output signal at 7 T for the DUT parallel oriented to B_0 was -2.7% ± 1.9% for the LDMOS and -1.1% ± 1.0% for the HEMT. For the DUT oriented perpendicularly to B_0, the change was -3.6% ± 2.2% for the LDMOS and +0.7% ± 0.9% for the HEMT. In both orientations the LDMOS output signal decreased more than the HEMT signal.

Following these results, the experiment was repeated and focused only on the HEMT device. The results from the repeated experiment in **Fig 4** show a generally larger decrease in output signal with increasing magnetic field strength, including a signal decrease at 5 T that requires further investigation. The decrease in output signal at 5 T is the largest signal change in both parallel and perpendicular orientation, decreasing by -1.9% ± 0.4% and -2.8% ± 0.5% in parallel and perpendicular orientation to B_0 respectively.

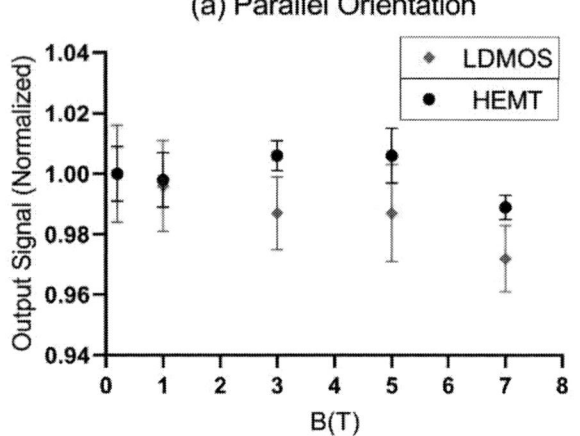

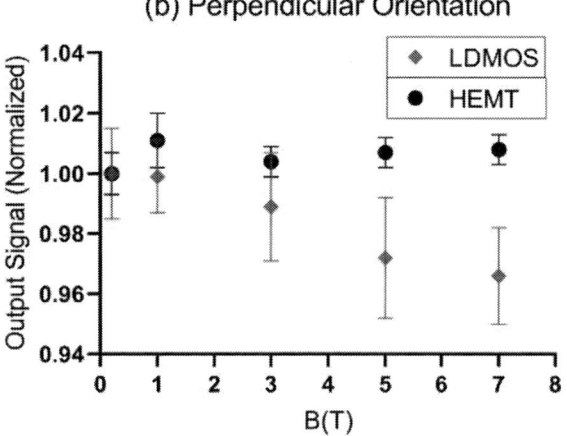

Fig. 3. *(a)* LDMOS and HEMT normalized output signal (proportional to I_D) with device oriented parallel to B_0 and *(b)* oriented perpendicularly.

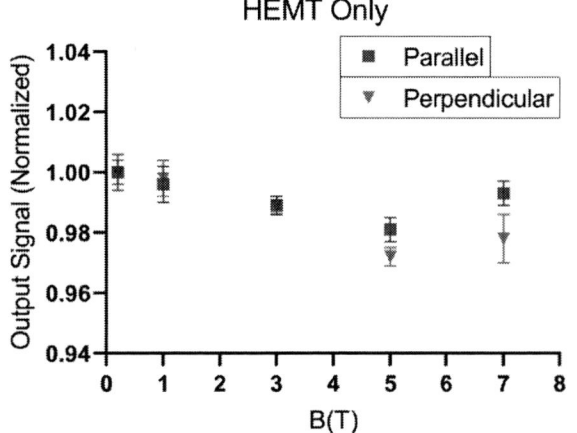

Fig. 4. HEMT normalized output signal oriented parallel and perpendicularly to B_0 from second experiment.

At 7 T, the decrease is less severe with a signal change of -0.7% ± 0.7% and -2.2% ± 0.9% for the parallel and perpendicular orientation respectively.

B. Low Temperature Perpendicular Orientation Experiment

Preliminary experimental results in **Fig 5** comparing device operation at room temperature and at 77 K indicate that low temperatures do increase device output signal, as expected. However, no reduction in output signal was observed for the device at 77 K when moving the DUT from 0.2 T to 5 T. In fact, output signal at 5 T increased +2.9% ± 0.7% from the signal at 0.2 T. Due to the difficulties of this preliminary experiment setup, the data points at 3 T and 7 T were not recorded due to user error and frost formation on the test circuit, respectively. It is worth noting that the HEMT continued to operate under these extreme conditions, and even on the bench after frost thawed and electronics dried. The 300 K experimental data reported in **Fig 5** shows a decrease in output signal from 0.2 T to 7 T of -3.3% ± 0.4% and another decrease of the same magnitude at 1 T. The latter was observed in both experiments at room and cryogenic temperatures.

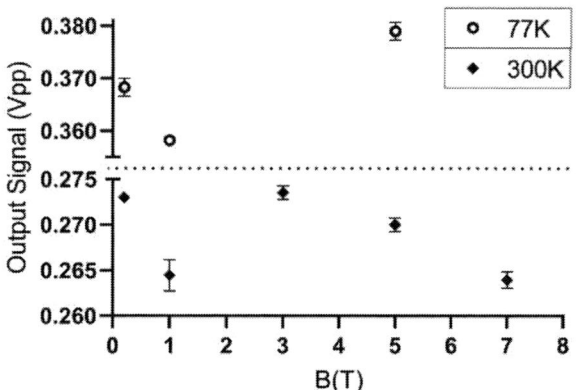

Fig. 5. Comparison of GaN HEMT output signal reported as a peak-to-peak voltage (proportional to I_D) at cryogenic temperature and room temperature. The device was oriented perpendicular to B_0.

V. DISCUSSION

The data presented here confirms that the performance of the DUTs is less sensitive to the magnetic field when located in a parallel orientation to B_0. These findings agree with both Lorentz force theory and with experimental data presented in [6] and [7], and appear to confirm the assumption that the active channel in these commercially packaged devices is parallel to the package bonding surface. The results show the Si-based device to be more sensitive to the magnetic field than the III-V device which contradict results presented elsewhere [6], reiterating that results are device specific. Therefore in this analysis device comparisons cannot be exclusively based on semiconductor materials. Importantly, the effective electron mass of GaAs is less than 1/3 of that of GaN and nearly 1/14 than that of Si [8]. This suggests that, given similar device geometries, GaAs-devices would be more sensitive to the magnetic field than either GaN or Si.

Comparing the output signal decrease for the perpendicularly oriented HEMT at 7 T in **Figs 3(b)** and **4,** there is a discrepancy in that **Fig 3(b)** shows an increase at 7 T of 0.7% ± 0.9% and **Fig 4** shows a decrease of - -3.3% ± 0.4%. The cause of this discrepancy may be that as the DUT is moved through the magnetic field gradient of the MRI scanner, the coaxial cable transmitting the output signal must be moved to accommodate the changing distance between the test circuit and the oscilloscope. This may be the same cause of the unexpected signal decreases at 5 T in **Fig 4** and at 1 T in **Fig 5.**

The data acquired in the low temperature experiment with the HEMT perpendicularly oriented to the field (**Fig 5**) did not show any trend. Instead, we recorded a signal decrease of - 2.7% ± 0.5% at 1 T and a signal *increase* of +2.9% ± 0.7% at 5 T. Because of the frost formation on the circuit, this experiment was interrupted, and data was recorded for a single sweep of the DUT across the specified magnetic field values. Further experiments may be necessary for more accurate assessment of the device output current at each field value. In addition, for the devices tested in this work, the amplification of the Lorentz force effect on carriers at 77 K may not be significant enough to observe differences from results obtained at 300 K. Noteworthy, the GaN HEMT displayed excellent robustness as it continued to operate inside the strong magnetic field while submerged in liquid nitrogen, far outside the datasheet specifications. Future work is planned to refine and repeat the cryogenic experiment inside B_0 to verify preliminary results.

VI. Conclusions

The EPC8009 GaN HEMT and the MRFE6VS252 LDMOS exhibited good robustness inside strong magnetic fields up to 7 T. Device output current was less impacted by magnetic fields when oriented parallel to B_0. These results confirm the suitability of GaN HEMTs for use in on-coil pTx MRI technology. Parallel orientation between the device and B_0 is easily realized in this application, minimizing their sensitivity to the magnetic field. Preliminary low-temperature experiments were performed to determine if amplified Lorentz forces would produce an observable difference from room temperature results, but no conclusive differences were observed.

Acknowledgment

The authors thank Steve Dodd and the Section on Instrumentation at the National Institute of Mental Health (National Institutes of Health). This research was supported by the Intramural Research Program of the National Institute of Neurological Disorders and Stroke (National Institutes of Health).

References

[1] J. H. Duyn, "The future of ultra-high field MRI and fMRI for study of the human brain.," *Neuroimage*, vol. 62, no. 2, pp. 1241–1248, Aug. 2012, doi: 10.1016/j.neuroimage.2011.10.065.

[2] P.-F. Van de Moortele *et al.*, "B(1) destructive interferences and spatial phase patterns at 7 T with a head transceiver array coil.," *Magn Reson Med*, vol. 54, no. 6, pp. 1503–1518, Dec. 2005, doi: 10.1002/mrm.20708.

[3] Y. Zhu, "Parallel excitation with an array of transmit coils.," *Magn Reson Med*, vol. 51, no. 4, pp. 775–784, Apr. 2004, doi: 10.1002/mrm.20011.

[4] N. Gudino *et al.*, "Optically controlled on-coil amplifier with RF monitoring feedback.," *Magn Reson Med*, vol. 79, no. 5, pp. 2833–2841, May 2018, doi: 10.1002/mrm.26916.

[5] N. Gudino, J. A. de Zwart, and J. H. Duyn, "Eight-channel parallel transmit-receive system for 7 T MRI with optically controlled and monitored on-coil current-mode RF amplifiers," *Magnetic Resonance in Medicine*, vol. 84, no. 6, pp. 3494–3501, Dec. 2020, doi: 10.1002/mrm.28392.

[6] R. L. Lagore, B. R. Roberts, C. Possanzini, C. Saylor, B. G. Fallone, and N. De Zanche, "A system for automated noise parameter measurements on MR preamplifiers and application to high B(0) fields.," *NMR Biomed*, vol. 27, no. 8, pp. 926–938, Aug. 2014, doi: 10.1002/nbm.3138.

[7] C. Possanzini and M. Boutelje, "Influence of magnetic field on preamplifiers using GaAs FET technology," *Proceedings of the 16th Annual Meeting of ISMRM*, Jan. 2008.

[8] S. M. Sze, *Semiconductor devices : physics and technology / S.M. Sze*. New York: Wiley, 2002.

[9] R. S. Popovic, *Hall Effect Devices*. in Series in Sensors. CRC Press, 2003. [Online]. Available: https://books.google.com/books?id=_H5n-5sO5BAC

[10] A. Endoh, I. Watanabe, Y. Yamashita, T. Mimura, and T. Matsui, "Effect of temperature on cryogenic characteristics of AlGaN/GaN MIS-HEMTs," *physica status solidi c*, vol. 6, no. S2, pp. S964–S967, Jun. 2009, doi: 10.1002/pssc.200880803.

[11] K. Fukuda, J. Hattori, H. Asai, J. Yaita, and J. Kotani, "Temperature-dependent mobility modeling of GaN HEMTs by cellular automaton method," in *2021 International Conference on Simulation of Semiconductor Processes and Devices (SISPAD)*, Sep. 2021, pp. 40–43. doi: 10.1109/SISPAD54002.2021.9592574.

[12] J. Shen *et al.*, "High-mobility n−-GaN drift layer grown on Si substrates," *Applied Physics Letters*, vol. 118, no. 22, p. 222106, Jun. 2021, doi: 10.1063/5.0049133.

[13] U. S. P., R. Karthik, and P. Mallick, "Effect of dislocation scattering on electron mobility in GaN," *Natural Science*, vol. Volume 3, pp. 812–815, Jan. 2011, doi: 10.4236/ns.2011.39106.

[14] C. Ianniello *et al.*, "Synthesized tissue-equivalent dielectric phantoms using salt and polyvinylpyrrolidone solutions," *Magnetic Resonance in Medicine*, vol. 80, no. 1, pp. 413–419, Jul. 2018, doi: 10.1002/mrm.27005.

Switching Loss Reduction On Cascaded H-Bridge Converter With Diode Clamped Transformer Grounding Scheme

Zihan Gao
Department of Electrical Engineering and Computer Science
University of Tennessee
Knoxville, TN, USA
zgao15@vols.utk.edu

Ruirui Chen
Department of Electrical Engineering and Computer Science
University of Tennessee
Knoxville, TN, USA
rchen14@vols.utk.edu

Dingrui Li
Department of Electrical Engineering and Computer Science
University of Tennessee
Knoxville, TN, USA
dli35@vols.utk.edu

Fred Wang
Department of Electrical Engineering and Computer Science
University of Tennessee
Knoxville, TN, USA
Oak Ridge National Laboratory
Oak Ridge, TN, USA
fred.wang@utk.edu

Abstract—**Parasitic capacitance of dc/dc transformers interfacing the cascaded H-Bridge (CHB) converter can introduce extra switching losses. A switching loss reduction method with diode clamped grounding for dc/dc transformers connecting CHB is proposed in this paper. The transformer grounding is connected to diode clamping circuit on the low voltage dc link to partially cancel the voltage charging/discharging the parasitic capacitances. Simulation and test have been demonstrated that the proposed method has reduced 53% of the parasitics induced loss and ~6% of the total converter loss in the setup with one simple diode half-bridge clamping circuit added on each converter cell.**

Keywords—cascaded H-bridge (CHB) converters, transformer, grounding, switching loss

I. INTRODUCTION

Power electronic devices and converters play an important role in modern power systems [1-3]. As medium voltage (MV) SiC devices and magnetics technology have brought benefit on medium voltage power conversion, high power and high voltage converters using isolated dc/dc with dc/ac stages are more and more popular for power grid applications [4-9]. For converter systems using isolated dc/dc converters, the parasitics may become a concern on interference and losses as the operating and insulation voltages increase [10-13]. One of the critical parasitics is the common-mode (CM) parasitic capacitance of the MV dc/dc transformers and inductors [14]. Due to safety and noise suppression concerns, the transformer electric shielding or metal case is usually grounded to the earth (see Fig. 1), which connects the CM capacitance from the transformer MV winding to the ground. In this case, the low voltage (LV) winding or the area near the MV winding may not see high electric field, yet the voltage of MV winding can charge and discharge the CM or grounding capacitance through the grounding loop. Current spikes can be induced by charging and discharging the dc/dc transformer parasitic capacitance, which can be seen in Fig. 2, causing not only insulation and EMI concerns, but also extra

switching losses for CHB devices, which can account for over 15% of converter total loss [15]. Previous efforts have been made on CHB modulation to equivalent parasitics charging frequency, but the complex modulation can cause unbalanced switching losses on different devices [15]. The CM capacitance induced loss can also be limited by reducing the parasitics with different shielding strategies [16], but difficulties will be added on insulation design for the transformers and converters.

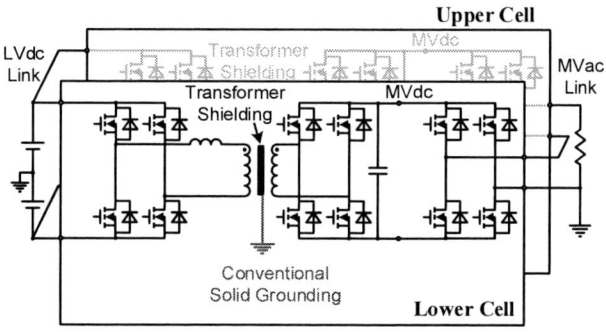

Fig. 1: Topology of converter using cells consisting of dc/dc and dc/ac, showing conventional transformer solid grounding.

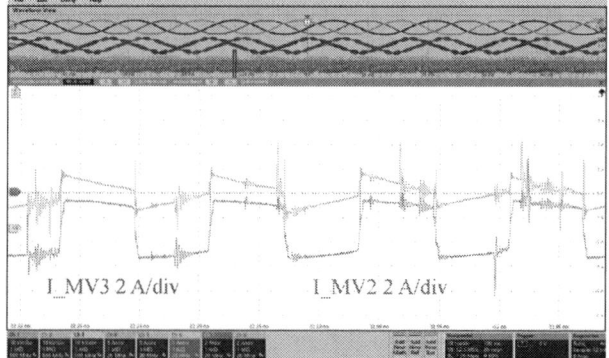

Fig. 2. Current spikes in transformers having high grounding capacitance [16].

This work was supported primarily by the Advanced Materials and Manufacturing Technologies Office (AMMTO), United States Department of Energy (DOE), under Awards no. DE-EE0008410 and DE-EE0009134.

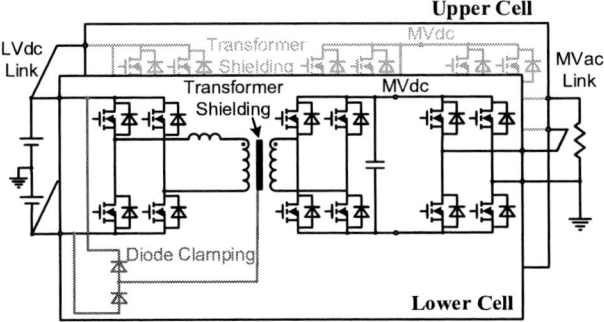

Fig. 3: Topology of converter using cells consisting of dc/dc and dc/ac, showing diode clamped transformer grounding.

In this paper, a diode clamped transformer grounding scheme is introduced to partially cancel the voltage stress across the transformer parasitic capacitance to the ground, and therefore, the losses caused by transformer capacitance can be reduced. Section II introduces the proposed grounding method. Simulation and test results are shown in Section III, along with discussions on influencing factors to the loss reduction. Section IV concludes the paper.

II. PROPOSED DIODE CLAMPED TRANSFORMER GROUNDING

A. Topology with Diode Clamped Grounding Circuit

First, the proposed topology with diode clamped grounding is shown in Fig. 3. Unlike the conventional topology in Fig. 1, the transformer shielding is not directly connected to the earth. However, the shielding is connected to a diode half-bridge network, which is linked to the LV dc link. Therefore, the potential of the transformer shielding is not solidly tied to the ground, but has a partially floating potential, which can cancel the CM voltage imposed on the MV winding by the CHB converter modulation.

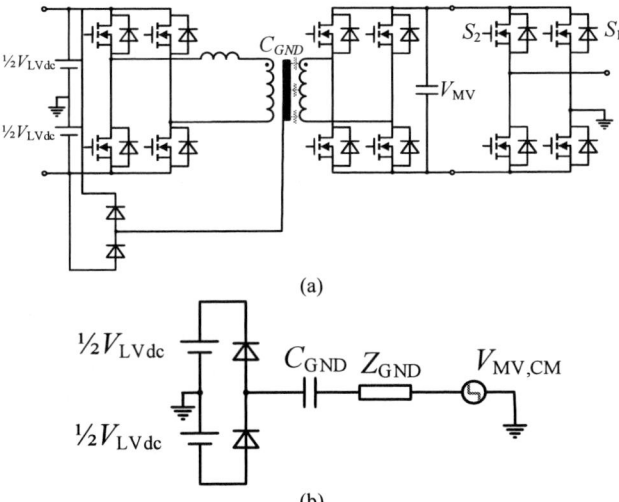

Fig. 4: Topology under study: (a) lowest cell with DAB and CHB converters using diode clamped circuit, (b) simplified CM loop.

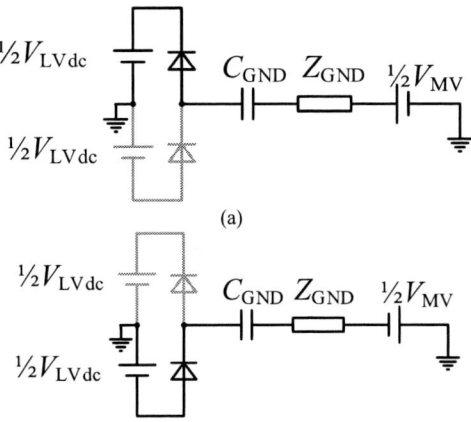

(a)

(b)

Fig. 5: Grounding parasitic capacitance charging and discharging modes with: (a) positive CM voltage, (b) negative CM voltage.

B. Operational Analysis of the Diode Grounding Circuit

As the MV dc/dc transformer is usually shielded between the LV and MV windings, only the MV side switching is studied. Also, with DAB converter acting with single phase-shift modulation, the DAB itself does not generate CM mode voltage. Therefore, the circuit loop concerning CM capacitance induced loss can be further simplified as the grounding CM capacitance directly connected to the CHB converter and diode clamping circuit, which is also grounded on the ac side, which is illustrated in Fig. 4.

Hence, the CM loop of CHB can be analyzed. As the CHB is grounded, for the lowest stage, the CM voltage of the MV dc side is

$$v_{MV,CM} = \frac{(1 - 2S_1)V_{MV}}{2} \tag{1}$$

where, S_1 is the switching function as in Fig. 4 [15]. From (1), the CM voltage of the MV dc side is flipping between half and negative half of the MV dc-link voltage. Therefore, as the CM voltage changes the polarity, the CM capacitance gets charged and discharged, through the diode clamping circuit. The two modes of the CM loop can be depicted as in Fig. 5.

From Fig. 6, as the CM voltage changes, the diode clamping circuit rectifies the charging/discharging current, causing voltage drop of half LV dc-link voltage on the circuit. Therefore, the voltage across the capacitance can be reduced by half of V_{LVdc}. Suppose the grounding impedance Z_{GND} is mainly resistive, the energy lost in the capacitance charging circuit is

$$E_{loss} = \frac{1}{2}C\Delta V^2 \tag{2}$$

Where, C is the grounding CM capacitance, ΔV the voltage change on the CM capacitance. For conventional solid grounded transformer, the total CM grounding induced loss for the lowest stage is [15]

$$P_{loss,solid} = CV_{MV}^2 f \tag{3}$$

979-8-3503-3714-3/23 $31.00 © 2023 IEEE

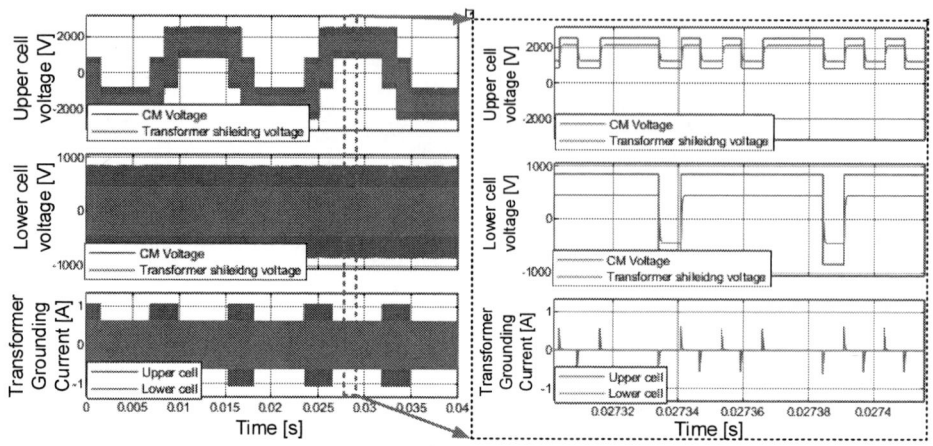

Fig. 6: Simulation results of diode clamped grounding, the transformer shielding potential swing for both upper and lower cell is reduced by half of LVdc link voltage (400 V).

Where, f is the equivalent switching frequency of the CHB converter. For the proposed diode clamping circuit, the loss can be reduced to

$$P_{loss,proposed} = C\Delta V^2 f = C(V_{MV} - V_{LVdc})^2 f \qquad (4)$$

By subtracting (4) with (3), the loss reduction for the lowest stage is

$$\Delta P_{loss} = C[V_{MV}^2 - (V_{MV} - V_{LVdc})^2]f \qquad (5)$$

Note that, the closer values V_{MV} and V_{LVdc} yield more effective loss reduction, and the analysis can be easily scale up to multiple cells cascaded by changing the switching frequency f to the sum of equivalent CM frequencies [15], and the Fig. 5 can also be valid with correct CM voltage levels.

III. SIMULATION AND TEST VALIDATION

As the effectiveness of the diode clamping circuit has been analyzed, simulation and experimental tests have been performed to get the proposed grounding method validated.

A. Simulation Verification

First, the simulation of a two-stage CHB converter with DAB transformers, as shown in Fig. 1 and 3, has been performed. The LV and MV dc-links are set to 800 V and 1600 V, and the waveforms of CM voltage, voltages across CM capacitances and charging currents are shown in Fig. 6. As the CHB device switches, the transformer parasitics see the common mode (CM) voltage abstracted by ½LVdc voltage (400 V) as in Fig. 6.

B. Test Verification

Then, tests have been performed with the same rating of the simulation. The photographs of the MV dc/dc transformer and the diode clamping circuits are shown in Fig. 7. The test waveforms are shown in Fig. 8. From the waveforms, the shielding layer potential is not constantly zero, floating between the positive and negative of half LV dc link. The power loss with constant resistive load up to 12 kW is shown in Fig. 9. Loss

reduction can be achieved constantly under variable output voltage by approximately 18 W, 53% of parasitic loss reduced (theoretically 34 W), and ~6% reduction on the total converter loss. Although the transformer potential swings ($\pm\frac{1}{2}V_{LVdc}$), the LV dc-link is typically below 1 kV and the margin on low

(a)

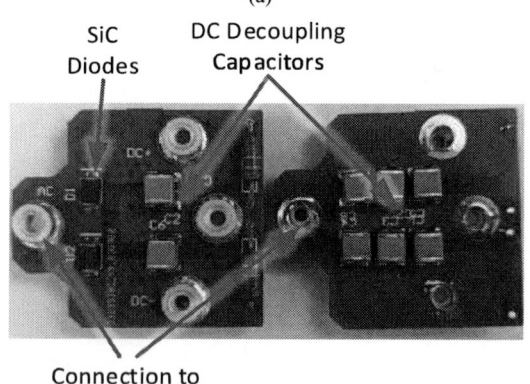

(b)

Fig. 7: Photography of (a) MV dc/dc transformer, (b) diode clamping circuit.

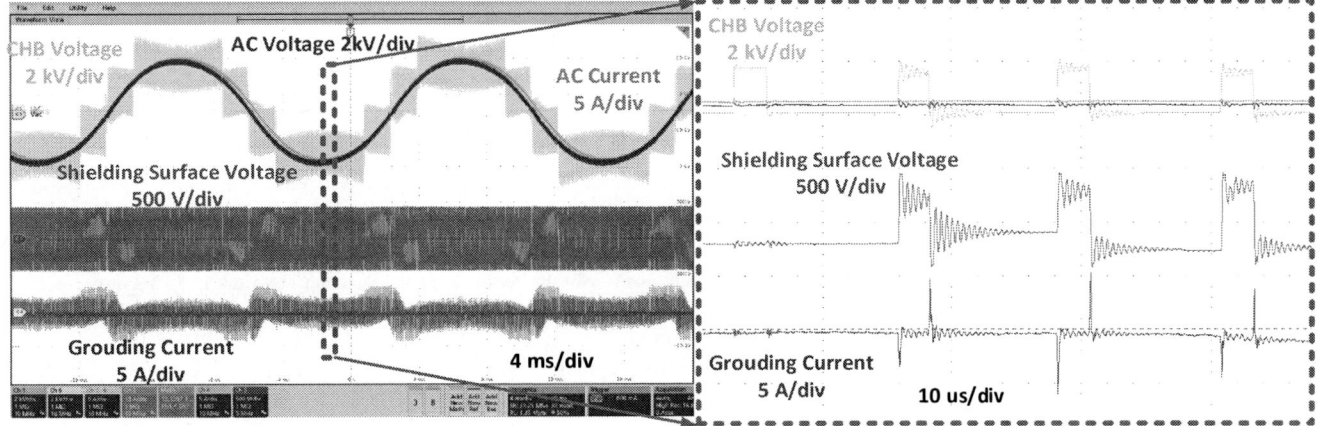

Fig. 8: Test waveforms for diode clamped grounding (with shielding voltage and grounding current for the upper cell).

voltage side insulation is usually high enough, so no further requirement on LV winding and power stage insulation is required.

C. Impact Factors on Loss Reduction

Since several different components can be found in the grounding loop, e.g., clamping diodes, grounding impedance, CHB devices, the effectiveness of the loss reduction can be influenced by multiple factors, too. The impact factors mainly include diode parasitics, grounding impedance, and modulation non-ideality, such as gate signal delays.

For the diodes, the non-idealities may cause extra conduction and switching losses. As the charging current is pulsating current with amplitude up to several amperes and short duration (<400 ns), and the diode forward voltage drop is also low compared to the CM voltage change, the conduction loss can be negligible. For the switching loss, as SiC Schottky diodes

are used, no reverse recovery loss is assumed, while the charges built-up in the junction capacitance should be considered. As the CHB MV devices hard switch, the diode junction capacitance discharges and causes CHB device switching loss. The switching energy loss for each diode is

$$E_{loss,SBD} = \int C(v_r) v_r dv_r \qquad (6)$$

where, the v_r is the diode reverse bias voltage during CHB device switching. As estimated during the test, the diode switching loss should be around 0.96 W.

The grounding parasitics can be complicated, which may be comprised of impedances of grounding wires, transformer shielding layer, and commutating loop of diode clamping boards. As the grounding loop should be low impedance required by the safety guidelines, only inductances are considered. To study the impact of total grounding impedance, a simulation sweep has been done to find out the loss variation on the grounding loop. Fig. 10 shows the loss results, from which the parasitic inductance increases from zero to 10 uH, while the loss is increased by 3 W. Hence, the inductance may impact the loss reduction, while as long as the grounding loop is controlled as short as possible, the loss impact is low.

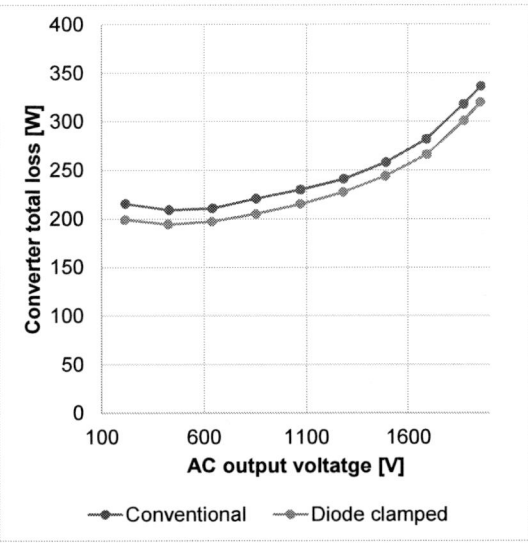

Fig. 9: Tested loss comparison between conventional and diode clamped grounding.

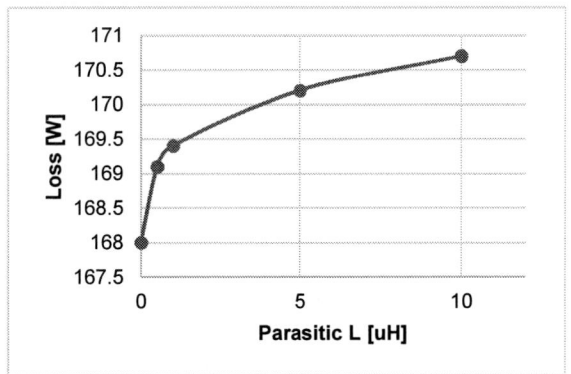

Fig. 10: Simulation sweep of grounding inductance and capacitance associated loss.

979-8-3503-3714-3/23 $31.00 © 2023 IEEE

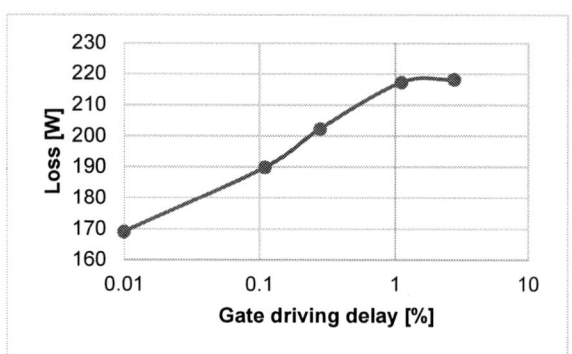

Fig. 11: Simulation sweep of gate driving delay and capacitance associated loss.

Modulation mismatches may cause mismatched CM voltages waveform, which can cause unexpected charging and discharging events on the CM capacitances. Fig. 11 shows the simulation sweep with gate driving delays. As shown in the figure, when gate driving delays up to 1% of the switching period, the loss can be increased by 30%, which can counterpart the proposed loss reduction. Therefore, the modulation and gate signaling may increase the CM capacitance induced loss, and the modulation delay and dead-time should be well managed.

IV. CONCLUSION

In this paper, a diode clamping circuit is proposed to reduce the CM parasitics induced switching loss on the CHB converters. The principle of loss reduction has been analyzed and validated through the simulation and test results. For the tested cases, the proposed method has reduced the parasitics associated loss up to 53%, which aligns with the analysis. The impact factors influencing the reduction effectiveness have also been discussed, among which the diode junction energy and gate signal delays may have high impact and should be carefully considered.

ACKNOWLEDGMENT

This work made use of the shared facilities of the Engineering Research Centre Program of the National Science Foundation (NSF) and the Department of Energy under NSF Award no. EEC-1041877. The authors also would like to acknowledge the contribution of Southern Company and Powerex.

REFERENCES

[1] Y. He *et al.*, "Control Development and Fault Current Commutation Test for the EDISON Hybrid Circuit Breaker," *IEEE Transactions on Power Electronics,* vol. 38, no. 7, pp. 8851-8865, July 2023, doi: 10.1109/TPEL.2023.3262605.

[2] Y. He, Y. Li, Q. Yang, R. B. Gonzatti, A. Taylor, and F. Peng, "Grid-connected converter without Interfacing Filters: Principle, Analysis and Implementation," in *2020 IEEE Energy Conversion Congress and Exposition (ECCE),* 11-15 Oct. 2020, pp. 4505-4511, doi: 10.1109/ECCE44975.2020.9235792.

[3] Y. He, Y. Li, B. Zhou, Y. Zou, and F. Z. Peng, "An Ultra-Fast Inrush-Current-Free Startup Method for Grid-tie Inverter without Voltage Sensors," in *2023 IEEE Applied Power Electronics Conference and Exposition (APEC),* 19-23 March 2023, pp. 2874-2880, doi: 10.1109/APEC43580.2023.10131356.

[4] H. Li, Z. Gao, and F. Wang, "Design and Demonstration of an 850 V dc to 13.8 kV ac 100 kW Three-phase Four-wire Power Conditioning System Converter Using 10 kV SiC MOSFETs," in *2023 IEEE Applied Power Electronics Conference and Exposition (APEC),* 19-23 March 2023, pp. 989-994, doi: 10.1109/APEC43580.2023.10131291.

[5] H. Li, Z. Gao, Z. Yang, C. Nie, and F. Wang, "A Medium Voltage Testbed for the Performance and Function Tests of a 13.8 kV Power Conditioning System Converter," in *2022 IEEE Applied Power Electronics Conference and Exposition (APEC),* 20-24 March 2022, pp. 757-763, doi: 10.1109/APEC43599.2022.9773522.

[6] Z. Li, E. Hsieh, Q. Li, and F. C. Lee, "High-Frequency Transformer Design With Medium-Voltage Insulation for Resonant Converter in Solid-State Transformer," *IEEE Transactions on Power Electronics,* vol. 38, no. 8, pp. 9917-9932, Aug. 2023, doi: 10.1109/TPEL.2023.3279030.

[7] M. Gao, L. Yi, and J. Moon, "Analysis, Modeling, and Validation of Cascaded Magnetics for Magnetic Energy Harvesting," in *2022 IEEE Energy Conversion Congress and Exposition (ECCE),* 9-13 Oct. 2022, pp. 1-7, doi: 10.1109/ECCE50734.2022.9947560.

[8] M. Gao, H. L. Herrera, and J. Moon, "Optimization of Core Size and Harvested Power for Magnetic Energy Harvesters based on Cascaded Magnetics," in *2023 IEEE Applied Power Electronics Conference and Exposition (APEC),* 19-23 March 2023, pp. 2926-2932, doi: 10.1109/APEC43580.2023.10131308.

[9] M. Gao, L. Yi, and J. Moon, "Mathematical Modeling and Validation of Saturating and Clampable Cascaded Magnetics for Magnetic Energy Harvesting," *IEEE Transactions on Power Electronics,* vol. 38, no. 3, pp. 3455-3468, March 2023, doi: 10.1109/TPEL.2022.3218725.

[10] J. E. Huber and J. W. Kolar, "Common-mode currents in multi-cell Solid-State Transformers," in *2014 International Power Electronics Conference (IPEC-Hiroshima 2014 - ECCE ASIA),* 18-21 May 2014, pp. 766-773, doi: 10.1109/IPEC.2014.6869674.

[11] Y. Lai, S. Wang, Y. Yang, Q. Huang, and Z. Ma, "Review on Modeling and Emissions from EMI Filters in Power Electronics: Inductors," in *2023 IEEE Symposium on Electromagnetic Compatibility & Signal/Power Integrity (EMC+SIPI),* 29 July-4 Aug. 2023, pp. 566-572, doi: 10.1109/EMCSIPI50001.2023.10241760.

[12] Z. Ma, Y. Lai, Y. Yang, Q. Huang, and S. Wang, "Review of Radiated EMI Modeling and Mitigation Techniques in Power Electronics Systems," in *2023 IEEE Applied Power Electronics Conference and Exposition (APEC),* 19-23 March 2023, pp. 1776-1783, doi: 10.1109/APEC43580.2023.10131552.

[13] Y. Yang, Q. Huang, Y. Lai, Z. Ma, Y. Liu, and S. Wang, "Analysis and Modeling of the Near Magnetic Field Distribution of Toroidal Inductors," in *2023 IEEE Symposium on Electromagnetic Compatibility & Signal/Power Integrity (EMC+SIPI),* 29 July-4 Aug. 2023, pp. 573-578, doi: 10.1109/EMCSIPI50001.2023.10241669.

[14] H. Li, P. Yao, Z. Gao, and F. Wang, "Medium Voltage Converter Inductor Insulation Design Considering Grid Requirements," *IEEE Journal of Emerging and Selected Topics in Power Electronics,* vol. 10, no. 2, pp. 2339-2350, November 2022, doi: 10.1109/JESTPE.2021.3131602.

[15] H. Li, Z. Gao, and F. Wang, "A PWM Strategy for Cascaded H-bridges to Reduce the Loss Caused by Parasitic Capacitances of Medium Voltage Dual Active Bridge Transformers," in *2022 IEEE Energy Conversion Congress and Exposition (ECCE),* 9-13 Oct. 2022, pp. 1-6, doi: 10.1109/ECCE50734.2022.9947550.

[16] Z. Gao, H. Li, and F. Wang, "A Medium-Voltage Transformer with Integrated Leakage Inductance for 10 kV SiC-Based Dual-Active-Bridge Converter," in *2022 IEEE 9th Workshop on Wide Bandgap Power Devices & Applications (WiPDA),* 7-9 Nov. 2022, pp. 221-226, doi: 10.1109/WiPDA56483.2022.9955258.

SiC MOSFETs performance modeling in Simulink Simscape environment

Jacopo Ferretti[1], Giacomo-Piero Schiapparelli[2], Enrico Sangiorgi[1] and Andrea Natale Tallarico[1]

[1]ARCES-DEI, University of Bologna, Campus of Cesena, Italy
{jacopo.ferretti, enrico.sangiorgi, a.tallarico}@unibo.it
[2]High Performance Engineering s.r.l., Italy
gpschiapparelli@hpe.eu

Abstract—**The paper proposes a model for Silicon-Carbide MOSFET, suitably defined to be simulated in the Simulink Simscape environment. The work details the model formalization, the tuning procedure based on the data available by the manufacturer's datasheet, and the simulation procedure. The model takes into account the role of temperature and ageing on the transistor's performance. Aging is implemented by drifting the drain-to-source ON-state resistance and/or the threshold voltage. The SiC MOSFET model is first validated on a commercial power module by comparing its outcomes with the manufacturer's measurements. Then, it is adopted to simulate a full-bridge inverter, assessing the effects of transistors' ageing on the current capability, limited by thermal effects, and on the efficiency degradation.**

Index Terms—**SiC MOSFET modelling, ageing, self-heating, power electronics simulation, inverter.**

I. INTRODUCTION

The electric vehicles market is growing rapidly, increasing the interest in Power Electronic Converters (PECs) [1], including their reliability assessment. The latter is important to achieve regulatory compliance and increase the quality level of the power circuits. The reliability of a PEC can be assessed by evaluating each subcomponent's lifetime, especially the most susceptible to ageing, such as DC Link capacitors and semiconductor-based power switches. The latter are responsible for converter failures for 72 % of cases [2]. The power transistor based on Wide Bandgap (WBG) semiconductors, such as Silicon Carbide (SiC) and Gallium Nitride (GaN) [3], represents one of the best choices in terms of robustness, efficiency, and circuit area factor, since it provides an excellent trade-off between blocking voltage, ON-resistance and switching frequency. To date, the most adopted WBG technology for the high power demand of the automotive industry is the SiC [4], widely appreciated for the traction inverter.

However, being a relatively emerging technology, it might be less reliable than its silicon competitor. As a matter of fact, the reliability investigation of SiC MOSFETs represents a topic of interest for the research community [5]. Failure modes are associated with internal and inaccessible microscopic ageing mechanisms, making them challenging to be monitored directly. One of the most adopted strategies to predict the device failure consists in the monitoring of the drain-to-source ON-state resistance (R_{ds_on}) [6] and the threshold voltage (V_{th}) drift [6]–[9]. Usually, a drift of $\sim$20 % is identified as a

failure for the power component [10]. However, this approach does not account for the role and effects of component ageing on the efficiency and maximum current capability of the converter. Indeed, the effects of a 20% drift could be different according to the specific application requirements. Thus, a tool aimed to assess the power losses of the converter, taking into account the ageing of the power MOSFET and the related thermal behavior, is of paramount importance. Simulation of aged SiC power MOSFETs has been proposed in [11]; nevertheless, it requires a wide amount of information often unavailable in the datasheet.

This paper proposes an approach aimed at providing a SiC-MOSFET model to be implemented in Simulink-Simscape simulator, able to account for the ageing and the electro-thermal behavior, and relying only on the typical set of information provided by the manufacturer's datasheet. First, the modelling assumptions are summarized; then, the parameter's tuning procedure is detailed and validated for a specific use case, i.e., the STMicroelectronics ACEPACK DRIVE SiC power module [12], a three-phase full-bridge SiC MOSFET power module representing state-of-the-art in Electric Vehicle (EV). Finally, the effect of the power module ageing is assessed in the case of a single-phase full bridge inverter.

The paper is structured as follows: section II reports the electrical and thermal models; section III shows the fitting procedure adopted to tune the model parameters against the experimental data available by the datasheet; section IV details the simulation environments; section V discusses the simulation results, concluding in section VI with the main achievements of this work.

II. SiC MOSFET MODEL

The equivalent circuit of the SiC MOSFET is shown in Fig. 1. It is modelled through a controlled current source, a set of parasitic elements [13], and a thermal network for taking into account self-heating effects.

A. Thermal Network

The Cauer thermal network has been implemented to simulate the MOSFET junction temperature (T_j) [15]. The related parameters, provided by the datasheet, are considered constant for the useful time of the converter, since the possible ageing of the thermal network is beyond the scope of this paper.

979-8-3503-3714-3/23 $31.00 © 2023 IEEE

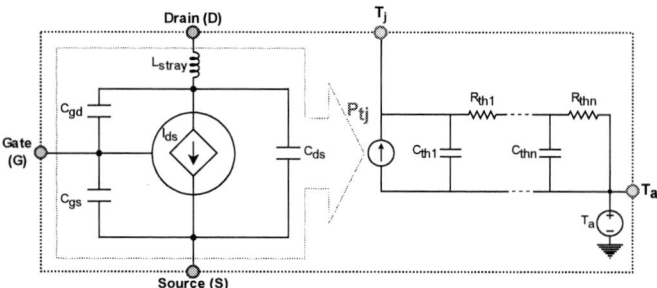

Fig. 1: MOSFET electrical and thermal model used in simulation [14], [15].

B. Parasitic elements

The parasitic elements are modelled by the drain-to-source (C_{ds}), gate-to-source (C_{gs}) and gate-to-drain (C_{gd}) capacitances, and the stray inductance L_{stray}. As a model assumption, L_{stray} is constant, while the capacitances depend only on the drain-to-source voltage V_{ds}.

C. Drain to source current model

The current controlled source, which generates the drain-to-source current I_{ds} can be obtained using different approximation models, *e.g.*, *i)* the Shichman-Hodges, *ii)* the Grove-Frohman, *iii)* empirical models and *iv)* surface-potential-based model [16], [17]. The first three models, typically used in SPICE simulators, are characterized by different operating regions modelled with specific equations and with increasing complexity and number of required parameters; while the latter, surface potential-based, relies on the charge equation. The Shichman-Hodges model based on the Gradual Channel Approximation has been adopted in this study as reported in (1)–(5), since it provides a good trade-off between number of required parameters and accuracy [16].

The I_{ds} current is expressed by a piecewise function f defined in four different domains (D_{off}, D_{lin}, D_{sat} and D_{rev}) and depends on gate-to-source voltage V_{gs}, drain-to-source voltage V_{ds}, and temperature T.

$$I_{ds} = f(V_{gs}, V_{ds}, T) =$$
$$= \begin{cases} f_{off}(V_{gs}, V_{ds}, T) & \text{for } I_{ds} \in D_{off} \\ f_{lin}(V_{gs}, V_{ds}, T) & \text{for } I_{ds} \in D_{lin} \\ f_{sat}(V_{gs}, V_{ds}, T) & \text{for } I_{ds} \in D_{sat} \\ f_{rev}(V_{gs}, V_{ds}, T) & \text{for } I_{ds} \in D_{rev} \end{cases} \quad (1)$$

Off-region

$$\begin{cases} f_{off} = 0 \\ D_{off} = \{V_{gs} < V_{th} \cap V_{ds} > 0\} \end{cases} \quad (2)$$

Linear region

$$\begin{cases} f_{lin} = K(T, V_{gs}) \left((V_{gs} - V_{th}(T)) V_{ds} - \frac{V_{ds}^2}{2} \right) \\ D_{lin} = \{V_{gs} \geq V_{th} \cap (0 < V_{ds} < V_{gs} - V_{th}) \} \end{cases} \quad (3)$$

Saturation region

$$\begin{cases} f_{sat} = \frac{K(T, V_{gs})}{2} \left((V_{gs} - V_{th}(T))^2 \right) \\ D_{sat} = \{V_{gs} \geq V_{th} \cap (V_{ds} \geq V_{gs} - V_{th}) \} \end{cases} \quad (4)$$

Reverse conduction region

$$\begin{cases} f_{rev} = \begin{cases} f_{diode}(V_{ds}, T) = I_s(T) \left(e^{\frac{-V_{ds}}{V_T(T)}} - 1 \right) \\ \qquad \text{for } V_{gs} < V_{th} \\ f_{diode}(V_{ds}, T) + f_{lin}(V_{ds}, V_{gs}, T) \\ \qquad \text{for } V_{gs} \geq V_{th} \cap V_{ds} > -(V_{gs} - V_{th}) \\ f_{diode}(V_{ds}, T) + f_{sat}(V_{ds}, V_{gs}, T) \\ \qquad \text{for } V_{gs} \geq V_{th} \cap V_{ds} \leq -(V_{gs} - V_{th}) \end{cases} \\ D_{rev} = \{V_{ds} < 0\} \end{cases}$$
$$(5)$$

Each domains defined in (2)–(5) identifies a specific operating condition of the MOSFET. In the OFF-state region, there is no current conduction. The linear and saturation regions refer to the forward conduction condition characterized by two temperature and ageing-dependent parameters, the transconductance (K) and threshold voltage (V_{th}).

Finally, the reverse conduction region identifies a condition in which a negative voltage is applied between the drain and source terminals. However, it can occur with V_{gs} values lower or higher than V_{th}. For $V_{gs} < V_{th}$, the I_{ds} current is completely ascribed to the antiparallel diode; for $V_{gs} \geq V_{th}$, I_{ds} is determined by the parallel conduction of diode and transistor (channel formed). I_s and V_T in the diode model (5) are the temperature-dependent saturation current and thermal voltage, respectively. However, it is worth noting that, in our case, the model (5) has not been adopted to reproduce the reverse conduction regime since the related curves are already available by the datasheet for the whole admissible operating range of V_{gs}, V_{ds} and T.

C.1 Temperature dependence

The temperature-dependent K and V_{th} have been modelled as reported in (6) and (7), respectively [18].

$$V_{th}(T) = V_{th}(T_0) + \alpha(T - T_0) \quad (6)$$

$$K(T, V_{gs}) = K(T_0, V_{gs}) \left(\frac{T}{T_0} \right)^{\beta(V_{gs})} \quad (7)$$

T_0 is the reference (ambient) temperature, provided as an input for the determination of both V_{th} and K, while T represents the temperature of the device. α and $\beta(V_{gs})$ are fitting parameters obtainable from the MOSFET datasheet as detailed in Section III.

C.2 Ageing dependence

The effect of the MOSFET ageing on the static and dynamic performance of the power circuit is reproduced and investigated by drifting the K and V_{th} parameters from their fresh value [7]. The modelling of time-dependent aging mechanisms is out of the scope of this paper, its effect is only investigated by performing different simulations with different levels (time-independent) of degradation.

III. MODEL FITTING

A. Thermal Network values

As anticipated in section II-A, the Cauer thermal network has been implemented by adopting values provided by the datasheet. As the power module under consideration in this paper is intended for liquid cooling [12], the ambient temperature T_a depicted in Fig. 1 is regarded as the coolant temperature ($T_{coolant}$).

B. Parasitic element values

The parasitic elements of the MOSFET equivalent circuit in Fig. 1 have been assigned as follows: L_{stray} is assumed to be a constant value as reported in the datasheet; the parasitic capacitances, available in the form of input C_{iss}, reverse transfer C_{rss} and output C_{oss} capacitances, and provided as a function of V_{ds}, are calculated as $C_{gd}(V_{ds}) = C_{rss}(V_{ds})$, $C_{ds}(V_{ds}) = C_{oss}(V_{ds}) - C_{rss}(V_{ds})$ and $C_{gs}(V_{ds}) = C_{iss}(V_{ds}) - C_{rss}(V_{ds})$. It is worth noting that such parameters are not affected by ageing in the present model.

C. Drain to source current fitting

To accurately reproduce the experimental I_{ds}, a careful adoption of the two parameters $K(V_{gs}, T)$ and $V_{th}(T)$ is needed. $V_{th}(T)$ is calculated from the transfer characteristics, whereas $K(V_{gs}, T)$ is finely tuned against the output characteristics by adopting a fitting algorithm detailed in the following. Both transfer and output characteristics are provided by the datasheet.

An example of a data set obtained from the I-V curves reported in the datasheet is shown in Fig. 2. In particular, the blue points represent the available data from the I-V characteristics, while the red points depict the unavailable ones. The proposed approach aims at *i)* tuning the models (2)–(5) on the known (blue) points, determining the K and V_{th} parameters; *ii)* adopting the tuned models to determine the unknown regions of the I-V map (red points in Fig. 2).

D_d represents the domain of the known data, and $f_l : D_d \rightarrow I_{ds}$ retrieves values from this domain. The subscript d refers to the datasheet-related domain. The fitting procedure aims at determining the model parameters for the domain D_d to compute the missing I-V values for the extended domain D, with $D_d \subset D$. The latter is defined as follows:

$$
\begin{aligned}
D = \{(V_{gs}, V_{ds}, T) \in \mathbb{R}^4 | V_{gs} \in [-5\,\mathrm{V}, 20\,\mathrm{V}], \\
V_{ds} \in [-6\,\mathrm{V}, 1200\,\mathrm{V}], T \in [25\,^\circ\mathrm{C}, 175\,^\circ\mathrm{C}]\}
\end{aligned}
\tag{8}
$$

The fitting algorithm reported in (9) determines for each j, with $j = 1, ..., \mathrm{card}(\{V_{gs}\}_d)$, and for each temperature $T \in \{T\}_d$, the value of K_j that ensures the minimum of the objective function. The latter is defined as the sum of the normalized error between I-V values reported in the datasheet

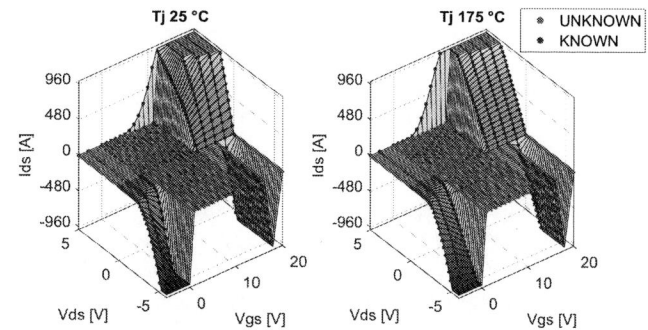

Fig. 2: Current-Voltage map obtained from the datasheet for 25 °C and 175 °C [12].

($I_{ds,i}^*$) and the corresponding computed ones ($I_{ds,i}$) by the tuned model.

$$
\min_{K_j} \sum_{i=1}^{N} \left| \frac{I_{ds,i}^* - I_{ds,i}}{I_{ds,i}^*} \right| \tag{9}
$$

s.t.

$$
\begin{aligned}
N &= \mathrm{card}(D_d) \\
I_{ds}^* &= f_l(V_{ds}, V_{gs}, T) && \forall I_{ds}^* \in D_d \\
I_{ds} &= f(V_{gs}, V_{ds}, T, V_{th}, K) = \\
&= \begin{cases} f_{lin}(\cdot) & \forall I_{ds} \in D_{lin} \cap D_d \\ f_{sat}(\cdot) & \forall I_{ds} \in D_{sat} \cap D_d \end{cases} \\
D_d &= \{(V_{gs}, V_{ds}, T) \in \mathbb{R}^4 | V_{gs} \in \{V_{gs}\}_d, \\
& \quad V_{ds} \in \{V_{ds}\}_d, T \in \{T\}_d\} \\
V_{th} &= V_{th}(T, t_0)
\end{aligned}
$$

After obtaining the optimal set of K_j values (denoted as K_j^*), equations (2), (3), and (4) are used to build up the whole current map in the domain D, as reported in Fig. 3. A measure of the fitting accuracy is given in terms of mean value μ and standard deviation σ of the error function computed to obtain the optimal set K_j^*. In particular, their value is $\mu = -0.5\,\%$ and $\sigma = 3.7\,\%$ in the case of $T = 25\,^\circ\mathrm{C}$ and $\mu = 0.2\,\%$ $\sigma = 2.9\,\%$ for $T = 175\,^\circ\mathrm{C}$.

C.1 Temperature dependence equations fitting

Equations (6) and (7) describe the temperature dependence of V_{th} and K, respectively, relying on α and β parameters, which can be simply calculated by as reported in (10) and (11) respectively.

$$
\alpha = \frac{V_{th}(T) - V_{th}(T_0)}{T - T_0}, \tag{10}
$$

$$
\beta(V_{gs}) = \frac{\log\left(\frac{K_j^*(V_{gs}, T_0)}{K_j^*(V_{gs}, T)}\right)}{\log\left(\frac{T_0}{T}\right)}. \tag{11}
$$

In particular, α can be directly evaluated from the datasheet available data (I_{ds}-V_{gs} at 25 °C and 175 °C), while β has to be computed from the results of the fitting algorithm (K_j^*) at the given temperatures.

It is important to highlight that K_j^* is undefined in the OFF-state region ($V_{gs} < V_{th}$), and V_{th} varies with temperature. This situation introduces a critical issue in eq. (11). There could be a possible scenario where, for the same V_{gs} value, the device may operate in OFF-state or conduction region depending on the temperature value. In such case, it is not possible to compute β for that specific V_{gs} value, since K_j^* is undefined for one of the two temperatures. Therefore, a suitably defined transformation $\mathcal{T}$ is required. The latter is a function that translates the K_j^* with respect to the domain boundary, obtained by defining a new variable $V_{ov} = V_{gs} - V_{th}$. In particular, the value of β is obtained from the transformed values of K_j^* by applying $\mathcal{T}$ to the optimal set K_j^* (12), computing the values of $\beta(V_{ov})$ (13), and calculating $\beta(V_{gs})$ with the inverse transformation function $\mathcal{T}^{-1}$ (14).

$$\mathcal{T}(K_j^*(V_{gs}, T)) = K_j^*(V_{gs} - V_{th}, T) = K_j^*(V_{ov}, T) \quad (12)$$

$$\beta(V_{ov}) = \frac{\log\left(\frac{K_j^*(T_0, V_{ov})}{K_j^*(T, V_{ov})}\right)}{\log\left(\frac{T_0}{T}\right)} \quad (13)$$

$$\beta(V_{gs}) = \mathcal{T}^{-1}(\beta(V_{ov})) = \beta(V_{ov} + V_{th}) \quad (14)$$

IV. MODEL IMPLEMENTATION

Once all the model parameters K, V_{th}, α and β have been fitted against the partial I-V curves provided in the datasheet, the model (1)–(4) can be used to evaluate the missing ones over the whole domain D, for any $I_{ds} = f(V_{ds}, V_{gs}, T)$ as reported in Fig. 3. Moreover, to take into account the transistor ageing, the values of K and V_{th} are arbitrarily changed within a specific range. It is worth noting that V_{th} is a direct ageing indicator, while K, being proportional to electron mobility [19], is linked to the variation of R_{ds_on}. A lower K corresponds to a higher R_{ds_on}, representing another ageing indicator.

The so-computed current values are implemented in the Simscape-Simulink environment as a Lookup Table (LUT). For any possible simulation case, a new three-dimensional LUT ($I_{ds} = f(V_{ds}, V_{gs}, T)$) is computed using fresh or aged values of K and V_{th}. The LUTs, as well as the parasitic elements and the thermal network, are used to simulate the real-time electrical behavior of the MOSFET, including switching energy and junction temperature.

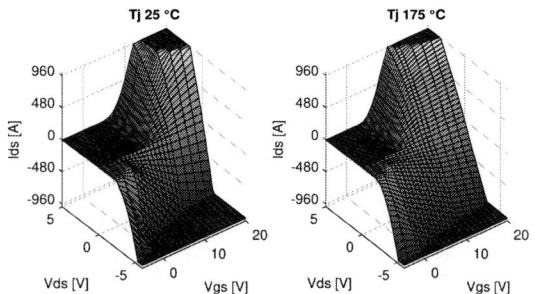

Fig. 3: Current-Voltage map obtained by adopting the fitting algorithm in the case of 25 °C and 175 °C.

V. SIMULATION RESULTS

A. Test circuit 1: model validation

The proposed model has been firstly validated with the test circuit shown in Fig. 4, implemented to evaluate the switching losses and to have a comparison with experiments reported in the datasheet. The DC source has been modelled as an ideal voltage generator, the load as a RL network, the high-side SiC MOSFET (S1) is used as a freewheeling diode, whereas the low-side one (S2) is the device under test. The parameters adopted in the simulation are summarized in Table I.

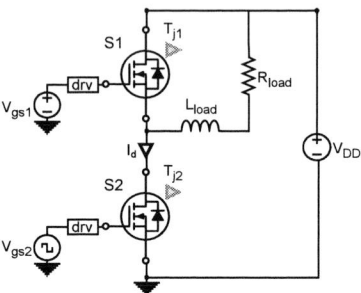

Fig. 4: Single-switch model used in the simulation to evaluate the switching losses.

TABLE I: Single-switch test parameters used in the simulation.

Parameter	Symbol	Value
ON gate resistor	R_{gs_on}	$10\,\Omega$
Off gate resistor	R_{gs_off}	$8.2\,\Omega$
Gate-source voltage S1	V_{gs_1}	$-5\,V$
Gate-source voltage S2	V_{gs_2}	$-5/18\,V$ (pulsed)
DC bus voltage	V_{DD}	$850\,V$
Load current	I_d	$[120\,A\ ...\ 960\,A]$
Load inductor	L	$1\,mH$
Load resistor	R	V_{DD}/I_d
Coolant temperature	$T_{coolant}$	$25\,°C$

As for real tests on power devices, the single-switch test circuit (Fig. 4) is adopted to simulate the SiC MOSFET switching energy as a function of the drain current, as shown in Fig. 5. It can be observed as the simulated switching energy (green circles), which is the sum of the turn-on and turn-off energy (E_{on}, E_{off}), accurately reproduces the experimental one reported in the datasheet (green line), with a mean error $= -0.42\,mJ$, a mean absolute percentage error $= -0.04\,\%$ and a root mean squared error $= 4.94\,mJ$.

Moreover, the switching energy has been evaluated under different aging conditions, revealing the R_{ds_on} drift (ΔK in Fig. 5) as the most impacting parameter for the switching losses.

B. Test circuit 2: full-bridge inverter simulation

A single-phase full-bridge inverter is adopted to assess the performance degradation given by the ageing of the power

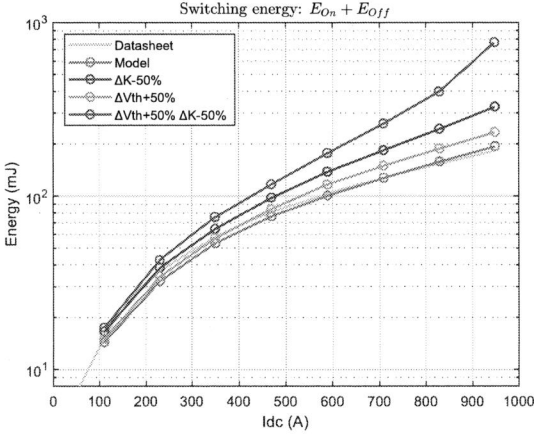

Fig. 5: Total switching energy VS load current with different ageing conditions.

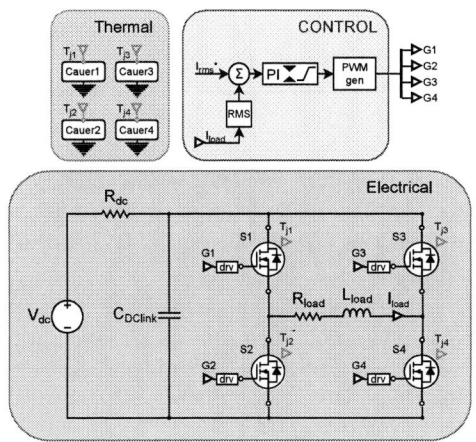

Fig. 6: Single-phase full-bridge inverter model used in the simulation.

SiC MOSFETs. The converter topology is shown in Fig. 6; it consists of a DC source modelled by an ideal voltage generator with a series resistor, a DC link capacitor, and a RL load. The four SiC MOSFETs (S1, S2, S3, S4) are controlled by a PID and a PWM generator which controls the RMS value of the load current. Circuit parameters are summarized in Table II.

Degraded conditions are simulated for each combination of K and V_{th} drift in the range 0% to -50% and 0% to 50%, respectively. For each simulation, the maximum current capability (which is limited by the maximum reachable junction temperature $T_j = 175\,^{\circ}\mathrm{C}$) and the circuit efficiency are considered.

Fig. 7 shows the T_j of one of the SiC MOSFET while operating in the full-bridge inverter, as a function of the transistor aging level. It is possible to note two aspects: *i)* the model is able to take into account the self-heating of the MOSFET, T_j increases during the operating time of the inverter reaching a steady state for longer times; *ii)* the

TABLE II: Single-phase full-bridge inverter parameters used in simulation.

Parameter	Symbol	Value
Switching frequency	f_{sw}	$20\,\mathrm{kHz}$
Output current frequency	f_{sin}	$1\,\mathrm{kHz}$
Fundamental sample time	Td_{ctrl}	$1 \times 10^{-6}\,\mathrm{s}$
Gate-source voltage	V_{gs}	$-5/18\,\mathrm{V}$
ON gate resistor	R_{gs_on}	$10\,\Omega$
Off gate resistor	R_{gs_off}	$8.2\,\Omega$
Dead Time	T_d	$500\,\mathrm{ns}$
DC-Bus voltage	V_{dc}	$850\,\mathrm{V}$
DC-link cap capacitance	C	$500\,\mu\mathrm{F}$
Load inductor	L	$100\,\mu\mathrm{H}$
Load resistor	R	$0.6\,\Omega$
Coolant temperature	$T_{coolant}$	$65\,^{\circ}\mathrm{C}$

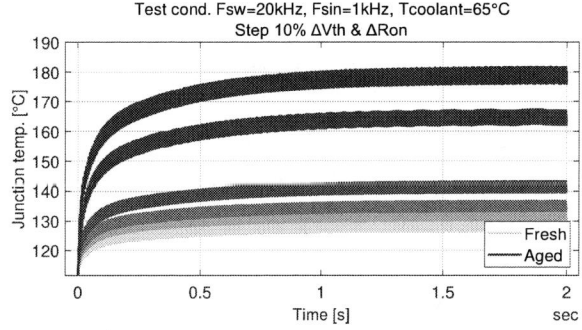

Fig. 7: Simulated junction temperature during the single-phase full-bridge inverter operation with different transistors ageing levels.

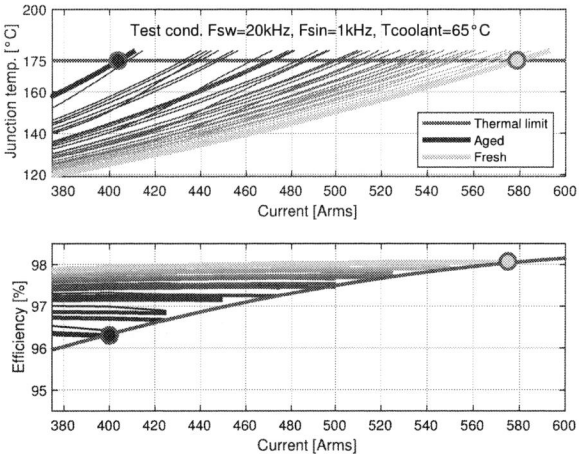

Fig. 8: Simulated junction temperature (top) and circuit efficiency (bottom) vs the output current of the single-phase full-bridge inverter as a function of different transistor ageing levels.

impact of transistor aging on the T_j, which increases with the MOSFET degradation. Fig. 8 (top) reports T_j as a function of the inverter output current for different levels of transistor aging. In particular, for a fixed current the T_j significantly increases with the level of degradation, making the transistor more prone to premature failures. On the contrary, the circuit efficiency is not particularly affected by transistors aging, $\sim 1.5\,\%$ of reduction by considering a $50\,\%$ of drift for both R_{ds_on} and V_{th} (Fig. 8, bottom).

Finally, Fig. 9 shows the impact of transistor aging (ΔR_{ds_on} and ΔV_{th}) on the maximum output current that can be supplied by a single-phase full-bridge inverter (simulated by Simulink), without exceeding the maximum junction temperature of the transistors, $i.e.$, $175\,°C$. As already mentioned, it is possible to observe that the variation of R_{ds_on} on (or K) has a larger impact on the degradation performance of the MOSFET compared to the variation of V_{th}.

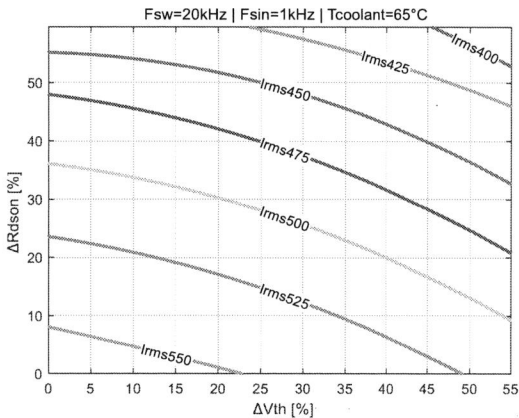

Fig. 9: Thermally limited ($175\,°C$) aging curves for different output currents of the Simulink-simulated single-phase full-bridge inverter.

VI. CONCLUSION

A Matlab-based approach able to build the transistor I-V-T maps (lookup tables), including the operating regions not reported in the datasheet, has been implemented. It turned out to be accurate as the adoption of the maps in the Simscape simulation highlighted similar switching losses reported in the datasheet. The proposed model allows power circuits simulations in Simulink environment by taking into account, since the early stage of design, the role of the transistor aging and temperature on the performance of the component and power circuit. The transistor aging did not show a significant impact on the circuit efficiency, rather it represents an accelerating factor for transistor failure due to temperature increase and a constraining factor for the maximum current capabilities. Finally, such an approach is computationally efficient, as the simulation time of 15 output current periods of the single-phase full-bridge takes about two minutes.

REFERENCES

[1] E. Robles, A. Matallana, I. Aretxabaleta, J. Andreu, M. Fernández, and J. L. Martín, "The role of power device technology in the electric vehicle powertrain," *International Journal of Energy Research*, vol. 46, no. 15, pp. 22222–22265, 2022.

[2] S. Yang, A. Bryant, P. Mawby, D. Xiang, L. Ran, and P. Tavner, "An industry-based survey of reliability in power electronic converters," *IEEE Transactions on Industry Applications*, vol. 47, no. 3, pp. 1441–1451, 2011.

[3] I. Husain, B. Ozpineci, M. S. Islam, E. Gurpinar, G.-J. Su, W. Yu, S. Chowdhury, L. Xue, D. Rahman, and R. Sahu, "Electric drive technology trends, challenges, and opportunities for future electric vehicles," *Proceedings of the IEEE*, vol. 109, no. 6, pp. 1039–1059, 2021.

[4] A. Elasser and T. Chow, "Silicon carbide benefits and advantages for power electronics circuits and systems," *Proceedings of the IEEE*, vol. 90, no. 6, pp. 969–986, 2002.

[5] J. Wang and X. Jiang, "Review and analysis of sic mosfets' ruggedness and reliability," *IET Power Electronics*, vol. 13, no. 3, pp. 445–455, 2020.

[6] S. Dusmez, H. Duran, and B. Akin, "Remaining useful lifetime estimation for thermally stressed power mosfets based on on-state resistance variation," *IEEE Transactions on Industry Applications*, vol. 52, no. 3, pp. 2554–2563, 2016.

[7] Z. Ni, X. Lyu, O. P. Yadav, B. N. Singh, S. Zheng, and D. Cao, "Overview of real-time lifetime prediction and extension for sic power converters," *IEEE Transactions on Power Electronics*, vol. 35, no. 8, pp. 7765–7794, 2020.

[8] S. Pu, F. Yang, B. T. Vankayalapati, and B. Akin, "Aging mechanisms and accelerated lifetime tests for sic mosfets: An overview," *IEEE Journal of Emerging and Selected Topics in Power Electronics*, vol. 10, no. 1, pp. 1232–1254, 2022.

[9] T. Santini, S. Morand, M. Fouladirad, L. Phung, F. Miller, B. Foucher, A. Grall, and B. Allard, "Accelerated degradation data of sic mosfets for lifetime and remaining useful life assessment," *Microelectronics Reliability*, vol. 54, no. 9, pp. 1718–1723, 2014. SI: ESREF 2014.

[10] A. Wintrich, U. Nicolai, W. Tursky, and T. Reimann, *Application Manual Power Semiconductors*. SEMIKRON International GmbH, 2 ed., 2015.

[11] L. Ceccarelli, A. Bahman, and F. Iannuzzo, "Impact of device aging in the compact electro-thermal modeling of sic power mosfets," *Microelectronics Reliability*, vol. 100-101, p. 113336, 2019.

[12] STMicroelectronics, "ADP480120W3-L: Automotive-grade acepack drive power module." https://www.st.com/en/power-modules-and-ipm/adp480120w3-l.html.

[13] Li, Jinyuan, Cui, Meiting, Du, Yujie, Ke, Junji, and Zhao, Zhibin, "Influence of parasitic inductances on switching performance of sic mosfet," *E3S Web Conf.*, vol. 64, p. 04005, 2018.

[14] TOISHIBA, "Electrical characteristics of mosfets (dynamic characteristics ciss/crss/coss)." https://toshiba.semicon-storage.com/eu/semiconductor/knowledge/faq/mosfet/electrical-characteristics-of-mosfetsdynamic-characteristics-cis.html.

[15] M. März and P. Nance, "Thermal modeling of power electronic systems," *Infineon Technologies AG Munich*, vol. 2, 2000.

[16] T. H. N. Yuan Taur, *Fundamentals of Modern VLSI Devices*. Cambridge University Press, 2 ed., 2013.

[17] T. Patel, "Comparison of level 1, 2 and 3 mosfet's," 12 2014.

[18] Mathworks, "Simscape, n-channel mosfet." https://www.mathworks.com/help/sps/ref/nchannelmosfet.html.

[19] A. S. Sedra and K. C. Smith, *Microelectronic Circuits*. Oxford University Press, fifth ed., 2004.

Exploring the Impact of Implant Temperature and a Novel Aluminum Ion Source on the Electrical Performance of 4H-SiC PiN Diodes

Justin Lynch[a], Ryota Wada[b], Nobuhiro Tokoro[b], Takashi Kuroi[b], and Woongje Sung[a]

[a] College of Nanotechnology Science and Engineering, University at Albany, State University of New York, Albany, NY 12203 USA

(jmlynch@albany.edu)

[b] Nissin Ion Equipment Co., LTD., Kyoto, Japan

Abstract— **This work compares the electrical performance of 4H-SiC PiN diodes fabricated using various P+ ion implantation conditions. For optimal device electrical performance, the use of high temperature implantation is necessary for high energy and high dose implants to achieve full lattice damage recovery. Furthermore, a new Aluminum (Al) ion source is introduced, significantly improving ion beam current, reducing beam glitches, and increasing source lifetime. It is noteworthy that the new ion source unintentionally introduces a small fraction of Chlorine (Cl) doping to the Al implants. The electrical performance of the fabricated PiN diodes using this new ion source is also examined to validate its effectiveness and understand the impact of Cl on device performance. Preliminary electrical results indicate that the use of the new ion source has no discernible effect on the static electrical performance of the fabricated diodes.**

Keywords—4H-SiC, ion implantation, PiN diodes

I. INTRODUCTION

The use of Silicon Carbide (SiC) as a power device substrate is becoming increasingly preferred over its Silicon (Si) counterpart. Due to its inert material properties, including a wide bandgap and large thermal conductivity, SiC power devices can offer robust and reliable operation with decreased conduction losses for applications 600V and above [1,2]. As SiC devices are prevalent due to the potential improvements they offer, there are still challenges that must be overcome to improve device efficiency and fabrication techniques. One key area of SiC device fabrication is ion implantation.

Ion implantation is crucial in fabricating SiC and other wide bandgap material power devices. Due to the low diffusivity of dopant atoms in SiC, ion implantation is used to create the necessary junctions for power device operation [3]. Ion implantations must be conducted to selectively form both high doping concentration and deep junctions for effective and efficient device operation. These deep and high concentration junctions have recently been shown to significantly improve device reliability and performance [4,5]. To form these junctions, the use of high energy and high dose implantation is traditionally required and must be performed at elevated temperatures to ensure a reduction in basal plane dislocation formation and proper device performance [6,7]. It is imperative to investigate further the effects of ion implantation temperature on device performance, as it is a cause of increased fabrication costs and the need for specialized high temperature capable ion implantation tools.

In addition to the temperature control ion implanters must have, given the current growing demand for wide bandgap power devices, it is vital to have reliable and high-current ion beam sources to enhance throughput and reduce fabrication costs. To address this need, Nissin Ion Equipment Co., LTD. is introducing the Indirect Hot Cathode-Flow Through Source (IHC-FTS) as a new ion source for Al ion implantations (see Fig. 1), the most commonly used p-type dopant for SiC. This innovative ion source offers an industry-leading Al beam current greater than 7mA, significantly reduces extraction glitches common in traditional Al ion sources, and dramatically improves the source lifetime to over 600 hours. An unintended consequence of this new source is the unintentional addition of a very low concentration of ~100ppm Cl to the Al implant.

In this work, two separate investigations of the effect of ion implantation variations on device performance are presented. First, given the limited available literature on the impact of Cl addition on power device performance, PiN diodes were fabricated with Cl intentionally added to the P+ anode to verify the use of the new IHC-FTS ion source for future power device fabrication. Additionally, a look at the impact of implantation temperature on high dose and high energy implanted PiN diodes is presented. It has been shown that high temperature (500°C – 600°C) is beneficial for these implants. Still, this work offers an investigation for the use of "medium" temperature (300°C) ion implantations, which provide the potential reduction of the fabrication thermal budget.

II. PROFILE DESIGN AND IMPLANT CONDITIONS

To investigate the effect of ion implantation variations on device performance, PiN diodes were fabricated on 4H-SiC substrates. To further explore the impact of Cl on device performance, PiN diodes were fabricated with aluminum implants to form the P+ anodes and implanted with the Gradient aluminum profile, which was simulated with SYNOPSYS Process simulations and is shown in Fig. 2. An overview of the PiN diode splits used for this investigation is shown in Table 1. Two ion sources were used for this ion implantation: the traditional ion source (without Cl) to form the w/o Cl_standard source diode split and the new IHC-FTS source (with unintentional Cl introduction) to create the Unintentional Cl_FTS split. In addition, diodes were also fabricated using the same gradient aluminum profile but with the addition of Chlorine implants at higher concentrations than the unintentionally added chlorine, to thoroughly investigate the effect of Cl on the device. The added chlorine implants

979-8-3503-3714-3/23 $31.00 © 2023 IEEE

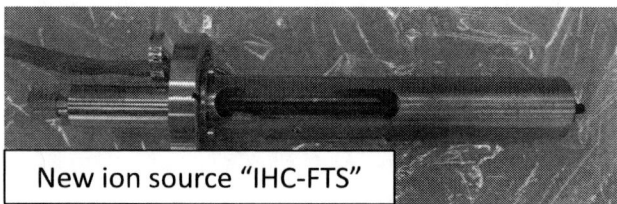

New ion source "IHC-FTS"

Figure 1: Image of the new ion source.

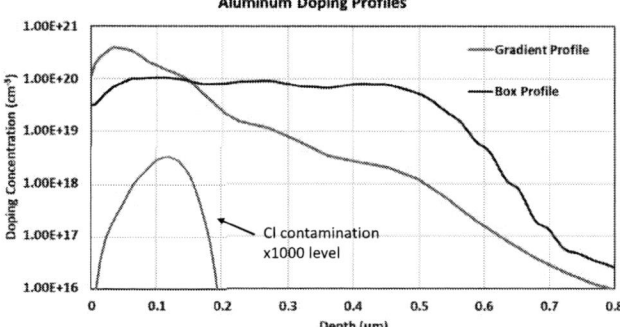

Figure 2: Doping profiles for the P+ anode of the fabricated PiN diodes. The Gradient profile was used to investigate the effect of Cl on device performance. The total dose for both implantation profiles shown was $4*10^{15}$cm^{-2}.

PiN Split Nickname	Implanted Aluminum Condition
w/o Cl_standard source	Conventional ion souce. No chlorine contamination.
Unintentional Cl_FTS	New IHC-FTS ion source. Unintentinal chlorine addition.
x10 Cl	Conventional ion souce. Intentinal chlorine addition 10 times more than Unintentional Cl_FTS split.
x1000 Cl	Conventional ion souce. Intentinal chlorine addition 1000 times more than Unintentional Cl_FTS split.

Table 1: Description of PiN diode splits used to investigate the effect of chlorine on device performance.

were conducted to add a Cl profile ten times larger (10x) and a thousand times larger (1000x) than what was found in the Unintentional Cl_FTS split. The 1000x Cl profile can be seen in Fig. 2, with a peak doping concentration of approximately $3*10^{18}$cm^{-3} just below the SiC surface. This chlorine profile's depth and concentration were verified using SIMS.

For the investigation of the effect of ion implantation temperature on device performance, aluminum was implanted as a p-type dopant to form the P+ anodes of the fabricated PiN diodes. Two aluminum profiles were simulated with SYNOPSYS Process simulations, as shown in Fig. 2, and were used for the P+ anode profiles of the diodes. The names of the profiles, Gradient and Box, refer to their shape. The implant temperature was varied between room temperature (RT), 300°C, and 500°C. Typical ion implantations are performed at 500°C - 600°C for commercial SiC fabrication. Two distinct shapes can be seen for the profiles shown in Fig. 2. For the Gradient profile, an extremely high surface concentration is present, followed by a steady decline of aluminum

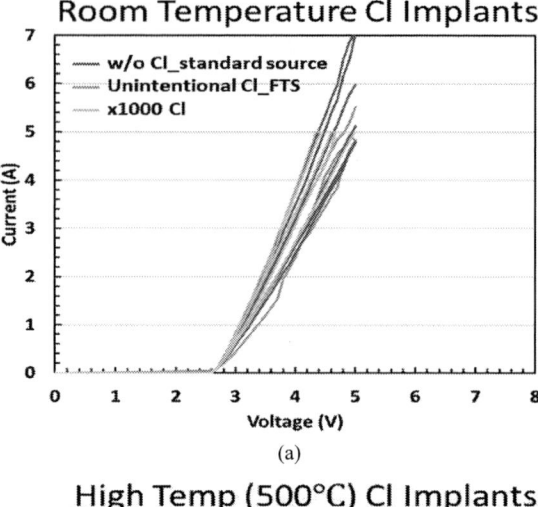

(a)

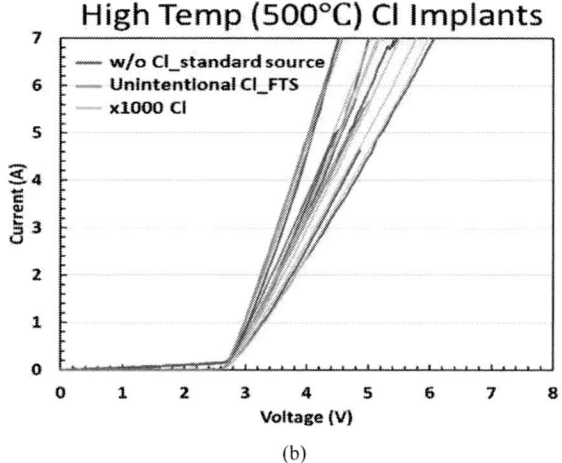

(b)

Figure 3: Forward IV characteristics of the PiN diodes fabricated with various Cl conditions at both (a) room temperature (RT) and (b) high temperature (500°C). The 10x Cl introduced case is not shown but exhibits similar characteristics.

concentration, resulting in a gradient shape. This is a typical implant profile in SiC power devices and allows for the formation of superior ohmic metal contacts on the surface of the SiC substrate. In the other profile present, the Box aluminum profile, a consistently high doping concentration of aluminum, around $1*10^{20}$cm^{-3}, is present from the wafer surface to a considerable depth of 0.5μm. High energy and high dose implantations must be conducted to create this profile shape to provide aluminum at increased depths. As previously mentioned, these implants are known to induce damage and defects into the SiC substrate, which can negatively affect device performance if not cured.

III. ELECTRICAL RESULTS

A. Chlorine Investigation

Fig. 3 illustrates the forward IV characteristics of PiN diodes fabricated with the improved IHC-FTS aluminum ion source (Unintentional Cl_FTS), with the traditional aluminum ion source (w/oCL_standard source), and the conventional ion source with the addition of 1000x Chlorine at RT (Fig. 3a) and

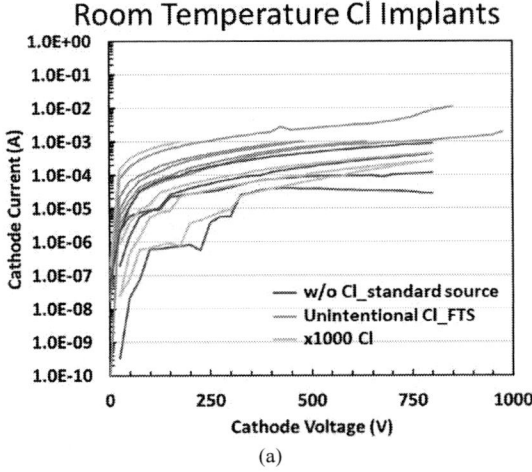

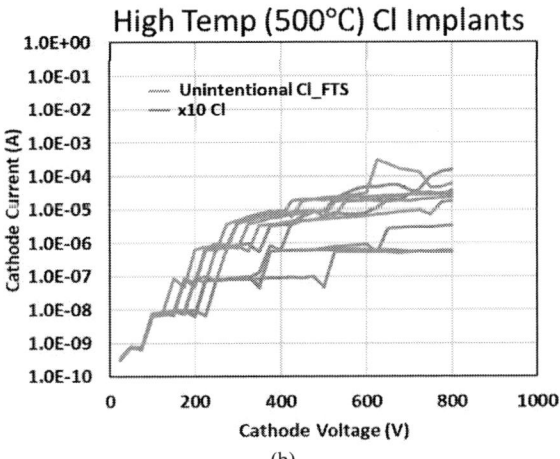

Figure 4: Blocking characteristics of the PiN diodes fabricated with various Cl conditions at both (a) room temperature and (b) high temperature (500°C).

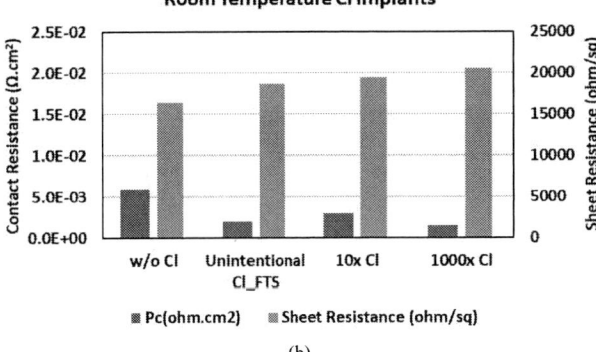

Figure 5: Contact and sheet resistance values for the various Cl implantations at (a) 500°C and (b) RT.

500°C (Fig. 3b). The figures demonstrate that the unintentional addition of Cl from the new IHC-FTS source, as well as the intentional addition of Cl up to x1000 compared to FTS, does not affect device performance. Similarly, the blocking characteristics at both RT (Fig. 4a) and 500°C (Fig. 4b) show that introducing Cl does not significantly impact performance. It is worth noting that at RT, the devices exhibit considerably higher leakage current compared to 500°C. The contact and sheet resistance values obtained from TLM structures, utilizing Al implantation with varying levels of introduced Cl, are displayed at 500°C (Fig. 5a) and RT (Fig. 5b). Once again, the introduced Cl has a negligible effect on the resistance values shown. Initial electrical measurements indicate that the introduction of Cl from the IHC-FTS Al ion source does not significantly affect the electrical performance of the fabricated devices.

B. Implant Temperature and Profile Investigation

In addition to investigating the effect of Cl on device performance, we also examined the impact of ion implant temperature on device performance using the Gradient and Box aluminum profiles implanted at various temperatures. For the Gradient aluminum profiles, both the forward IV characteristics and blocking characteristics (Fig. 6a & Fig. 6b) show the slight benefit of using elevated temperatures during ion implantation. When a high temperature implant is used, the forward voltage drops of the diodes are lower, and the leakage current in the blocking characteristics is decreased. This indicates that though the SiC cannot fully recover with RT ion implantation, proper device performance can still be achieved with the hill shape profile. For the Box aluminum profiles, both the forward IV characteristics and blocking characteristics (Fig. 7a & Fig. 7b) show significant improvement in device performance when using elevated temperatures during ion implantation. For the RT ion implanted Box aluminum profile diodes, a substantial increase in forward voltage drops is seen compared to the 300°C and 500°C implanted diodes. Additionally, no blocking capability is demonstrated by the room temperature implanted Box aluminum profile diodes. These fabricated devices demonstrate the advantages of implementing high-dose and high-energy implants, the implants required to form the box profile at elevated temperatures, as evidenced by lower forward voltage drops and leakage current. From the electrical results, despite the similar forward operation of diodes implanted at 300°C compared to the 500°C case, a slightly higher leakage current is present during the off-state of the diodes. Elevated temperatures of at least 500°C provide the best performance of high energy and

979-8-3503-3714-3/23 $31.00 © 2023 IEEE   187

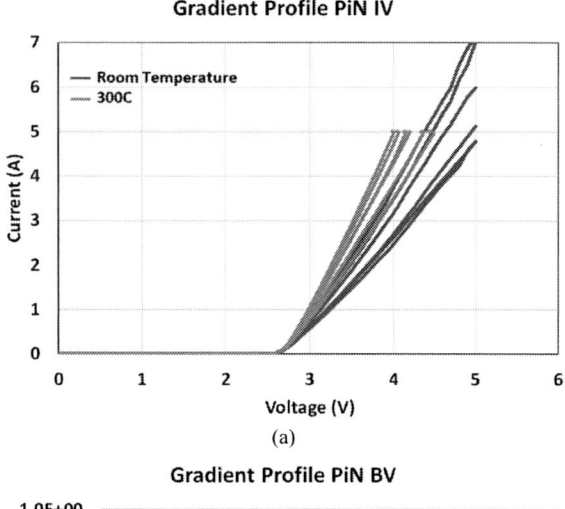

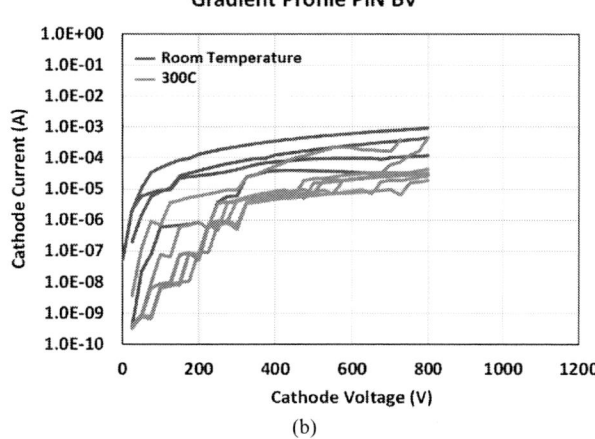

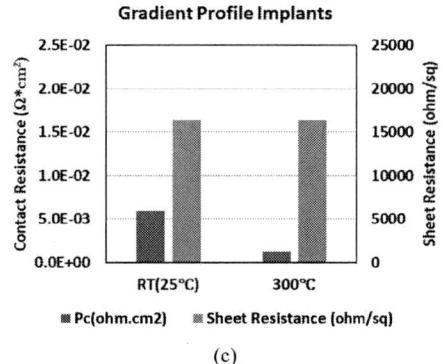

Figure 6: Forward IV curves (a), blocking characteristics (b), contact and sheet resistance values (c) of PiN diodes fabricated with the Gradient aluminum profile at various temperatures.

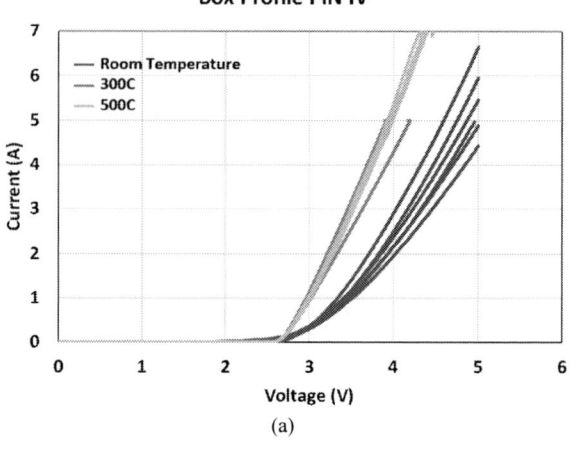

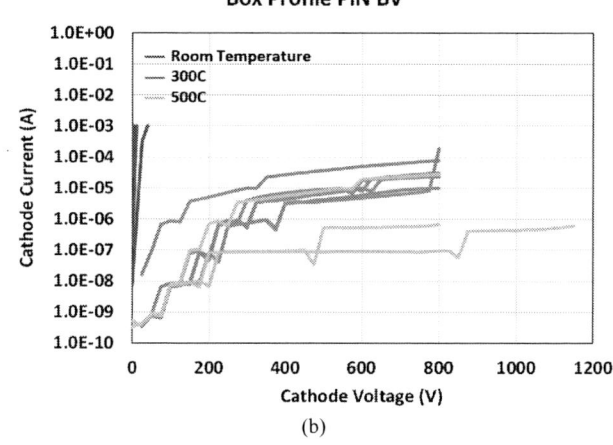

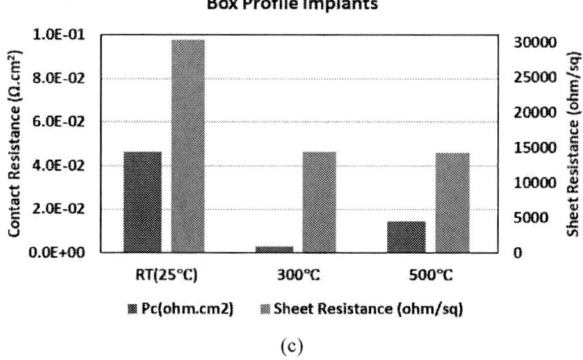

Figure 7: Forward IV curves (a), blocking characteristics (b), contact and sheet resistance values (c) of PiN diodes fabricated with the Box aluminum profile at various temperatures.

high dose ion implants. The contact and sheet resistance values obtained from TLM structures utilizing the Gradient and Box aluminum implantations at various temperatures are displayed in Fig. 6c and Fig. 7c. These results again reaffirm that for the box profile, the use of elevated implantation temperatures is needed as a significant reduction of sheet resistance and contact resistance is shown for the 300°C and 500°C profiles. Interestingly, for the Gradient aluminum profile, which does not use high dose and high energy implantations, no sheet or contact resistance improvement is seen with the use of 300°C

compared to RT, indicating implant temperature has a limited effect on implant profiles that do not use high energy and high dose implants.

IV. CONCLUSIONS

This work compared the electrical performance of 4H-SiC PiN diodes fabricated under various P+ implant conditions. First, a new Aluminum (Al) ion source, the IHC-FTS, is introduced with the potential for improving ion implanter tool operation. Devices fabricated using this new ion source were demonstrated, showing the potential of using an ICH-FTS source in power device fabrication. Additionally, the impact of chlorine on SiC PiN diode device was presented, indicating that the small Chlorine profile unintentionally implanted with this new source does not affect device performance. Lastly, a comparison of PiN diodes fabricated with the Gradient and Box aluminum profiles was presented, showing the impact of ion implantation temperature on fabricated device performance. Electrical results indicate that high temperature implantation is beneficial for device performance of standard power device implant profiles and is necessary for high-energy and high-dose implants to achieve full lattice damage recovery and efficient device operation.

ACKNOWLEDGMENT (Heading 5)

The authors would like to acknowledge the help of Emran Ashik and Dr. Bongmook Lee for their contributions and help to complete this work.

REFERENCES

[1] B. J. Baliga, Fundamentals of Power Semiconductor Devices. New York, NY: Springer, 2008

[2] J. S. Glaser, J. J Nasadoski, P. A. Losee, A. S. Kashyap, K. S. Matocha, J. L. Garrett, and L. D. Stevanovic, "Direct comparison of silicon and silicon carbide power transistors in high-frequency hard-switched applications," 2011 Twenty-Sixth Annual IEEE Applied Power Electronics Conference and Exposition (APEC), Fort Worth, TX, USA, 2011, pp. 1049-1056, doi: 10.1109/APEC.2011.5744724.

[3] T. Kimoto, " Material science and device physics in SiC technology for high-voltage power devices." Japanese Journal of Applied Physics, vol. 54, no. 4, 2015, p. 040103, https://doi.org/10.7567/jjap.54.040103.

[4] D. Kim and W. Sung, "Improved Short-Circuit Ruggedness for 1.2kV 4H-SiC MOSFET Using a Deep P-Well Implemented by Channeling Implantation," in IEEE Electron Device Letters, vol. 42, no. 12, pp. 1822-1825, Dec. 2021, doi: 10.1109/LED.2021.3123289.

[5] D. Kim, N. Yun, A. J. Morgan, and W. Sung, "The Effect of Deep JFET and P-Well Implant of 1.2kV 4H-SiC MOSFETs," in IEEE Journal of the Electron Devices Society, vol. 10, pp. 989-995, 2022, doi: 10.1109/JEDS.2022.3218689.

[6] A. Agarwal, H. Fatima, S.Haney, S.-H. Ryu, "A New Degradation Mechanism in High-Voltage SiC Power Mosfets," IEEE Electron Device Letters vol. 28, no. 7, pp.587-589, 2007. doi.: 10.1109/LED.2007.897861.

[7] S. A. Mancini, S. Y. Jang, D. Kim, and W. Sung, "Static Performance and Reliability of 4H-SiC Diodes with P+ Regions Formed by Various Profiles and Temperatures," 2022 IEEE International Reliability Physics Symposium (IRPS), 2022, pp. P62-1-P62-6, doi: 10.1109/IRPS48227.2022.9764538.

Multi-MHz Auto-Resonant Power Oscillator in a 650 V GaN-on-SOI Technology for Compact Wireless Power Transfer Systems

Manuel Rueß, Dominik Koch and Ingmar Kallfass

Institute of Robust Power Semiconductor Systems, University of Stuttgart, Germany

(manuel.ruess@ilh.uni-stuttgart.de)

Abstract—In this research, we demonstrate a 60 W, 60 V, 3 MHz **auto-resonant integrated power oscillator implemented on IMEC's 650 V GaN-on-SOI MPW process. The monolithic integrated circuit, with dimensions of 2.5 × 2.5 mm², incorporates the switching cell of the so-called Royer-Circuit which can be used both in transmitter (inverter) and receiver (rectifier) circuits for WPT systems. Designed for power levels of 100 W, the chip can handle maximum input voltages up to 150 V and switching frequencies in a single-digit MHz range. The integrated design as well as MHz switching frequencies allow for a very compact and simple WPT system. This paper describes the operation and functionality of the Royer converter including the processed GaN-Royer IC and addresses important aspects of the integration of the circuit. The operation and performance of the designed IC is demonstrated in operation at resonant frequencies of up to 3 MHz and output powers up to 100 W at an input voltage of maximum 60 V. The whole WPT system, including a Royer converter as transmitter and receiver reaches efficiencies between DC in and DC out up to $\eta = 90\%$. The limitations and further optimizations for such a converter system are discussed at the end of the paper.**

Index Terms—power electronics, wireless power transfer, gallium nitride, wide-bandgap semiconductors, DC-DC power converters, inductive charging, resonant charging, energy transmission

I. INTRODUCTION

Wireless power transmission (WPT) is increasingly gaining interest in battery-powered applications such as in the field of electrical vehicles, medical implants, robotics or mobile phones. In applications where space is limited, the so-called Royer converter offers a cost-effective and very efficient solution. This converter is an auto-resonant topology, which operates in its own resonant frequency and converts DC voltage into AC voltage. This circuit was developed and first patented by George Howard and Richard Louis Bright in 1954 [1]. In [2], this principle is further improved by an optimized driving of the power transistors. Due to the auto-resonant concept, the Royer converter does not require any additional control or drive circuitry. In addition, such a converter topology is bidirectional and can be used as an inverter or rectifier. The power class is scalable by appropriate selection of the required power transistors and is typically used at resonant frequencies in the range of several 100 kHz [3].

In this paper, the version presented in [2] is even further improved by using gallium nitride (GaN) high electron mobility transistors (HEMTs). This allows to further increase

the achievable resonant frequencies and efficiencies of the Royer circuit [4]. Especially in resonant topologies which are used in WPT systems, GaN HEMTs show their full potential and outperform the counterpart sillicon (Si) metal oxide semiconductor field-effect transistors (MOSFETs) by a significant extent [5], [6]. Additional, the core of the circuit consisting of four transistors is further being integrated onto a single chip using IMEC's 650 V GaN-on-SOI (sillicon-on-insulator) technology [7]. This enables further improvement by increasing integration density and performance of the Royer converter. With these improvements, the overall losses of the system are reduced, while the achievable switching frequencies can be increased even further. The higher frequencies allow the required passive components to be reduced, which additionally increases the power density. In the first chapter, the properties and functionality of the Royer circuit are described. Following this, the GaN-Royer integrated circuit (IC) is introduced and important aspects in the design of the circuit are addressed. The functionality is demonstrated in a WPT system consisting of two Royer converters. We further investigate the performance of the system at various resonant frequencies and output powers. Finally, limitations and possible optimization potentials of the GaN-Royer IC are discussed.

II. ROYER CIRCUIT AND MODE OF OPERATION

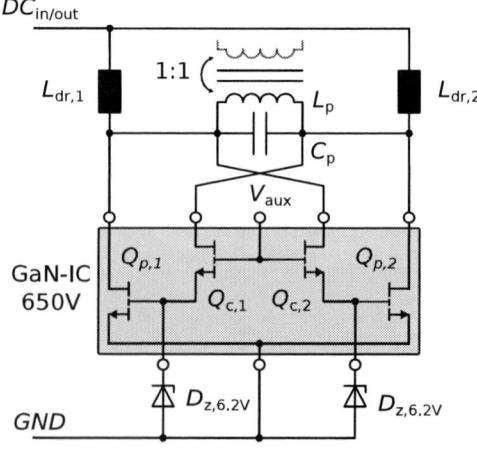

Fig. 1. Electrical schematic of a Royer converter acting as an inverter. The monolithic integrated part of the presented IC is marked.

The electrical structure of the applied circuit is shown in figure 1. The circuit consists of a resonant tank including the coil L_p, the capacitor C_p, the control transistors $Q_{c,1/2}$, the power transistors $Q_{p,1/2}$, and the chokes $L_{dr,1/2}$.

The coil L_p acts as the primary side of the transformer system and forms an oscillating circuit with the capacitor C_p. The resonant frequency $\omega_0 = 1/\sqrt{L_p \cdot C_p}$ of this parallel oscillator is also the transmission frequency of the WPT system. The power transistors and their control transistors build the core of the circuit. Here, $Q_{p,1}$ and $Q_{p,2}$ alternately carry the active current, which is transmitted to the receiver side. The control transistors allow an efficient gate drive of the power transistors and secure the gates from higher voltages. $L_{dr,1/2}$ form constant current sources when an input voltage $V_{DC,in}$ is applied. To be able to guarantee this, they must be a minimum of ten times larger than L_p [3].

During operation, a sinusoidal voltage V_{Lp} is formed. The maximum amplitude $\hat{v}_{Lp}$ is a π-fold of $V_{DC,in}$ [3]. During switching, a half-wave with corresponding amplitude is applied across the power transistors, depending on which transistor is active. Therefore, the breakdown voltage V_{bds} of all four transistors must be higher than the maximum amplitude $\hat{v}_{Lp}$. To protect the gates of the power transistors against the high voltages, control transistors are used. A constant voltage V_{aux} is required at the gate of $Q_{c,1}$ and $Q_{c,2}$. In the voltage-free state, they are permanently switched on. When the voltage at the drain rises they slowly close their channel and disconnect the power transistor gate from the resonant tank. V_{aux} can be provided from $V_{DC,in}$ using a simple Zener diode circuit. The voltage at the gate of the power transistors depends directly on the threshold voltage of the control transistors $V_{th,c}$ and the applied voltage V_{aux}. As a result of the cross-coupling, it is guaranteed that only one of the power transistors is active at any time of operation. Switching will always take place close to the zero-crossing of the sinusoidal wave. Therefore, the Royer converter uses zero-voltage switching (ZVS). A more detailed analytical discussion of the circuit can be found in [8].

III. GaN Royer IC Design

The improved Royer circuit in [2] is using discrete MOSFETs as switches. This solution is very efficient at low frequencies in the kHz range [8]–[10]. At higher frequencies in the MHz range, MOSFETs reach their limits due to their threshold voltage and parasitic capacitances. By using discrete GaN HEMTs with comparatively low threshold voltage and parasitic capacitances, the circuit can achieve higher frequencies with similar efficiency ratings [4].

In this paper, an attempt was made to integrate the core of the circuit, consisting of the four transistors $Q_{p,1/2}$ and $Q_{c,1/2}$, into a single die. On the one hand, integration of the circuit on a single chip can greatly reduce unwanted parasitic elements. For example, parasitic inductances in the gate circuit can have a strong influence on the switching behavior at higher frequencies. On the other hand, the complexity of the entire converter can be further reduced by using a single chip. In

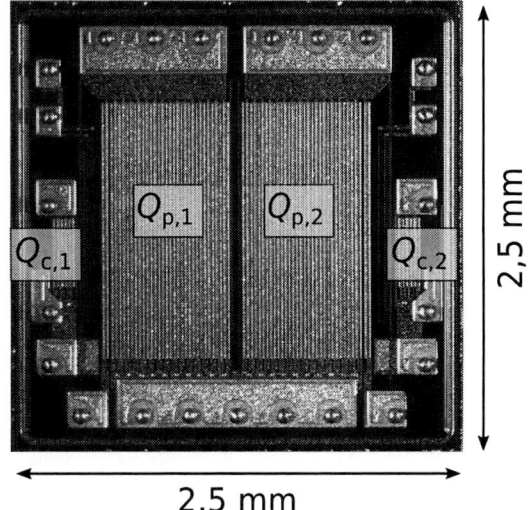

Fig. 2. Design of the introduced and fabricated GaN-Royer IC, which contains the power transistors and control transistors.

figure 2, we present the processed $2.5 \times 2.5\,\mathrm{mm}^2$ GaN-Royer IC, which comprises the power transistors in a common source topology and the control transistors connected directly to the gates of the power transistors. This IC was fabricated by IMEC using their 650 V GaN-on-SOI multi-project wafer (MPW) process [7]. The devices used in this design are p-GaN HEMTs operating in enhancement mode (e-mode). The insulating SOI layer enables a direct connection of the substrate to the heat sink, which allows an optimal thermal management. In order to cover a wide input voltage range up to $V_{DC,in} = 150\,\mathrm{V}$ as well as to protect the device from possible voltage overshoots that may occur during power-up, a breakdown voltage of $V_{bds} = 650\,\mathrm{V}$ has been selected.

For the dimensioning of the transistors in a Royer circuit there are following aspects which have to be taken into account:

- $R_{DS,on}$ **of the power transistors:** Responsible for the conduction losses in the Royer system depending on the load current. Due to the cross-coupling, the voltage drop must not be greater than the threshold voltage of the power transistor. If the voltage drop is too large, the opposite power transistor is switched on.
- $R_{DS,on}$ **of the control transistors:** The control transistors charge or discharge the gate of the power transistors. The speed is influenced by their $R_{DS,on}$.
- **Capacitive voltage divider at the power gate:** The drain-to-source capacitance $C_{oss,c}$ of the control transistor and the gate-to-source capacitance $C_{iss,p}$ of the power transistor form a capacitive voltage divider. The voltage at the gate of the power transistor increases depending on the voltage in the resonant circuit.

The aim of this design is to achieve a power class of up to 100 W and possible switching frequencies of up to 5 MHz. The data of the respective transistors are listed in table I. The

TABLE I
MEASURED STATIC TRANSISTOR PARAMETERS IN THE PRESENTED GaN-ROYER IC

Transistor	Q_p	Q_c
V_{bds}	650 V	650 V
V_{th}	1.8 V	2.1 V
$R_{DS,on}$ [1]	0.36 Ω	13 Ω
C_{iss} [2]	64.2 pF	3.8 pF
C_{oss} [2]	24.3 pF	3.2 pF
C_{rss} [2]	2.1 pF	1.4 pF

[1] measured at $V_{GS} = 6\,V$
[2] measured at $V_{DS} = 400\,V$

power class is directly connected to the $R_{DS,on}$ and the V_{th} of the power transistors. The corresponding size of $Q_{p,1/2}$ allows to scale the power class of the IC. Therefore a resistance of $R_{DS,on} = 0.36\,\Omega$ and $V_{th} = 1.8\,V$ allows load currents of up to 3 A. In this case, a safety margin must be subtracted to ensure that the voltage drop across the power transistor is still small enough to guarantee safe operation even if the resistance increases due to rising temperature, for example. The control transistors were connected directly to the gate of a power transistor to minimize the parasitic inductance. The gate can additionally be contacted externally via a pad, for example to limit the voltage with the help of a Zener diode in order to better protect the power transistor. Here, the additional capacitance helps to reduce the effect of the capacitive voltage divider. The cross-coupling connection between the two halves has to be implemented externally on the circuit carrier. This can be realized in the first two metall layers of the PCB.

IV. ROYER WPT SYSTEM - FUNCTIONALITY AND PERFORMANCE TESTS

The functionality of the IC is tested using a WPT system. The setup can be seen in figure 3. Here, one Royer circuit acts as an inverter and one as a rectifier. Two PCB coils are used to transfer the energy over a distance of 2 cm. The design parameters of the WPT system can be found in table II. The IC is mounted on the backside of the board and is directly connected to a heat sink. This backside mounting solution allows to place the resonant capacitance directly on top of the chip to minimize parasitic inductances. Discrete Zener diodes

TABLE II
DESIGN PARAMETERS OF THE USED WPT SYSTEM INCLUDING A ROYER CONVERTER AS TRANSMITTER AND RECIEVER

Parameter	Values
Input voltage	$V_{DC,in} = 60\,V$
Auxillary voltage	$V_{aux} = 7\,V$
Input Chokes	$L_{dr} = 47\,\mu H$
Resonance coil	$L_p = 3.84\,\mu H$
Resonance capacitance:	C0G/NP0 500 V
@ 1.8 MHz	3*560 pF
@ 2.3 MHz	2*560 pF
@ 3.1 MHz	1*560 pF
Q-factor resonance coil	Q(3 MHz) = 250
Coupling factor	k = 0.28

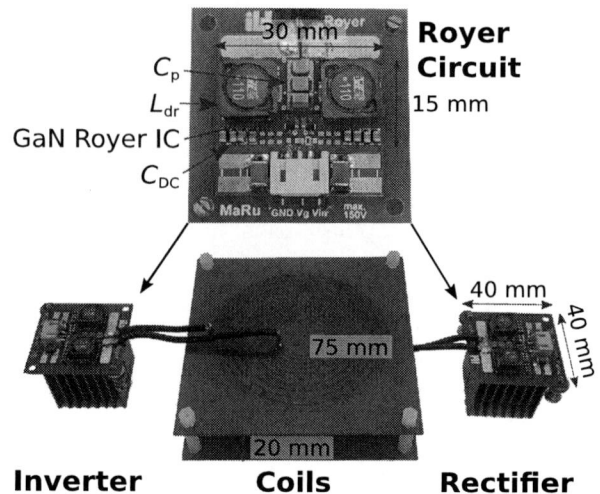

Fig. 3. Configuration of the WPT system used, consisting of two Royer systems and the coil system, which consists of PCB coils.

with a voltage of 6.2 V are added. The converter has an overall size of $4 \times 4\,cm^2$ including additional space for probes and mounting holes. The actual Royer switching cell has a size of $3 \times 1.5\,cm^2$.

Figure 4 shows and proves the operation and functionality of the GaN-Royer IC at an input voltage $V_{DC,in} = 60\,V$ and a load resistance of $R_{load} = 50\,\Omega$. With two 560 pF ceramic capacitors the resulting resonant frequency is $f_{res} = 2.139\,MHz$. The voltage curves of the drain-to-source voltages $V_{DS,Qp1/2}$ of the power transistors $Q_{p,1}$ and $Q_{p,2}$ as well as their gate-to-source voltages $V_{GS,Qp1/2}$ are shown in figure 4 for inverter and rectifier operation respectively. A maximum amplitude of $\hat{v}_{DS,Qp1/2} = 188.5\,V$ is applied. Due to the energy transfer from the primary side to the secondary side, the sinusoidal voltage is slightly deformed. This is due to the DC component of the load current in the system.

In this operation point the base of the gate voltages turns out to be $V_{GS,Qp1/2,base} = 0.55\,V$. This offset is positive for the inverter and negative for the rectifier, because the current direction is opposite. With an input current of $I_{DC,in} = 1.18\,A$, a deviation from the expected value $V_{GS,base} = R_{DS,on} \cdot I_{DC,in} = 0.420\,V$ can be seen. This indicates a higher resistance during operation, which is possible due to temperature increase or chip mounting problems. The increase of $V_{GS,Qp1/2}$ by the voltage divider is still within safe operating range and allows the gate to rise up to an voltage of $V_{GS,Qp1/2} = 5.7\,V$.

Figure 5 shows an enlarged view of a switching transition in the inverter. A delay in the falling edge of the gate voltage $V_{GS,2}$ is visible, compared to the drain voltage $V_{DS,1}$. This delay causes the gate to discharge more slowly and delays the switching operation. Additional signs of a delayed switching are the undershoot of the voltage $V_{DS,1}$ and the dip in the rising edge of $V_{DS,2}$. This leads to the fact that the switching transition does not have a pure ZVS behavior and

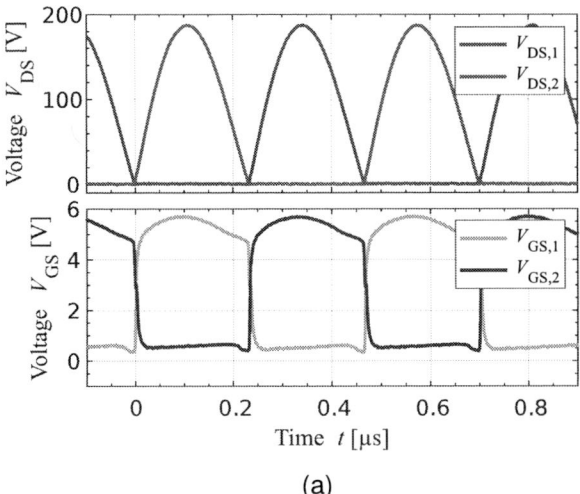

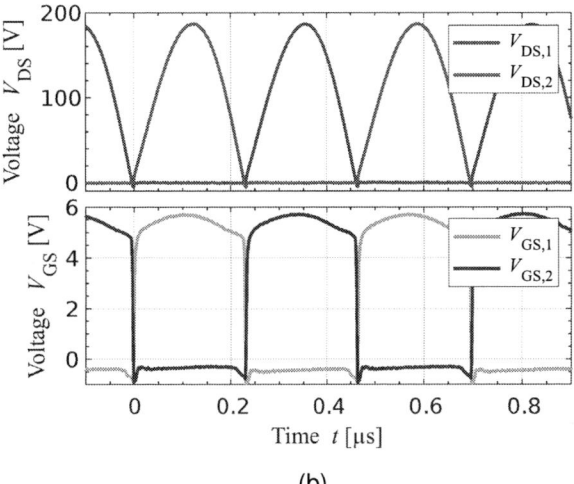

(a)　　　　　　　　　　　　　　　　　　(b)

Fig. 4. Drain-to-source voltage $V_{DS,Qp1/2}$ and gate-to-source voltage $V_{GS,Qp1/2}$ of power transistors $Q_{p,1}$ and $Q_{p,2}$ in inverter mode (a) and rectifier mode (b) at an input voltage of $V_{DC,in} = 60\,V$ a resulting resonant frequency of $2.2\,MHz$ and a load resistance of $50\,\Omega$, resulting in an output power of $P_{out} = 60\,W$.

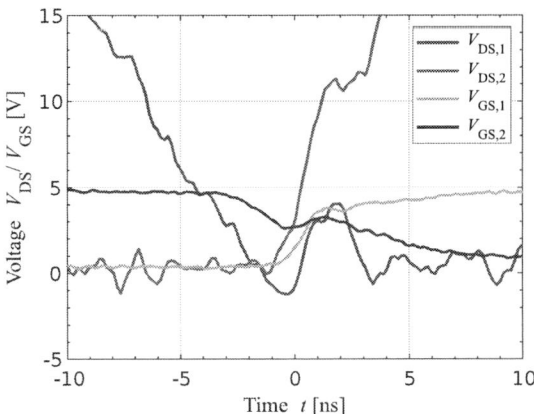

Fig. 5. Enlarged switching transition at the power transistor of the inverter at $V_{DC,in} = 60\,V$ and $f_{res} = 2.2\,MHz$.

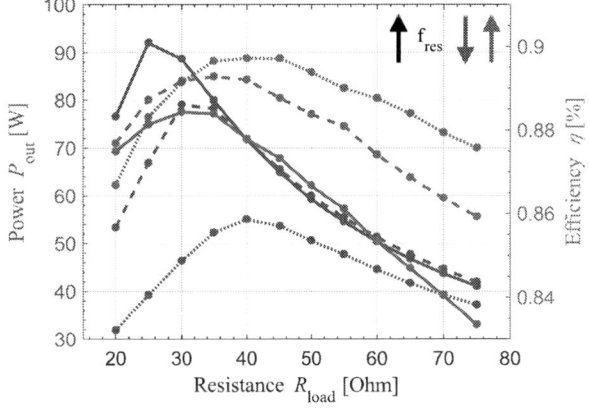

Fig. 6. Maximum transmittable power and achieved efficiency of the WPT system at different resonant frequencies and load points.

the control transistor are slowing down the operation. As a result, switching losses increase and additional losses occur in the resonant tank, since it does not oscillate at its desired resonant frequency. This effect is more noticeable at increased frequencies.

The resulting efficiencies at three different resonant frequencies are shown in figure 6. System efficiency is determined from DC in to DC out including inverter, rectifier and coil losses. The maximum power of $P_{out} = 92\,W$ could be transmitted at a frequency $f_{res} = 1.9\,MHz$ with an efficiency of $\eta = 88\,\%$, proving the ability to transfer power levels up to $100\,W$. At a frequency $f_{res} = 3.07\,MHz$ and a power level $P_{out} = 55\,W$ the system reaches the highest measured efficiency of $\eta = 90\,\%$, mainly limited by the quality factor of the used coil system. With an optimal coil design, tuned

to the desired frequency higher efficiencies and output powers are possible.

Compared to previous work, this work can be classified according to table III. Here, Royer converter systems are compared. [8], [10] are systems using Si-MOSFETs and [4] is a previous work of the authors using a combination of GaN and Si transistors. With a resonant frequency of $3.07\,MHz$ and an efficiency of $90\,\%$ at a transmitted power of $55\,W$, the performance of the converter could be greatly improved using the presented GaN-Royer IC. Compared to [4], the resonant frequency could be increased by a factor of 6 with same efficiency and output power on a $3 \times 1.5\,cm^2$ swithing cell. An example of a WPT system with a controlled resonant frequency is presented in [11]. Gu et al. achieve an efficiency of $95\,\%$ with their approach for a $6.78\,MHz$ WPT system with

979-8-3503-3714-3/23 $31.00 © 2023 IEEE

TABLE III
PERFORMANCE COMPARISON WITH ROYER-BASED WPT SYSTEMS AND A STATE-OF-THE-ART WPT SYSTEM.

WPT-System: Reference:	Si Royer [10]	High Power Royer [8]	GaN Royer [4]	Controlled Frequency [11]	This work
Input voltage $V_{DC,in}$ [V]	36	780	30	120	60
Frequency f_{reso} [MHz]	0.25	0.1	0.5	6.78	3.07
Coil L_p [µH]	13.7	618.5	10.2	0.16	3.84
Coupling k	0.33	0.35	0.4	0.15	0.28
Power P_{out} [W]	98	375	59	300	55
Efficiency η [%]	89	92.7	94.1	94.6	90

300 W transmitted power. Gate driving losses are excluded from their study. High Q-factor, Q = 700, self-oscillating coils were used here. The size of the inverter is $7 \times 7\,\text{cm}^2$. This shows that the Royer circuit with a perfectly designed coil system can compete with such a controlled switched converter. However, one of the biggest disadvantages of the Royer converter is the load-dependent resonant frequency. Depending on the load resistance, the natural frequency of the system shifts. Detailed information on this behavior is explained in more detail in [9]. For the operation of a self-oscillating Royer converter itself, the frequency change is not a problem. However, it becomes a major problem when a specific frequency band must be maintained, such as the ISM band of 6.78 MHz and 13.56 MHz that are commonly used for such resonant charging.

CONCLUSION AND FUTURE WORK

In this work, a $2.5 \times 2.5\,\text{mm}^2$ GaN Royer IC is presented, which is processed in IMEC's 650 V GaN-on-SOI technology. With this IC, a complete WPT system is setup, which includes inverter, rectifier and coil system. On a size of $3 \times 1.5\,\text{cm}^2$ the required Royer switching cell could be realized. The functionality was proven in this work up to voltages of 60 V. At a resonant frequency of 3.07 MHz a maximum efficiency of 90 % could be achieved with a transferred power of 55 W. At the same time, the frequency could be increased by a factor of 6 for the same efficiency compared to previous designs. The Royer IC thus offers a very efficient and cost-effective alternative for systems that do not have any restrictions for transmission frequencies. Especially for lower power systems with limited installation space. In the following work the optimization of the coil system as well as the comparison to typical topologies in WPT systems will be investigated in more detail.

ACKNOWLEDGMENT

This work is funded by the German Research Foundation (DFG) as part of the 3D-CeraGaN project (No.: 462828009). Furthermore, the work was supported by ASCENT+ as well as IMEC. This made the processing and fabrication of the GaN-Royer IC possible. Many thanks to the responsible persons at ASCENT+ and IMEC who supported and supervised the project during this time.

REFERENCES

[1] Bright, Richard L., and G. H. Royer. "Electrical inverter circuits." Google Patents (1957).

[2] Rehrmann, Jörg, and Bernd Scott. "MOSFET/IGBT-Oszillatorschaltung für parallelgespeiste Leistungsoszillatoren." German, DE202007011745 (2007).

[3] M. Maier, D. Maier, M. Zimmer and N. Parspour, "A novel self oscillating power electronics for contactless energy transfer and frequency shift keying modulation," 2016 International Symposium on Power Electronics, Electrical Drives, Automation and Motion (SPEEDAM), Capri, Italy, 2016, pp. 67-72, doi: 10.1109/SPEEDAM.2016.7525952.

[4] D. Koch, M. Rueß, D. Maier and I. Kallfass, "Optimization of Self-Oscillating Power Converter Based on GaN HEMTs for Wireless Power Transfer," 2021 IEEE 8th Workshop on Wide Bandgap Power Devices and Applications (WiPDA), Redondo Beach, CA, USA, 2021, pp. 164-169, doi: 10.1109/WiPDA49284.2021.9645129.

[5] D. Reusch and J. Strydom, "Evaluation of Gallium Nitride Transistors in High Frequency Resonant and Soft-Switching DC–DC Converters," in IEEE Transactions on Power Electronics, vol. 30, no. 9, pp. 5151-5158, Sept. 2015, doi: 10.1109/TPEL.2014.2364799.

[6] A. Lidow, "The Path Forward for GaN Power Devices," 2020 IEEE Workshop on Wide Bandgap Power Devices and Applications in Asia (WiPDA Asia), Suita, Japan, 2020, pp. 1-3, doi: 10.1109/WiPDAAsia49671.2020.9360274.

[7] X. Li et al., "200 V Enhancement-Mode p-GaN HEMTs Fabricated on 200 mm GaN-on-SOI With Trench Isolation for Monolithic Integration," in IEEE Electron Device Letters, vol. 38, no. 7, pp. 918-921, July 2017, doi: 10.1109/LED.2017.2703304.

[8] D. Maier, J. Noeren, N. Parspour and C. Lauer, "A Novel Power Electronics for Contactless Inductive Energy Transfer Systems," 2018 IEEE 18th International Power Electronics and Motion Control Conference (PEMC), Budapest, Hungary, 2018, pp. 40-45, doi: 10.1109/EPEPEMC.2018.8521752.

[9] D. Maier, J. Heinrich, M. Zimmer, M. Maier and N. Parspour, "Contribution to the System Design of Contactless Energy Transfer Systems," in IEEE Transactions on Industry Applications, vol. 55, no. 1, pp. 316-326, Jan.-Feb. 2019, doi: 10.1109/TIA.2018.2866247.

[10] J. Heinrich, P. Präg, N. Parspour and D. Maier, "Efficiency Factor Calculation for Contactless Energy Transfer Systems," 2019 IEEE Wireless Power Transfer Conference (WPTC), London, UK, 2019, pp. 130-135, doi: 10.1109/WPTC45513.2019.9055684.

[11] L. Gu, G. Zulauf, A. Stein, P. A. Kyaw, T. Chen and J. M. Rivas-Davila, "Design and Optimization of 6.78 MHz Wireless Power Transfer with Self-Resonant Coils," 2020 IEEE 21st Workshop on Control and Modeling for Power Electronics (COMPEL), Aalborg, Denmark, 2020, pp. 1-5, doi: 10.1109/COMPEL49091.2020.9265744.

979-8-3503-3714-3/23 $31.00 © 2023 IEEE

Ultra-Wideband Surface Current Sensor Topology for Wide-Bandgap Power Electronics Applications

Ali Parsa Sirat
Design Center
East West Manufacturing
Asheville, United States
aparsasirat@ewmfg.com

Hossein Niakan
ECE Department
University of North Carolina at Charlotte
Charlotte, United States
hniakan@charlotte.edu

Babak Parkhideh
ECE Department
University of North Carolina at Charlotte
Charlotte, United States
bparkhideh@charlotte.edu

Abstract— Through the development of wide-bandgap semiconductor devices, power electronics systems are enabled to operate at higher switching frequencies, requiring minimally invasive current sensing solutions to allow real-time monitoring and control of the system. The majority of current sensors available on a commercial level, however, do not possess many of the properties needed for high-frequency wide-bandgap converters (physical dimensions, bandwidth, noise immunity, isolation, and circuit invasion). While clamp-shaped current transducers (such as current transformers and non-invasive toroidal/helical Hall-effect or Rogowski coil sensors) are more suitable for traditional vertical power modules and switches, they may not be appropriate for lateral wide-bandgap power devices with very tight layouts and ultra-high switching speeds. Surface-mounted Rogowski coils or lateral current sensors would typically be more appropriate for this application, allowing a seamless design integration that is required for compact high-frequency power electronics. In this paper, a Rogowski coil-based embedded lateral current sensor is proposed, manufactured, and tested to demonstrate its ability to measure the switching current of lateral power switches, also known as on-trace current sensing.

Keywords— surface current sensing, switching current measurement, Rogowski coil, GaN.

I. INTRODUCTION

Wide-bandgap (WBG) semiconductors can enhance the operational conditions in power electronics systems. They are more efficient, have lower switching losses, and are increasingly used in applications such as electric vehicles and renewable energy sources [1-6]. Nevertheless, they require minimally intrusive sub-systems for monitoring and controlling the operation. The current sensing sub-system is one of the major bottlenecks in high-frequency power converters. Many current sensors available on the market lack the properties necessary for high-frequency WBG converters (such as adequate size, bandwidth (BW), immunity to EMI noise, isolation, and circuit invasion) [6, 7]. Shunt resistors and coreless Hall-effect IC are examples of small-size current sensors that require current to pass through them to detect current [6-8]. Due to the required changes in the optimal layout of WBG power converters and added parasitic elements, it is not possible to monitor switch or capacitor currents in high-frequency converters with such sensors [4-7]. A large clamp-shaped current sensor or other toroidal sensor may also introduce parasitic elements due to the

length of the traces and connectors. Furthermore, any magnetic-core-based current sensor, such as a current transformer, is likely to result in excessive parasitic inductance near switches and decoupling capacitors. Magneto-resistive or micro-fluxgate current sensors can resolve the problem, but they possess a limited detection bandwidth [8].

On the other hand, the layout design guidelines for GaN high electron mobility transistors (HEMT) from their manufacturers such as GaN Systems [9], EPC [10], Infineon [11], etc. emphasize the importance of keeping low parasitic values on printed circuit board (PCB) designs (especially the loop inductance of the traces) for better switching performance. In general, GaN HEMTs can be switched at very high switching speeds (MHz range), which results in high di/dt and di/dt during switching transients, causing voltage spikes and ringing due to parasitic inductances in the power- and driving loops. Consequently, currents or voltages ring at the switching moment, resulting in higher switching losses, insulation failures (due to voltage overshoots), and high EMI emissions. Operating at such high switching speeds, GaN HEMTs are susceptible to excess parasitic inductances, which prevents these devices from fully utilizing their superior high-frequency advantages without minimizing layout loop inductances. Such effects are detrimental to the system's power quality, leading to reduced efficiency, increased energy costs, and shortened equipment life [12-16].

A GaN-based layout with tight current traces and a small loop inductance implies the importance of non-invasive current sensing [6, 7], which is essential to ensure the reliable operation of GaN-based power electronics. GaN switch-current monitoring may be achieved through electrically contactless surface and on-trace current sensing, as long as they do not introduce parasitic insertion and have sufficient BW and accuracy. Such sensing principles have been investigated previously using embedded pick-up coils [17-20], but they either altered the exiting switch-current path or were unable to provide a general formula or level of accuracy.

Rogowski-based surface/internal current sensors are demonstrated to overcome all the problems mentioned above associated with the use of current sensors in WBG converters. Specifically, this paper provides a brief design and prototype of such lateral pick-up Rogowski coils that are non-invasive, have a high detection BW, are EMI-immune, and are accurate at measuring surface AC currents in GaN-based converters. The Rogowski coils were successfully tested in the laboratory and

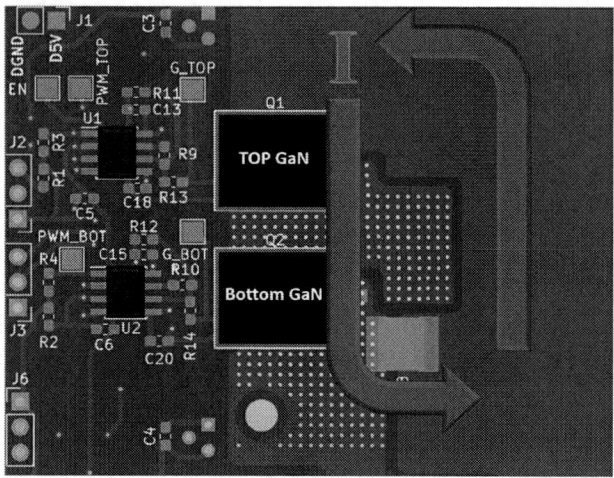

(a)

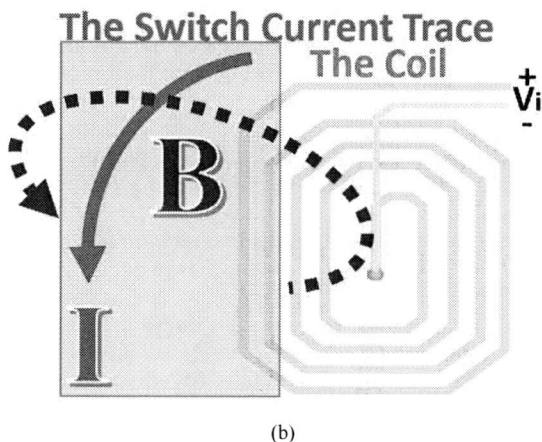

(b)

Fig. 1. A typical GaN half-bridge layout (a) and a single internal spiral coil placed on the top-switch power trace (b).

their performance was found to be satisfactory, which demonstrates the potential of this type of current sensors for use in WBG converters.

II. THE COIL DESIGN PARAMETERS

Based on Faraday's law of induction, variable (AC) currents (I) cause an AC magnetic field (B) that can generate a voltage at any nearby coil's terminals (Fig. 1(b)). A coil's induced voltage depends on its mutual inductance with the current carrying conductor as well as the frequency and amplitude of the observed AC current. This induced voltage can be useful in detecting the presence of an AC current in a circuit. As long as the current amplitude and frequency components remain unchanged, increasing M may increase the magnitude of the induced voltage in order to improve the resolution of the current sensing. Thus, it is essential to optimize the parameters associated with mutual inductance (M) when designing a coil. Faraday's law can be expressed in two simplified forms:

$$V_{coil} = -M \frac{dI}{dt} \quad \text{(V)} \tag{1}$$

$$V_{coil} = -N \frac{d\Phi}{dt} \quad \text{(V)} \tag{2}$$

N is the turns number of the coil and Φ is the magnetic flux passing through the coil. As a simplified formulation, we can assume that the magnetic field is uniformly distributed throughout the coil's cross-sectional area. Hence, the magnetic flux can be obtained from (3):

$$\Phi \approx BA \quad \text{(wb)} \tag{3}$$

In the simplified formulation, A refers to the average area of the coil, while B refers to the magnetic flux density at the coil's center. Ampere's law can be modified to find the flux density at the center of the coil:

$$\oint_C B.dl = \mu_0 I \tag{4}$$

$$B = \frac{\mu_0 I}{2\pi r} \quad \text{(T)} \tag{5}$$

in which r can be considered as the distance of the coil's center from the effective current (I) path, and μ_0 is the magnetic permeability of air (or PCB). By combining (1-5), mutual inductance between the coil and the current-carrying conductor can be found using (6):

$$M \approx \frac{N\mu_0 A}{2\pi r} \quad \text{(H)} \tag{6}$$

Thus, increasing the number of turns and the coil area as well as reducing the coil distance from the current-carrying conductor can increase the mutual inductance of the coil and, consequently, the overall sensitivity of the current measurement. Having a single coil as depicted in Fig. 1 poses the risk of the presence of external AC magnetic fields around that area which may result in inaccurate readings and noisy output. Fig. 2 illustrates an alternative solution to this issue by placing two identical coils on either side of the current's effective path. In this way, the AC magnetic fields they observe from the target trace are opposite polarity, whereas the external AC magnetic fields going through them are almost identical. In other words, this differential-induced voltage (or AC current) measurement will enhance the reading from the target trace while canceling

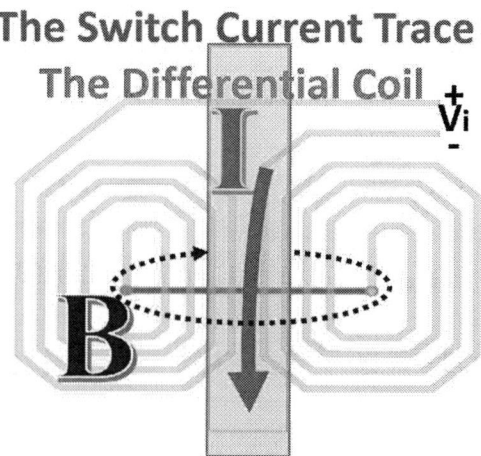

Fig. 2. A pair of spiral coils under differential surface current/magnetic-field sensing.

out most external magnetic fields. This configuration has a mutual inductance approximately double that obtained from (6).

Referring back to (6), a larger coil area and an increase in the number of turns, as well as a reduction in the distance between the coil and the current-carrying conductor, may increase the mutual inductance of the coil and, consequently, the sensitivity and the lower frequency band of the current measurement. However, identifying all parasitic *RLC* elements in a coil is essential to determining the accurate sensitivity and BW of the coil, in which self-inductance and stray capacitance of the coil are the major factors that contribute to dampening sensitivity and BW, respectively. A detailed description of calculating the parasitic values of a coil can be found in [5, 6, 23, 24], but it can also be inferred that the factors increasing mutual inductance can also increase parasitic inductance and capacitance. Although a greater number of turns can result in a higher mutual inductance, it also increases the self-inductance by a power of two. Another example would be placing the coil closer to the current-carrying conductor, but the increasing coupling between the windings and the conductor might result in a higher parasitic capacitance, and therefore a lower sensing BW. Hence, the trade-off between coil sensitivity and its parasitic effects needs to be taken into consideration when designing and implementing such coils for current sensing.

The equivalent circuit of a coil terminated with a burden resistor of R_t is shown in Fig. 3. As shown in the same figure, the burden resistor was used to alter the bandwidth and sensitivity of the coil. A coil without a magnetic core is generally unable to achieve high mutual inductance, so its lower frequency band is a few MHz or a few hundred kHz. A solution that can be used to make them work at lower frequencies is to employ an OPAMP-based analog integrator to flatten the gain slope of the coil, which is +20dB/dec. Detailed explanations of integrator calculations and implementation will be presented in the next section.

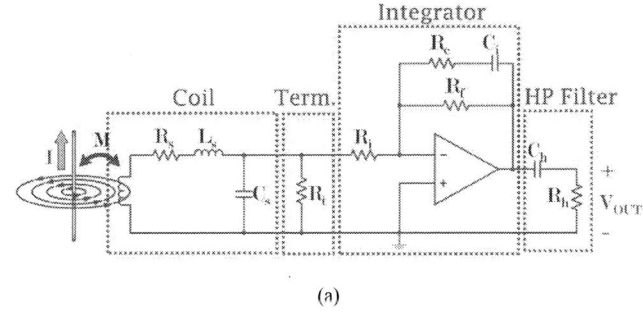

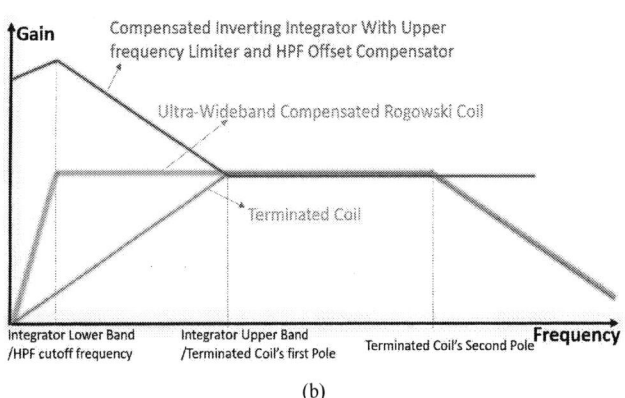

(b)

Fig. 4. Ultra-wideband Rogowski coil topology: the circuitry and its gain-frequency Bode plot [6].

III. EXTENDING THE SENSING LOWER FREQUENCY BAND USING A CONTROLLED-BAND INVERTING INTEGRATOR

A PCB-embedded coil lacks a magnetic core, so its mutual inductance is generally not sufficient to function passively (as a current transformer). By utilizing the integrated differentiating frequency region, an integrator circuit could increase the lower-band frequency gain (Fig. 4). Op-Amp-based analog integrators can provide enough gain at low frequencies while having a flexible frequency range (in contrast to passive integrators) and do not suffer from HF noise as much as digital integrators [24]. By adjusting the integration order of the amplifier, the desired gain and frequency response characteristics can be achieved, in which the gain of the amplifier will be sufficient for driving PCB-embedded coils and the circuit will be tuned to achieve the desired frequency response.

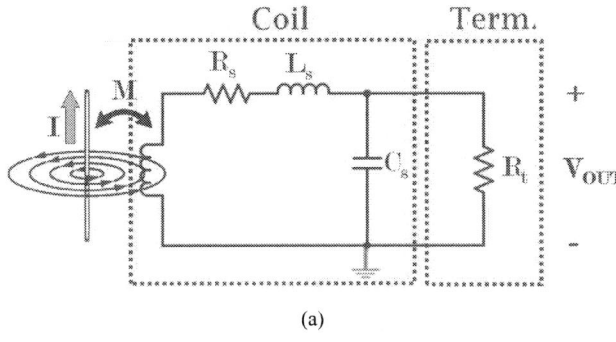

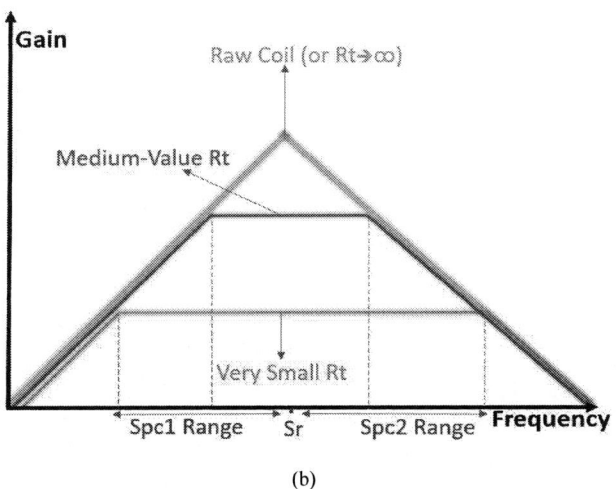

(b)

Fig. 3. Coil self-integration: simplified equivalent circuit plus a burden resistor for *RL* passive integration (a) and their overall gain-frequency response (b) [6].

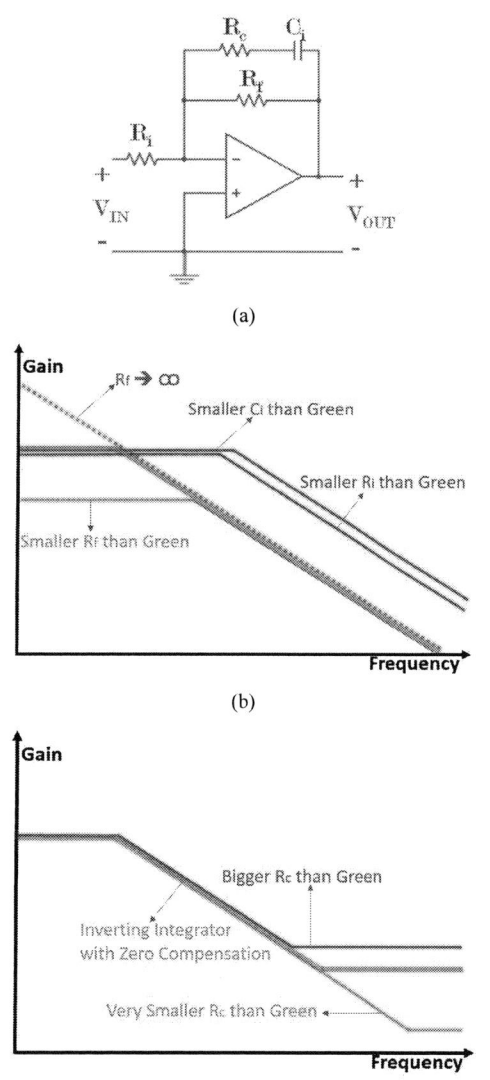

(a)

(b)

(c)

Fig. 5. Frequency-controlled inverting integrator topology: the circuit (a), the frequency response with a shorted Rc (b), and the frequency response with a non-zero Rc (c) [6].

consideration one important point; non-inverting integrators have a gain that is always greater than one, indicating that they are ineffective at ultra-high frequencies, whereas inverting integrators have a gain that is lower than one [6]. Inverting integrators can also have their upper-frequency band controlled by placing a resistor in series with their capacitor (Fig. 5). There is a complete set of formulations for adjusting the gain and frequency corners of the coil and the integrator in [5, 6].

As a result of Op-Amp non-idealities, integrators based on Op-Amps can suffer from undesirable DC voltage offsets at their output caused by bias currents and their resulting input offset voltages. It is shown in Fig. 5(b) that due to the finite feedback resistor of analog integrators, integration only occurs at frequencies higher than a specific frequency.

IV. Prototyping and Experimental Results

A. Sensor Prototyping

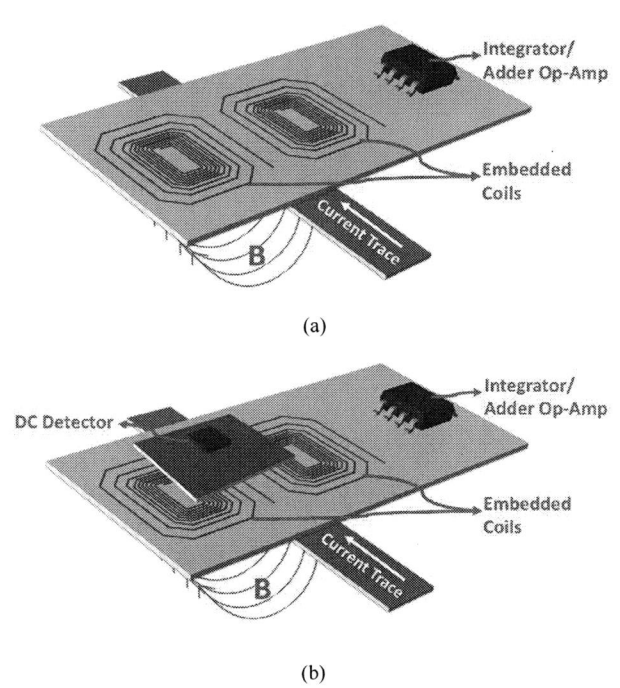

(a)

(b)

Fig. 6. Surface current sensor realizations: AC detector (a) and DC and AC detector combination (b).

When choosing between an inverting or non-inverting integrator configuration, it is important to take into

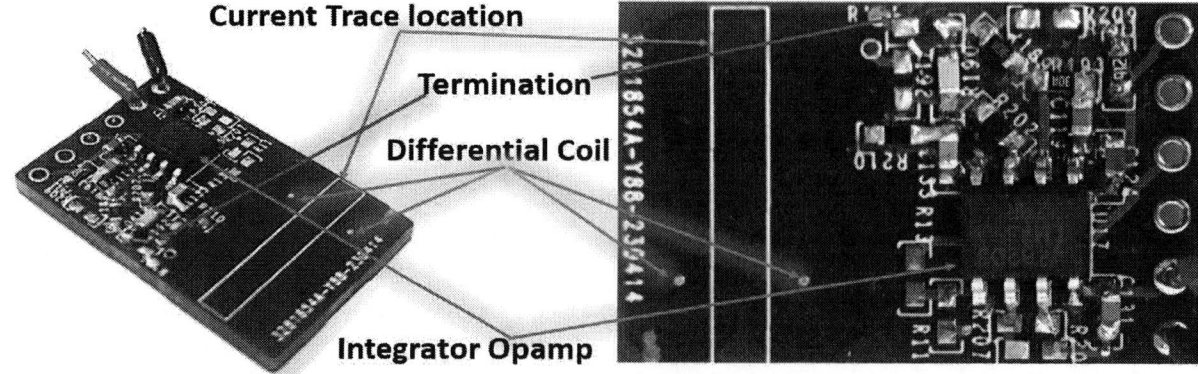

Fig. 7. Prototyped AC current sensor.

979-8-3503-3714-3/23 $31.00 © 2023 IEEE

To eliminate the undesirable Op-Amp offset, a passive high-pass filter with a cutoff frequency below the integration band can be used at the Op-Amp output terminal (Fig. 4(a)).

As shown in Fig. 6(a), a single piece of PCB can be used to implement the coil shown in Fig. 2 along with the circuitry shown in Fig. 4(b). As a result of its inductive sensing nature, this topology is unable to detect DC and very low-frequency current components. Fig. 6(b) illustrates how a contactless DC detector can be used, such as magneto-resistive sensors or micro-fluxgate magnetometers, for currents having interrupted DC values, such as the current flowing through an inductor in a buck or boost converter. As shown in Fig. 7, a surface current

Fig. 8. The GaN half-bridge for surface current sensing.

sensor based on Fig. 2, 4(b), and 6(b) has been constructed. Using a resistor smaller than one ohm for coil termination allows the integrator to use a lower BW opamp to provide better square-wave signal conditioning [6]. The LM6172 has been used as the integrator opamp, which is a voltage feedback opamp with ultra-low distortion and a high slew rate.

B. Switching Current Measurement

When Fourier transforms of square or trapezoidal signals are used to calculate switch-current waveforms, a sensor with the highest BW is needed. A GaN half-bridge (shown in Fig. 8) was modified for a double pulse test setup to measure the top switch current with the prototyped sensor. In Fig. 9, to determine accuracy, the sensor results are compared with those from a DC-50MHz current probe (Tektronix TCP305A) that captures frequency components up to tens of megahertz. The scope captures show Tektronix TCP305A probes in magenta and AC surface current sensors in blue waveforms. According to the results, the sensor captured frequency components of ultra-fast rising switch currents, even faster than the 50MHz current probe.

CONCLUSION

A non-invasive ultrafast current sensor has been constructed and tested for extremely fast surface switch-current measurements, especially in high-frequency applications. Because both schemes are small and compensate for each

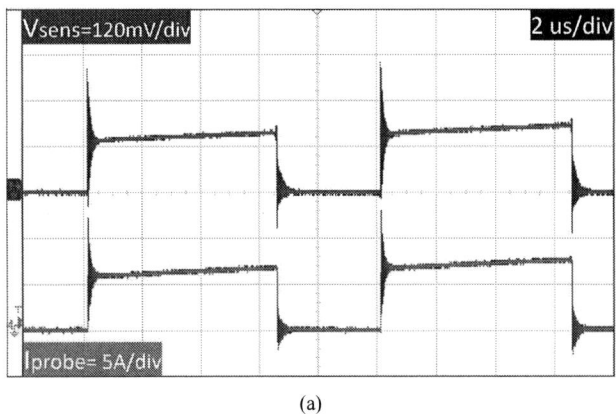

(a)

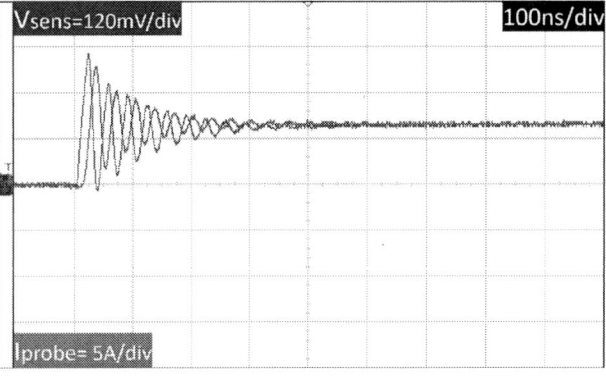

(b)

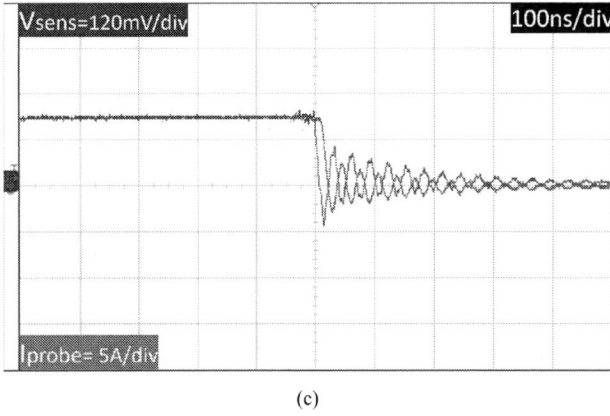

(c)

Fig. 9. Switching current pulse and transients: a) two last pulses out of 8 pulses b) rising moment of the last pulse, and c) falling instant of that pulse. The magenta is a DC-50 MHz (Tektronix TCP305A) current probe, and the blue is the output of the prototyped sensor.

other's lack of bandwidth, the sensor can be easily integrated into high-power applications due to its small size and high bandwidth. Compared to a commercial high-bandwidth current probe, the experimental results demonstrate clean, fast, and accurate results, confirming the capability of the proposed current sensor to be integrated into high-power WBG converters, offering a small footprint and low cost.

REFERENCES

[1] A. Morya, M. Moosavi, M. C. Gardner and H. A. Toliyat, "Applications of Wide Bandgap (WBG) devices in AC electric drives: A technology status review," 2017 IEEE International Electric Machines and Drives Conference (IEMDC), Miami, FL, USA, 2017, pp. 1-8, doi: 10.1109/IEMDC.2017.8002280.

[2] R.A. Ghaderloo, Y. Shen, C. Singhabahu, R. Resalayyan, and A. Khaligh, "Dead-time Compensation Method for Bus-clamping Modulated Voltage Source Inverter, " in 2023 arXiv preprint arXiv:2307.11868.

[3] A. Parsa Sirat, H. Niakan, D. Evans, J. Gafford and B. Parkhideh, "Ultra-Wideband Unidirectional Reset-Less Rogowski Coil Switch Current Sensor Topology for High-Frequency DC-DC Power Converters," 2023 IEEE Applied Power Electronics Conference and Exposition (APEC), Orlando, FL, USA, 2023, pp. 1662-1669, doi: 10.1109/APEC43580.2023.10131463.

[4] R. Asrar Ghaderloo, A. Parsa Sirat, and A. Shoulaie. "A High Frequency Active Clamp Forward Converter with Coreless Transformer." arXiv preprint arXiv:2307.12804 (2023).

[5] A. Parsa Sirat, C. Roy, D. Evans, J. Gafford and B. Parkhideh, "In-Situ Ultrafast Sensing Techniques for Prognostics and Protection of SiC Devices," 2022 IEEE 9th Workshop on Wide Bandgap Power Devices & Applications (WiPDA), Redondo Beach, CA, USA, 2022, pp. 142-147, doi: 10.1109/WiPDA56483.2022.9955254.

[6] A. Parsa Sirat, 2023. Ultra-Wideband Contactless Current Sensors for Power Electronics Applications (Doctoral dissertation, The University of North Carolina at Charlotte).

[7] A. Parsa Sirat and B. Parkhideh, "Current Sensor Integration Issues with Wide-Bandgap Power Converters," Sensors (2023), 23(14), 6481.

[8] Z. Xin, H. Li, Q. Liu and P. C. Loh, "A Review of Megahertz Current Sensors for Megahertz Power Converters," in IEEE Transactions on Power Electronics, vol. 37, no. 6, pp. 6720-6738, June 2022, doi: 10.1109/TPEL.2021.3136871.

[9] GaN Systems. (2023). PCB Thermal Design Guide for GaN Enhancement Mode Power Transistors. [Online]. Available: http://www.gansystems.com/whitepapers.php.

[10] EPC. (2023). Gen4 eGaN FETs. [Online]. Available: http://epcco.com/epc/Products/eGaNFETs/Gen4eGaNFETs.aspx.

[11] Infineon. (2023). Infineon and Panasonic Will Establish Dual Sourcing for Normally-Off 600 V GaN Power Devices. [Online]. Available: http://www.infineon.com/cms/en/aboutinfineon/press/pressreleases/2015/INFPMM201503-041.html.

[12] H. Niakan, A. Parsa Sirat and B. Parkhideh, "Current Distribution Monitoring of Paralleled GaN HEMTs," 2020 IEEE Energy Conversion Congress and Exposition (ECCE), Detroit, MI, USA, 2020, pp. 2865-2870, doi: 10.1109/ECCE44975.2020.9235770.

[13] S. Lu, T. Zhao, Z. Zhang, K. D. T. Ngo, R. Burgos and G. -Q. Lu, "Low Parasitic-Inductance Packaging of a 650 V/150 A Half-Bridge Module Using Enhancement-Mode Gallium-Nitride High Electron Mobility Transistors," in IEEE Transactions on Industrial Electronics, vol. 70, no. 1, pp. 344-351, Jan. 2023, doi: 10.1109/TIE.2022.3148750.

[14] P Z. Qi, Y. Pei, L. Wang, Q. Yang and K. Wang, "A Highly Integrated PCB Embedded GaN Full-Bridge Module With Ultralow Parasitic Inductance," in IEEE Transactions on Power Electronics, vol. 37, no. 4, pp. 4161-4173, April 2022, doi: 10.1109/TPEL.2021.3128694.

[15] Z. Ma, Y. Lai, Y. Yang, Q. Huang and S. Wang, "Review of Radiated EMI Modeling and Mitigation Techniques in Power Electronics Systems," 2023 IEEE Applied Power Electronics Conference and Exposition (APEC), Orlando, FL, USA, 2023, pp. 1776-1783, doi: 10.1109/APEC43580.2023.10131552.

[16] KASHYAP, N., 2022. Addressing GaN Converter Challenges: False Turn-On Issues & Switching Loss Modelling (Doctoral dissertation).

[17] Kim, U.J., Song, M.S. and Kim, R.Y., 2020. PCB-based current sensor design for sensing switch current of a nonmodular GaN power semiconductor. Energies, 13(19), p.5161.

[18] K. Wang, X. Yang, H. Li, L. Wang and P. Jain, "A High-Bandwidth Integrated Current Measurement for Detecting Switching Current of Fast GaN Devices," in IEEE Transactions on Power Electronics, vol. 33, no. 7, pp. 6199-6210, July 2018, doi: 10.1109/TPEL.2017.2749249.

[19] M. Niliyan, M. Mohamadian and A. Y. Varjani, "One step model predictive control of five level ANPC permanent magnet motor drive," 2016 7th Power Electronics and Drive Systems Technologies Conference (PEDSTC), Tehran, Iran, 2016, pp. 620-625.

[20] P. T. N. Kishore, S. K. Pramanick and S. S. Nag, "Development of a PCB Embedded High Bandwidth Coil Based Current Sensor Suitable for Characterizing GaN Devices," 2023 11th International Conference on Power Electronics and ECCE Asia (ICPE 2023 - ECCE Asia), Jeju Island, Korea, Republic of, 2023, pp. 99-104, doi: 10.23919/ICPE2023-ECCEAsia54778.2023.10213475.

[21] M. A. Dashtaki, H. Nafisi, A. Khorsandi, M. Hojabri, and E. Pouresmaeil, "Dual Two-Level Voltage Source Inverter Virtual Inertia Emulation: A Comparative Study," *Energies*, vol. 14, no. 4, p. 1160, Feb. 2021, doi: 10.3390/en14041160.

[22] M. Salehi, M. Shahabadini, H. Iman-Eini, and M. Liserre, "Predictive control of grid-connected modified-chb with reserve batteries in photovoltaic application under asymmetric operating condition," IEEE Transactions on Industrial Electronics, vol. 69, no. 9, pp. 9019–9028, 2021.

[23] A. Parsa Sirat, H. Niakan, M. Campo, J. De La Rosa Garcia and B. Parkhideh, "An All-Passive Compound Current Sensor for Fast Switching Current Monitoring," 2022 IEEE Energy Conversion Congress and Exposition (ECCE), Detroit, MI, USA, 2022, pp. 01-07, doi: 10.1109/ECCE50734.2022.9947615.

[24] A. Parsa Sirat, H. Niakan, C. Roy and B. Parkhideh, "Rogowski-Pair Sensor for High-Speed Switch Current Measurements without Reset Requirement," 2022 IEEE Energy Conversion Congress and Exposition (ECCE), Detroit, MI, USA, 2022, pp. 1-8, doi: 10.1109/ECCE50734.2022.9948065.

[25] M. Salehi, M. Khodabandeh and M. Amirabadi, "Zeta-Based AC-Link Universal Converter," 2022 IEEE Energy Conversion Congress and Exposition (ECCE), Detroit, MI, USA, 2022, pp. 1-7, doi: 10.1109/ECCE50734.2022.9947744.

[26] A. Pourghorban, D. Maity, "Target defense against a sequentially arriving cooperative intruder team," Proc. SPIE 12544, Open Architecture/Open Business Model Net-Centric Systems and Defense Transformation 2023, 1254409 (12 June 2023); https://doi.org/10.1117/12.2663418.

Electro-thermal Design of MV SiC JFET Based Solid State Circuit Breakers

Sima Azizi Aghdam, Mohammed Agamy
Dept. of Electrical and Computer Engineering
University at Albany, SUNY
Albany NY, USA
saziziaghdam@albany.edu, magamy@albany.edu

Zhongda Li and Peter Losee
Qorvo Power Device Solutions
Qorvo
Princeton NJ, USA
zhongda.li@qorvo.com, pete.losee@qorvo.com

Abstract—This paper discusses the electro-thermal performance of solid-state circuit breakers (SSCBs) employing 1.7kV and 3.3kV SiC JFETs in medium voltage (MV) applications. It develops SiC JFET based circuit models that can be used to evaluate their performance in various SSCBs applications. Also, it studies the challenges associated with high conduction losses and elevated junction temperatures (T_J) of power devices in SSCB application during faults. Based on this analysis, new arrangement of power devices in SSCB are introduced. Given the critical nature of device failure resulting from elevated T_J level, it is crucial to conduct a comprehensive electro-thermal analysis and consider both electrical and thermal constraints during the design process. Extensive simulations examine the performance of scaled voltage/current of SSCB using SiC JFET under various condition such as ambient temperature, current levels, and SSCB response times, particularly in severe fault scenarios. Results show how the proposed design procedure enables proper sizing the semiconductor devices of the SSCB to handle high fault currents and enable longer withstand times so that these breakers can be incorporated in grid protection coordination schemes.

Index Terms—Solid-state circuit breaker, SiC JFET, Short-circuit fault, Electro-thermal analysis.

I. INTRODUCTION

The increasing demand for more efficient and reliable electrical systems has led to a growing interest in the development of advanced technologies for controlling, monitoring, and protecting of these systems. Regarding the protection of the power system, the conventional electromechanical circuit breakers have been replaced by solid-state circuit breakers (SSCBs) [1]. SSCBs that rely on the use of power semiconductor to isolate fault currents, have emerged as a promising alternative to traditional mechanical circuit breakers as they offer numerous advantages such as faster response times, arc-free interruption, improved fault detection, and enhanced control capabilities [1]- [2]. For implementations of SSCB, different power devices such as IGBTs, thyristors, SiC MOSFET and SiC JFETs, etc, have been used [3]. Compared to SiC MOSFETs, SiC JFET provides distinct advantages such as lower on-resistance resulting from high-channel density, long-term stability of the pinch-OFF voltage even at high temperatures, and more ruggedness and reliability attributed to the absence of a gate oxide layer [4] and when placed in cascode the resulting switch has a lower input and reverse transfer capacitance than

This work was supported in part by PowerAmerica - Task no. BP6-3.27.

a MOSFET. Hence, in this work, SiC JFET is used for SSCB implementation. Although SSCBs bring significant advantages to the system, they cause high conduction losses resulting in the increased thermal stress on power devices [5], particularly in fault conditions. During instances of short-circuit faults, the SSCB turn-off process typically generates high di/dt, leading to significant over-voltage across the power switch, due to the presence of line inductance in power systems. These conditions have the potential to exceed the device rating, resulting in failure. Therefore, electrical and thermal considerations in the SSCB design are required. To implement the thermal modeling of SiC JFET, among different approaches Finite Element Method (FEM) approach is used. The thermal model using Cauer models which is "RC" thermal networks of the equivalent thermal impedance is utilized. Developing this model can be effective to evaluate the performance of various SSCBs applications [6]. Based on these models, the impact of different parameters such as response time of breaker, the level of current and the ambient temperature on the conduction losses and the T_J has been studied. Moreover, in this paper, the impact of on-resistance of power devices such as 1.7kV and 3.3kV were discussed. As the $R_{ds,on}$, as well as die and package thermal resistance of JFETs directly influence the device maximum current handling capacity and overall power dissipation, a new arrangement was proposed. In this configuration, in addition to series connection of SiC JFETs, each switch position is composed of several die connected in parallel. This arrangement enables SSCB to operate in higher voltage and current. The rest of paper is as follows: Section II discusses the SSCB design in the application of MV and high current. Section III discusses the electro-thermal analysis and thermal modeling of SiC JFET. Section IV presents the results. Finally Section V outlines the conclusion.

II. MV SOLID-STATE CIRCUIT BREAKER USING SiC JFETs

Solid-state circuit breakers provide fast responses, arc-free current interruption and low peak fault current during faults [2], making them valuable assets for the evolving power systems of the future. Given that SSCBs are predominantly used in higher voltage and DC systems, careful attention must be given to factors such as voltage and current ratings, reliability, temperature tolerance, and safety mechanisms dur-

979-8-3503-3714-3/23 $31.00 © 2023 IEEE 201

ing the design phase. When incorporating SSCBs into MV power systems, the adjustment of voltage and current levels is necessary. To achieve this, techniques like supercascode and parallel connection of power devices will be utilized [7].

A. Implementation of MV SSCB

Wide bandgap (WBG) semiconductors such as SiC JFET provide superior performance in terms of efficiency, power density, and high-voltage-blocking capability compared to Si devices [8]. However, the availability of high voltage switches, suitable for MV applications is limited. To employ these devices in MV or HV applications, various designs have been utilized [4]. One promising approach is supercascode (SC), wherein multiple low-voltage, high-current (LV-HC) SiC JFETs are connected in series with balancing elements to manage voltage distribution within the string. For enabling a normally-off state in SiC JFETs, Si MOSFETs are integrated into the circuit configuration. The SC operation is based on the concept of self-triggered, sequential semiconductor behavior, explained in more detail in [4]. The approach adopted for this study, as illustrated in Fig 1, is introduced by Gao et al. [9]. This design comprises **N** SiC JFETs (U_1-U_N), with the balancing networks implemented using RCD components, which are avalanche diodes (D), resistors (R_{diode}), and capacitors ($C_1 - C_N$). An additional component, R_L, is responsible for providing a bias current to the diodes, while R_{g1} to R_{gN} control the rise and fall times of the JFETs. The MV SSCB is designed for a 5kV(rms) nominal operation, leading to a peak withstand requirement of 8kV+10%. Details of the design of the corresponding supercascode, based on the principles presented in [9], are summarized in Table I.

B. Implementation of high current SSCB

To design a SSCB with a robust current-handling capacity, it is imperative to account for the increase in T_J under different conditions. During normal operation, the device temperature gradually rises, and settles at a level corresponding to the conduction loss and cooling network in use. In severe scenarios, when the T_J surpasses a critical threshold, the device may enter a state of thermal runaway. There are two primary failure mechanisms associated with SiC power devices. Firstly, when T_J exceeds the intrinsic temperature of SiC, it results in a significant current increase. Secondly, if the temperature surpasses the specified limits of the ohmic contact metal or the passivation/packaging materials, it can lead to device failure, even at relatively lower temperatures [10]- [11]. To enable the handling of high currents in this context, the drain-to-source resistance ($R_{ds,on}$) is reduced by inserting multiple power devices in parallel, since $R_{ds,on}$ in JFETs has a direct impact on the device maximum current-handling capability and overall power dissipation. Therefore, the total on state resistance of **K** power devices in parallel will be ($R_{ds,on}/K$), which results in lower conduction loss. Further, the thermal impedance will be reduced due to the increase in die area. These two factors result in a lower temperature rise and thus

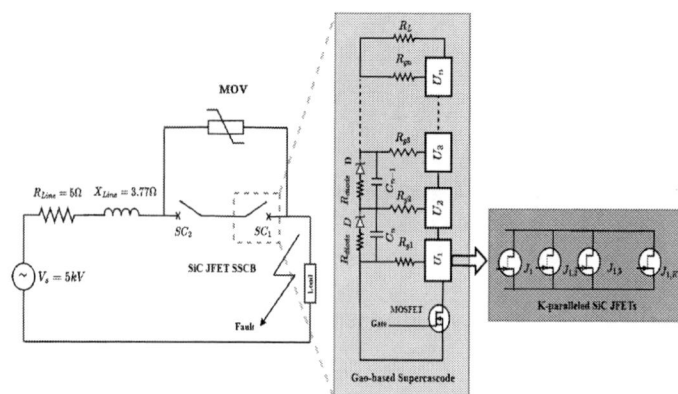

Fig. 1: Medium voltage AC SSCB schematic using SiC JFET in supercascode connection

a lower settling point for T_J, which enables the SSCB to withstand larger current for longer time.

TABLE I: Supercascode parameters

Common parameters for Both Power Devices	
Parameters	**Value**
Gate resistor of MOSFET	R_{on}=1.5 Ω,R_{off}=5 Ω
Gate resistor of JFET	R_g=10Ω
R_{diode}	0.01 Ω
R_L	50k Ω
Diode	AU1PM
1.7kV SiC JFET	
Power device voltage and current rating	1.7 kV, 100A
Number of JFETs in series	N=10
Capacitance	$C_1 = 600pF$, $C_2 = 1200pF, \cdots, C_9 = 5400pF$
3.3kV SiC JFET	
Power device voltage and current rating	3.3 kV, 100A
Number of JFETs in series	N=5
Capacitance	$C_1 = 750pF$, $C_2 = 1500pF, \cdots, C_4 = 3000pF$

III. ELECTRO-THERMAL MODELING OF SiC JFET

To design a high efficient, high power density and reliable SSCB, electro-thermal analysis of power devices is critical, which are discussed in the following subsections.

A. SiC JFET Electrical Characteristics

To electrically characterize two different power devices including 1.7kV and 3.3kV SiC JFET, isothermal electrical SiC JFET SPICE models were developed using the subcircuit method. The model consists of a standard JFET model at its core and the other circuit elements around it to model the drift resistance and capacitance behaviors of the SiC JFET. The model parameters were determined in a combination of TCAD and experimental measurements, and good matching of I-V and C-V between the SPICE model and TCAD/measurement were achieved, as shown in Fig. 2.

979-8-3503-3714-3/23 $31.00 © 2023 IEEE

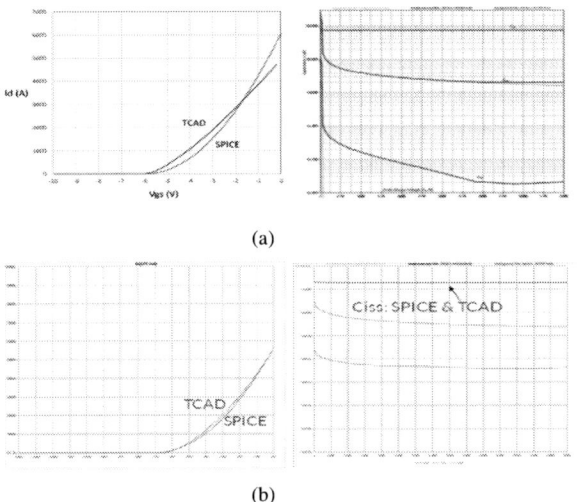

(a)

(b)

Fig. 2: Model fit of static characteristics, transfer I-V and C-V of (a) 1.7kV SiC JFET, (b) 3.3kV SiC JFET

B. SiC JFET Thermal characteristics

To estimate the transient response of the JFETs, practical packaging assumptions needs to be considered. For each of the device, a representative material stack from die to baseplate is outlined in Table II. For the 1.7kV package assumptions, the study used data from measurements taken on prototype discrete and power modules developed separately. A heating pulse to the device under test (DUT) is applied and the temperature using a Temperature Sensitive Electrical Parameter (TSEP) is monitored, and thermal resistance values are measured. To accurately assess the transient thermal response, a specialized Transient Dual Interface Method (TDIM) was employed. The resulting data of both thermal resistance values for the 1.7kV and 3.3kV prototype modules is showed in Fig.3. Along with the traditional junction-to-case impedance ($Z_{th}, J - c$), the extracted thermal impedance from junction to bottom of SiC die and to the bottom of the DBC is also presented. According to the results in Fig.3, one can ascertain that the time scale critical for absorbing heat energy into the SiC die is 100μs or less, into the DBC is 10ms or less and into the 5mm thick baseplate is 100ms to 1second, respectively. Based on the transient thermal impedance curves, RC thermal (Cauer) network [6] to model the device behavior is obtained, Fig4.

TABLE II: Assumed material stack for SiC JFETs

Voltage	1.7kV	3.3kV
Die	150μm SiC	200μm SiC
Die Attach	50μm Ag Sinter	50μm Ag Sinter
Ceramic	250μm Si3N4 AMB	630μm AlN DBC
Solder	200μm SaC Solder	200μm SaC Solder
Baseplate	5mm AlSiC	5mm AlSiC

C. SiC JFET safe operation area(SOA)

SOA curve of SiC JFETs is defined as shown in Fig. 5, based on the obtained Electro-Thermal SPICE models,. These

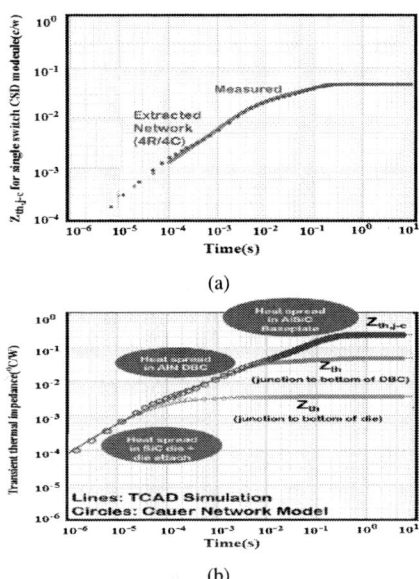

(a)

(b)

Fig. 3: Measured steady-state and transient thermal impedance of (a) 1.7kV SiC JFET, (b) 3.3kV SiC JFET

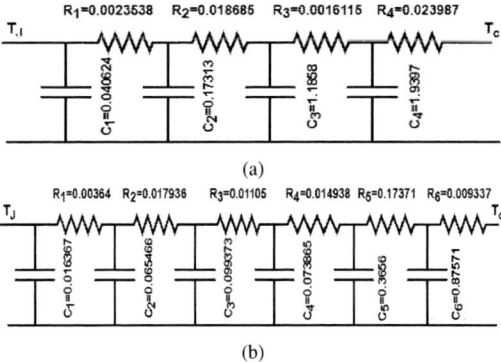

(a)

(b)

Fig. 4: Extracted Cauer network of (a) 1.7kV, (b) 3.3kV SiC JFET

curves can help to estimate the maximum trip current and time in SSCB. In this section, the case temperature (T_c) is assumed to be T_c=80 °C, the maximum junction temperature is $T_{J,max}$=175°C and a square current pulse is used. From the figure, the maximum dc current of the JFETs subject to these case temperatures are 92A, 134A, for 1.7kV and 3.3kV, respectively. The ET-SPICE model predicts the safe current versus pulse width defined in the shaded region.

IV. RESULT AND DISCUSSION

To evaluate the performance of an MV AC SSCB utilizing 1.7kV and 3.3kV SiC JFETs, the power system depicted in Fig.1 is employed, and simulations are conducted in LtSpice. This study investigates two distinct scenarios. In the first scenario, short-circuit faults occur when the line current is at its maximum, allowing us to assess the SSCB capability to manage high currents. The second scenario examines extended delays between fault occurrence and breaker activation,

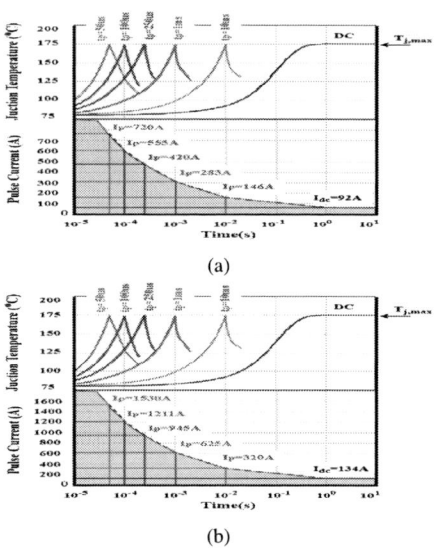

(a)

(b)

Fig. 5: SOA curves based on forward bias of (a) 1.7kV SiC JFET, (b) 3.3kV SiC JFET

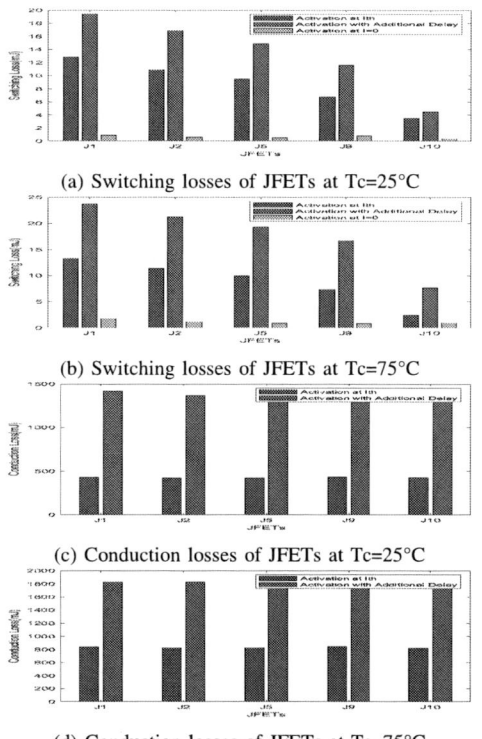

(a) Switching losses of JFETs at Tc=25°C

(b) Switching losses of JFETs at Tc=75°C

(c) Conduction losses of JFETs at Tc=25°C

(d) Conduction losses of JFETs at Tc=75°C

Fig. 6: Power losses of 1.7kV SiC JFETs in SSCB application

with a particular focus on assessing the breaker's response time limitations. In scenario I, three different cases will be discussed including breaker activation (a) at threshold current (I_{th}), (b) with a delayed response subsequent to reaching the I_{th}, (c) at zero-current crossing ($I = 0$). In the scenario II, zero-current crossing case ($I^* = 0$) is repeated. In case (a), I_{th} is set at three times the nominal current (I_n), which is

366A, and the corresponding time for activation is $370\mu s$ after the fault occurs. In case (b), a delay equals to $480\mu s$ after reaching I_{th} will be considered, resulting in a total response time of $850\mu s$ from fault occurrence. In case (c), the breaker activates when the current reaches zero for the first time, which occurs $5.4ms$ after the fault. In Scenario II, the breaker is activated $8.92ms$ after the fault occurrence, indicating a longer activation time. Electro-thermal analyses are conducted for two different case temperatures (T_c) including 25°C and 75°C. In the 1.7kV-based implementation, although the SC requires 10 devices, for simplicity, results from 5 of these switch locations are presented. These five devices are the lower pair (U_1 and U_2), the middle unit (U_5), and the upper pair (U_9 and U_{10}). Electrical analyses depicted in Fig. 6 and Fig. 7 present conduction losses and switching losses for both 1.7kV and 3.3kV-based SSCBs. As expected, an elevation in junction temperature (T_J) has negligible effect on switching losses, but it leads to a substantial increase in conduction losses. In situations involving zero-current crossing, conduction losses surge dramatically, resulting in predicted device failure due to thermal runaway. The thermal analysis results in TableIII-IV confirm this. For the 1.7kV JFET, under normal operating conditions, the T_J reaches 80°C at T_c = 25°C and 159°C at T_c= 75°C . These values soar to 152.8°C and 296.8°C during short-circuit faults when the breaker disconnects the system upon reaching the I_{th}. Unacceptable temperature levels are prominently marked in red within Table III-IV to indicate thermal runaway. To ensure reliable operation of the AC SSCB using the 1.7kV JEFT with $R_{ds,on}$=11mΩ, each switch location in the supercascode arrangement will be composed of 7 parallel SiC JFET devices, leading to a total $R_{ds,on}$ to $1.57m\Omega$ per set. This also leads to a reduction of the assumed thermal resistance by a factor of 7 (from 0.435°C/W to 0.062°C/W). Consequently, the AC SSCB can sustain higher current loads over an extended duration without device failure or temperature instability, with the maximum junction temperature $T_{J,max}$ capped at 151°C in the most adverse conditions. Similar analyses are conducted for the 3.3kV device implementation. During normal operation, T_J reaches 44°C and 100°C for T_c=25 °C and 75°C , respectively. These figures increase during fault scenarios to 52.9°C and 111°C when the breaker interrupts the circuit at the I_{th}. Comparison of 1.7kV device and 3.3 kV device shows, in $T_c = 75°C$, the $T_{J,max}$ in 3.3kV-based design experienced lower T_J than 1.7kV-based design, (111°C vs. 296.8°C), when the breaker disconnects the circuit at threshold current. This can be attributed to the larger die area for the 3.3kV device, which is designed for a similar on state resistance to 1.7kV; this larger area provides a lower thermal impedance and thus lower temperature rise. Even with an extended delay, 3.3kV-based design can operate with a peak T_J of 130°C at T_c=75°C. However, similar to 1.7kV-based design, 3.3 kV device with $R_{ds,on} = 10m\Omega$ is incapable of functioning in zero-current crossing scenarios. The number of devices that needs to be connected in parallel for 3.3kV is 5 devices, resulting in a reduction of the assumed thermal resistance by a factor of 5 (from 0.231°C/W to 0.046°C/W).

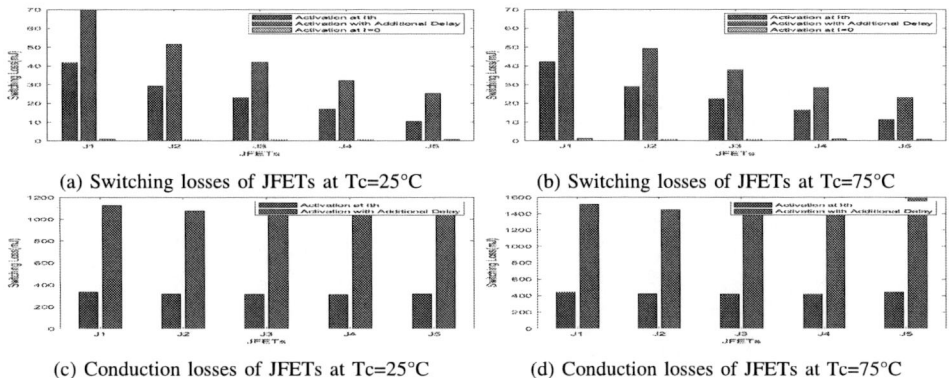

(a) Switching losses of JFETs at Tc=25°C (b) Switching losses of JFETs at Tc=75°C

(c) Conduction losses of JFETs at Tc=25°C (d) Conduction losses of JFETs at Tc=75°C

Fig. 7: Power losses of 3.3kV SiC JFETs in SSCB application

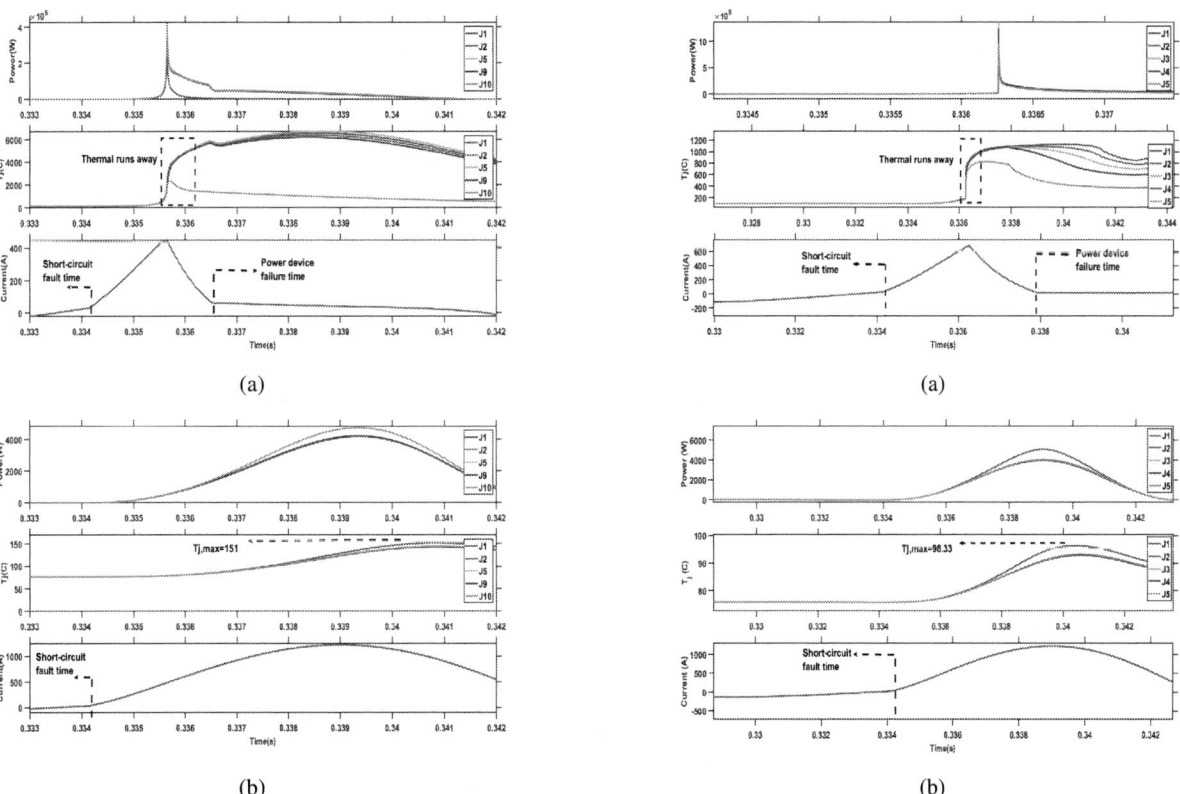

Fig. 8: Dissipation power, junction temperature, line current of (a)$R_{ds,on} = 11m\Omega$ and (b)$R_{ds,on} = 1.57m\Omega$ per 1.7kV JFET

Fig. 9: Dissipation power, junction temperature, line current of (a)$R_{ds,on} = 10m\Omega$ and (b)$R_{ds,on} = 2m\Omega$ per 3.3kV JFET

The transient thermal behavior of a SSCB for a 5kV grid voltage in zero-crossing breaking scenario using 1.7kV JFETs with a single device per switch location vs. 7 parallel devices per switch location is shown in Fig. 8. Similar scenario is applied to a breaker using 3.3kV JFET with 5 parallel devices per switch location is shown in Fig. 9. Therefore, the total number of SiC JFETs necessary for implementation based on the 3.3kV JFET is lower compared to the 1.7kV JFET, with a configuration of 5 sets, each containing 5 devices, as opposed to 10 sets of 7 devices.

V. CONCLUSION

In this paper the electro-thermal modeling and design of a MV AC SSCB employing 1.7kV and 3.3kV SiC JFETs have been investigated. The analysis revealed that the SSCB time response is constrained, and under circumstances like zero-current crossings, the power devices can experience failure due to elevated junction temperatures. Consequently, the solution involved reducing the on-state resistance and thermal impedance of each device by incorporating multiple devices

TABLE III: Simulated maximum T_J in $^{\circ}C$ under different SSCB breaking scenarios for 1.7kV SiC JFETs

JFETs		$T_c = 25\,^{\circ}\text{C}$		$T_c = 75\,^{\circ}\text{C}$	
		$R_{ds,on} = 11m\Omega$	$R_{ds,on} = 1.57m\Omega$	$R_{ds,on} = 11m\Omega$	$R_{ds,on} = 1.57m\Omega$
J1	I_{th}	152.8	30.8	296.8	80
	delay	244.4	34.1	Thermal runaway	82
	I=0	Thermal runaway	46.8	Thermal runaway	102.1
	I*=0	Thermal runaway	83	Thermal runaway	151
J5	I_{th}	150.8	27	294.5	77.1
	delay	235.4	28.2	Thermal runaway	78.6
	I=0	Thermal runaway	44.8	Thermal runaway	100.1
	I*=0	Thermal runaway	74.8	Thermal runaway	142.5
J10	I_{th}	150.6	26.9	292.1	77.1
	delay	235.4	28.4	Thermal runaway	78.6
	I=0	Thermal runaway	45.3	Thermal runaway	100.5
	I*=0	Thermal runaway	75.8	Thermal runaway	143.2

TABLE IV: Simulated maximum T_J in $^{\circ}C$ under different SSCB breaking scenarios for 3.3kV SiC JFETs

JFETs		$T_c = 25\,^{\circ}\text{C}$		$T_c = 75\,^{\circ}\text{C}$	
		$R_{ds,on} = 10m\Omega$	$R_{ds,on} = 2m\Omega$	$R_{ds,on} = 10m\Omega$	$R_{ds,on} = 2m\Omega$
J1	I_{th}	52.9	25.9	111.1	76.1
	delay	66.4	26.4	130.1	76.7
	I=0	Thermal runaway	31.3	Thermal runaway	83
	I*=0	Thermal runaway	42	Thermal runaway	96.3
J3	I_{th}	51.4	25.26	110.4	76.1
	delay	63.9	26.3	127.1	76.6
	I=0	Thermal runaway	31.2	Thermal runaway	82.5
	I*=0	Thermal runaway	39.6	Thermal runaway	92.9
J5	I_{th}	51.3	25.2	111.1	76.1
	delay	64.7	26.3	129.8	76.6
	I=0	Thermal runaway	31.3	Thermal runaway	82.6
	I*=0	Thermal runaway	39.9	Thermal runaway	93.3

in parallel. To achieve this, 7 power devices were necessary for the 1.7kV case, while 5 power devices were required for the 3.3kV configuration. As a result, the circuit breaker can sustain high voltage and current levels for extended duration, making it suitable for integration into distribution networks necessitating relay coordination. The 3.3kV configuration required fewer devices and exhibited lower junction temperatures under normal and fault conditions since the device is designed with a larger area to achieve similar on-state resistance to 1.7kV device.

REFERENCES

[1] R. Rodrigues, Y. Du, A. Antoniazzi, and P. Cairoli, "A review of solid-state circuit breakers," *IEEE Transactions on Power Electronics*, vol. PP, pp. 1–1, 06 2020.

[2] X. Song, P. Cairoli, Y. Du, and A. Antoniazzi, "A review of thyristor based dc solid-state circuit breakers," *IEEE Open Journal of Power Electronics*, vol. 2, pp. 659–672, 2021.

[3] P. Cairoli, L. Qi, C. Tschida, V. R. Ramanan, L. Raciti, and A. Antoniazzi, "High current solid state circuit breaker for dc shipboard power systems," in *2019 IEEE Electric Ship Technologies Symposium (ESTS)*, 2019, pp. 468–476.

[4] L. Gill, L. A. G. Rodriguez, J. Mueller, and J. Neely, "A comparative study of sic jfet super-cascode topologies," in *2021 IEEE Energy Conversion Congress and Exposition (ECCE)*, 2021, pp. 1741–1748.

[5] A. Giannakis and D. Peftitsis, "Electro-thermal design of a solid-state mvdc circuit breaker," in *2019 10th International Conference on Power Electronics and ECCE Asia (ICPE 2019 - ECCE Asia)*, 2019, pp. 1–8.

[6] R. Kunzi, "Thermal design of power electronic circuits," 2016. [Online]. Available: https://api.semanticscholar.org/CorpusID:119280390

[7] X. Lyu, H. Li, Z. Ma, B. Hu, and J. Wang, "Dynamic voltage balancing for the high-voltage sic super-cascode power switch," *IEEE Journal of Emerging and Selected Topics in Power Electronics*, vol. 7, no. 3, pp. 1566–1573, 2019.

[8] J. Millán, P. Godignon, X. Perpiñà, A. Pérez-Tomás, and J. Rebollo, "A survey of wide bandgap power semiconductor devices," *IEEE Transactions on Power Electronics*, vol. 29, no. 5, pp. 2155–2163, 2014.

[9] B. Gao, A. Morgan, Y. Xu, X. Zhao, B. Ballard, and D. C. Hopkins, "6.5kv sic jfet-based super cascode power module with high avalanche energy handling capability," in *2018 IEEE 6th Workshop on Wide Bandgap Power Devices and Applications (WiPDA)*, 2018, pp. 319–322.

[10] Y. Zhang, M. Tang, Q. Song, X. Tang, H. Lv, and S. Liu, "High temperature characterization of normally-on 4h-sic junction field-effect transistor," *Superlattices and Microstructures*, vol. 99, pp. 113–117, 2016.

[11] S. Wang, Z. Song, P. Fu, K. Wang, X. Xu, W. Tong, and Z. Wang, "Thermal analysis of water-cooled heat sink for solid-state circuit breaker based on igcts in parallel," *IEEE Transactions on Components, Packaging and Manufacturing Technology*, vol. 9, no. 3, pp. 483–488, 2019.

979-8-3503-3714-3/23 $31.00 © 2023 IEEE

Highly-Integrated, Low-Noise, Dual-Output GaN DC/DC for GaN Solid State Power Amplifier Supplies in Space Applications

Dominik Koch, Jeremy Nuzzo, Michael Bosch, Manuel Rueß, Dominik Wrana, Benjamin Schoch
and Ingmar Kallfass

Institute of Robust Power Semiconductor Systems, University of Stuttgart, Germany (dominik.koch@ilh.uni-stuttgart.de)

Abstract—In this work a highly-integrated, 24 V input, dual-output (5 to 18 V up to 1 A and −1.3 V up to 100 mA), low-noise GaN-based DC/DC supply for the gate and drain bias of a GaN solid state power amplifier for E- and W-band applications in space ór phased-array applications is presented. By using two fast-switching GaN half-bridges as the main element of two highly-integrated DC/DC converters for gate and drain bias supply, a ≈50 % higher efficiency and an adjustable drain voltage (e.g. for orbital tracking) can be achieved compared to state-of-the-art solutions (LDOs only). The required drain supply of the GaN SSPA (5 to 18 V with currents up to 1 A) is generated by a 70 mΩ, 100 V monolithic GaN half-bridge based buck-converter switching at 3.1 MHz with high gate-resistors (62 Ω/10 Ω), extended filtering (resulting in an ac voltage noise of below 5 mV RMS for most bias points) and efficiencies above 80 % for the whole output current range. The gate supply (−1.3 V, ≤100 mA) is provided by a multi-stage approach to achieve efficient conversion and low noise: In the first stage the battery voltage is down converted to 5 V in a 3.1 MHz buck-converter ($\eta \geq 90$ %) with two 3.3 Ω, 65 V GaN transistors, to optimize the efficiency of the second stage. The second stage (switched capacitor, 2 MHz circuit, $\eta \geq 80$ %) converts the voltage to −2 V. A final LDO stage converts the −5 V to −1.3 V with 0.8 µV RMS-noise, to avoid instabilities of the SSPA due to gate bias oscillations. The converter is not actively cooled and does not exceed a local hotspot temperature of 85 °C at any operation point.

Index Terms—gallium nitride, DC/DC converter, performance evaluation, low-voltage, low-noise, multi-stage converter, bias supply, SSPA, E-band amplifier

I. INTRODUCTION

For space and phased-array applications, efficiency and thermal aspects of the supply of GaN SSPAs, which typically suffer from low power added efficiency (PAE), are critical since they consume the major amount of power [1]–[4] in an RF-frontend. For example, the GaN-based SSPA with an output power above 1 W at E-/W-band millimeter wave frequencies and a low power added efficiency (PAE) of ≤20 % [1], [5] requires an input power of 8.4 W, which is more than 70 % of the necessary RF-frontend power shown in Fig. 1. Typically, the supply voltages are generated with low dropout regulators (LDO) [6], [7], which have high losses in down-converting high input voltages above 20 V to the required drain and especially gate bias voltages with low-noise to

avoid instabilities and oscillations. Furthermore, LDOs requires complex approaches to adjust the output voltage with sufficient speed [8]. Therefore, in this work two 3.1 MHz GaN DC/DC stages are used to improve the efficiency significantly, while maintaining a high power density (≥ 125 W/in³ for both gate and drain bias) and a low noise output (ac voltage noise below 5 mV RMS) only achievable by GaN devices, since they are able to achieve switching frequencies up to 10 MHz and therefore enable much efficient and compact converters in comparison to Silicon (Si) devices. In addition, a variable drain voltage can be implemented to enable environmental (slow, e.g. in highly dynamic down-/uplinks in low-earth orbit cubesats or terrestrial internet backbones with varying weather conditions) or envelope-tracking (fast, supply-modulation [9], [10]), necessary to adjust the (required) RF output power for efficiency improvement [11]–[14]. Again, only a fast switching GaN-based half-bridge can enable a fast load response and therefore tracking frequency. Several approaches with switched mode power supplies units (PSU) are presented in [10]–[16], where mainly the drain bias is optimized, since it requires significantly more power than the gate bias, or they require expensive chip area [10]. Therefore, in this work a single-input/dual-output DC/DC converter based on two discrete GaN-based buck converters with high switching frequency (3.1 MHz) for gate and drain bias supply and an output power of up to 18 W is presented. To achieve a compact design together with low output noise advanced filtering, high gate resistor values and a multi-stage approach for the gate bias is suggested. First, the single-input/dual-output converter design is presented in section II. In section III, the evaluation of the drain bias supply in terms of efficiency, noise behavior and temperature is evaluated and information on the gate supply are illustrated. Finally, a conclusion and the outlook on future work are given.

II. 24 V INPUT, DUAL-OUTPUT GaN BASED DC/DC CONVERTER DESIGN AND SETUP FOR LOW-NOISE APPLICATIONS

As mentioned in the introduction, the purpose of this work is to increase the PAE of a GaN SSPA, which is the most power-consuming component of an E-/W-band trans-

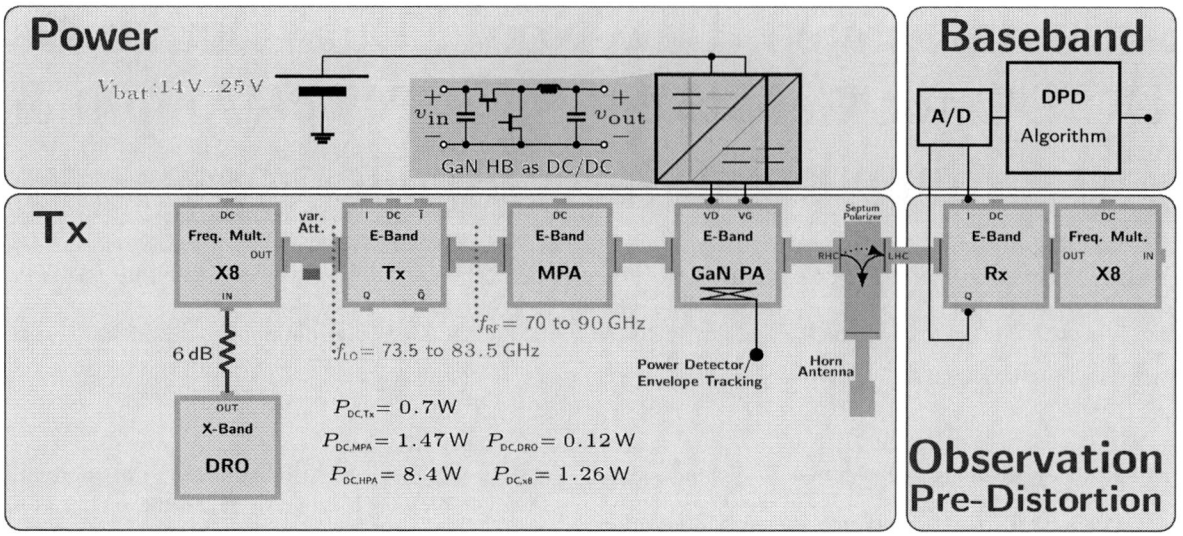

Fig. 1. Block diagram of the E-band transmitter chain consisting of the RF-components (dielectric resonator oscillator (DRO), eight-times frequency multiplier (x8), transmitter (Tx), medium power amplifier (MPA) and GaN power amplifier (PA) for the transmitter and receiver (Rx) and x8 for the receiver (based on [1]) and the baseband block for frequency down-conversion of an envelope signal as mutual input for the control of the DC/DC) and the power part. For the drain and the gate of the GaN PA, two DC/DC converters (which utilize a 3.1 MHz GaN half-bridge as central DC/DC stage) are used to convert the battery voltage (14 to 25 V) to 8 to 15 V for the drain voltage and -1.3 V for gate voltage. The gate voltage supply is further utilizing a switched capacitor IC (converts 5 V to -5 V) and a low-noise LDO for a low-noise gate bias. The power PCB is shown in Fig. 2

mitter (Tx) frontend as shown in Fig. 2. The GaN SSPA is consuming 8.4 W, which is $\approx 70\%$ of the whole transmitter chain power (11.95 W). To minimize the power consumption, the LDO-based gate and drain bias supply is replaced by switch-mode power supply units (PSU), where simple GaN DC/DC converters acts as main power electronics building block [7]. The GaN DC/DC converters are based on low-voltage GaN devices in combination with an analog control IC, which allows a maximum switching frequency f_{sw} of around 3.1 MHz, resulting in small output filter values for a compact design. The drain bias supply is using a 70 mΩ, 100 V monolithic GaN half-bridge, together with a multi-stage output filtering and reduced voltage slew rates due to high gate resistances of $R_{g,off}$: 62 Ω for the turn-off and $R_{g,on}$: 10 Ω for the turn-on. The resulting drawbacks in terms of efficiency are comparably low in comparison to the advantage in the output noise (more details in [7]). The input voltage range V_{in} is 18 to 24 V with an output voltage range V_{out} between 5 to 18 V with currents up to 1000 mA.

The gate bias supply is based on a similar DC/DC conversion stage with two discrete 3.3 Ω, 65 V GaN transistors for high- and low-side. Other parameters as e.g. gate-resistors are identical to the drain supply. After the first voltage down-conversion stage from the input voltage V_{in} to 5 V, a switched capacitor converter is converting the voltage to -5 V, providing a galvanic isolation, necessary for the single-input, dual-output converter. The switched capacitor converter is switching at 2 MHz, resulting in a small flying capacitor value of 1 µF. In the final stage, a low-noise LDO is providing the necessary gate bias supply v_G of -1.3 V with an RMS voltage noise of 0.8 µV. The gate bias supply is supporting currents up

to 100 mA. The details of the used components (transistors, magnetics and ICs) are indicated in Table I. The converter is shown in Fig. 2, designed on a six-layer PCB with 34 mm edge length fitting in a split-blockwaveguide module (cf. [1]) and is passively cooled (in air: convection, in vacuum: radiation), resulting in a power density of around 125 W/in^3 for the dual-

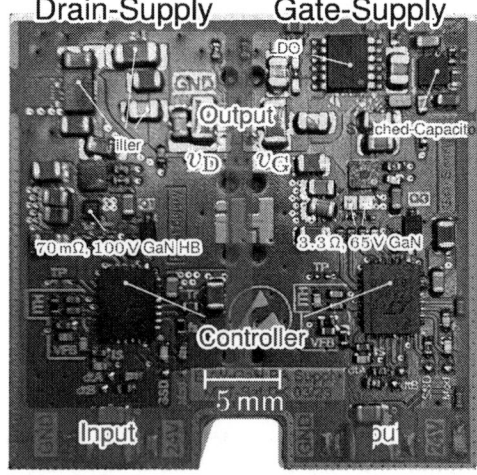

Fig. 2. Top-side view of the dual-output DC/DC PCB containing the drain (8 to 20 V up to 850 mA) and the gate supply (-1.3 V up to 100 mA) based on GaN half-bridges switching at 3.1 MHz.

979-8-3503-3714-3/23 $31.00 © 2023 IEEE

TABLE I
PARAMETERS OF THE DESIGNED 24 V INPUT, DUAL-OUTPUT, 3.1 MHz GaN DC/DC CONVERTER.

Drain-Supply V_D			
Monolithic GaN half-bridge			
$V_{ds,max}$	100 V		
$R_{ds,on}$	55 mΩ to 70 mΩ		
$V_{gs,max}$	−4 V/6 V		
High-side transistor		**Low-side transistor**	
C_{ISS}	79 pF	C_{ISS}	79 pF
C_{OSS}	52 pF	C_{OSS}	61 pF
C_{RSS}	0.5 pF	C_{RSS}	0.5 pF
$R_{g,int}$	1.3 Ω	$R_{g,int}$	1.5 Ω
Q_{OSS}	3960 pC	Q_{OSS}	4680 pC
Output filter			
Coil			
Inductance	3.3 µH	DCR (max.)	228 mΩ
SRF	55 MHz	I_{sat} (30 % drop)	1.05 A
Gate-Supply V_G			
GaN transistor for LS and HS			
$V_{ds,max}$	65 V	C_{ISS}	7 pF
$R_{ds,on}$	1.7 Ω to 3.3 Ω	C_{OSS}	2 pF
$V_{gs,max}$	−4 V/5.75 V	C_{RSS}	0.013 pF
Q_{OSS}	114 pC	$R_{g,int}$	4.8 Ω
Coil			
Inductance	22 µH	DCR (max.)	3.06 Ω
SRF	55 MHz	I_{sat} (30 % drop)	0.240 A
Second Stage			
Low-Noise, Regulated, Switched-Capacitor Voltage Inverter			
2 MHz switching		Efficiency ≤ 91 %	
V_{in} : 2.7 to 5.5 V		V_{out} : −2.7 to −5.5 V	
Third Stage			
Ultralow Noise, Ultrahigh PSRR Negative Linear Regulator			
RMS noise: 0.8 µV (10 Hz to 100 kHz)		I_{out} ≤ 500 mA	
PSRR 74 dB (1 MHz)		V_{out} : −1.8 V	

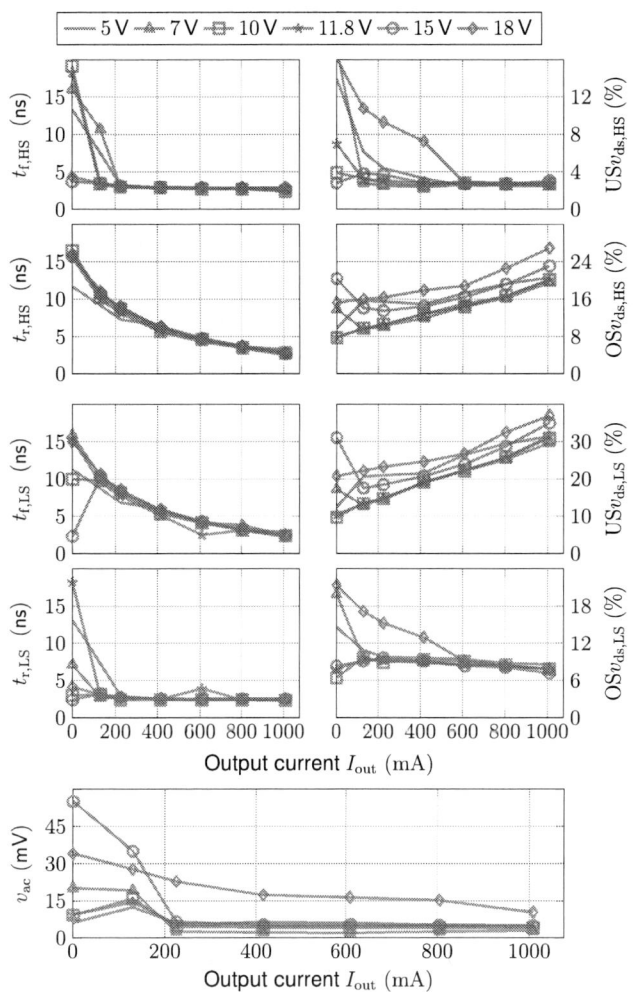

Fig. 3. Rise- and fall times $t_{r,f}$ and the corresponding drain-source voltage under- and overshoots $\mathrm{US}v_{ds}$ and $\mathrm{OS}v_{ds}$ for high-side and low-side transistors HS and LS over the output current I_{out} and in dependence of the output drain-bias voltages V_D. In the bottom the measured ac voltage amplitude $\hat{v}_{ac}$ at different I_{out} and V_D.

output converter. Considering, that the gate supply requires 50 % of the area with a maximum power contribution of 0.7 % to the converter, the drain bias supply has a power density of ≈220 W/in³, which is comparable to converters in similar power classes [7].

Due to packaging constraints, a component placement only on one layer is possible, which is an additional boundary condition for the power density.

III. DRAIN BIAS AND GATE BIAS SUPPLY EVALUATION

To achieve a minimum ac noise at the output for a stable operation of the GaN SSPA together with a reasonable efficiency, the drain bias supply of the designed DC/DC converter is evaluated in this section. Furthermore, a simulated and analytical loss breakdown of the three-stage gate bias supply approach is presented and discussed. The drain bias supply is evaluated at different drain voltage operation points v_D between 5 V (low RF power) and 18 V (high RF power) with a nominal drain bias supply voltage of 11.8 V. The output current I_{out} is swept between 0 mA (not occurring in

application) and 1 A. Fig. 3 is showing the rise- and fall times $t_{r,f}$ for both the high-side and low-side transistors as well as the corresponding under- and overshoot values (US/OS) for the respective drain-source voltages in dependence of v_D and I_{out}. It is shown, that the rise and fall times for all switching transients are reducing with increasing currents and saturating at approximately 2.5 ns at 1 A independent of the gate bias voltage. Some differences in the rise-and fall times at 0 mA can be attributed to the smart near zero dead-time setting of the control IC, where the dead-time is as small as possible (1 V threshold of the opposite voltage transition) and always below 20 ns. Since the voltage rise and fall times at 0 A are only hard-switching, the voltage detection differs between the different voltages due to the steep slope of the drain-source voltages. The voltage over- and undershoots are mainly inversely proportional to the rise- and fall-times, as can be seen, for example, in the undershoots of the low-side drain-source voltage $USv_{\mathrm{ds,LS}}$ at 0 A. The high-side overshoot $OSv_{\mathrm{ds,HS}}$ and the low-side undershoot $USv_{\mathrm{ds,LS}}$ as well as the high-side undershoot $USv_{\mathrm{ds,HS}}$ and the low-side overshoot $OSv_{\mathrm{ds,LS}}$ are proportional to each other, since they occur at the same switching event. While for the low-side turn-off, the overshoot saturates far below 10 %, the low-side turn-on could further be optimized, by increasing e.g. the high-side turn-off gate-resistor values $R_{\mathrm{g,off,HS}}$. Nevertheless, the bottom part of Fig. 3 indicates the measured ac voltage amplitude $\hat{v}_{\mathrm{ac}}$ for the different bias points. Expect for the drain bias voltage V_{D} of 18 V, where the effective output capacitor values are decreasing [17], the voltage amplitude $\hat{v}_{\mathrm{ac}}$ achieves values of around 5 mV at I_{out} around 200 mA, which is sufficient for the application and similar to prior work [7].

Beside the noise behavior of the converter, the efficiency η is investigated in Fig. 4, since the performance should significantly be improved in comparison to the LDO-based solution. The peak efficiency for the nominal operation point of $V_{\mathrm{D}} = 11.8$ V is 89.2 % at 800 mA with a nearly flat behavior in the current range between 400 to 1000 mA. The highest efficiencies η of around 93 % are achieved at $V_{\mathrm{D}} = 18$ V and is never below 70 % (low power: 5 V with 200 mA) for all real bias points that occur. Through a detailed investigation of the gate-resistor values [7], a further efficiency enhancement together with similar or improved noise behavior will be investigated in future work. Since the converter is integrated

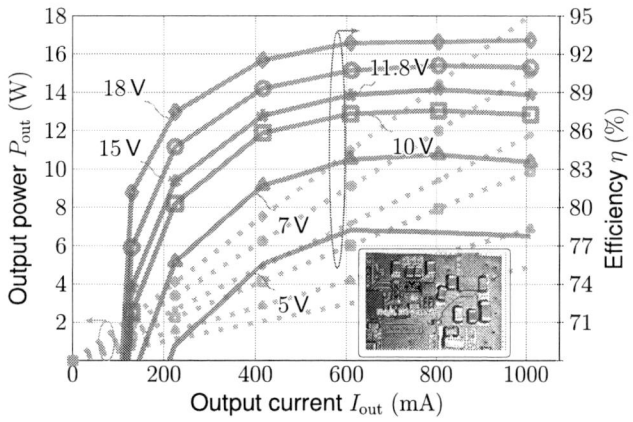

Fig. 4. Efficiency η and output power P_{out} of the drain-supply in dependence of the bias drain voltage V_{D} and the output current I_{out}. The peak efficiency of $\approx$92.5 % is achieved at 600 mA at 18 V. The gate-resistors for turn-off are $R_{\mathrm{g,on}} = 10\,\Omega$ and $R_{\mathrm{g,off}} = 62\,\Omega$ for turn-on, respectively.

in a split-blockwaveguide module, no active cooling can be utilized, resulting in a strictly passively cooled converter. To prove a safe temperature operation, a thermal image of the full power operation (drain bias supply: 18 V at 1 A and gate bias supply at -1.8 V and 100 mA) is shown in Fig. 5. For the maximum operation point, the hotspot temperature is occurring at the GaN transistors for the gate-supply, due to the high gate-resistor values in combination with the small duty-cycle and therefore high-switched currents (forced continuous

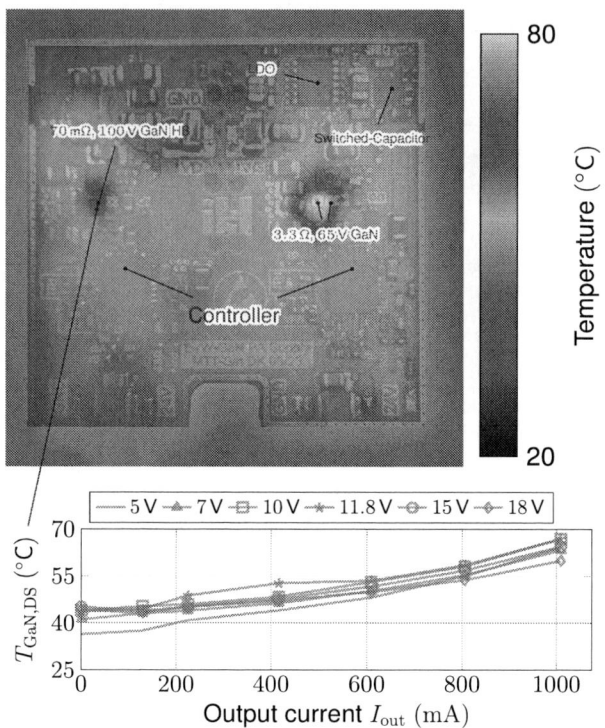

Fig. 5. Thermal image of the supply from Fig. 2 at full output power (drain bias supply: 18 V at 1 A and gate bias supply at -1.8 V and 100 mA). Bottom part: Temperature of the drain-supply monolithic half-bridge $T_{\mathrm{GaN,DS}}$ in dependence of the bias drain voltage V_{D} and the output current I_{out} measured at an ambient temperature of 22 °C. The temperature increase is mainly current dependent and shows no dependence on the drain bias voltage.

operation) and third-quadrant conduction. Since 100 mA of gate bias current is the worst case and the temperature is below 90 °C, the converter is still in a safe-operation area and the gate bias supply can be optimized in future work. In the later application, the converter is mounted into the split-blockwaveguide module [7], where conducted heat transfer to the split-block module with a significant thermal mass is affecting the ambient temperature due to the high loss power of the SSPA but simultaneous thermal convection or radiation (space) through the module. To keep the gate bias supply in a safe operation area, the ambient temperature should not exceed 60 °C. The temperature increase $T_{\mathrm{GaN,DS}}$ of the GaN monolithic half-bridge for the drain bias supply is shown in the bottom part of Fig. 5 and is a strict function of the output current I_{out} and not dependent on the drain bias voltage $_{\mathrm{D}}$. The maximum temperature is around 65 to 70 °C. Beside the drain bias supply, the gate bias supply is also investigated in this work in comparison to the state-of-the-art: Since the high battery voltage V_{in} of above 20 V has to be down-converted to below 2 V, a single-stage LDO is either not available or is resulting in very poor efficiencies of below 10 % (calculated at the example of the used LDO). As shown

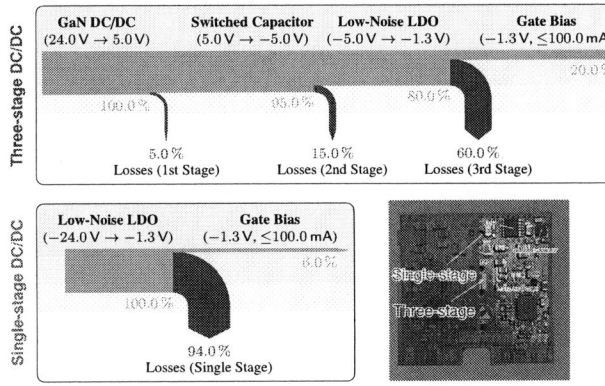

Fig. 6. Simulated and analytical loss breakdown of the single-stage LDO-based DC/DC conversion directly from (−) 24 V to −1.3 V and for the three-stage approach with GaN DC/DC converter, switched capacitor and low-noise LDO used in this work. The single-stage approach requires a separate supply with reversed polarity, which is neglected in this calculation (functionality of the switched capacitor in the three-stage approach).

in Fig. 6 a three-stage approach with individual efficiencies of the first GaN DC/DC $\eta_{\mathrm{GaN,DCDC}} = 95\,\%$, the second stage (switched capacitor) $\eta_{\mathrm{swCap}} = 95\,\%$ and the last LDO-stage $\eta_{\mathrm{LDO}} \approx 25\,\%$ is still more than three times efficient ($\eta_{\mathrm{threeStage}} = 20\,\%$) compared to the single-stage approach with an efficiency $\eta_{\mathrm{singleStage}}$ of $6\,\%$. If the necessity of a galvanic isolation is taken into account, the efficiency of the single-stage approach is further decreasing or requires a second supply with inversed polarity. Nevertheless, due to the low overall power (loss) contribution to the whole converter, a design choice between efficiency (three-stage approach) or power density (single-stage approach) is possible and further dependent on the given design and input supply constraints.

CONCLUSION AND FUTURE WORK

In this work a highly-integrated ($\geq 125\,\mathrm{W/in}^3$ for both gate and drain bias), 24 V input, dual-output (8 to 20 V up to 1 A and $-1.3\,\mathrm{V}$ up to 100 mA), low-noise ($\hat{v}_{\mathrm{ac}}: \leq 5\,\mathrm{mV}$) GaN-based DC/DC supply for the gate and drain bias supply of a GaN solid state power amplifier for E- and W-band applications in space ór phased-array applications is presented. Utilizing two different (discrete and monolithic) GaN half-bridges as the core components in two DC/DC converters, one dedicated to gate bias supply and the other to drain bias supply, can yield an enhancement in efficiency, approximately 50 % higher than current solutions, which rely solely on LDOs. Furthermore, this setup allows for the flexibility of adjusting the drain voltage, enabling functionalities like orbital tracking. The GaN SSPA's required drain bias supply, spanning from 5 to 18 V with currents of up to 1000 mA, is generated through a carefully designed system. This system features a monolithic GaN half-bridge operating at a 3.1 MHz and employs high gate-resistors ($62\,\Omega/10\,\Omega$) for optimized noise to efficiency performance. Extended filtering has been added to reduce AC voltage noise to below 5 mV RMS for most

bias points. The system maintains efficiencies exceeding 80 % across the entire application output current range. For the gate supply, which demands a voltage of $-1.3\,\mathrm{V}$ and a current below 100 mA, a multi-stage approach is employed to ensure efficient conversion and minimal noise. In the first stage, the input battery voltage is reduced to 5 V using a 3.1 MHz buck-converter with two $3.3\,\Omega$, 65 V GaN transistors, achieving an efficiency of at least 90 %. The second stage utilizes a switched capacitor circuit operating at 2 MHz, achieving an efficiency of at least 80 % and converting the voltage to $-5\,\mathrm{V}$. Finally, a LDO stage is providing $-1.3\,\mathrm{V}$ with a low noise level of $0.5\,\mu\mathrm{V}$ RMS, necessary for a stable SSPA operation. The entire converter operates without active cooling and stays below a maximum local hotspot temperature of $90\,°\mathrm{C}$ across all operational conditions. In future work, the efficiency to noise behavior of the drain bias supply as well as the gate bias supply will further be investigated and optimized. Furthermore, the gate bias supply could further be improved by a decrease of the first intermediate voltage to 2.7 V (minimum operation voltage of the switched capacitor converter) to further reduce the dropout voltage at the LDO. In combination with improved gate-resistor values and adapted filter inductance, this could result in a more efficient operation of the first GaN DC/DC stage and a colder operation temperature. Finally, a joint operation with the mentioned GaN SSPA will be investigated.

ACKNOWLEDGMENT

TThis work has been supported in the frame of the MTT-Sat Challenge project "W-Band GaN PA with Pre-Distortion and Efficiency Enhancement Based on GaN DC/DC Variable Power Supply for Space Applications".

REFERENCES

[1] B. Schoch, D. Wrana, R. Henneberger, S. Wagner, E. Ture, A. Tessmann, and I. Kallfass, "E-band transmitter with 3 w complex modulated signal output power performance," in *2021 51st European Microwave Conference (EuMC)*, 2022, pp. 458–461.

[2] V. Prabhu, Vengadarajan, and V. John, "Realisation of dc-dc converters for active phased array radar applications," in *2022 IEEE International Conference on Power Electronics, Drives and Energy Systems (PEDES)*, 2022, pp. 1–6.

[3] R. Tong and D. Dancila, "Compact and highly efficient lumped push-pull power amplifier at kilowatt level with quasi-static drain supply modulation," in *2020 IEEE/MTT-S International Microwave Symposium (IMS)*, 2020, pp. 29–32.

[4] R. Giofrè, P. Colantonio, L. Gonzalez, L. Cabria, and F. De Arriba, "A 300w complete gan solid state power amplifier for positioning system satellite payloads," in *2016 IEEE MTT-S International Microwave Symposium (IMS)*, 2016, pp. 1–3.

[5] D. Schwantuschke, R. Henneberger, S. Wagner, A. Tessmann, I. Kallfass, P. Brückner, R. Quay, and O. Ambacher, "GaN-based E-Band Power Amplifier Modules," in *2016 46th European Microwave Conference (EuMC)*, 2016, pp. 564–567.

[6] Y. Wang, Y. Lu, Q. Pan, Z. Hou, L. Wu, W. Ki, and C. P. Yue, "A 3-mW 25-Gb/s CMOS Transimpedance Amplifier with Fully Integrated Low-Dropout Regulator for 100GbE Systems," in *2020 IEEE Radio Frequency Integrated Circuits Symposium (RFIC)*, 2020, pp. 275–278.

[7] D. Koch, D. Wrana, B. Schoch, and I. Kallfass, "Low-noise, 24 v, 1 a, 2.1 mhz gan dc/dc converter for variable power supply of a gan-based solid-state power amplifier," in *2022 IEEE Applied Power Electronics Conference and Exposition (APEC)*, 2022, pp. 1208–1213.

[8] R. Wu, Y. Tsukui, R. Minami, K. Okada, and A. Matsuzawa, "A 0.7 V-to-1.0V 10.1 dBm-to-13.2 dBm 60-GHz Power Amplifier using Digitally-assisted LDO considering HCI issues," in *2012 IEEE Asian Solid State Circuits Conference (A-SSCC)*, 2012, pp. 269–272.

[9] N. Wolff, S. Chevtchenko, A. Wentzel, O. Bengtsson, and W. Heinrich, "Switch-type modulators and pas for efficient transmitters in the 5g wireless infrastructure," in *2018 IEEE MTT-S International Microwave Workshop Series on 5G Hardware and System Technologies (IMWS-5G)*, 2018, pp. 1–3.

[10] S. Bhardwaj, S. Moallemi, and J. Kitchen, "A review of hybrid supply modulators in cmos technologies for envelope tracking pas," *IEEE Transactions on Power Electronics*, vol. 38, no. 5, pp. 6036–6062, 2023.

[11] P. Zhou, X. Ruan, N. Liu, and Y. Wang, "A series-parallel-form switch-linear hybrid envelope tracking power supply with two multilevel converters sharing a voltage-level provider," *IEEE Transactions on Power Electronics*, vol. 38, no. 1, pp. 593–603, 2023.

[12] X. Liu, J. Jiang, C. Huang, and P. K. T. Mok, "Design techniques for high-efficiency envelope-tracking supply modulator for 5th generation communication," *IEEE Transactions on Circuits and Systems II: Express Briefs*, vol. 69, no. 6, pp. 2586–2591, 2022.

[13] J. Rodriguez, J. R. Garcia-Mere;, D. G. Aller, and J. Sebastia;n, "Pulsewidth modulated three-level buck converter based on stacking switch-cells for high power envelope tracking applications," *IEEE Transactions on Power Electronics*, vol. 37, no. 5, pp. 5786–5800, 2022.

[14] C. Nogales, Z. Popović, and G. Lasser, "A 600-w enhancement-mode gan multi-level dynamic converter for supply modulated pas," in *2021 51st European Microwave Conference (EuMC)*, 2022, pp. 902–905.

[15] W. Yan, Y. Wang, Z. Wang, Y. Jin, and G. Shi, "The Design of Power Amplifier Switch Mode Power Supply for Handset Applications," in *2015 IEEE International Conference on Cyber Technology in Automation, Control, and Intelligent Systems (CYBER)*, 2015, pp. 1394–1398.

[16] W. -T. Lin, Z. -Y. Lin, C. -H. Liu, K. -H. Chen, Y. -H. Lin, J. -R. Lin, and T. -Y. Tsai, "A 20MHz Low Dropout Controlled Current Sensor for Constant On-Time Based Envelop Tacking Supply Modulator for Radio Frequency Power Amplifier," in *2018 IEEE International Symposium on Circuits and Systems (ISCAS)*, 2018, pp. 1–4.

[17] J. Weimer, D. Koch, and I. Kallfass, "Compact half-bridge module for a charger application utilizing gan power devices with integrated driver," in *PCIM Europe 2022; International Exhibition and Conference for Power Electronics, Intelligent Motion, Renewable Energy and Energy Management*, 2022, pp. 1–8.

Gate Driver with Dynamic Drive Strength on High-temperature CMOS Process for Heterogeneous Integration inside the SiC Power Module

Asif Faruque
*Department of Electrical
Engineering & Computer Science
University of Arkansas*
Fayetteville, AR 72701, USA
kafaruqu@uark.edu

Ayesha Hassan
*Department of Electrical
Engineering & Computer Science
University of Arkansas*
Fayetteville, AR 72701, USA
ah111@uark.edu

Yuyang Wang
*Department of Electrical
Engineering & Computer Science
University of Arkansas*
Fayetteville, AR 72701, USA
yuyangw@uark.edu

*H. Alan Mantooth
Department of Electrical
Engineering & Computer Science
University of Arkansas*
Fayetteville, AR 72701, USA
mantooth@uark.edu

Abstract— **This paper presents the design, integration technique, and test results of a non-isolated single-channel gate driver for heterogeneous integration inside a SiC power module. In order to survive at the anticipated junction temperature, the driver is built on a 180 nm silicon-on-insulator (SOI) CMOS technology that can run safely up to 175 °C. Multiple pull-up (PMOS) slices and pull-down (NMOS) slices arranged in parallel serve as the current sourcing and sinking components of the gate driver. Through a controller block, each slice can be independently turned on or off, allowing the gate driver's driving strength to be adjusted. Such dynamic drive strength can be crucial to optimizing the SiC power die's voltage overshoot and switching loss in power converter applications. Additionally, a die temperature monitoring circuit is integrated inside the gate drive chip. The packaging, module integration techniques, and power switching test results using the designed gate driver are also presented in this work.**

Keywords—Gate driver IC, heterogenous integration, high-temperature electronics, SiC power module, SOI CMOS process.

I. INTRODUCTION

The movements toward electrification in transportation systems like electric aircraft and vehicles and hybrid electric vehicles pose a serious challenge to future power electronic converters [1] – [3]. SiC devices can be a promising candidate to meet these challenges, featuring normally off-state, lower conduction and switching loss, expanded operating temperature capabilities, higher thermal conductivity, and breakdown voltage [4] – [8]. However, the connection parasitics between the module housing and the gate driver must be kept to a minimum in order to prevent oscillation across the power devices, which is likely to result from quicker switching inside the SiC power module. The impacts mentioned can be significantly diminished by integrating a gate driver inside the module. Also, by eliminating the need for aggressive active and passive cooling systems for such a heterogeneously integrated system, higher power-to-volume

and power-to-weight ratios can be achieved. Conventional silicon-based gate drivers fail to survive at such a high junction temperature in SiC modules. Therefore, research is underway to integrate high-temperature gate drivers with SiC power modules. In [9] – [14], high-temperature gate drivers on silicon-on-insulator (SOI) process with high-temperature passive components are integrated into the power package to demonstrate the concept. However, they either lack the dynamic drive strength feature or require a complex interface with the controller unit to make that happen. In [15], drivers with up to 4 A fixed sourcing and sinking current capacity are reported. This work presents the design and test results of a high-current, non-isolated single-channel gate driver on a high-temperature process for heterogeneous integration inside a SiC power module with a dynamically adjustable drive strength feature. Additionally, a die temperature monitoring circuit is integrated inside the gate drive chip since the die is expected to be placed in close proximity to the SiC power die and exposed to significant passive heating. The following sections briefly describe the design techniques and key test results for the fabricated gate driver. The packaging and module integration techniques of the driver are also briefly discussed. A successful demonstration of double pulse tests (DPT) conducted on a commercial half-bridge SiC power module up to 800 V and 200 A using this module-integrable SOI gate driver is also presented towards the end.

II. GATE DRIVER DESIGN

The gate driver design is implemented using the 180 nm silicon-on-insulator (SOI) CMOS process from XFAB, which can operate safely up to 175 °C and maybe even above this value to survive at the projected junction temperature inside the module. The following subsections briefly describe the top-level building blocks of the gate driver without elaborating much on the transistor-level circuit design and present some key test results.

The work in this paper is supported by the Department of Energy under Grant DE- EE0008707.

A. The Driver Core

As the current sourcing and sinking components, the gate driver core is made up of four pull-up (PMOS) slices and four pull-down (NMOS) drive slices arranged in parallel. Each slice can be triggered individually through a CMOS buffer chain, as shown in Fig. 1, ensuring a variable drive strength feature for the gate driver. The input capacitances of these drive slices are significant due to their relatively larger width, which allows them to source and sink a considerable amount of current. Hence, the length of the CMOS buffer chain and transistor sizing inside the chain are optimized to reduce the loading effect on the pre-driving circuits, thereby reducing the PWM input-to-gate driver output propagation delay. Fig. 2 shows the schematics of a single pull-up slice and a single pull-down slice, along with their CMOS buffer chains. The positive and negative voltage rails for both the chain and the drive slice are +15 V and -3 V, respectively. Each chain receives either +15 V or -3 V from the controller unit and activates or deactivates the drive slice accordingly.

B. Controller Units

Two sperate controller units generate the switching signals for the driver core's CMOS buffer chains to operate the pull-up and pull-down drive slices. Fig. 3 shows the block-level schematic of the pull-up controller unit. A similar architecture exists for the pull-down controller unit. The first block inside the controller unit is the flash analog-to-digital converter (ADC). Fig. 4 shows the schematic of the flash ADC. The positive and negative power supplies for the ADC are 0 V and -3 V, respectively. Depending on the magnitude of the analog control voltage it receives, the output pins C1 through C4 stay high (0V) or low (-3V).

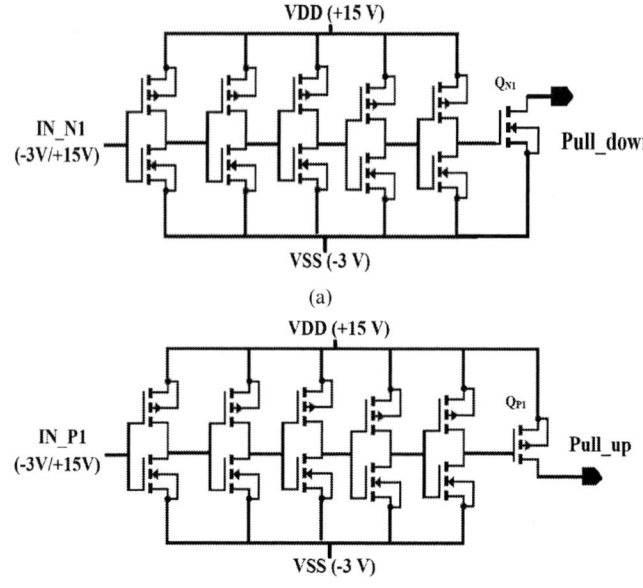

Fig. 2. (a) A single pull-down, (b) A single pull-up slice driven by CMOS buffer chain

Table I summarizes the ranges for the control voltages and the status of the pins for each range. The next block inside the controller unit is the logic block. The logic block synchronizes the PWM input signal (-3V or 0V) with the output of the ADC pins to ensure the activated PMOS and NMOS drive slices are switching in a complementary fashion, whereas the non-activated drive slices maintain their non-switching states during the gate driver operation. For example, if the output of pin C1 is high, for both the pull-up and pull-down controller units, N1 is switching in phase with the PWM input signal (-3V or 0V). If C1 is low, N1 is held at -3V for the pull-up controller unit and at 0V for the pull-down controller unit. Similar logic states appear for the other pins as well. As to be noted, the logic levels for all the signals generated up to this point are -3V or 0V, whereas the logic levels for the driver core are -3V or +15V. The level shifter, the third block inside the controller unit, does this voltage conversion. These outputs are then passed into the driver core to activate or deactivate the desired numbers of pull-up and pull-down drive slices. Thus, using only two input control voltages, the drive strength for the gate driver can be adjusted.

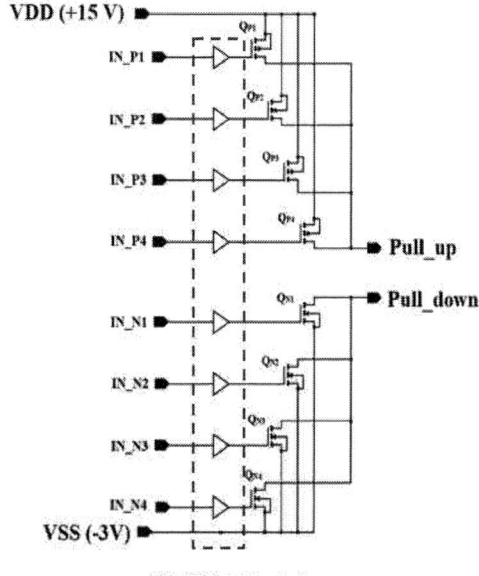

Fig. 1. Gate driver functional schematic

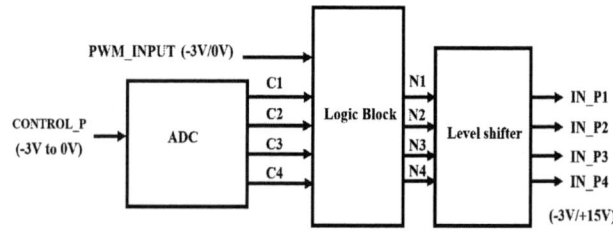

Fig.3. Block level schematic of the pull-up controller unit

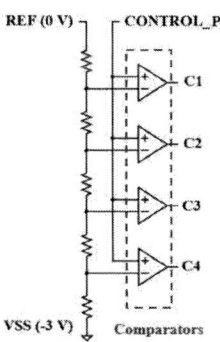

Fig. 4. Schematic of the flash analog to digital converter (ADC)

TABLE I. CONTROL VOLTAGES AND STATUS OF THE PINS

Control voltage	C1	C2	C3	C4
< -2	Low	Low	Low	Low
> -2 & <-1.7	Low	Low	Low	High
> -1.7 & <-1.4	Low	Low	High	High
> -1.4 & <-1.1	Low	High	High	High
> -1.1	High	High	High	High

C. Temperature Monitoring Circuit

Fig. 5 shows the schematic of the temperature monitoring circuit integrated with the gate driver on the same die. A p-type MOSFET is biased by a resistor divider network with its source connected to the positive supply voltage. A resistor is connected to the drain of the device, and the output is taken at this connection port. In this particular process, both the device and the resistor have a negative temperature coefficient, as the simulation results in Fig. 6 suggest. So, as we go higher in temperature, both the device current and the resistance decrease, causing a multiple-order voltage drop at the output node. Thus, by noting the output voltage, we can monitor the gate driver die temperature.

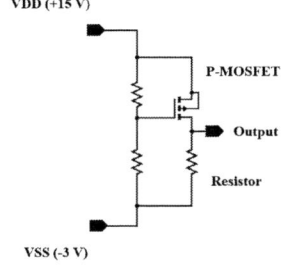

Fig. 5. Schematic of the temperature monitoring circuit

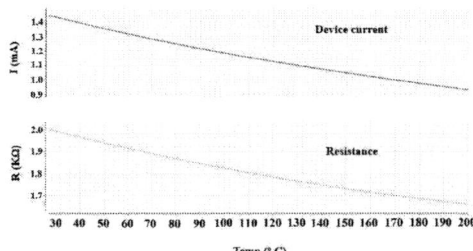

Fig. 6. Simulation results showing the degradation of device current and resistance with increasing temperature

III. DIE FABRICATION AND TEST RESULTS

All the designs and simulations are carried out in the Cadence Virtuoso environment. After being satisfied with the parasitic extracted (PEX) simulation results, the physical designs are laid down. Fig. 7 (a) shows the layout view of the gate driver, showing the relative positions of the driver core and the controller units. Fig. 7(b) shows the die micrograph of the fabricated gate driver die with the functionalities of the pads briefly outlined in Table II.

Fig. 8 shows the input-output response of the gate driver under a continuous pulse train, whereas Table III summarizes the gate driver's key performance figures both at room temperature and at an elevated temperature of 175 °C. All the measurements were carried out on the probe station by putting the die on a hot chuck. The gate driver is characterized by a 16 nF load capacitance, consistent with the expected capacitance to be seen by the driver inside the module, where two CPM3-1200-0013A power die are placed in parallel (an illustrative power module that the gate drive aims to drive). The driver's performance degrades a bit as we go up in temperature. This is due to the decreased carrier mobility at higher temperatures, an intrinsic property of the SOI process. However, the robust continuous operation of this driver even at 175 °C puts it much ahead of its silicon (Si) counterparts.

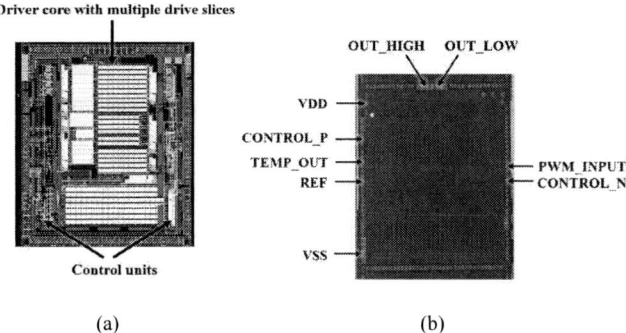

(a)   (b)

Fig. 7. (a) Layout view of the gate driver (b) micrograph of the fabricated die showing its pads

TABLE II. GATE DRIVER PADS AND THEIR FUNCTIONALITIES

Pad name	Type	Functionality
VDD (+15 V)	Power supply	Positive supply voltage
REF (0 V)	Power supply	Reference
VSS (-3 V)	Power supply	Negative supply voltage
PWM_INPUT	Input	PWM input signal into the gate driver
CONTROL_P	Input	Analog control voltage for pull-up drive strength variation
CONTROL_N	Input	Analog control voltage for pull-down drive strength variation
OUT_HIGH	Output	Pull_up output node
OUT_LOW	Output	Pull_down output node

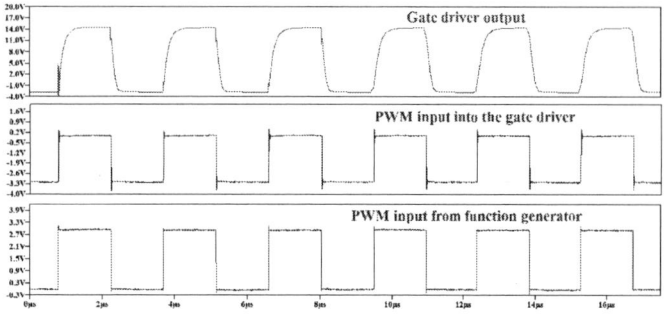

Fig. 8. Input-output response of the gate driver under continuous pulse train at 175 ℃

TABLE III. PERFORMANCE FIGURES OF THE GATE DRIVER

Test Condition	Parameter	Measured Value (25℃)	Measured Value (175℃)
$R_g = 2\,\Omega$, $C_{load} = 16$ nF Gate driver in full drive strength	Output voltage swing	+ 15 V / - 3 V	+ 15 V / - 3 V
	Peak source / sink current	4.15 A / 6.38 A	4.01 A / 6.06 A
	Rise time	272 ns	298 ns
	Fall time	163 ns	179 ns

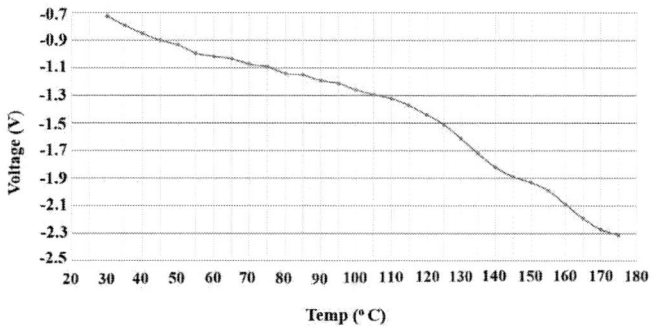

Fig. 9. Output voltage of the die temperature monitoring circuit

Fig. 9 shows the output voltage of the die temperature monitoring circuit. As expected, as we go up in temperature, the voltage drops quite reasonably. A lookup table (LUT) can be constructed to monitor the gate driver die temperature.

IV. PACKAGING AND MODULE INTEGRATION

Fig. 10 shows the bare die micrograph of the fabricated SOI gate driver. The die area is approximately 5.4 mm by 3.3 mm. Fig. 11 shows the top view of the power module-integrable PCB accommodating the gate driver die along with the power supply board. High-temperature non-conductive epoxy is used to attach the die to the PCB pad, whereas high-temperature Rogers 4350B material is used for PCB fabrication with all its pad electroless nickel immersion gold (ENIG) platted to facilitate 25 µm Al wire bonds. Connections are made directly from the die pad to the exposed pads on the

PCB. The input and output signals for the gate driver are routed through the PCB traces. The purpose of the power supply board is to provide an isolated power supply for the driver. It also contains PWM input signal isolators and isolation buffers with unity gain for the analog input control voltages for variable drive strength.

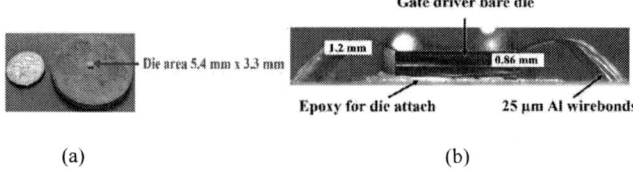

(a)   (b)

Fig. 10. (a) Gate driver die size comparison, (b) Close view of the wirebonded die

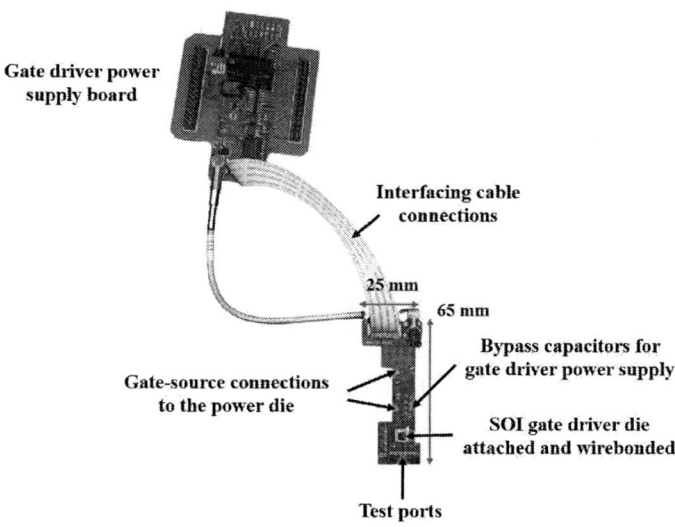

Fig. 11. Module-integrable gate driver PCB along with the power supply board

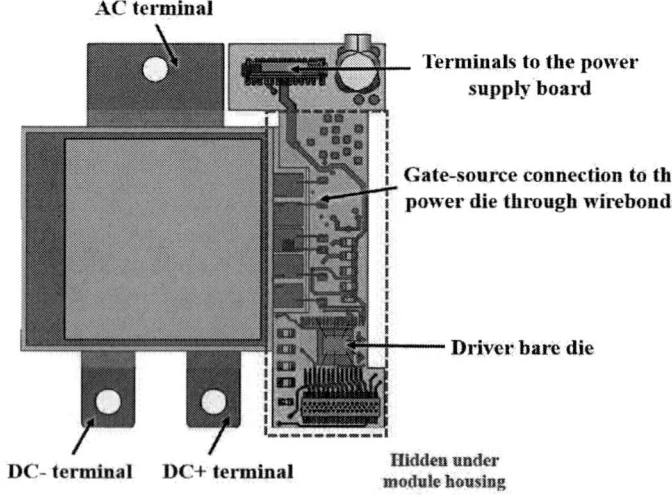

Fig. 12. Conceptual integration of the gate driver inside a SiC power module.

Fig. 12 shows the conceptual diagram of a half-bridge SiC power module where this module-integrable driver PCB can be integrated. Gate-source connections to the power die can be made directly through wirebonds. The PCB can then be safely hidden under the module housing. For the sake of simplicity, only the bottom switch substrate of the power module and the driving PCB are shown. A similar configuration can be assembled for the top switch on the other side (left side).

V. POWER SWITCHING TEST RESULTS

In order to validate the driving capability of the bare die gate driver, the module-integrable PCB is used to drive a commercial SiC module manufactured by Wolfspeed. Connections are made from the gate-source pads of the PCB to the gate-source pins of the half bridge module by smaller jumper wires. A successful double pulse test (DPT) is demonstrated for both the bottom and top switching positions up to 800 V and 200 A, as shown in Fig. 13 and Fig. 14. This validates the capabilities of the power supply board and module-integrable PCB to drive a SiC power module.

The dynamic drive strength feature of the gate driver is also validated. By changing the analog input control voltages for the pull-up (CP) and pull-down (CN) drive slices, the source and sink currents of the gate driver can be adjusted. These changes are reflected in the varying rise times (RT) and fall times (FT) of the gate driver, as shown in Fig. 15 and Fig. 16.

Fig. 17 and Fig. 18 show how the turn-on and turn-off switching dv/dt of the SiC modules are affected by the varying source and sink currents of the gate driver. Fig. 19 summarizes the varying switching energy of the module at a test condition of 800 V and 100 A at room temperature. Thus, by changing the magnitude of only two input control voltages, the switching speed of the module can be dynamically adjusted. Such adjustments can be crucial to optimize the switching loss and voltage/current overshoots of the SiC module in power electronic converter applications.

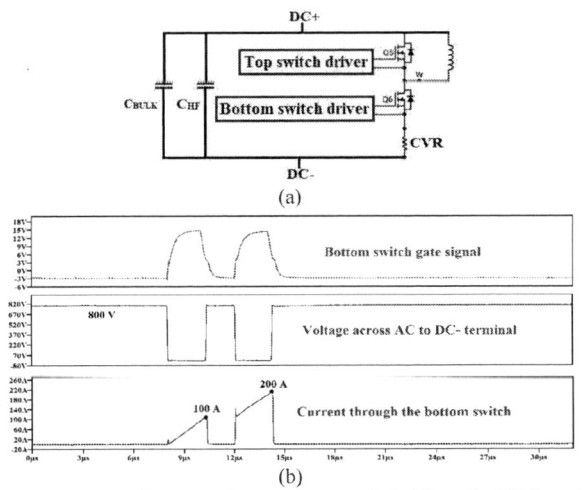

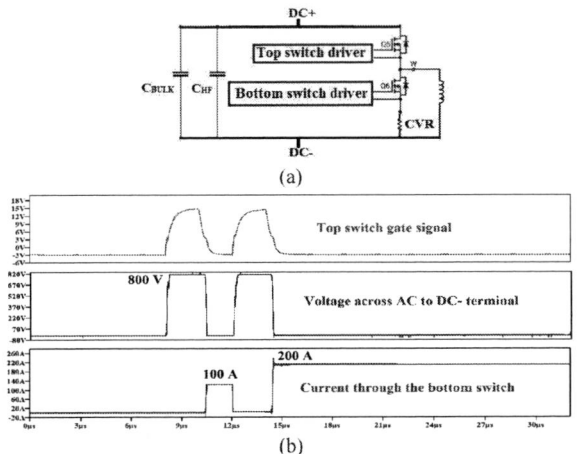

Fig. 13. (a) DPT configuration for the bottom switch driver, (b) DPT results up to 800 V, 200 A

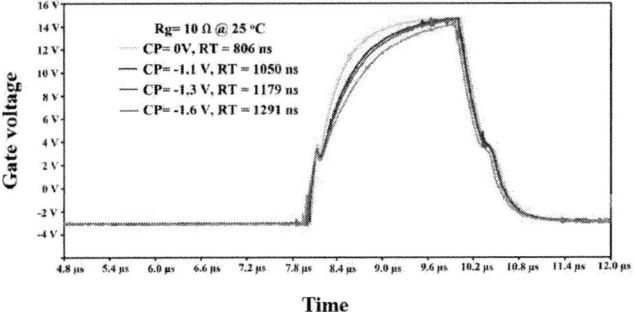

Fig. 15. Varying rise times of the gate driver with pull-up input control voltage (CP)

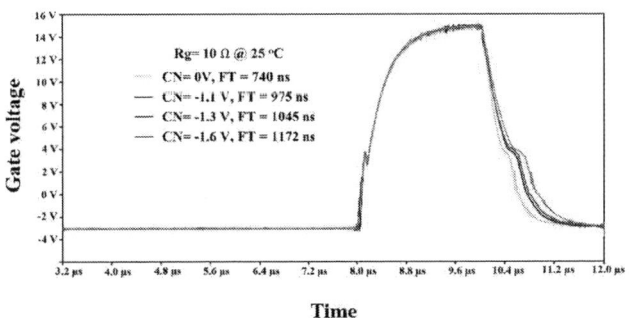

Fig. 16. Varying fall times of the gate driver with pull-down input control voltage (CN)

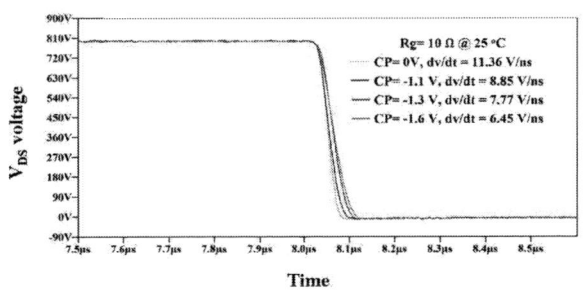

Fig. 14. (a) DPT configuration for the top switch driver, (b) DPT results up to 800 V, 200 A

Fig. 17. Varying turn-on dv/dt of the SiC module with gate driver's pull-up input control voltage (CP)

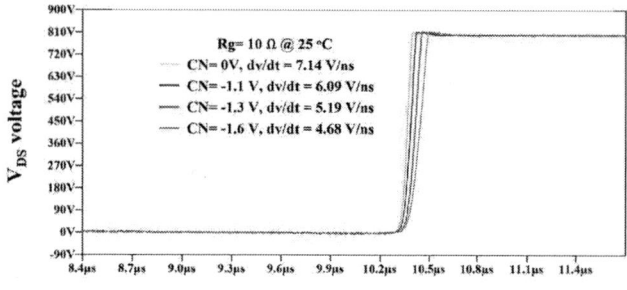

Fig. 18. Varying turn-off dv/dt of the SiC module with gate driver's pull-down input control voltage (CN)

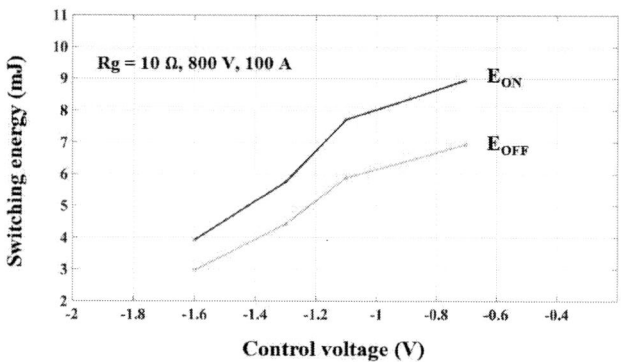

Fig. 19. Varying turn-on and turn-off switching energies of the module with gate driver's input control voltages.

VI. CONCLUSION

In this work, the design technique and test results of a high-temperature, non-isolated single-channel gate driver on a 180 nm SOI CMOS process for heterogeneous integration inside a SiC power module are presented. Two separate analog control voltages can control the driver's active pull-up and pull-down strengths. Such heterogeneous integration not only increases the system reliability by reducing the gate loop inductance but also helps improve the power density by integrating the gate driver and the auxiliary temperature monitoring circuitry into a single chip. Additionally, the SOI process gives them endurance at the anticipated elevated ambient temperature as they are positioned so near the power die inside the module.

REFERENCES

[1] J. Biela, M. Schweizer, S. Wafer, and J. W. Kolar, "SiC versus Si evaluation of potentials for performance improvement of inverter and DC -DC converter systems by SiC power semiconductors," IEEE Trans. Ind. Electron., vol. 58, no. 7, pp. 2872–2882, Jul. 2011.

[2] R. Wang, D. Boroyevich, P. Ning, Z. Wang, F. Wang, P. Mattavelli, K. Ngo, and K. Rajashekara, "A high-temperature SiC three-phase AC-DC converter design for > 100° C ambient temperature," IEEE Trans. Power Electron., vol. 28, no. 1, pp. 555–572, Jan. 2013.

[3] K. Rajashekara, "Present status and future trends in electric vehicle propulsion technologies," IEEE J. Emerg. Select. Topics Power Electron., vol. 1, no. 1, pp. 3–10, Mar. 2013.

[4] T. Funaki, J. C. Balda, J. Junghans, A. S. Kashyap, H. A. Mantooth, F. Barlow, T. Kimoto, and T. Hikihara, "Power conversion with SiC devices at extremely high ambient temperatures," IEEE Trans. Power Electron., vol. 22, no. 4, pp. 1321–1329, Jul. 2007.

[5] H. Zhang and L. M. Tolbert, "Efficiency impact of SiC power electronics for modern wind turbine full scale frequency converter," IEEE Trans. Ind. Electron., vol. 58, no. 1, pp. 21–28, Jan. 2011.

[6] R. A. Wood and T. E. Salem, "Evaluation of a 1200-V, 800-A all-SiC dual module," IEEE Trans. Power Electron., vol. 26, no. 9, pp. 2504–2511, Sep. 2011.

[7] I. Josifovic, J. Popovic-Gerber, and J. Ferreira, "Improving SiC JFET switching behavior under influence of circuit parasitics," IEEE Trans. Power Electron., vol. 27, no. 8, pp. 3843–3854, Aug. 2012.

[8] X. Wu, S. Cheng, Q. Xiao, and K. Sheng, "A 3600 V/80 A series-parallel connected silicon carbide MOSFETs module with single external gate driver," IEEE Trans. Power Electron., vol. 29, no. 5, pp. 2296–2306, May 2014.

[9] J. Hornberger, E. Cilio, R. Schupbach, A. Lostetter, and H. A. Mantooth, "A high-temperature multichip power module (MCPM) inverter utilizing silicon carbide (SiC) and silicon on insulator (SOI) electronics," in Proc. IEEE Power Electron. Spec. Conf., 2006, pp. 1–7.

[10] A. Lostetter, J. Hornberger, B. McPherson, B. Reese, R. Shaw, R. Schupbach, B. Rowden, H. A. Mantooth, J. Balda, T. Otsuka, K. Okumura, and M. Miura, "High-temperature silicon carbide and silicon on insulator based integrated power modules," in Proc. IEEE Vehicle Power Propulsion Conf., 2009, pp. 1032–1035.

[11] J. Hornberger, S. Mounce, R. Schupbach, B. McPherson, H. Mustain, H. A. Mantooth, W. Brown, and A. Lostetter, "High-temperature integration of silicon carbide (SiC) and silicon-on-insulator (SOI) electronics in multichip power modules (MCPMs)," in Proc. Eur. Conf. Power Electron. Appl., 2005, pp. 1–10.

[12] J. Hornberger, E. Cilio, B. McPherson, R. Schupbach, A. Lostetter, and H. A. Mantooth, "A fully integrated 300° C, 4 kW, 3-phase, SiC motor drive module," in Proc. IEEE Power Electron. Spec. Conf., 2007, pp. 1048– 1053.

[13] J. Valle-Mayorga, C. P. Gutshall, K. M. Phan, I. Escorcia-Carranza, H. A. Mantooth, B. Reese, M. Schupbach, and A. Lostetter, "High- temperature silicon-on-insulator gate driver for SiC-FET power modules," IEEE Trans. Power Electron., vol. 27, no. 11, pp. 4417–4424, Nov. 2012.

[14] Cissoid, HADES: High-Reliability, High-Temperature Half-Bridge Isolated Gate-Driver, Datasheet. (2011). [Online]. Available: http://www.cissoid.com/.

[15] Z. Wang et al., "A high temperature silicon carbide mosfet power module with integrated silicon-on-insulator-based gate drive," in IEEE Transactions on Power Electronics, vol. 30, no. 3, pp. 1432-1445, March 2015, doi: 10.1109/TPEL.2014.2321174.

Analysis of 10 kV SiC MOSFET Module Baseplate Parasitic Capacitance Impact on Switching Loss

Ruirui Chen[1], Min Lin[1], Dingrui Li[1], Zihan Gao[1], Fred Wang[1,2], Hua Bai[1], and Leon M. Tolbert[1]

[1]Min H. Kao Department of Electrical Engineering and Computer Science, The University of Tennessee, Knoxville, TN, USA

[2]Oak Ridge National Laboratory, Oak Ridge, TN, USA

rchen14@vols.utk.edu

Abstract — The parasitic capacitance of a power module baseplate provides a path for displacement current during the switching transient of the power module, which causes extra switching loss and EMI noise. It becomes an increasing concern especially for emerging 10 kV SiC MOSFET modules due to their high voltage and fast switching speed. This paper presents an in-depth analysis of 10 kV SiC MOSFET module baseplate parasitic capacitance impact on switching loss. The turn-on switching energy is divided into four parts, and the module baseplate parasitic capacitance impact on each part of the energy is analyzed. In addition to a constant capacitive energy (i.e., $\frac{1}{2}CV^2$), the module baseplate parasitic capacitance also introduces extra V-I overlap energy during voltage fall stage of the turn-on switching transient because it increases voltage fall time. This portion of loss is modeled and quantified. The module baseplate parasitic capacitance impact on switching loss is experimentally evaluated at 6 kV dc voltage and 0 to 50 A currents. This 10 kV SiC MOSFET module baseplate capacitance contributed loss is in the 5 - 10% range.

Keywords — 10 kV SiC MOSFET, parasitic capacitance, switching loss model.

I. INTRODUCTION

High voltage (HV, >3.3 kV) silicon carbide (SiC) power semiconductor devices show higher blocking voltage and faster switching speed compared to their silicon (Si) IGBT counterparts [1-3]. As a result, the HV SiC MOSFET based converters show great benefit in efficiency, size, weight, and control bandwidth compared to Si IGBT based converters [4-8].

The parasitic capacitance of module baseplate provides a path for displacement current during the switching transient of the power module, which causes extra switching loss and EMI noise. This becomes an increasing concern especially for emerging 10 kV SiC MOSFET modules due to their high voltage and fast switching speed.

In [7], the importance of parasitic capacitance impact on the loss of a 10 kV SiC MOSFET based 100 kW 13.8 kV converter has been stated and analyzed. In [9], the parasitic capacitance introduced energy in the MV double pulse test (DPT) setup is studied. However, the switching energy is estimated directly using the measured parasitic capacitance current. This approach neglects the parasitic capacitance's impact on slowing device switching transient which causes

extra V-I overlap energy. In [10], the switching energy of two custom packaged 10 kV single-die SiC MOSFET modules with different baseplate parasitic capacitances are analyzed in detail. However, the parasitic capacitance introduced extra V-I overlap energy is also not quantified.

This paper presents in-depth analysis of the module baseplate parasitic capacitance impact on switching loss of the Wolfspeed 10 kV 100 A multi-die XHV-9 SiC MOSFET half-bridge module as shown in Fig. 1. Section II provides analysis of the module baseplate parasitic capacitance impact on the switching energy. A model is proposed to estimate the module baseplate parasitic capacitance contributed extra V-I overlap energy. Section III presents experimental evaluation of the module baseplate parasitic capacitance impact on switching loss of this 10 kV SiC MOSFET module. Section VI concludes this paper.

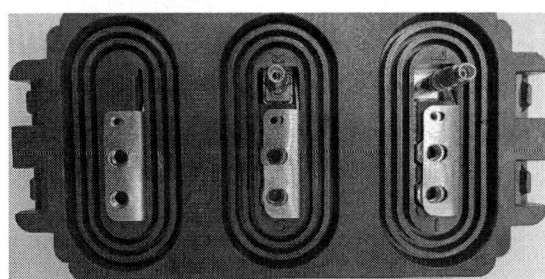

Fig. 1. Wolfspeed 10 kV XHV-9 SiC MOSFET power module.

II. MODELING OF BASEPLATE PARASITIC CAPACITANCE IMPACT ON SWITCHING ENERGY

A. Baseplate Parasitic Capacitance Impact on Turn-on Switching Transient

The turn-on switching transient of a SiC MOSFET can be divided into four stages: (1) turn-on delay stage, (2) current rise stage, (3) voltage fall stage, and (4) ringing stage. Fig. 2 shows the turn-on waveform of the 10 kV SiC MOSFET module conducted at 6 kV 40 A with the four stages labeled. The turn-on delay stage and ringing stage will not generate energy while the current rise stage and voltage fall stage will generate energy loss.

979-8-3503-3714-3/23 $31.00 © 2023 IEEE

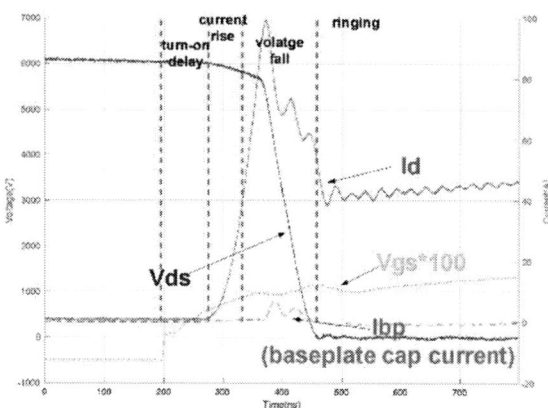

Fig. 2. Turn-on waveform of the 10 kV module conducted at 6 kV 40 A.

Fig. 3(a) describes the module baseplate parasitic capacitances of the studied 10 kV SiC MOSFET module. C_{ph} represents the module positive power terminal (module high side device drain terminal) to baseplate parasitic capacitance. C_{mh} and C_{gdmh} denotes the module middle power terminal (switching node) to baseplate parasitic capacitance and the module high side device gate driver to baseplate parasitic capacitance, respectively. C_{nh} and C_{gdnh} represents the module negative power terminal (module low side device source terminal) to baseplate parasitic capacitance and the module low side device gate driver to baseplate parasitic capacitance, respectively. Note that C_{gdmh} and C_{gdnh} are very small (e.g., < 1 pF) and can be neglected.

The baseplate is connected to the module negative power terminal. This helps to limit the switching node injected displacement current to within the module during switching transient and thus mitigating EMI noise. In the DPT setup, the module negative power terminal is grounded. Z_{gnd} denotes the impedance of the connection wire from baseplate to module negative power terminal.

Fig. 3 (b) shows the equivalent impedance network. Since Z_{gnd} is very small, the potential of the module baseplate is almost equal to ground potential. The baseplate displacement current during switching transient is mainly determined by C_{mh} and C_{gdmh} which is the parasitic capacitance between the switching node and baseplate. This parasitic capacitance is called baseplate capacitance C_{bp} for simplification. The measured C_{bp} of the studied 10 kV SiC MOSFET module is 55 pF.

During voltage fall stage, the current paths are described as Fig. 4. i_{rr} denotes the high side device body diode reverse recovery current. i_{c_H} represents the high side device junction capacitance charging current. I_L is load current. i_{ch_L}, i_{C_L}, and i_{bp} are the low side device channel current, junction capacitance discharging current, and baseplate capacitance

discharging current, respectively. Note that the DPT inductor parasitic capacitance is not shown in Fig. 4. It is in parallel with the high side device junction capacitance. If the DPT inductor parasitic capacitance is very small, its impact can be neglected.

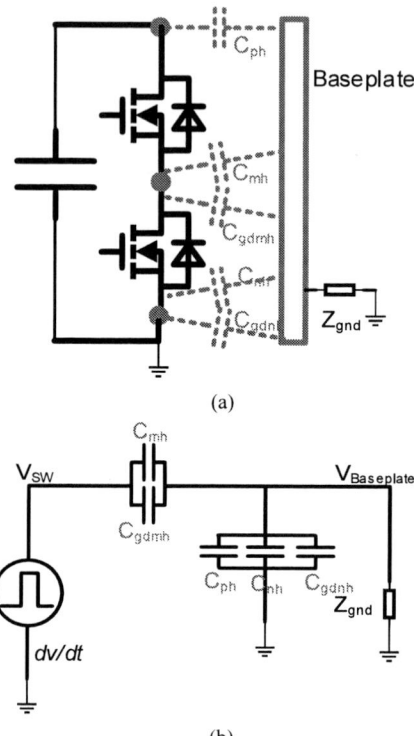

(a)

(b)

Fig. 3. (a) Baseplate related parasitic capacitances of the 10 kV half bridge module. (b) Equivalent impedance network.

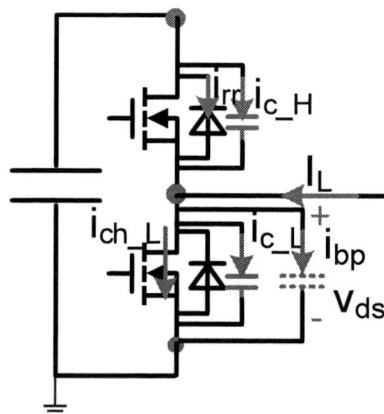

Fig. 4. Current paths during voltage fall stage of turn-on switching transient.

To help better understand baseplate capacitance impact on switching loss, the turn-on energy is separated into four parts.

(1) E_{on1}: V-I overlap energy during the current rise stage. The calculation of E_{on1} is expressed as

979-8-3503-3714-3/23 $31.00 © 2023 IEEE

$$E_{on1} = \int_{t_{cr1}}^{t_{cr2}} v_{ds} i_d dt \qquad (1)$$

where t_{cr1} and t_{cr2} represents the start and end time of the current rise stage, respectively. v_{ds} and i_d are the measured low side device drain-source voltage and drain current, respectively.

(2) E_{on2}: Constant load current contributed V-I overlap energy during the voltage fall stage. The calculation of E_{on2} is expressed as

$$E_{on2} = \int_{t_{vf1}}^{t_{vf2}} v_{ds} I_L dt \qquad (2)$$

where t_{vf1} and t_{vf2} represents the start and end time of voltage fall stage, respectively.

(3) E_{on3}: High side device junction capacitance charging current and body diode reverse recovery current contributed energy during the voltage fall stage. Based on the direction defined in Fig. 4, the calculation of E_{on3} is expressed as

$$E_{on3} = \int_{t_{vf1}}^{t_{vf2}} v_{ds}(i_d - I_L) dt \qquad (3)$$

(4) E_{on4}: Baseplate capacitance discharging energy during the voltage fall stage. The calculation of E_{on4} is expressed as

$$E_{on4} = \int_{t_{vf1}}^{t_{vf2}} v_{ds} i_{bp} dt = \frac{1}{2} C_{bp} V_{dc}^2 \qquad (4)$$

E_{on1} is not impacted by the baseplate capacitance. During current rise stage, voltage v_{ds} is equal to dc bus voltage if neglecting the voltage drop on loop inductance. i_d is equal to the channel current which is determined by gate voltage. The gate voltage slew rate is determined by input capacitance which is not impacted by the baseplate capacitance.

E_{on2} is impacted by the baseplate capacitance. During voltage fall stage, baseplate capacitance discharge process will slow the voltage fall process. Although I_L is constant, v_{ds} is impacted by the baseplate capacitance. Thus, extra constant load current contributed V-I overlap energy is introduced, which is called E_{on2_extra}. Modeling of this part of loss is discussed in the next section.

The baseplate capacitance impact on E_{on3} is neglectable although the baseplate capacitance will impact the voltage fall process of voltage fall stage . E_{on3} contains two parts as

$$E_{on3} = E_{qoss} + E_{rr} \qquad (5)$$

where E_{qoss} represents the high side device junction capacitance discharging current generated energy on low side device channel. E_{rr} is the high side device body diode reverse recovery current generated energy.

E_{qoss} is a capacitive energy and keeps constant for a given dc voltage since the stored charge on the capacitor only charges with the dc voltage. The change of voltage fall time will only result in the change in current magnitude that the parasitic capacitance are being charged.

Diode reverse recovery current is mainly determined by di_F/dt. i_{rr} can be considered not impacted by baseplate capacitance since the low side device channel current slew rate (which is the same as high side device body diode current slew rate) during current rise stage is determined by gate voltage which is not impacted by baseplate capacitance. E_{rr} could still be impacted by baseplate capacitance which need further investigation. However, for the 10 kV SiC MOSFET, E_{rr} is very small compared with E_{qoss}. Therefore, E_{on3} can be considered not impacted by baseplate capacitance.

The calculation of E_{rr} is briefly introduced. Based on (3) and (5), E_{rr} can be obtained if E_{qoss} is known. E_{qoss} can be estimated by

$$E_{qoss} = \int_0^{V_{dc}} (V_{dc} - v_{ds}) C_{oss} dv_{ds} \qquad (6)$$

where C_{oss} is the device junction/output capacitance. Another simple and more accurate method to estimate E_{qoss} is to use the first pulse of DPT calculation. As the first pulse turn-on transient is actually zero current turn-on and there is no load current contributed V-I overlap energy and device body diode reverse recovery energy. Then E_{qoss} is calculated by

$$E_{qoss} = E_{pulse1} - E_{C_{bp}} \qquad (7)$$

where E_{pulse1} is the total turn-on energy of the first pulse and $E_{C_{bp}}$ is the baseplate capacitance energy (i.e., $\frac{1}{2} C_{bp} V_{dc}^2$).

As an example, at 6 kV 40 A test points, the calculated E_{on3}, E_{qoss}, and E_{rr} are 13.63 mJ, 12.54 mJ, and 1.09 mJ, respectively. E_{rr} is only 8% of E_{on3}.

E_{on4} is a constant capacitive energy and is determined by baseplate capacitance and the given dc voltage.

B. Baseplate Capacitance Introduced Extra V-I overlap Energy Modeling

To quantify the baseplate capacitance introduced extra V-I overlap energy, how the baseplate capacitance impact v_{ds} during voltage fall stage need to be determined. Two cases – one with baseplate capacitance and another without baseplate capacitance – are investigated.

For the case with baseplate capacitance, based on Fig. 4, the low side device channel current i_{ch_L} can be expressed as

$$i_{ch\_L} = i_{c\_H} + i_{rr} + I_L - i_{c\_L} - i_{bp} \qquad (8)$$

For the case without baseplate parasitic capacitance, the module lower device channel current i'_{ch_L} can be expressed as

$$i'_{ch\_L} = i'_{c\_H} + i'_{rr} + I_L - i'_{c\_L} \tag{9}$$

Based on the discussion in last section, i_{ch_L} and i'_{ch_L} can be considered equal. i_{rr} and i'_{rr} can be also considered equal.

Defining

$$i_{eq}(t) = i_{ch\_L} - I_L - i_{rr} \tag{10}$$
$$i'_{eq}(t) = i'_{ch\_L} - I_L - i'_{rr} \tag{11}$$

$i_{eq}(t)$ and $i'_{eq}(t)$ can be also considered equal for the two cases.

Based on (8)-(11), the equivalent circuit for the two cases is drawn as Fig. 5. C_{H+L} represents the switching node capacitance. The low side device junction capacitance C_L, high side device junction capacitance C_H (as a function of low side device voltage), and the total switching node capacitance C_{H+L} are shown in Fig. 6. $v_{ds}(t)$ and $v'_{ds}(t)$ represents the low side device drain-source voltages for the cases with and without baseplate parasitic capacitance, respectively. Since the current sources in the two cases are the same, the existence of baseplate current reduces the current flowing through C_{H+L} slowing the voltage fall transient and leading to higher constant load current V-I overlap loss.

$v'_{ds}(t)$ Using the backward Euler method, (12)-(15) are obtained.

$$i = \frac{dq}{dt} = \frac{d(C(v) \cdot v)}{dt} \tag{12}$$
$$i_{k+1} = C(v_k)\frac{v_{k+1} - v_k}{\Delta T} \tag{13}$$
$$C_{H+L}(v'_k)\frac{v'_{k+1} - v'_k}{\Delta T} = (C_{H+L}(v_k) + C_{bp})\frac{v_{k+1} - v_k}{\Delta T} \tag{14}$$
$$v'_{k+1} = \frac{(C_{H+L}(v_k) + C_{bp})}{C_{H+L}(v'_k)}(v_{k+1} - v_k) + v'_k \tag{15}$$

Thus, $v'_{ds}(t)$ can be estimated using measured $v_{ds}(t)$ for voltage fall stage based (15). Then, baseplate capacitance introduced extra constant load current V-I overlap E_{on2_extra} is calculated as

$$E_{on2\_extra} = \int_{t_{vf1}}^{t_{vf2}} (v'_{ds} - v_{ds})I_L dt \tag{16}$$

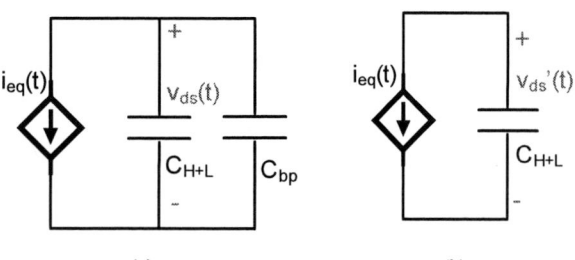

(a)   (b)

Fig. 5. Equivalent circuit during voltage fall stage. (a) With baseplate parasitic capacitance. (b) Without baseplate parasitic capacitance.

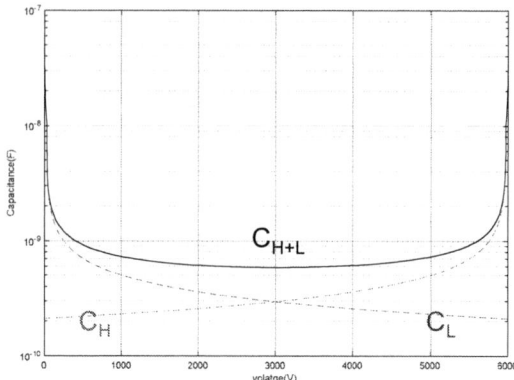

Fig. 6. Curves for C_L, C_H, and C_{H+L}

C. Turn-off Loss Discussion

The turn-off switching transient of a SiC MOSFET can also be divided into four stages: (1) turn-off delay stage, (2) voltage rise stage, (3) current fall stage, and (4) ringing stage. The analysis on baseplate capacitance impact on turn-off loss will be very similar to that of turn-on loss and is not repeated here.

Note that the baseplate capacitance charging current will not flow through the device channel and will not introduce joule heating during the turn-off switching transient. This part of energy is stored and dissipated during the next turn-on switching transient.

The channel current portion V-I overlap contributed energy will increase as the baseplate capacitance slows the voltage rise stage. However, the channel current will rapidly decrease to zero during the turn-off switching transient of this 10 kV SiC module, and this part of loss is very small. The turn-off switching transient is dominated by load current charging/discharging the junction capacitances process and this part of loss is not impacted by the baseplate capacitance.

Therefore, because the channel current portion V-I overlap energy is quite a small portion of total turn-off energy, the baseplate capacitance impact on turn-off loss is considerably small and does not have significant impact on the total turn-off loss. Note that the turn-off loss is also very small compared with turn-on loss of this 10 kV SiC MOSFET module.

III. EXPERIMENTAL EVALUATION

Fig. 7 shows the DPT hardware setup with the studied 10 kV SiC module. DPT at 6 kV dc bus voltage and different load currents from 0 to 50 A are conducted. The turn-on waveforms at 6 kV 40 A test point is shown in Fig. 2. Based on the models developed in Section II A, the defined four parts of turn-on switching energy are shown in Fig. 8. E_{on2} and E_{on3} are the two dominant parts of turn-on switching energy. E_{on1} and E_{on2} are load dependent while E_{on3} and E_{on4} are almost load independent.

Fig. 7. 10 kV SiC module based DPT test platform.

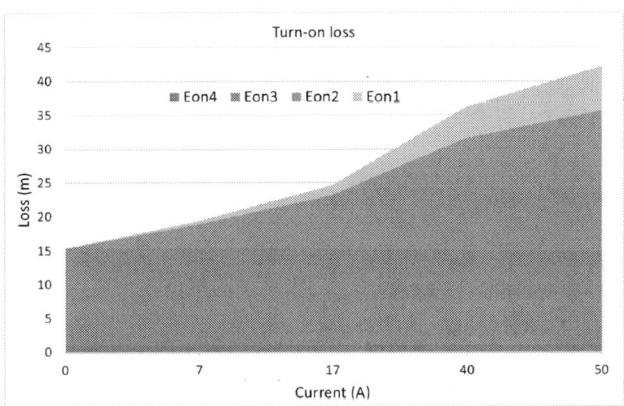

Fig. 8. The four parts of turn-on switching energy of the 10 kV SiC module at 6 kV voltage and different load currents.

Based on the models developed in Section II B, $v'_{ds}(t)$ is estimated based on the tested $v_{ds}(t)$ at 6 kV and different load currents. The 6 kV 40 A test point waveforms are shown in Fig. 9. Due to the existence of baseplate capacitance, $v_{ds}(t)$ is slowed compared with $v'_{ds}(t)$ during the voltage fall stage and extra constant load current contributed V-I overlap energy is introduced. At the 6 kV 40 A test point, the total turn on energy E_{on} is 35.77 mJ and E_{on1}, E_{on2}, E_{on3}, and E_{on4} are 4.63 mJ, 16.54 mJ, 13.63 mJ, and 0.97 mJ, respectively. The baseplate parasitic capacitance introduced extra constant load current contributed V-I overlap loss E_{on2_extra} is 1.29 mJ.

Fig. 10 summarizes the two portion of loss (E_{on4} and E_{on2_extra}) induced by the baseplate parasitic capacitance at 6 kV voltage and different loads. E_{on2} is almost load independent while E_{on2_extra} is load dependent. In general, the baseplate capacitance contributed loss is in the 5 - 10% range of the total switching loss at various loads (0 to 50 A) for this 10 kV SiC module.

Fig. 11 shows the turn-off energy at 6 kV voltage and different load currents. E_{oss} is the estimated device junction capacitances charging/discharging contributed energy during turn-off switching transient, and E_{off1} is the remaining V-I overlap energy.

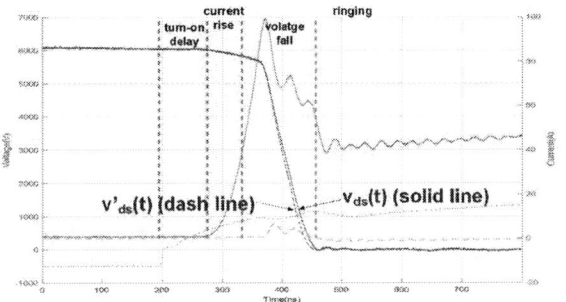

Fig. 9. Calculated $v'_{ds}(t)$ based on $v_{ds}(t)$ at 6 kV 40 A point.

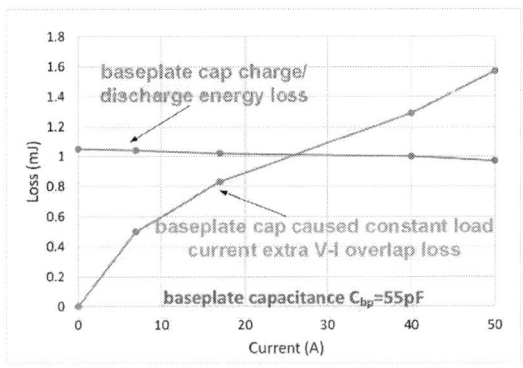

Fig. 10. Summary of baseplate capacitance introduced switching energy E_{on4} and E_{on2_extra} at 6 kV voltage and different load currents.

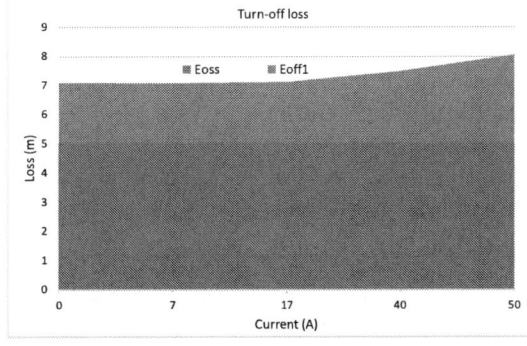

Fig. 11. Turn-off energy at 6 kV voltage and different load currents.

IV. CONCLUSION

This paper presents an in-depth analysis of 10 kV SiC MOSFET module baseplate parasitic capacitance impact on switching loss. The turn-on switching energy is divided into four parts, and the module baseplate parasitic capacitance impact on each part of the energy is analyzed. The module baseplate parasitic capacitance will introduce not only a constant capacitive energy (E_{on4}) but also an extra constant load current contributed V-I overlap energy (E_{on2_extra}) during the voltage fall stage of turn-on switching transient because it slows the voltage fall transient. This portion of loss is quantified and experimentally evaluated at 6 kV voltage and different load currents. In general, this 10 kV SiC MOSFET module baseplate parasitic capacitance contributed switching loss is in the 5 - 10% range. The developed models and experimental evaluation provide better understanding of 10 kV SiC module baseplate impact on switching loss and accurate estimation of 10 kV SiC module based MV converter loss.

ACKNOWLEDGMENT

This work was supported primarily by the AMMTO office, United States Department of Energy, under Award Number DE-EE0009134. This work made use of the Engineering Research Center Shared Facilities supported by the Engineering Research Center Program of the National Science Foundation and DOE under NSF Award Number EEC-1041877 and the CURENT Industry Partnership Program.

REFERENCES

[1] J. Casady *et al.*, "New generation 10kV SiC power MOSFET and diodes for industrial applications," in *Proc. Int. Exhib. Conf. Power Electron., Intell. Motion, Renewable Energy Manage.*, 2015, pp. 1–8.

[2] D. Johannesson, M. Nawaz, and K. Ilves, "Assessment of 10 kV, 100 A silicon carbide MOSFET power modules," *IEEE Trans. Power Electron.*, vol. 33, no. 6, pp. 5215-5225, Jun. 2018.

[3] R. Chen, M. Lin, X. Huang, F. Wang and L. M. Tolbert, "An improved turn-on switching transient model of 10 kV SiC MOSFET," *IEEE Applied Power Electronics Conference and Exposition (APEC)*, Houston, TX, USA, 2022, pp. 1-8.

[4] S. Madhusoodhanan et al., "Solid-state transformer and MV grid tie applications enabled by 15 kV SiC IGBTs and 10 kV SiC MOSFETs based multilevel converters," *IEEE Trans. Ind. Appl.*, vol. 51, no. 4, pp. 3343– 3360, Jul./Aug. 2015.

[5] A. Huang, L. Wang, Q. Tian, Q. Zhu, D. Chen, and W. Yu, "Medium voltage solid state transformers based on 15 kV SiC MOSFET and JBS diode," in *Proc. IEEE 42nd Annu. Conf. Ind. Electron. Soc.*, 2016, pp. 6996–7002.

[6] S. Parashar, A. Kumar and S. Bhattacharya, "High power medium voltage converters enabled by high voltage SiC power devices," in *Proc. International Power Electronics Conference (IPEC-Niigata 2018 - ECCE Asia)*, 2018, pp. 3993-4000.

[7] R. Chen et al., "10 kV SiC MOSFET based medium voltage power conditioning system for asynchronous microgrids," in *IEEE Access*, vol. 10, pp. 73294-73308, 2022.

[8] S. Ji et al., "Medium voltage (13.8 kV) transformer-less grid-connected DC/AC converter design and demonstration using 10 kV SiC MOSFETs," *IEEE Energy Conversion Congress and Exposition (ECCE)*, Baltimore, MD, USA, 2019, pp. 1953-1959.

[9] H. Li, Z. Gao, R. Chen and F. Wang, "Improved double pulse test for accurate dynamic characterization of medium voltage SiC devices," in *IEEE Transactions on Power Electronics*, vol. 38, no. 2, pp. 1779-1790, Feb. 2023.

[10] D. N. Dalal *et al.*, "Impact of power module parasitic capacitances on medium-voltage SiC MOSFETs switching transients," in *IEEE Journal of Emerging and Selected Topics in Power Electronics*, vol. 8, no. 1, pp. 298-310, Mar. 2020.

A Si IGBT Circuit Breaker for Protection of 10 kV SiC MOSFET Power Module

Ruirui Chen[1], Min Lin[1], Dingrui Li[1], Zihan Gao[1], Fred Wang[1,2], Hua Bai[1], and Leon M. Tolbert[1]

[1]Min H. Kao Department of Electrical Engineering and Computer Science, The University of Tennessee, Knoxville, TN, USA

[2]Oak Ridge National Laboratory, Oak Ridge, TN, USA

rchen14@vols.utk.edu

Abstract — **This paper presents a low cost Si IGBT circuit breaker for protection of the Wolfspeed 10 kV XHV-9 half bridge SiC MOSFET module. Based on the 10 kV SiC module short circuit (SC) protection requirement, both current limiting and desaturation protection are needed for this IGBT breaker. Voltages balancing for series-connected IGBTs and effective current limiting of this IGBT breaker during SC transient are discussed. Detailed modes transition between the 10 kV SiC module and the IGBT breaker during SC transient is analyzed. Demonstration of the 10 kV SiC module with the IGBT breaker SC test and continuous power test at 6 kV voltage are presented.**

Keywords — 10 kV SiC MOSFET, circuit breaker, IGBT, short circuit fault.

I. INTRODUCTION

The emerging 10 kV SiC MOSFET is drawing increased attention in medium and high voltage power converter applications due to the benefits in efficiency, size, and control bandwidth compared to Si IGBT based converters [1-4].

Due to the high cost of the 10 kV SiC MOSFET module at the time being, even though the 10 kV SiC module may has its own protection, it is usually necessary to add a circuit breaker to serve as an additional layer of protection when conducting characterization test of the 10 kV SiC module during the early stage of converter development.

In [5], the 6.5 kV Si IGBT module is inserted to a 10 kV 120 A SiC MOSFET module based power electronics building block (PEBB). However, such IGBT has long turn-off delay time (usually 6-8 μs) and which is not fast enough to turn-off the circuit until short-circuit fault current increased to very high. In [6], series-connected Si IGBTs are used to protect a 10 kV 20 A SiC MOSFET discrete device based double pulse test (DPT) circuit. However, the detailed design and experimental demonstration are not presented.

This paper presents a low-cost Si IGBT based circuit breaker for protection of the Wolfspeed 10 kV 100 A multi-die XHV-9 half-bridge SiC MOSFET module. Section II illustrates the design considerations of the Si IGBT breaker. Section III analyzes the modes transition for the 10 kV SiC module and the IGBT breaker during short circuit (SC) transient. Section VI demonstrates experimental results and section V concludes this paper.

II. SI IGBT CIRCUIT BREAKER DESIGN CONSIDERATION

A. 10 kV SiC Module Protection Requirements

The short circuit current of this 10 kV 100 A SiC multi-die power module increases fast and could exceed 1 kA in less than 1 μs due to the high voltage and low loop inductance. High current during short circuit transient may cause module degradation or even module failure. It is preferred to limit short circuit current to a couple of hundred amps during short circuit event. However, the desaturation protection of this 10 kV SiC power module is hard to achieve < 1 μs response time due to the blanking time and noise immunity requirements. Also, the commercial Si IGBTs with a couple of kilovolts voltage rating is hard to achieve < 2 μs protection response time. Thus, the IGBT breaker needs to have current limiting function and should limit the short circuit current to desired value (e.g., a couple of hundred amps) before the circuit turn-off by desaturation protection of either the 10 kV SiC module or the IGBT breaker. The current limiting function can be realized by tuning the IGBT gate voltage.

B. Voltage Balancing of Series-Connected IGBTs

In this study case, three 3.6 kV 125 A Si discrete IGBTs (IXBX50N360HV) are series-connected to achieve 10 kV voltage rating. Extensive dynamic voltage balancing methods for series connected IGBTs has been proposed in literature, which generally requires complex circuit and control design. Different from the switching device in power converter, metal oxide varistor (MOV) or transient voltage suppression (TVS) diode is required to be paralleled with switching device to absorb the energy in loop inductance for circuit breaker applications. MOV or TVS diode will clamp the voltages of series-connected switching devices if severe voltage unbalance occurs during the breaker turn-off transient. Thus, for circuit breaker applications, voltage balancing of series-connected switching devices is not a concern due to the existence of MOV or TVS diode. In this study case, TVS diode is used due to its lower clamping factor and smaller size compared with MOV. Considering 6 kV dc bus voltage for the 10 kV SiC module application, six series-connected 350 V TVS diodes are used and paralleled with each of the IGBT.

C. Effective Current Limiting of IGBT

As discussed above, the current limiting function is realized by tuning IGBT gate voltage. Fig. 1 shows the tested output curve for the selected 3.6 kV IGBT. To limit the 10 kV module short circuit current to within 300 A, the gate voltage of the IGBT is selected to be 12 V.

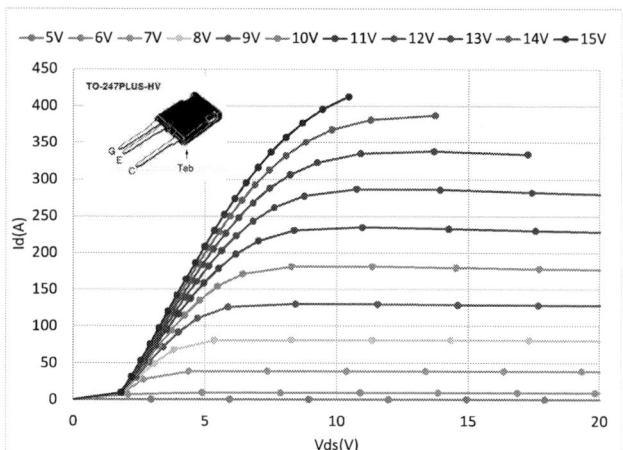

Fig. 1. Tested output curve of the selected IGBT.

To achieve effective current limiting during SC transient, the internal gate voltage of the IGBT needs to be kept stable and close to the designed value. The impact of common-source inductance and Miller capacitance on IGBT internal gate voltage during SC transient is briefly discussed.

The detailed mode transitions of the 10 kV SiC module SC with IGBT breaker inserted will be discussed in the next section. In general, the IGBT breaker will experience current rise (di/dt) period and voltage rise (dv/dt) period. During the di/dt period, as shown in Fig. 2(a), the common source inductance will be induced with a positive voltage. As a result, the internal gate voltage (voltage on C_{GE}) will be reduced which helps limiting SC current. Impact of common source inductance on internal gate voltage can be controlled by gate resistance. When gate resistance is small, internal gate voltage reduction is large. When gate resistance is large, internal gate reduction is small. During the dv/dt period, the current flowing through the Miller Capacitance will increase the internal gate voltage. The common source inductance will be induced with a negative voltage and will also increase the internal gate voltage.

To suppress the internal gate voltage fluctuation of the IGBT during SC transient, a clamping circuit is added to the gate driver as shown in Fig. 3. The main component is the capacitor C_{Clamp} stabling the gate voltage to V_{Clamp} during SC transient, which should be put close to the gate terminal. Resistance of R1 should be small to achieve effective gate voltage clamping.

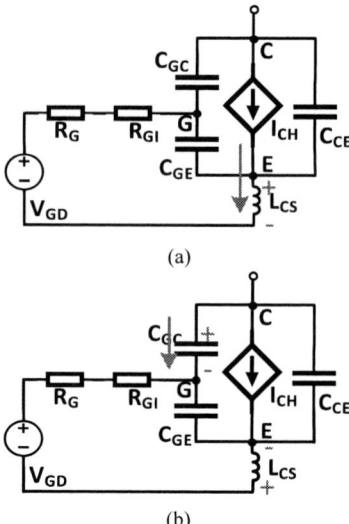

(a)

(b)

Fig. 2. Illustration of common source inductance and Miller capacitance impact on internal gate voltage. (a) di/dt period. (b) dv/dt period.

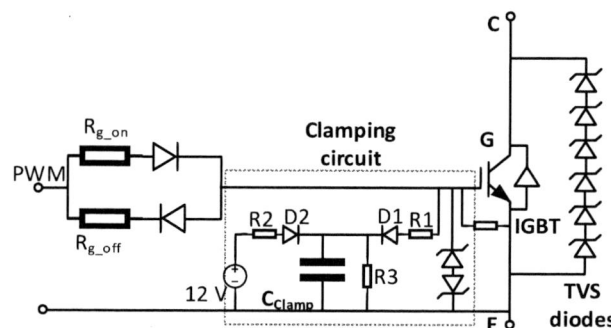

Fig. 3. Gate voltage clamping circuit for the IGBT.

III. MODES TRANSITION DURING SHORT-CIRCUIT TRANSIENT

Detailed modes transition during SC transient of the 10 kV SiC module with the IGBT breaker are analyzed in this section. For simplification, the 10 kV SiC module is called DUT, and one IGBT is used to represents the three series-connected IGBTs of this IGBT breaker in this section. The simplified circuit scheme is shown as Fig. 4. Considering the hard switching fault (HSF) of the 10 kV SiC, there will be five substages during SC transient.

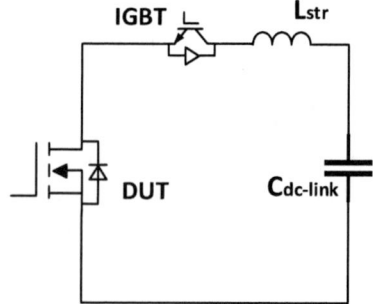

Fig. 4. Simplified circuit scheme.

Substage 1: DUT turn-on delay stage. During this stage, IGBT breaker is in on-state. DUT withstands the DC voltage and gate voltage increases from V_{EE} to V_{TH_DUT}.

For IGBT

$$v_{CE} \approx 0 \tag{1}$$

$$v_{GE} = V_{Clamp} \tag{2}$$

For DUT

$$v_{DS} = V_{DC} \tag{3}$$

$$i_{G\_DUT} = (C_{GS} + C_{GD})\frac{dv_{GS}}{dt} \tag{4}$$

$$v_{GS} + R_G i_{G\_DUT} = V_{GDH} \tag{5}$$

Substage 2: SC current rise stage. During this stage, the IGBT is still in on-state. The DUT works in saturation region, SC current increases and is determined by DUT gate voltage. The equivalent circuit is shown as Fig. 5.

For IGBT,

$$v_{CE} \approx 0 \tag{6}$$

$$v_{GE} + R_{GI}i_G + (L_{GI} + L_{EI})\frac{di_G}{dt} + L_{EI}\frac{di_{SC}}{dt} = V_{Clamp} \tag{7}$$

For DUT,

$$i_{G\_DUT} = C_{GS}\frac{dv_{GS}}{dt} + C_{GD}\frac{d(v_{GS} - v_{DS})}{dt} \tag{8}$$

$$v_{GS} + R_G i_{G\_DUT} = V_{GDH} \tag{9}$$

$$i_{SC} = i_{CH\_DUT} + C_{DS}\frac{dv_{DS}}{dt} - C_{GD}\frac{d(v_{GS} - v_{DS})}{dt} \tag{10}$$

$$i_{CH\_DUT} = g_{fs\_DUT}(v_{GS} - V_{TH\_DUT}) \tag{11}$$

$$v_{DS} + L_{loop}\frac{di_{SC}}{dt} = V_{DC} \tag{12}$$

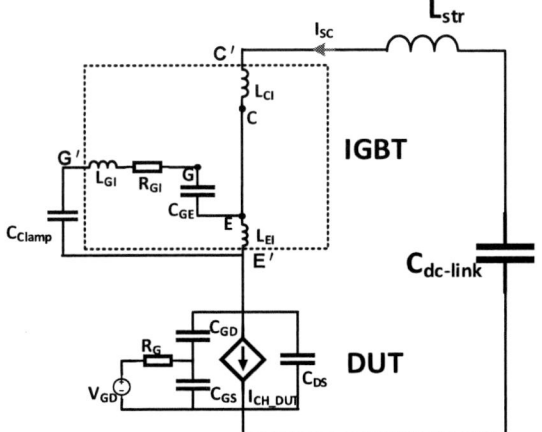

Fig. 5. Equivalent circuit for substage 2 during SC.

Substage 3: Voltage transition stage. During this stage, IGBT enters saturation region and v_{CE} increases while DUT voltage v_{DS} decreases. The equivalent circuit is shown as Fig. 6.

For IGBT,

$$i_G = C_{GE}\frac{dv_{GE}}{dt} + C_{GC}\frac{d(v_{GE} - v_{CE})}{dt} \tag{13}$$

$$v_{GE} + R_{GI}i_G + (L_{GI} + L_{EI})\frac{di_G}{dt} + L_{EI}\frac{di_{SC}}{dt} = V_{Clamp} \tag{14}$$

$$i_{SC} = C_{CE}\frac{dv_{CE}}{dt} - C_{CC}\frac{d(v_{GE} - v_{CE})}{dt} + g_{fs}(v_{GE} - V_{TH}) \tag{15}$$

For DUT,

$$i_{G\_DUT} = C_{GS}\frac{dv_{GS}}{dt} + C_{GD}\frac{d(v_{GS} - v_{DS})}{dt} \tag{16}$$

$$v_{GS} + R_G i_{G\_DUT} = V_{GDH} \tag{17}$$

$$i_{SC} = i_{CH\_DUT} + C_{DS}\frac{dv_{DS}}{dt} - C_{GD}\frac{d(v_{GS} - v_{DS})}{dt} \tag{18}$$

$$i_{CG\_DUT} = g_{fs\_DUT}(v_{GS} - V_{TH\_DUT}) \tag{19}$$

and

$$v_{DS} + v_{CE} + L_{loop}\frac{di_{SC}}{dt} = V_{DC} \tag{20}$$

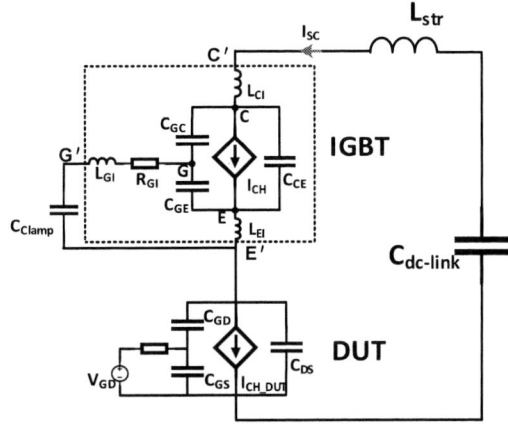

Fig. 6. Equivalent circuit for substage 3 during SC.

Substage 4: SC current slow changing stage. During this stage, DUT is in on-state. IGBT works in saturation region. SC current could increase or decrease depending on IGBT internal v_{GE}. The equivalent circuit is shown as Fig. 7.

For IGBT,

$$i_G = C_{GE}\frac{dv_{GE}}{dt} + C_{GC}\frac{d(v_{GE} - v_{CE})}{dt} \tag{21}$$

$$v_{GE} + R_{GI}i_G + (L_{GI} + L_{EI})\frac{di_G}{dt} + L_{EI}\frac{di_{SC}}{dt} = V_{Clamp} \tag{22}$$

$$i_{SC} = C_{CE} \frac{dv_{CE}}{dt} - C_{CC} \frac{d(v_{GE} - v_{CE})}{dt} + g_{fs}(v_{GE} - V_{TH}) \quad (23)$$

$$v_{CE} + L_{loop} \frac{di_{SC}}{dt} = V_{DC} \quad (24)$$

For DUT,

$$v_{DS} \approx 0 \quad (25)$$

$$i_{G\_DUT} = (C_{GS} + C_{GD}) \frac{dv_{GS}}{dt} \quad (26)$$

$$v_{GS} + R_G i_{G\_DUT} = V_{GDH} \quad (27)$$

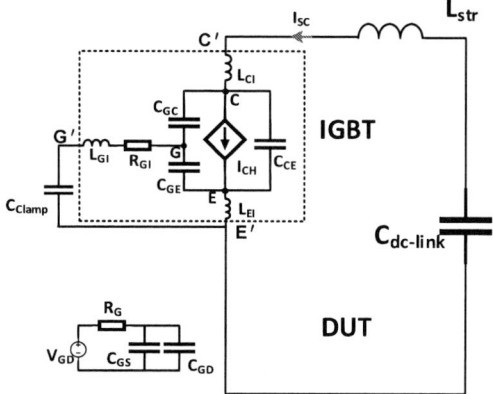

Fig. 7. Equivalent circuit for substage 4 during SC.

Substage 5: SC current holding stage. During this stage, IGBT internal gate voltage and SC current keep constant.

For IGBT,

$$v_{GE} = V_{Clamp} \quad (28)$$
$$v_{CE} = V_{DC} \quad (29)$$
$$i_{SC} = g_{fs}(v_{GE} - V_{TH}) \quad (30)$$

For DUT,

$$v_{DS} \approx 0 \quad (31)$$

$$v_{GS} = V_{GDH} \quad (32)$$

From above analysis, during SC transient, the DUT changes from saturation mode to on-state mode while the IGBT changes from on-state mode to saturation mode. The short circuit energy is transferred from the DUT to the IGBT and SC withstand time of the DUT is equivalently increased.

IV. EXPERIMENTAL RESULTS

A. IGBT Breaker Standalone Tests

Fig. 8 shows the hardware of the IGBT breaker. Fig. 9 shows the DPT circuit scheme and experimental waveforms of the IGBT breaker conducted at 6.3 kV and 40 A. The 10 kV SiC MOSFET body diode is used as the freewheeling

diode in the setup. The measure one IGBT collector-emitter voltage (v_{ce1}) is one third of total IGBT breaker voltage (v_{bus}), during the turn-off switching transient and steady state, indicating good voltage balancing of the three IGBTs.

Fig. 8. IGBT breaker hardware.

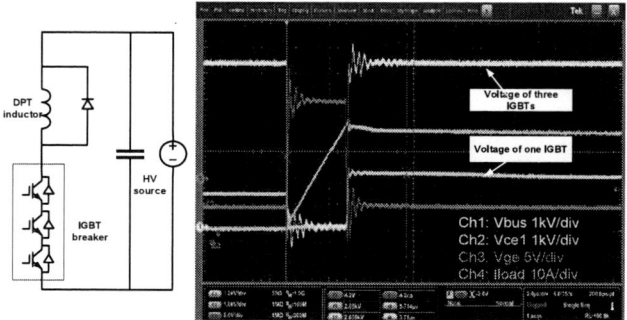

Fig. 9. IGBT breaker DPT waveform at 6 kV 40 A.

Fig. 10 shows the IGBT breaker SC test waveforms conducted at 6 kV dc voltage and the SC current is limited to below 300 A. The response time is around 2 μs.

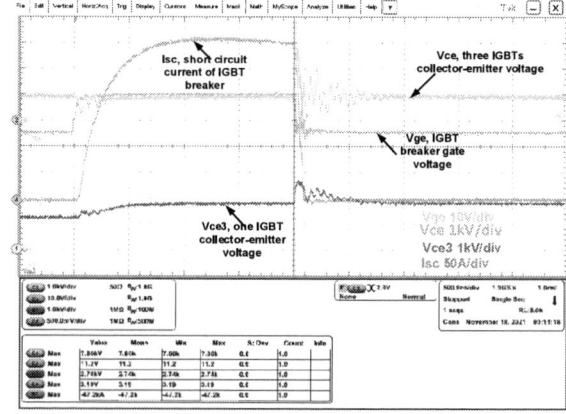

Fig. 10. The IGBT breaker SC test waveforms at 6 kV.

B. 10 kV SiC Module SC Test with the IGBT Breaker

Fig. 11 shows the circuit scheme and the experimental waveforms of the 10 kV SiC module SC test with the IGBT

979-8-3503-3714-3/23 $31.00 © 2023 IEEE

breaker inserted. The high side device in the module is bypassed for this test.

During the SC transient, first, the SC current increasing fast together with the 10 kV SiC device drains-source voltage decrease from dc bus voltage to almost zero while the IGBT breaker collector-emitter voltage increases from zero to almost dc bus voltage (substages 2 and 3). The short circuit energy is transferred from the 10 kV SiC device to the IGBT breaker. Then, the SC current increasing slew rate is reduced and controlled by the IGBT internal gate voltage (substage 4). Next, before the SC reaches steady state (substage 5), the 10 kV SiC device desaturation triggered and soft turn off at around 1 μs. The peak SC current is also limited to < 300 A. Since the 10 kV SiC device desaturation protection response time is faster than that of the IGBT breaker, the IGBT breaker mainly provides current limiting function during the SC transient. The SC withstand time of the 10 kV SiC device is equivalently increased.

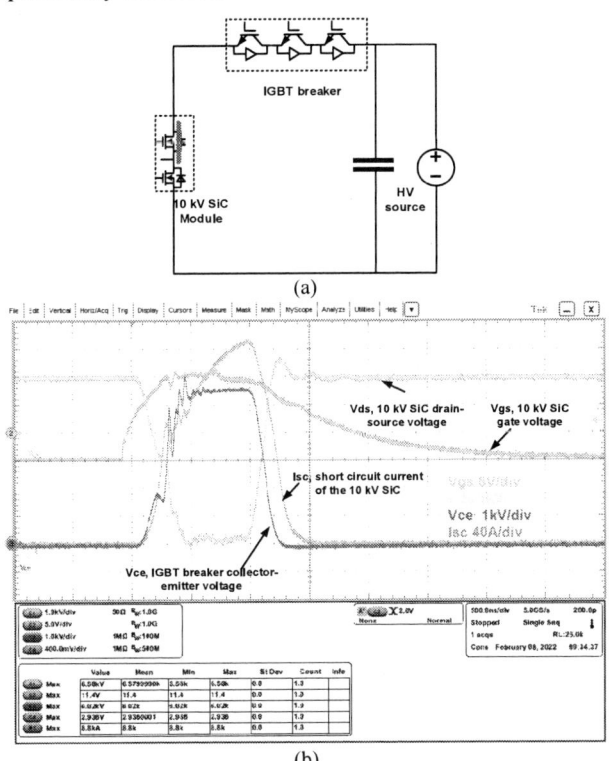

(a)

(b)

Fig. 11. 10 kV SiC module SC test at 6 kV voltage with IGBT breaker inserted. (a) Circuit scheme. (b) Waveforms.

C. 10 kV SiC Module Continuous Power Test

The designed IGBT breaker is mainly used for pulses tests of the 10 kV SiC module during the early stage of converter design. If used for continuous power test, the loss of the GBT needs to be considered.

Considering a half-bridge converter, Fig. 12 shows two configurations when the IGBT breaker is inserted into the converter. In Fig. 12(a), only the ripple current flows through

the IGBT breaker and the loss is very small. In Fig. 12(b), the main current flows through the IGBT breaker and the loss is a concern although it can reliably detect and clear overcurrent faults.

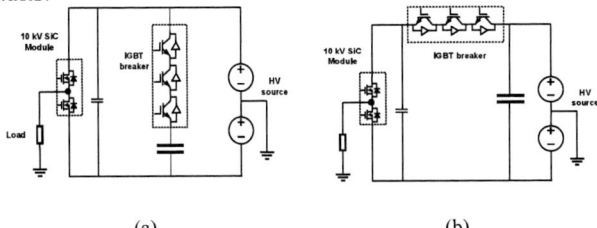

(a)   (b)

Fig. 12. IGBT breaker configuration. (a) Inserted into energy storage capacitor path. (b) Inserted into main current conduction path.

The continuous power tests of the 10 kV SiC module is conducted with the IGBT breaker inserted. The experiment setup and waveforms at 6 kV dc bus voltage and around 50 A peak output current are shown in Fig. 13 (a) and Fig. 13(b) respectively. At this test point, the loss of the 10 kV SiC module and the IGBT breaker are evaluated and summarized in Table 1.

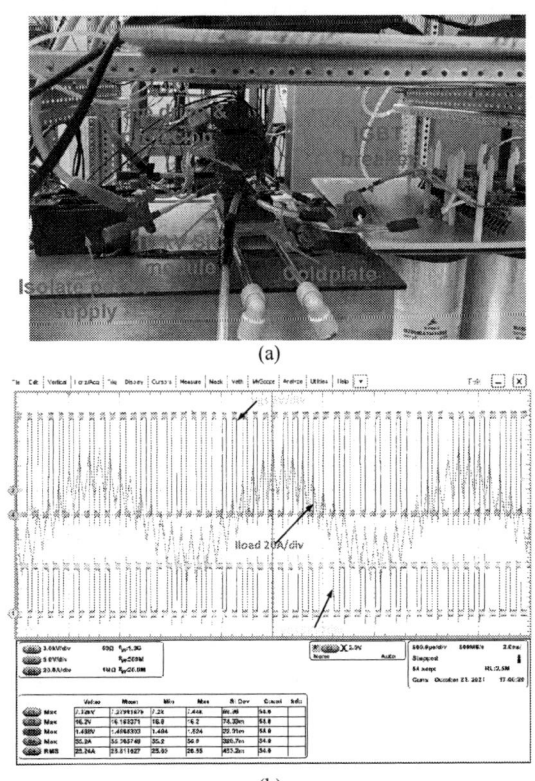

(a)

(b)

Fig. 13. 10 kV SiC module continuous power test at 6 kV bus voltage and 50 A peak output current. (a) Hardware setup. (b) Waveforms.

Table I. Loss at 6 kV dc voltage 50 A peak current case

	10 kV SiC module loss	IGBT breaker loss
Configuration (a)	432.6W	18.6W×3
Configuration (b)	433W	~2.5W×3

V. CONCLUSION

This paper presents a low cost Si IGBT circuit breaker for protection of the Wolfspeed 10 kV XHV-9 half bridge SiC MOSFET module. The 10 kV SiC module protection requirements, series-connected IGBTs voltage balancing, and effective current limiting of the IGBT breaker during SC transient are discussed. Detailed modes transition of the 10 kV SiC module and the IGBT breaker during SC transient are analyzed. Experimental demonstration of the SC test and continuous power test of the 10 kV SiC module with the IGBT breaker at 6 kV voltage are presented and verifies the effectiveness of the proposed IGBT breaker for 10 kV SiC module protection.

ACKNOWLEDGMENT

This work was supported primarily by the AMMTO office, United States Department of Energy, under Award Number DE-EE0009134. This work made use of the Engineering Research Center Shared Facilities supported by the Engineering Research Center Program of the National Science Foundation and DOE under NSF Award Number EEC-1041877 and the CURENT Industry Partnership Program.

REFERENCES

[1] S. Madhusoodhanan et al., "Solid-state transformer and MV grid tie applications enabled by 15 kV SiC IGBTs and 10 kV SiC MOSFETs based multilevel converters," *IEEE Trans. Ind. Appl.*, vol. 51, no. 4, pp. 3343–3360, Jul./Aug. 2015.

[2] A. Huang, L. Wang, Q. Tian, Q. Zhu, D. Chen, and W. Yu, "Medium voltage solid state transformers based on 15 kV SiC MOSFET and JBS diode," in *Proc. IEEE 42nd Annu. Conf. Ind. Electron. Soc.*, 2016, pp. 6996–7002.

[3] S. Parashar, A. Kumar and S. Bhattacharya, "High power medium voltage converters enabled by high voltage SiC power devices," in *Proc. International Power Electronics Conference (IPEC-Niigata 2018 - ECCE Asia)*, 2018, pp. 3993-4000.

[4] R. Chen et al., "10 kV SiC MOSFET based medium voltage power conditioning system for asynchronous microgrids," in *IEEE Access*, vol. 10, pp. 73294-73308, 2022.

[5] C. DiMarino, I. Cvetkovic, Z. Shen, R. Burgos and D. Boroyevich, "10 kV, 120 A SiC MOSFET modules for a power electronics building block (PEBB)," *IEEE Workshop on Wide Bandgap Power Devices and Applications*, Knoxville, 2014, pp. 55-58.

[6] S. Ji, S. Zheng, F. Wang and L. M. Tolbert, "Temperature-dependent characterization, modeling, and switching speed-limitation analysis of third-generation 10-kV SiC MOSFET," *IEEE Trans. on Power Electronics*, vol. 33, no. 5, pp. 4317-4327, May 2018.

Planar Implantation Edge Termination for Vertical GaN Power Devices

Yifan Wang[a]*, Ming Xiao[a], Matthew Porter[a], Ruizhe Zhang[a], Qihao Song[a], Albert Lu[b], Nathan Yee[b], Hiu Yung Wong[b], Yuhao Zhang[a]*

[a]Center for Power Electronics Systems (CPES), Virginia Tech, Blacksburg, USA
[b]Department of Electrical Engineering, San Jose State University, San Jose, USA.
*yifanwang@vt.edu, yhzhang@vt.edu

Abstract—Edge termination plays a crucial role in achieving near-ideal avalanche breakdown in power semiconductor devices. In this paper, two edge termination designs, one GR (guard ring) [1], the other USAB-JTE (ultra-small-angle bevel junction termination extension) [2] that utilize planar ion implantation are developed and studied. The fabrication process only has a single implantation step that does not need precise control over the depth. Isolation is also done by the same process, avoiding the need of etching and possible etch-induced damages to the devices. Comprehensive characterization, including static I-V test and avalanche circuit test are conducted to confirm the avalanche breakdown capability of both devices. It is found both the GR design and the JTE design achieved an efficiency over 83% and a positive temperature coefficient of breakdown voltage, suggesting avalanche breakdown capability. The JTE design specifically shows robust avalanche breakdown behavior to pass high avalanche current under UIS (unclamped inductive switching) test. Finally, a comprehensive comparison between these two designs and other vertical GaN device edge terminations is performed, showing that these two designs are promising as the building blocks for vertical GaN devices.

Index Terms—power electronics, power semiconductor devices, gallium nitride, edge termination, avalanche capability, ion implantation

I. INTRODUCTION

Gallium nitride vertical power devices is deemed as a strong candidate for high-voltage, high power applications [3]–[5]. Edge termination is essential for managing electric field crowding at the edge of device active area, achieving the desired breakdown voltage, and improving the device robustness. Many edge terminations have been studied already, including beveled field plate, mesa, JTE, guard rings and their combinations. Most of the designs reported so far either have complicated fabrication process, or do not show robust avalanche breakdown mechanism. In this work, the two methods - GR and USAB-JTE [1], [2] all utilize single step ion implantation to accomplish both the isolation and edge termination. Their fabrication does not require precise control of the implantation depth and obviates the need for GaN etching, enabling a great process latitude and less fabrication complexity. Meanwhile, both the designs are able to achieve positive temperature coefficients, a strong indicator

This work was supported in part by National Science Foundation under Grant No. ECCS-2134374, and in part by the Power Management Consortium of the Center for Power Electronics Systems.

for avalanche breakdown. High efficiencies are also obtained at about 83% for USAB-JTE design and 88% for GR design.

The paper is organized as follows. Section II presents the guard ring design structures, fabrication process and static I-V characterization. Section III presents the JTE design, fabrication, and device I-V characteristics. Section IV presents the dynamic UIS circuit test setup and resutls. Section V presents the benchmark and comparsion. Section VI concludes and summarizes the paper.

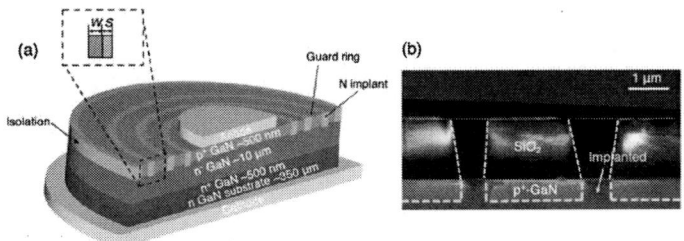

Fig. 1. (a) 3D cross-section view of vertical GaN PN diode with guard ring edge termination. "W" and "S" stand for guard ring width and spacing. (b) Cross-sectional scanning electron microscopy image showing the contrast between ion-implanted region and protected region.

II. GUARD RING DESIGN AND FABRICATION

A. GR Device Structure

Fig. 1 shows the 3D schematic of the guard ring structure. W and S refer to the width and spacing, where the width is unimplanted p-GaN region and spacing is the implanted isolation region. W and S vary from 1 μm to 2.5 μm and 1 μm to 2 μm, respectively. The active area is surrounded by the implanted rings. The wafer consists of a 500 nm thick n+-GaN ([Si] = 2×10^{18}cm^{-3}) layer, a 10 μm thick n−-GaN ([Si] = 1×10^{16}cm^{-3}) drift layer, a 500 nm thick p+-GaN ([Mg] = 1×10^{19}cm^{-3}) layer and a 20 nm thick p++-GaN ([Mg] = 1×10^{20}cm^{-3}), all grown by MOCVD by Enkris Semiconductor Inc. on the 2-inch GaN substrates supplied by Nanowin Science and Technology Co., Ltd. The net donor concentration (N$_D$-N$_A$) is extracted to be 8×10^{15}cm^{-3} by C-V measurements.

979-8-3503-3714-3/23 $31.00 © 2023 IEEE

B. GR Device Fabrication

Fig. 2 shows the fabrication process. The process started from first depositing a thin layer of Cr onto the surface of p^{++}-GaN for protection, followed by deposition of thick SiO_2 by plasma-enhanced chemical vapor deposition (PECVD) as the hard mask for implantation. Subsequently, lithography was done to pattern the SiO_2 layer for the following dry etch, creating openings for implantation. Nitrogen (N) was then implanted at 7° incidence. SRIM Monte-Carlo simulation was used to design the implantation profile, making sure the implanted region is fully compensated by having the defect concentration greater than the Mg concentration in p-GaN [7]. Fig. 1(b) shows SEM cross section of patterned SiO_2. A contrast is observed between the implanted region and unimplanted region. A uniform 400nm shrink of S is found due to the sloped sidewall created by non-vertical etching process. SiO_2 masks and Cr layer were removed in an HF rinse and Cr etchant after implantation. Ohmic contacts of both n-GaN and p-GaN were then deposited onto the device.

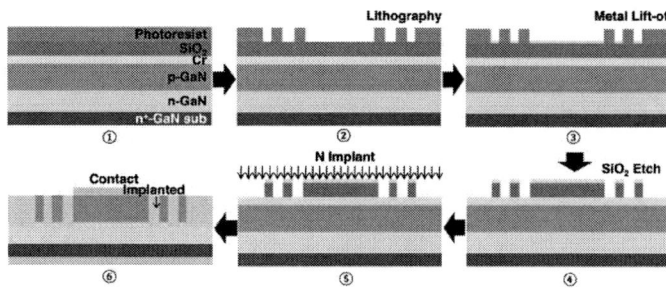

Fig. 2. Fabrication process of the GR edge terminations.

C. GR Device Static I-V Characterization

Fig. 3 shows the static I-V measurements for the fabricated GR devices. It is found that ring count and spacing are strongly related to the breakdown voltage while the width effect is relatively minor. Fig. 3(a) shows the reverse I-V characteristics of diodes with 4 and 16 rings, each with the practical S of 0.6, 1.1, and 1.6 μm and an identical W of 2.4 μm. The difference of breakdown voltage over different S proves the ring spacing effect on BV (breakdown voltage). Fig. 3(b) demonstrated that BV is not sensitive to W.

Fig. 3(c) shows the temperature-dependent reverse I-V characteristics of the vertical GaN PN diodes with only the implanted isolation (i.e., no guard ring) and those with 4, 8, and 16 guard rings ($S = 1.1$ μm and $W = 2.4$ μm). A clear trend of increased BV over increased spacing is observed. Positive temperature coefficient of breakdown voltage is demonstrated on all devices measured, suggesting a uniform avalanche breakdown capability. As ring count increases, the BV reaches saturation. Fig. 3(d) shows that 24-ring designs is comparable to 16-ring designs. The box plot in Fig. 3(e) further proves the BV trend and saturation.

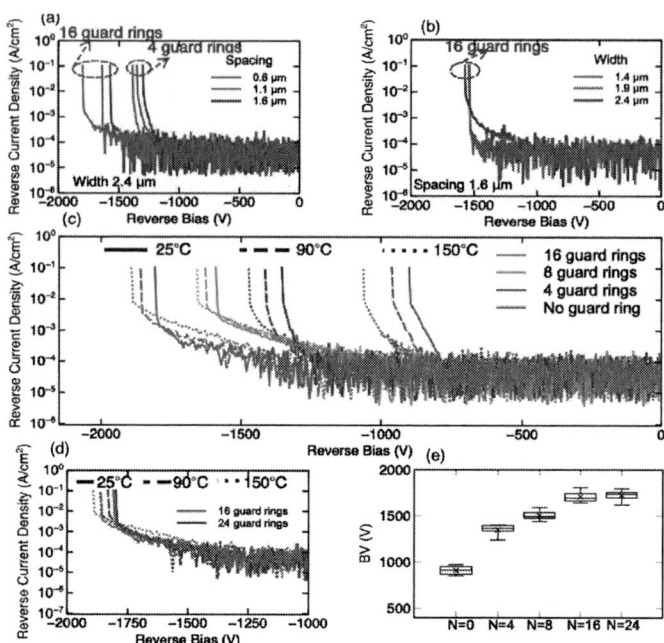

Fig. 3. (a) Reverse I-V characteristics for 4-guard-ring and 16-guard-ring devices of fixed $W = 2.4$ μm, and varying S of 0.6, 1.1, and 1.6 μm. (b) Reverse I-V characteristics for 16-guard-ring devices with varying W of 1.4, 1.9, and 2.4 μm, and fixed $S = 1.6$ μm. (c) Temperature-dependent reverse I-V characteristics of devices with no guard ring and 4, 8 and 16 guard rings (all with S of 1.1 μm and W of 2.4 μm). For each devices, measured data at three temperatures, i.e., 25, 90 and 150 °C, are shown. (d) Reverse I-V for 16-guard-ring and 24-guard-ring devices with same $W = 2.4$ μm, $S = 1.1$ μm. (e) Box plot of BV distribution for devices with different number of rings and fixed $W = 2.4$ μm, $S = 1.1$ μm.

III. JTE DESIGN AND FABRICATION

A. Device Structure

Fig. 4(a) shows the 3D schematic of the JTE design, the ultra-small-angle bevel JTE (USAB-JTE). The uncompensated p-GaN at the device edge exhibits a wedge shape with a large ratio between the JTE width W and thickness T. The wafer used is the same as the ones in guard ring designs. Fig. 4(b) shows the 2d cross-section of the edge termination as well as the electric field peak over different $W:T$ ratio. It is clear that as the bevel angle becomes smaller, electric field peak is reduced.

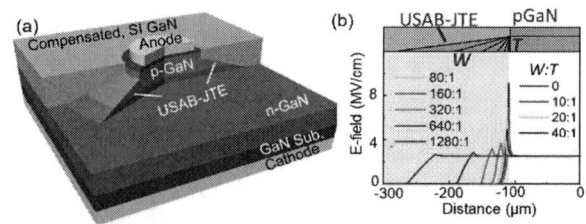

Fig. 4. (a) 3D schematic of the JTE design. (b) 2D schematic and simulated electric field with different bevel angle.

B. JTE Device Fabrication

In Fig. 5(a), an overview of the primary fabrication steps is demonstrated. A 480 nm Al_2O_3 layer is deposited through

979-8-3503-3714-3/23 $31.00 © 2023 IEEE

atomic layer deposition (ALD) at 300 °C. Subsequently, a bilayer photoresist (PR) is employed for lithography, consisting of 2 μm PMGI and 3 μm AZ3330F. Following a bake step at 200 °C, the sample is submerged in a 2.38% TMAH solution to completely etch away the Al_2O_3 in regions not covered by the PR. This results in the formation of a significant lateral undercut beneath the PR, as the etch rate of PMGI is approximately 750 nm/min, significantly faster than that of Al_2O_3, which is roughly 1.2 nm/min. Consequently, a small bevel angle is formed along the Al_2O_3 sidewall, and the magnitude of this bevel angle is determined by the ratio of the lateral (PMGI) and vertical (Al_2O_3) etch rates.

This patterned and etched Al_2O_3 is then used as the mask for nitrogen (N) implantation. Similar to the guard ring design, SRIM Monte-Carlo simulation was used to determine the energy and dose for full and partial compensation; 190 nm Al_2O_3 mask would give a compensation depth of 400 nm while an unmasked region would give a compensation depth of 610 nm, enough for compensating the whole p-GaN layer (500 nm). As shown in Fig. 4(c), the small bevel angle created in the Al_2O_3 mask is transferred to the uncompensated p-GaN, forming a USAB-JTE with T of ~120 nm and W of ~75 μm.

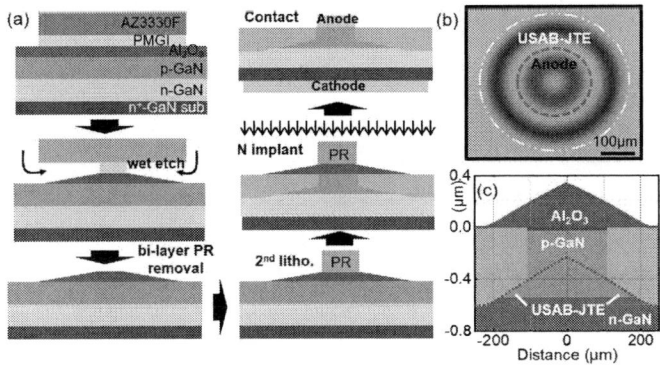

Fig. 5. (a) Schematics of the fabrication steps of the JTE design. (b) Top-view microscopic image of the fabricated USAB Al_2O_3 mask. (c) Surface profile of Al_2O_3 measured by Dektak (solid red line) and the simulated profile of the uncompensated p-GaN (dashed line) from Monte Carlo simulations.

C. JTE Device Static I-V Characterization

Fig. 6 shows the reverse I-V characteristics of the devices with the USAB-JTE design and the ones with only the isolation. The USAB-JTE is able to reach a 83% efficiency, about 1700V breakdown voltage, a large boost compared to the diodes without this edge termination. The breakdowns of both devices have a positive temperature coefficient and are non-destructive, suggesting an impact ionization induced avalanche breakdown.

IV. DYNAMIC UIS CIRCUIT TEST

A. UIS Test Principle and Setup

In static I-V test, low compliance of current is set, restricted by the curve tracer upper power limit. Dynamic UIS circuit test is then commonly used to determining whether the device can pass high I_{AVA} (avalanche current) and dissipate energy,

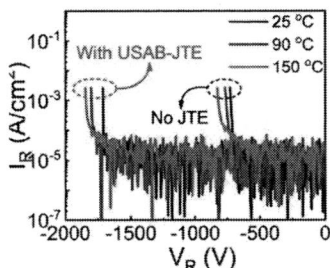

Fig. 6. Reverse I-V characteristics of the diodes with and without the JTE at various temperatures.

which is more essential in determine the avalanche breakdown capability desired for power electronics applications. Fig. 7(a) and 7(b) shows the on-wafer test setup. During the UIS test, the transistor is first turned ON to charge the inductor. Device is then turned-OFF, and the inductive energy is dissipated all through the DUT (device under test), forcing it to withstand high voltage and high current. DUT with the true avalanche capability will be able to clamp the voltage at the breakdown and dissipate large amount of energy through impact ionization.

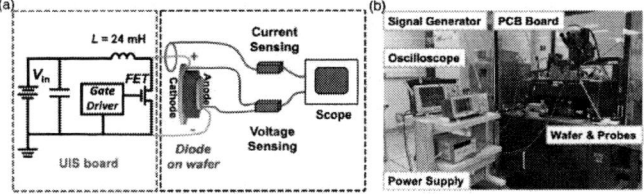

Fig. 7. (a) Schematic and (b) photo of the UIS test setup for on-wafer devices.

B. JTE and GR Device UIS Test

Fig. 8 shows the voltage and current waveforms of the diodes with and without the USAB-JTE in the UIS tests with an increased turn-ON time (t_{ON}) and inductive energy. Classic avalanche waveforms with the voltage clamped at the BV is observed. The current goes down to zero after energy dissipattion. At 1.7 kV, the I_{AVA} is over 1100 A/cm^2 is observed.

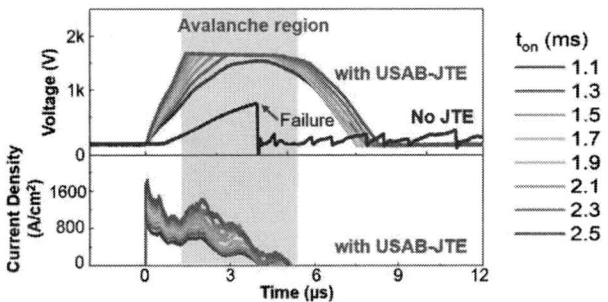

Fig. 8. UIS waveforms of the diodes with and without JTE under various t_{ON}.

979-8-3503-3714-3/23 $31.00 © 2023 IEEE   233

Fig. 9 shows the UIS test waveforms for GR device. Although the GR devices is able to achieve over 1700 V breakdown voltage as well in dynamic test, the device does not clamp at the BV, instead, capacitive charging phase is observed and ultimately a destructive failure is shown [6]. The behavior suggests that although positive temperature coefficient is observed in static I-V test, it is not necessarily a indicator of a robust full ion impact ionization. The impact ionization could happen locally, restricting the amount of energy the device can withstand [8]. In the guard ring case, the impact ionization might happen only on the outer ring instead of the main junction, limiting the energy dissipation capability. The further optimization of the GR design to enable the robust avalanche will be investigated in our future work.

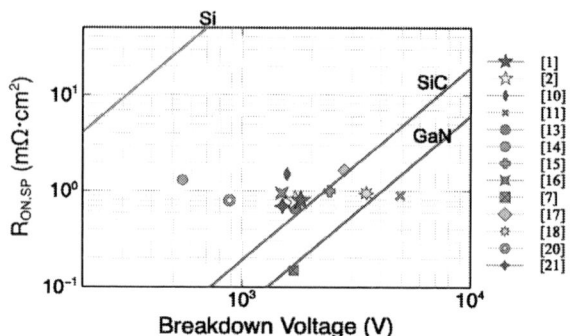

Fig. 10. Benchmark of differential $R_{ON,SP}$ vs. Breakdown voltage for the reported GaN PN diodes.

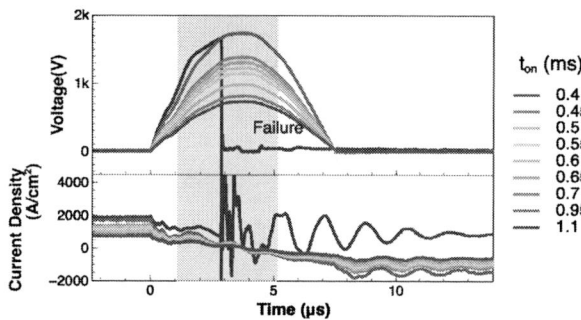

Fig. 9. UIS waveforms of the diodes with GR under various t_{ON}.

V. BENCHMARK AND TABLE

In this section, we compare the guard ring and JTE edge terminations to vertical GaN PN diodes reported in the literature and compare the performance of our fabricated devices with them. Fig. 10 benchmarks the tradeoff between the differential $R_{ON,SP}$ and BV. The performance of our devices are comparable to the state of the art of vertical GaN PN diodes with a similar voltage rating. Table. I compares the key metrics (avalanche capability, efficiency, fabrication complexity) between our edge termination designs and other reported GaN termination technologies. Efficiency in this table is defined as the practical BV over the 1D prallel-plane BV limit. The 1D parallel-plane BV limit is defined as $BV_{pp} = E_c t - q N_D t^2 / 2\epsilon$, where t is the drift region thickness, ϵ is GaN's permittivity (10.4 used for c-axis) and N_D is the net donor concentration of the drift region. E_c is the avalanche critical E-field of GaN, and its dependence on N_D can be modeld by the following equation $E_c = 2.81/[1 - 0.205(N_D/10^{16})]$ [9]. The comparison shows the GR and JTE design advantages over other edge termination designs.

VI. CONCLUSION

In summary, two single planar step implantation edge terminations are demonstrated in this paper, the GR and USAB-JTE designs. Both design show a high efficiency of over 83%

TABLE I
COMPARISON OF THE STRUCTURE, AVALANCHE CAPABILITY, FABRICATION COMPLEXITY AND EFFICIENCY OF VERTICAL GANEDGE TERMINATION TECHNOLOGIES

Structure	Ref	Avalanche		Fabrication		Efficiency	
		Static	Circuit	GaN Etch	Implant Plasma	BV (kV)	BV/ BV_{pp}
GR	[1]	Yes	No	0	1	1.8	88%
USAB-JTE	[2]	Yes	Yes	0	1	1.7	83%
GR	[10]	Yes	N/A	0	1	1.575	90%
GR	[11]	N/A	N/A	0	1	4.92	N/A
GR	[12]	N/A	N/A	1	1	0.97	N/A
GR	[13]	N/A	N/A	0	1	1.7	N/A
JTE	[14]	N/A	N/A	4	0	0.55	N/A
JTE	[15]	Yes[1]	N/A	1	1	2.6	87%
JTE	[16]	N/A	N/A	0	1	1.48	N/A
JTE	[7]	N/A	N/A	1	1	1.68	N/A
FP	[17]	Yes	N/A	0	1	2.79	85%
FP	[18]	Yes	N/A	1	0	3.48	81%
Mesa	[19]	Yes	Yes	1	0	2.9	88%
Mesa	[20]	Yes	N/A	1	0	0.88	68%
Mesa	[21]	N/A	N/A	1	1	1.5	N/A

[1] Based on electroluminescence.

and positive temperature coefficients of BV. The JTE edge termination also demonstrates dynamic avalanche breakdown robustness to pass high avalanche current. The difference in the circuit test for the two designs shows that static I-V test is not sufficient for proving the avalanche breakdown robustness desirable for power electronics applications. Both designs are promising as the building blocks for high voltage GaN applications.

REFERENCES

[1] Y. Wang, M. Porter, M. Xiao, A. Lu, N. Yee, I. Kravchenko, B. Srijanto, K. Cheng, H. Wong, Y. Zhang, "Implanted Guard Ring Edge Termination with Avalanche Capability for Vertical GaN Devices," *IEEE Trans. Electron Devices*, 2023.

[2] M. Xiao, Y. Wang, R. Zhang, Q. Song, M. Porter, E. Carlson, K. Cheng, K. Ngo, and Y. Zhang, "Robust Avalanche in 1.7 kV Vertical GaN Diodes With a Single-Implant Bevel Edge Termination," *IEEE Electron Device Lett.*, vol. 44, pp. 1616-1619, October, 2023.

[3] Y. Zhang, F. Udrea and H. Wang, "Multidimensional device architectures for efficient power electronics," *Nat Electron*, vol. 5, pp. 723-734, Nov, 2022.

[4] J.P. Kozak, R. Zhang, M. Porter, Q. Song, J. Liu, B. Wang, R. Wang, W. Saito, and Y. Zhang, "Stability, Reliability, and Robustness of GaN Power Devices: A Review," *IEEE Trans. Power Electron.*, vol. 38, pp. 8442–8471, Jul, 2023.

[5] Y. Zhang and T. Palacios, "(Ultra)Wide-Bandgap Vertical Power Fin-FETs," *IEEE Trans. Electron Devices*, vol. 67, pp. 3960–3971, Oct, 2020.

[6] R. Zhang, J.P. Kozak, M. Xiao, J. Liu, and Y. Zhang, "Surge-Energy and Overvoltage Ruggedness of P-Gate GaN HEMTs," *IEEE Trans. Power Electron.*, vol 35, Dec, 2020, pp.13409–13419.

[7] J. Wang, L. Cao, J. Xie, E. Beam, R. McCarthy, C. Youtsey and P. Fay, "High voltage, high current GaN-on-GaN p-n diodes with partially compensated edge termination," *Appl. Phys. Lett.*, vol. 113, pp. 023502, July 2018.

[8] J. Liu, M. Xiao, R. Zhang, S. Pidaparthi, C. Drowley, L. Baubutr, A. Edwards, H. Cui, C. Coles and Y. Zhang, "Trap-Mediated Avalanche in Large-Area 1.2 kV Vertical GaN p-n Diodes," *IEEE Electron Device Lett.*, vol 41, Sep, 2020, pp.1328-1331.

[9] J. A. Cooper and D. T. Morisette "Performance Limits of Vertical Unipolar Power Devices in GaN and 4H-SiC," *IEEE Electron Device Lett.*, vol. 41., 2020, pp.892-895.

[10] V. Talesara, Y. Zhang, Z. Chen, H. Zhao and W. Lu,"Design and development of 1.5 kV vertical GaN pn diodes on HVPE substrate," *Journal of Materials Research*, vol. 36., Dec, 2021, pp.4919-4926.

[11] V. Talesara, Y. Zhang, V.G.T. Vangipuram, H. Zhao and W. Lu, "Vertical GaN-on-GaN pn power diodes with Baliga figure of merit of 27 GW/cm^2," *Appl. Phys. Lett.*, vol 122., Mar, 2023, pp.123501.

[12] M. Matys, T. Ishida, K.P. Nam, H. Sakurai, K. Kataoka, T. Narita, T. Uesugi, M. Bockowski, T. Nishimura, J. Suda and T. Kachi, "Design and demonstration of nearly-ideal edge termination for GaN p–n junction using Mg-implanted field limiting rings," *Appl. Phys. Express*, vol 14., Jul, 2021, pp.074002.

[13] H. Fu, K. Fu, S.R. Alugubelli, C.-Y. Cheng, X. Huang, H. Chen, T.-H. Yang, C. Yang, J. Zhou, J. Montes, X. Deng, X. Qi and S.M. Goodnick, "High Voltage Vertical GaN p-n Diodes With Hydrogen-Plasma Based Guard Rings," *IEEE Electron Device Lett.*, vol 41., Jan, 2020, pp.127–130.

[14] H.-S. Lee, Y. Zhang, Z. Chen, M.W. Rahman and H. Zhao, S. Rajan, "Design and Fabrication of Vertical GaN p-n Diode With Step-Etched Triple-Zone Junction Termination Extension," *IEEE Trans. Electron Devices*, vol 67., Sep, 2020, pp.3553-3557.

[15] J.R. Dickerson, A.A. Allerman, B.N. Bryant, A.J. Fischer, M.P. King, M.W. Moseley, A.M. Armstrong, R.J. Kaplar, I.C. Kizilyalli, O. Aktas and J.J. Wierer, "Vertical GaN Power Diodes With a Bilayer Edge Termination," *IEEE Trans. Electron Devices*, vol 63., Jan, 2016, pp.419–425.

[16] W. Lin, M. Wang, R. Yin, J. Wei, C.P. Wen, B. Xie, Y. Hao and B. Shen, "Hydrogen-Modulated Step Graded Junction Termination Extension in GaN Vertical p-n Diodes," *IEEE Electron Device Lett.*, vol 42., Aug, 2021, pp.1124–1127.

[17] Z. Bian, K. Zeng and S. Chowdhury, "2.8 kV Avalanche in Vertical GaN PN Diode Utilizing Field Plate on Hydrogen Passivated P-Layer," *IEEE Electron Device Lett.*, vol 43., Jan, 2022, pp.596–599.

[18] K. Nomoto, Z. Hu, B. Song, M. Zhu, M. Qi, R. Yan, V. Protasenko, E. Imhoff, J. Kuo, N. Kaneda, T. Mishima, T. Nakamura, D. Jena and H.G. Xing, "GaN-on-GaN p-n power diodes with 3.48 kV and 0.95 mΩ·cm^2: a record high figure-of-merit of 12.8 GW/cm^2," *IEEE International Electron Devices Meeting (IEDM)*, Dec, 2015, pp.9.7.1-9.7.4.

[19] B. Shankar, Z. Bian, K. Zeng, C. Meng, R.P. Martinez, S. Chowdhury, B. Gunning, J. Flicker, A. Binder, J.R. Dickerson and R. Kaplar, "Study of Avalanche Behavior in 3 kV GaN Vertical P-N Diode Under UIS Stress for Edge-termination Optimization," *IEEE International Reliability Physics Symposium (IRPS)*, Mar, 2022, pp.2B.2-1-2B.2-4.

[20] H. Fukushima, S. Usami, M. Ogura, Y. Ando, A. Tanaka, M. Deki, M. Kushimoto, S. Nitta, Y. Honda and H. Amano, "Vertical GaN p–n diode with deeply etched mesa and the capability of avalanche breakdown," *Appl. Phys. Express*, vol 12, Feb, 2019, pp.026502.

[21] D. Ji, S. Li, B. Ercan,C. Ren and S. Chowdhury, "Design and Fabrication of Ion-Implanted Moat Etch Termination Resulting in 0.7 mΩ · cm^2 1500V GaN Diodes," *IEEE Electron Device Lett.*, vol 41, Dec, 2020, pp.264-267.

979-8-3503-3714-3/23 $31.00 © 2023 IEEE

Output Capacitance Loss in Wide-bandgap and Superjunction Power Transistors: Impact of Switching Voltage and Current

Qihao Song, Qiang Li, and Yuhao Zhang
Center for Power Electronics Systems (CPES)
Virginia Tech, Blacksburg, Virginia, USA
Email: {qihao95, yhzhang}@vt.edu

Abstract— Output capacitance (C_{OSS}) loss (E_{DISS}) is a loss recently found to incur in a power semiconductor device when its C_{OSS} is charged and discharged, which ideally should be a lossless process. Recent studies have reported considerable E_{DISS} in various power devices when they are utilized in high-frequency soft-switching converters. However, the individual impact of switching voltage (V_{DS}) and current (I_{DS}) has not been fully revealed due to their coupled modulation. This work performs comprehensive experimental characterization of the E_{DISS} of SiC, GaN, and Si superjunction (SJ) transistors, and, for the first time, derives a unified model to describe their E_{DISS}. The individual impact of V_{DS} and I_{DS} on E_{DISS} is revealed for all devices. This model provides device users with quick and useful reference for calculating device E_{DISS} when designing high-frequency soft-switching converters.

Keywords—Gallium Nitride, Silicon Carbide, Wide bandgap Devices, Superjunction, Soft-switching, Output Capacitance Loss, Modeling

I. INTRODUCTION

A power device is expected to achieve zero loss in soft-switching operations. The switching loss is reduced or eliminated, allowing the devices to operate at high frequencies with no penalties [1]. The recent advanced wide-bandgap devices, capable of delivering high power at high frequencies, have enabled more compact and efficient power converters [2]. However, recently, dynamic C_{OSS} loss (E_{DISS}) has been reported in high-frequency soft-switching converters, which could be a bottleneck of this promise [3]. In zero-voltage-switching (ZVS), the device's C_{OSS} is charged and discharged once in each switching cycle, which is not a lossless process as supposed. For example, a GaN HEMT's E_{DISS} can account for up to 80% of its total loss in a soft-switching converter [4].

Accurate E_{DISS} measurement is critical yet challenging. Several approaches have been proposed in the literature to characterize E_{DISS}, which also unveiled E_{DISS}'s dependences on switching parameters such as V_{DS} and I_{DS}, as summarized in Table I. A Sawyer-Tower (ST) circuit is employed in [5], in

This work is supported in part by a PowerAmerica member-initiated project through the Office of Energy Efficiency and Renewable Energy (EERE), U.S. Department of Energy, under Award Number DE-EE0006521, and in part by the High-Density Integration (HDI) Consortium and the Power Management Consortium (PMC) of Center for Power Electronics Systems at Virginia Tech.

TABLE I. A COMPARISON OF PREVIOUSLY REPORTED APPROACHES WITH THE APPROACH IN THIS WORK FOR STUDYING THE C_{OSS} LOSS' DEPENDENCE ON CURRENT AND VOLTAGE

Method	Modulation		E_{DISS} Model	Notes
	V_{DS}	I_{DS}		
Sawyer-Tower (ST) [5]	Yes	No	No I_{DS} invlovled	Limited I_{DS} and V_{DS} range
Modified ST [7]	Yes	Yes	No V_{DS} invlovled	Only two different V_{DS} are measured
Nonlinear Resonance [8]	Yes	No	No I_{DS} invlovled	V_{DS} and I_{DS} change simutantously
Modified UIS (this work)	Yes	Yes	Both V_{DS} and I_{DS} invlovled	V_{DS} and I_{DS} are decoupled

which the device under test (DUT) is always OFF and functions as a passive C_{OSS}. A reference capacitor (C_{ref}) is put in series with the OFF-state DUT. With both voltage across DUT and C_{ref} measured, the E_{DISS} of the DUT can be extracted. However, the ranges of V_{DS} and I_{DS} are limited by the performance of the RF power amplifier [6], [7]. Also, I_{DS} is not tunable for the same $V_{DS(peak)}$, limiting the ST method's capability to probe the E_{DISS}'s current dependence. A modified ST method proposed in [8] allows for I_{DS} modulation, revealing a linear relationship between E_{DISS} and I_{DS} for some devices. Nevertheless, the V_{DS} dependence is unclear with this method. The E_{DISS} with V_{DS} dependence is modeled in [9], which proposes a non-linear (NR) method for E_{DISS} measurement. The method is based on an L-C resonance between a load inductor (L_{load}) and DUT's C_{OSS}. The V_{DS} modulation is achieved by simply tuning L_{load}. This makes V_{DS} and I_{DS} change simultaneously and is not able to separate the impact of V_{DS} and I_{DS}. Thus, a method with decoupled and wide-range V_{DS} and I_{DS} modulation is highly desired. A simple model of E_{DISS} with V_{DS} and I_{DS} dependence can provide device users with a quick and useful reference in the design of a high-frequency soft-switching power converter.

II. METHODOLOGY

A. DUT and Test Setup

Various commercial 600/650-V ~50mΩ-rated wide-bandgap and superjunction power transistors are characterized,

TABLE II. MAIN CHARACTERISTICS OF THE DUTs IN THIS WORK

Technology	Part Number	V_{DS} (V)	I_{DS} (A)	R_{ON} (mΩ)	$C_{OSS(er)}$ (pF)
Si SJ	IPP65R045 C7	650	46	40	146
SiC MOSFET	C3M00450 65J1	650	47	45	126
SiC cascode	UF3SC065 040B7S	650	43	42	146
P-gate HEMT	GS66508T	650	30	50	100
HD-GIT	IGOT60R0 70D1	600	31	55	80
GaN Direct-drive	LMG3410 R050	600	34	57	119
GaN cascode	TP65H050 WS	650	36	50	190

as listed in Table II. The method used in this work is a modified unclamped inductive switching (UIS) circuit based on [10], as shown in Fig. 1(a). Note that only the test results of GaN devices using this method have been reported previously; this work expands the test and modeling to Si and SiC devices. The device under test (DUT) is in series with a power supply (V_{bus}) as well as a L_{load}. Low-ESR ceramic capacitors (C_{ext}) are paralleled with the devices' C_{OSS}. Fig. 1(b) shows a photo of the setup with a TO-247 packaged device. Air-core Litz-wire-based inductors are used in the setup for L_{load}, as they are known to produce low and computable losses at high frequencies [7]. A commercial screw-in co-axial shunt resistor with resistance (R_{SHUNT}) of 0.1 Ω accurately measures the DUT drain-to-source current (I_{DS}).

B. C_{OSS} Loss Extraction

A typical test waveform is shown in Fig. 2(a). The test mainly contains three stages: in stage I, the DUT is ON, V_{bus} charges L_{load}; in stage II, the DUT is turned OFF, and L_{load} resonates with the DUT's C_{OSS} and C_{ext} for about a half resonance cycle (at a resonant frequency of f_R); in stage III, the DUT conducts reversely, and the resonance gradually dies

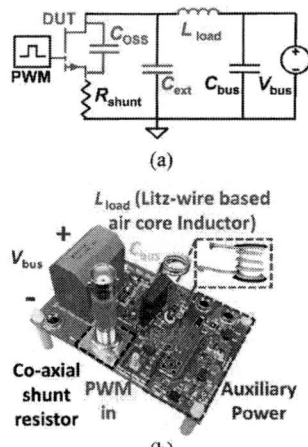

(a)

(b)

Fig. 1: (a) Schematic and (b) photo of a modified UIS test setup.

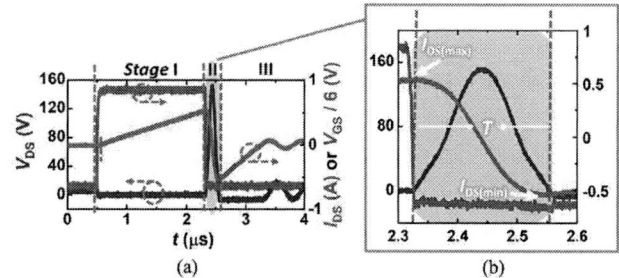

(a)   (b)

Fig. 2: (a) A typical UIS waveform. (b) Zoom-in picture of stage II (L-C resonance), where E_{DISS} is extracted.

down. The E_{DISS} is extracted from stage II, and the zoom-in image is shown in Fig. 2(b). For a lossless L_{load}-C_{OSS} resonance, the magnitude of the current (inductor current (I_L) or I_{DS}), i.e., I_{DS}'s peak ($I_{DS(max)}$) and valley ($I_{DS(min)}$) points, should be equal. The energy loss during the L-C resonance consists of E_{DISS} and loss incurred on other parasitic components (E_{other}) and is dominated by E_{DISS}.

$$E_{DISS} = 0.5\ L\ \left(I_{DS(max)}^2 - I_{DS(min)}^2\right) - E_{other} \qquad (1)$$

where L is L_{load}'s inductance. E_{other} is experimentally measured following the method in [11], proven to be marginal compared with E_{DISS}.

The potential impact factors for E_{DISS} in a single UIS pulse include the f_R (related to dv/dt), $V_{DS(peak)}$, $I_{DS(max)}$ and temperature. Therefore, to solely investigate the impact of $V_{DS(peak)}$ and $I_{DS(max)}$, f_R is fixed in the test (1 MHz for Si superjunction and 2 MHz for SiC and GaN devices), and the test is conducted at room temperature.

C. Modulation of Switching Voltage and Current

To probe the E_{DISS}'s dependence on $V_{DS(max)}$, other variables, e.g., $I_{DS(max)}$ and f_R, must be kept unchanged during the modulation process. This can be achieved by tuning C_{ext} and L. C_{ext}, realized by multiple MLCC surface-mount capacitors, achieves a marginal ESR, leading to a negligible loss addition for the system [12]. In this modified UIS circuit, f_R can be derived:

$$f_R = 1\ /\ \left[2\pi\sqrt{L'(C_{OSS} + C_{ext})}\right] \qquad (2)$$

where L' is the adjusted inductance as required to maintain the constant f_R after C_{ext} is added to the system.

$V_{DS(peak)}$ and $I_{DS(max)}$ can be expressed as:

$$V_{DS(peak)} = I_{DS(max)}\sqrt{L'/(C_{OSS} + C_{ext})} \qquad (3)$$

$$I_{DS(max)} = V_{DS(peak)}\sqrt{(C_{OSS} + C_{ext})/L'} \qquad (4)$$

Therefore, $V_{DS(peak)}$ can be increased by reducing C_{ext} and increasing L', at the same time optimizing the C_{ext} and L' based on (2) to maintain the f_R and $I_{DS(max)}$ unchanged. $I_{DS(max)}$ can be modulated in the opposite way. Fig. 3(a) and (b) show a series of UIS waveforms with various switching voltage (or current) but the constant current (or voltage) and f_R. Note that C_{ext} and L' are carefully selected to also compensate for the impact of C_{OSS}'s voltage non-linearity, enabling a superior f_R and dv/dt consistency over the fixed L'.

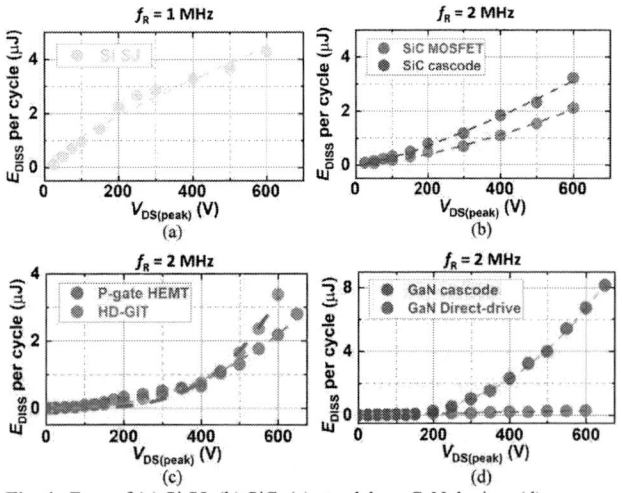

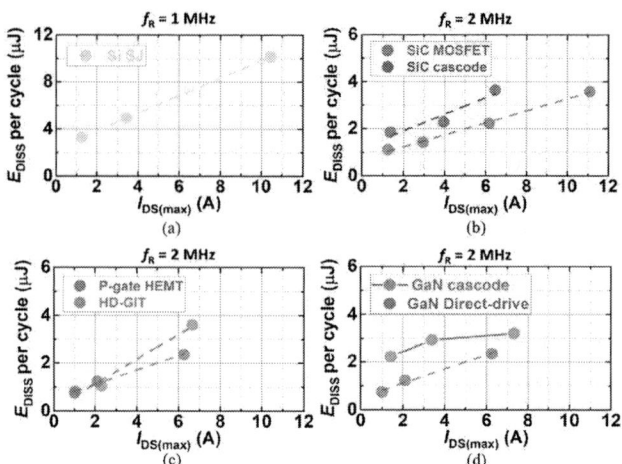

Fig. 3: UIS waveforms with various (a) switching voltage and (b) switching current with frequency unchanged.

Fig. 5: E_{DISS} of (a) Si SJ, (b) SiC, (c) standalone GaN devices (d) composite GaN devices under the $I_{\text{DS(max)}}$ up to 11 A.

III. Measurement Results and C_{OSS} Loss Model

A. C_{OSS} Loss Model

The E_{DISS} of various commercial 600/650-V ~50-mΩ-rated wide-bandgap and Si-SJ transistors are characterized under different $V_{\text{DS(peak)}}$ and $I_{\text{DS(max)}}$ (Fig. 4 and 5). In general, all the tested DUTs' E_{DISS} increases with V_{DS} and I_{DS}. All the DUTs' E_{DISS} can be modeled as power-law relationships with V_{DS}, as shown in Fig. 4. Si SJ's E_{DISS} has a weaker dependence on V_{DS} as signified by the smaller than unity power law coefficient. This is due to the nearly full depletion of superjunction pillars at high V_{DS}, leading to a minimal E_{DISS} [13]. All the tested DUTs' E_{DISS} increases nearly linearly with peak I_{DS} except for the GaN cascode (Fig. 5). Its E_{DISS} exhibits a relatively weaker dependence on I_{DS}, due to the complex interplay between GaN HEMT and Si MOSFET [14]–[16]. Therefore, for most SiC, GaN, and Si SJ devices, their E_{DISS} (in μJ) per C_{OSS} charging /discharging cycle as a function of V_{DS} (in V) and I_{DS} (in A) can be modeled by:

$$E_{\text{DISS}} = k[\alpha + \beta I_{\text{DS(max)}}]V_{\text{DS(peak)}}^{\gamma} \qquad (5)$$

The fitting parameters are shown in Table III.

B. Discussions

As mentioned above, besides $V_{\text{DS(peak)}}$ and $I_{\text{DS(max)}}$, E_{DISS} also shows dependence on temperature and f_{R} (related to dv/dt). The $E_{\text{DISS}} - f_{\text{R}}$ relationship is much more complex hence specific frequencies of interest are selected for modeling. The E_{DISS} of GaN HEMTs at different temperatures and f_{R} has been discussed in [11], [17], [18]. Fig. 6 shows the E_{DISS} of Si and SiC devices at different temperatures, indicating that the E_{DISS} of SiC MOSFET and SiC JFET cascode decreases at an elevated temperature while the E_{DISS} of Si SJ is nearly temperature-independent. Regardless of the complexity of various dependencies, E_{DISS} increases with $V_{\text{DS(peak)}}$ and $I_{\text{DS(max)}}$ in a power law and linear relation, respectively. This indicates the common roles of switching voltage and current in impacting the E_{DISS} of a wide variety of power transistors. The proposed model should give device users a useful and convenient reference of E_{DISS} in a high-frequency soft-switching converter.

Fig. 4: E_{DISS} of (a) Si SJ, (b) SiC, (c) standalone GaN devices (d) composite GaN devices under $V_{\text{DS(peak)}}$ from 0 to 600 V.

TABLE III. Fitting Parameters of the Tested DUTs (1 MHz for Si SJ and 2 MHz for SiC and GaN)

DUT	k	α	β	γ
Si SJ	1.13×10^{-8}	2.36	0.75	0.74
SiC MOSFET	1.41×10^{-10}	0.71	0.26	1.49
SiC cascode	3.37×10^{-10}	1.21	0.35	1.34
P-gate HEMT	1.45×10^{-11}	0.42	0.33	1.86
HD-GIT	0.95×10^{-15}	0.54	0.37	3.42
GaN Direct-drive	4.12×10^{-10}	0.03	0.30	1.20
GaN cascode	0.99×10^{-13}	2.18	0.15	2.69

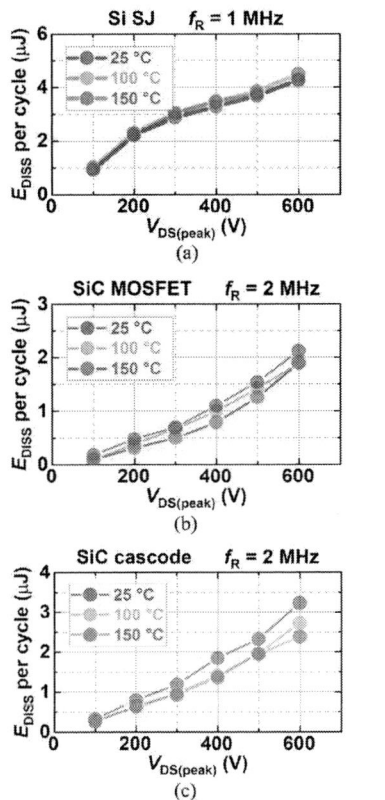

Fig. 6: E_{DISS} of (a) Si SJ, (b) SiC MOSFET, and (c) SiC cascode as a function of $V_{\text{DS(peak)}}$ at case temperature of 25°C, 100°C, and 150°C.

IV. CONCLUSIONS

E_{DISS} of power devices is critical in high-frequency soft-switching applications. E_{DISS} is found to be both V_{DS} and I_{DS}-related; prior E_{DISS} models did not fully capture these dependencies. This work utilizes the modified UIS method to experimentally characterize the E_{DISS} of GaN, SiC, and Si-superjunction devices over a wide V_{DS} and I_{DS} range. Through circuit-level component modulation, the impacts of V_{DS} and I_{DS} on E_{DISS} are separated, allowing for the derivation of a unified E_{DISS} model that is applicable to most of these devices. The results and model provide a useful reference for the design of high-frequency soft-switching converters.

REFERENCES

[1] X. Yu, J. Feng, and Q. Li, "A Planar Omnidirectional Wireless Power Transfer Platform for Portable Devices," in *2023 IEEE Applied Power Electronics Conference and Exposition (APEC)*, Orlando, FL, USA: IEEE, Mar. 2023, pp. 1654–1661. doi: 10.1109/APEC43580.2023.10131566.

[2] Y. Zhang, F. Udrea, and H. Wang, "Multidimensional device architectures for efficient power electronics," *Nat Electron*, vol. 5, no. 11, Art. no. 11, Nov. 2022, doi: 10.1038/s41928-022-00860-5.

[3] J. P. Kozak *et al.*, "Stability, Reliability, and Robustness of GaN Power Devices: A Review," *IEEE Transactions on Power Electronics*, vol. 38, no. 7, pp. 8442–8471, Jul. 2023, doi: 10.1109/TPEL.2023.3266365.

[4] A. Jafari *et al.*, "Comparison of Wide-Band-Gap Technologies for Soft-Switching Losses at High Frequencies," *IEEE Transactions on Power Electronics*, vol. 35, no. 12, pp. 12595–12600, Dec. 2020, doi: 10.1109/TPEL.2020.2990628.

[5] G. Zulauf, S. Park, W. Liang, K. N. Surakitbovorn, and J. Rivas-Davila, "COSS Losses in 600 V GaN Power Semiconductors in Soft-Switched, High- and Very-High-Frequency Power Converters," *IEEE Transactions on Power Electronics*, vol. 33, no. 12, pp. 10748–10763, Dec. 2018, doi: 10.1109/TPEL.2018.2800533.

[6] G. Zulauf, Z. Tong, J. D. Plummer, and J. M. Rivas-Davila, "Active Power Device Selection in High- and Very-High-Frequency Power Converters," *IEEE Transactions on Power Electronics*, vol. 34, no. 7, pp. 6818–6833, Jul. 2019, doi: 10.1109/TPEL.2018.2874420.

[7] M. Samizadeh Nikoo, A. Jafari, N. Perera, and E. Matioli, "Measurement of Large-Signal COSS and COSS Losses of Transistors Based on Nonlinear Resonance," *IEEE Transactions on Power Electronics*, vol. 35, no. 3, pp. 2242–2246, Mar. 2020, doi: 10.1109/TPEL.2019.2938922.

[8] D. Bura, T. Plum, J. Baringhaus, and R. W. De Doncker, "Hysteresis Losses in the Output Capacitance of Wide Bandgap and Superjunction Transistors," in *2018 20th European Conference on Power Electronics and Applications (EPE'18 ECCE Europe)*, Sep. 2018, p. P.1-P.9.

[9] M. S. Nikoo, A. Jafari, N. Perera, and E. Matioli, "New Insights on Output Capacitance Losses in Wide-Band-Gap Transistors," *IEEE Transactions on Power Electronics*, vol. 35, no. 7, pp. 6663–6667, Jul. 2020, doi: 10.1109/TPEL.2019.2958000.

[10] Q. Song, R. Zhang, J. P. Kozak, J. Liu, Q. Li, and Y. Zhang, "Robustness of Cascode GaN HEMTs in Unclamped Inductive Switching," *IEEE Transactions on Power Electronics*, pp. 1–1, 2021, doi: 10.1109/TPEL.2021.3122740.

[11] Q. Song, R. Zhang, Q. Li, and Y. Zhang, "Output Capacitance Loss of GaN HEMTs in Steady-State Switching," *IEEE Transactions on Power Electronics*, pp. 1–10, 2023, doi: 10.1109/TPEL.2023.3279308.

[12] D. Menzi, S. Ben-Yaakov, G. Zulauf, and J. W. Kolar, "ESR Modeling of Class II MLCC Large-Signal-Excitation Losses," *IEEE Trans. Power Electron.*, vol. 38, no. 5, pp. 5711–5715, May 2023, doi: 10.1109/TPEL.2023.3246095.

[13] J. Roig and F. Bauwens, "Origin of Anomalous C_{OSS} Hysteresis in Resonant Converters With Superjunction FETs," *IEEE Transactions on Electron Devices*, vol. 62, no. 9, pp. 3092–3094, Sep. 2015, doi: 10.1109/TED.2015.2455072.

[14] X. Huang, W. Du, F. C. Lee, Q. Li, and Z. Liu, "Avoiding Si MOSFET Avalanche and Achieving Zero-Voltage Switching for Cascode GaN Devices," *IEEE Transactions on Power Electronics*, vol. 31, no. 1, pp. 593–600, Jan. 2016, doi: 10.1109/TPEL.2015.2398856.

[15] Q. Song, R. Zhang, Q. Li, and Y. Zhang, "Origin of Soft-Switching Output Capacitance Loss in Cascode GaN HEMTs at High Frequencies," *IEEE Trans. Power Electron.*, vol. 38, no. 11, pp. 13561–13566, Nov. 2023, doi: 10.1109/TPEL.2023.3299977.

[16] Q. Song, R. Zhang, Q. Li, and Y. Zhang, "Investigation on Physical Origins of Output Capacitance Loss in Cascode GaN HEMTs," in *2023 IEEE Applied Power Electronics Conference and Exposition (APEC)*, Mar. 2023, pp. 651–655. doi: 10.1109/APEC43580.2023.10131161.

[17] Q. Song, R. Zhang, Q. Li, and Y. Zhang, "A Simple and Accurate Method to Characterize Output Capacitance Losses of GaN HEMTs," in *2022 IEEE Energy Conversion Congress and Exposition (ECCE)*, Oct. 2022, pp. 1–6. doi: 10.1109/ECCE50734.2022.9948018.

[18] Q. Song, R. Zhang, Q. Li, and Y. Zhang, "Impact of Conduction Current on Output Capacitance Loss in GaN HEMTs," in *2023 IEEE Applied Power Electronics Conference and Exposition (APEC)*, Orlando, FL, USA: IEEE, Mar. 2023, pp. 2533–2537. doi: 10.1109/APEC43580.2023.10131551.

979-8-3503-3714-3/23 $31.00 © 2023 IEEE

Comprehensive Investigation on Effects of Anti-Parallel Diodes in GaN-Based Converters

Kazuma Sakamoto, Yosuke Kato, Kenji Natori, Yukihiko Sato
Department of Electrical and Electronic Engineering, Chiba University, Chiba, Japan,
Email: {ksakamoto, y.kato1212, knatori}@chiba-u.jp, ysato@faculty.chiba-u.jp

Abstract—The voltage drop in GaN HEMTs during reverse conduction is larger than that of Si-MOSFETs, causing an increase in dead-time loss and a rise in the bootstrap capacitor voltage when a bootstrap circuit is applied. We propose connecting a low-side device in parallel with a Schottky barrier diode as a solution. This allows the bootstrap capacitor to be charged at the appropriate voltage because the voltage drop is lower. This also reduces dead-time losses but increases the losses due to capacitance across the connected diode terminals. These losses increase with frequency and are not negligible in ultra-high frequency switching operations such as 1MHz. This study examines the effect of using anti-parallel diodes for GaN devices.

Keywords—GaN, Reverse Conduction, Anti-Parallel Diodes, Dead-time, Bootstrap Circuit

I. INTRODUCTION (*HEADING 1*)

Gallium nitride (GaN) is suitable for high-voltage, high-frequency, and high-temperature applications because of its large critical electric field, high electron mobility, and good thermal conductivity compared to silicon (Si) [1], and many studies on the reliability of GaN power devices have been conducted [2]. In recent years, there have been examples of hard-switching operation of power converters at 1MHz [3-6].

Compared to Si-MOSFETs, GaN HEMTs have a larger voltage drop during reverse conduction, causing various problems when applied to power converters. The first is an increase in dead-time loss. The second is overcharging the bootstrap capacitor (BSC) [7]. Since the maximum allowable gate voltage of GaN HEMTs is usually lower than that of Si-MOSFETs, this is a serious problem in gate driving.

One solution to the above two problems is to connect a Schottky barrier diode (SBD) in anti-parallel with the low-side GaN devices, which has been reported to improve the dead-time loss and increase conversion efficiency [8, 9]. However, the overcharging of BSCs and the loss due to charging and discharging of the capacitance between terminals of SBDs have not been investigated, as has been confirmed in SiC [10]. This study aims to ascertain the effect of SBDs connected to GaN HEMTs as anti-parallel diodes.

II. REVERSE CONDUCTION CHARACTERISTICS OF GaN HEMTs

This section describes the reverse conduction characteristics of GaN HEMTs, the problems they cause, and how to solve them. This paper uses the GaN-based two-level synchronous rectifier buck converter shown in Fig.1 to verify the voltage drop phenomenon in GaN HEMTs during reverse conduction. The converter specifications and the devices used are shown in TABLE I and TABLE II, respectively.

A. Reverse conduction characteristics

GaN HEMTs are devices with a horizontal structure, unlike Si-MOSFETs. Therefore, GaN HEMTs do not have body diodes but have equivalent functionality when reverse-biased [11]. Usually, the voltage drop of the body diode of Si-MOSFETs is 0.6-0.8V, but the voltage drop of GaN HEMTs is about 2V in the case of GS61004B manufactured by GaN Systems, which is more than twice as high [12].

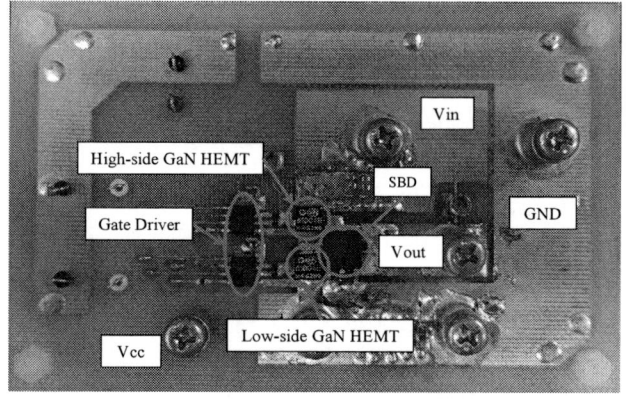

(a)

(b)

Fig. 1. GaN-based two-Level synchronous rectifier buck converter. (a) Front, (b) Back.

TABLE I.	CONVERTER SPECIFICATIONS	
Symbol	**Description**	**Value**
V_{in}	Input voltage	Up to 90V
I_L	Load current	Up to 9A
L	Inductor	150μH
V_{GS}	Gate–source voltage	6V (ON) and 0V (OFF)
f_{SW}	Switching frequency	100k–1MHz
t_d	Dead time	30–200ns
D	Duty cycle	50%

TABLE II.	DEVICES SPECIFICATIONS	
Device	**Manufacturer**	**Part Number**
GaN HEMT	GaN Systems	GS61004B [12]
Si-SBD	Vishay General Semiconductor	V10PW10C [13]
SiC-SBD	Infineon	IDD10SG60C [14]

B. Dead-Time Loss

One of the problems caused by the large voltage drop during reverse conduction is an increase in dead-time loss. When the load current conducts through the low-side device during the dead time, the voltage drop across the device causes the dead-time loss P_{dead}, expressed in the following (1).

$$P_{dead} = I_D * V_{rev} * t_d * f_{SW} \qquad (1)$$

where I_D is drain current, V_{rev} is reverse voltage drop of low-side devices during the dead time, t_d is dead time, and f_{SW} is switching frequency. From (1), P_{dead} is proportional to V_{rev} and f_{SW}, so the effect of dead-time loss is significant when the GaN-based power converters are driven at high frequencies.

C. Overcharge of Bootstrap Capacitor

The bootstrap method is widely employed as a simple strategy to realize isolated gate power supplies for floating devices. In this approach, the bootstrap capacitor is charged during turn-on periods of the low-side devices. In Fig.2(a), the potential of the high-side GaN source terminal is equal to GND, but in Fig.2(b), it drops by V_{rev} due to the reverse conduction operation of the low-side GaN during the dead time. Therefore, the voltage applied to BSC during the dead time can be calculated as (2).

$$V_{BSC} = V_{CC} - V_{Dboot} + V_{rev} \qquad (2)$$

where V_{BSC} is voltage applied to BSC, V_{CC} is supply voltage for gate drive, V_{Dboot} is forward voltage drop of the bootstrap diode, and V_{rev} is voltage drop during reverse conduction of low-side GaN. Since the voltage drop during reverse conduction of GaN is generally about 2V, the BSC is overcharged by 2V higher than the default voltage.

Accordingly, the gate-source voltage of high-side GaN rises to around 8V when V_{CC} is set to 6V, as shown in Fig.3.

D. Proposed Method

Both the increase in dead-time loss and the overcharging of the bootstrap capacitor are caused by the large voltage drop of the GaN HEMT during reverse conduction. Therefore, these problems can be solved by connecting SBDs in anti-parallel with the low-side GaN to reduce the reverse voltage drop. The dead-time loss can be reduced because V_{rev} in (1) becomes smaller. As shown in Fig.2(c), the voltage drop at the high-side GaN source terminal becomes $-V_{SBD}$, and overcharge can be suppressed. The investigation using the proposed method is described in sections 3 and 4.

III. IMPROVEMENT OF BOOTSTRAP CAPACITOR VOLTAGE CHARACTERISTICS

As shown in Fig.2(c), connecting the SBD in parallel with the low-side GaN reduces the source-drain voltage drop, resulting in a low V_{BSC}. This allows the high-side GaN to be driven at an appropriate voltage. In addition, the BSC voltage increases in proportion to the length of the dead time and the frequency. This section examines how the BSC voltage characteristics differ with and without connecting SBD.

A. Dead Time Characteristics

The longer the dead time, the longer the BSC charge time, and the smaller the BSC voltage drop during discharge. This causes the average BSC voltage to increase. Fig.4 shows the relationship between the length of the dead time and the BSC voltage. Without SBD, the voltage increases as the dead time increases, eventually rising to 8.4V. In contrast, with SBD, the voltage remains constant at around 5.8V regardless of the dead time length.

B. Frequency Characteristics

The average BSC voltage increases as the frequency increases because the time to release the electric charge becomes shorter. Fig.5 shows the relationship between the frequency and the BSC voltage. Without SBD, the voltage rises sharply between 100 and 400kHz, reaching 8V at 1MHz. In contrast, with SBD, the BSC voltage varies little, regardless of switching frequency.

C. Load Current Characteristics

The reverse conduction voltage V_{rev} of GaN HEMTs depends on the drain current I_D. Therefore, as I_D increases, V_{rev} increases, and the BSC voltage is expected to be higher from (2).

Fig.6(a) shows the results when the switching frequency is 100kHz. In this case, the voltage with and without SBD was only about 0.4V at maximum. This result indicates that the increase in V_{rev} due to the increase in I_D does not significantly affect the rise in BSC voltage.

Fig.6(b) shows the results when the switching frequency is 1MHz. Without SBD, the voltage at the BSC increased almost

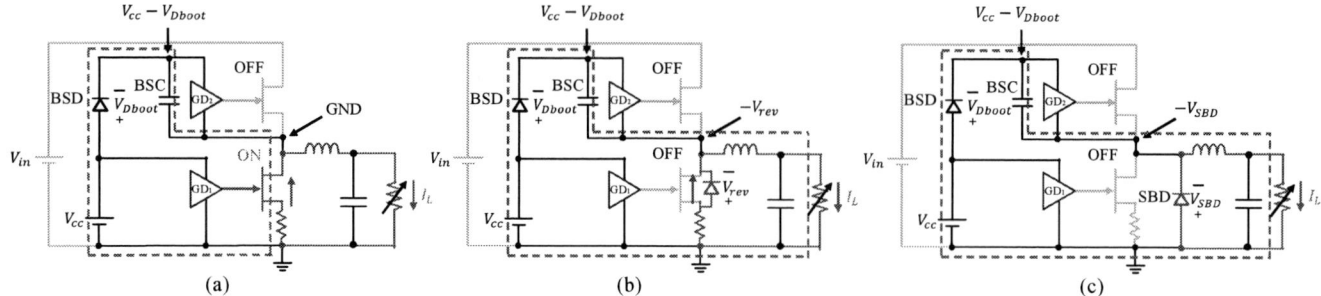

Fig. 2.   Operating mode of the two-level synchronous rectifier buck converter. (The Blue dotted line is the charging path of the BSC.) (a) Low-side GaN turn-on. (b) Dead time without SBD. (c) Dead time with SBD.

linearly, up to 8V. On the other hand, with SBD, the voltage did not increase as much as the case without SBD, reaching a maximum of only 6V. This result shows that when the switching frequency is high, the increase in V_{rev} due to the increase in I_D has a significant effect on the voltage of the BSC.

IV. CONVERTER EFFICIENCY MEASUREMENT

In this section, the losses that vary with the connection of the anti-parallel diode are investigated. It then compares the efficiency with and without the parallel connection of Si-SBD and SiC-SBD to the low-side GaN of the power converter in Fig.1.

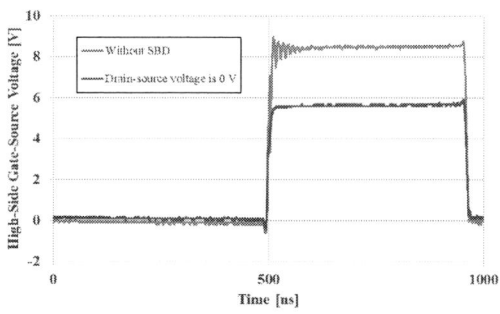

Fig. 3.   Increase in high-side gate-source voltage.

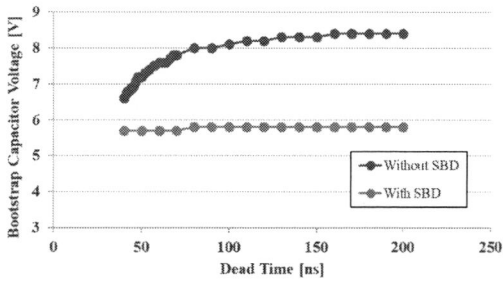

Fig. 4.   BSC voltage vs. dead time with and without Si-SBD. (V_{in} = 50V, I_L = 4A, f_{SW} = 100kHz.)

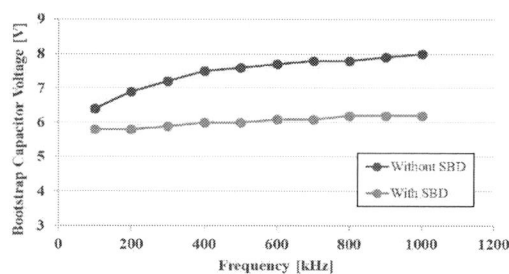

Fig. 5.   BSC voltage vs. frequency with and without Si-SBD. (V_{in} = 50V, I_L = 4A, t_d = 40ns.)

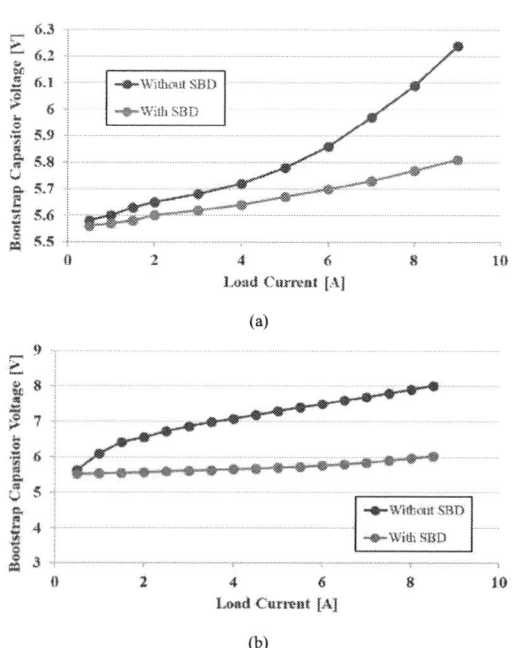

Fig. 6.   BSC voltage vs. load current with and without Si-SBD. (V_{in} = 50V, I_L = 4A, t_d = 40ns.) (a) f_{SW} = 100kHz, (b) f_{SW} = 1MHz.

979-8-3503-3714-3/23 $31.00 © 2023 IEEE   242

A. Losses Related to Anti-Parallel Diodes

Using SBDs as anti-parallel diodes in low-side GaN can reduce the dead-time loss because the voltage drop during reverse conduction is smaller. However, the output capacitance loss increases due to the additional capacitance between the terminals of the SBD. In general, it is known that not only GaN devices but also Si and SiC devices generate losses due to hysteresis during charging and discharging of parasitic capacitances such as output capacitance and capacitance between terminals [15]. It is also reported that the loss energy E_{qoss} due to output capacitance is noticeable in GaN devices instead of the absence of reverse recovery charge Q_{rr} [16]. Since the stored energy is proportional to the square of the input voltage, the higher the voltage, the larger the loss. The above losses are more noticeable in high-frequency operations because they increase proportionally to the switching frequency. The following section compares the efficiency by focusing on these losses.

B. Comparison at Low Voltage

Fig.7(a)-(d) compares the efficiency without SBD, with Si-SBD, and with SiC-SBD for input voltages from 10 to 40V.

First, the efficiency without SBD is the highest for all voltage ranges compared to the other two patterns at light loads of around 0.5A. In particular, Fig.7(b) shows that the efficiency is up to 3% higher than that with Si-SBD. However, as the load current increases, the efficiency begins to decrease, and at input voltages of 10-30V, the efficiency is the lowest among the three patterns. This is because the voltage drop during reverse conduction is larger than in the case with Si-SBD and with SiC-SBD.

Next, examining the case with Si-SBD, as shown in Fig.7(a), at an input voltage of 10V, the efficiency is highest at 1.5A or higher, and at 4.5A, the efficiency is 3% higher than that obtained without SBD. This is thought to be because the forward voltage of the Si-SBD is the lowest, which reduces the dead-time loss. However, there is no significant advantage above 20V; at 40V, the efficiency is lower than the other patterns at all current ranges. The efficiency is also lower than the two different patterns at light loads around 0.5A at all voltage ranges.

Finally, the efficiency with SiC-SBD is lower than that without SBD at light loads, as with Si-SBD. On the other hand, in all cases in Fig.7(a)-(d), the efficiency is higher than that without SBD at current ranges above 1A. The maximum efficiency is obtained at any voltage range among the three patterns.

C. Comparison at High Voltage

Fig.7(e)-(i) compares the efficiency without SBD, with Si-SBD, and with SiC-SBD for input voltages from 50 to 90V.

First, without SBD, the efficiency is above 97% at 2-4A in all cases shown in Fig.7(e)-(i). The maximum efficiency is obtained at current ranges of 3 to 4A when the efficiency is around 97.2 to 97.4%. However, due to the I_D dependence of the V_{rev}, the dead-time loss increases, and the efficiency decreases above 5A.

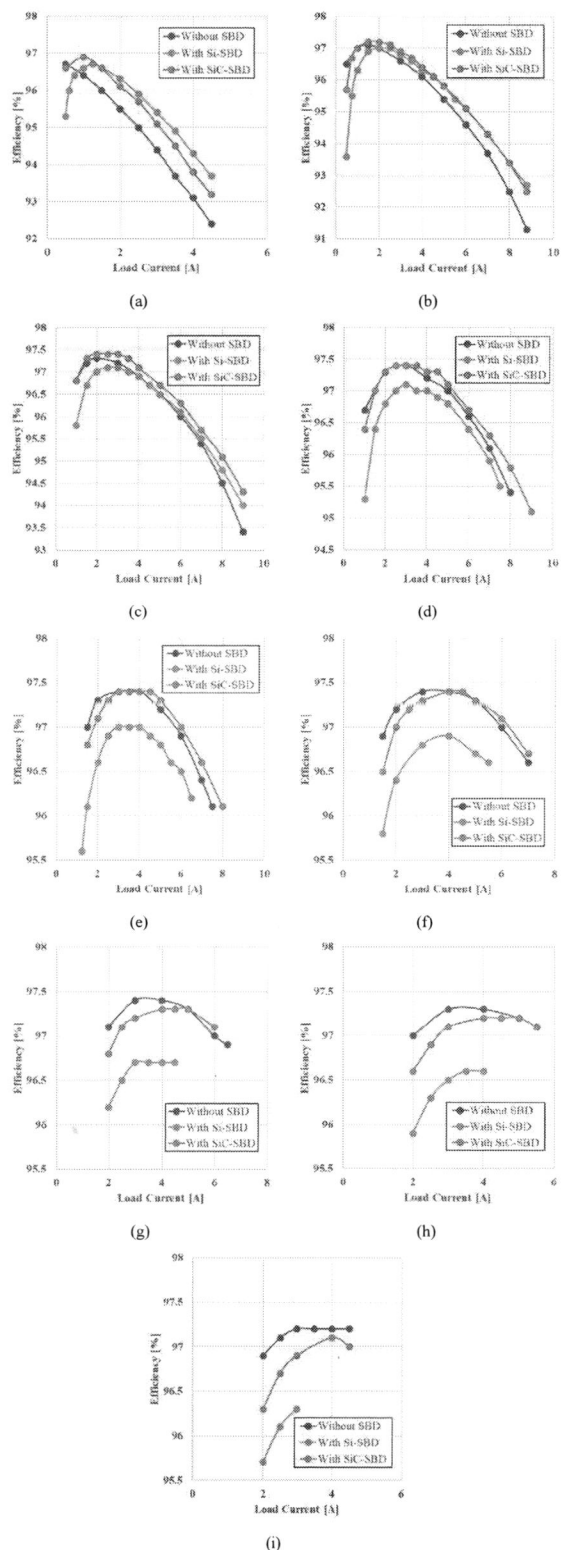

Fig. 7. Load current vs. efficiency characteristics at 10-90V input voltage. ($t_d = 30$ ns, $f_{SW} = 1$ MHz.) (a) 10V, (b) 20V, (c) 30V, (d) 40V, (e) 50V, (f) 60V, (g) 70V, (h) 80V, (i) 90V.

Next, the efficiency with Si-SBD decreases with increasing voltage. At no voltage or current range, the efficiency with Si-SBD exceeds that of the other patterns.

Finally, with SiC-SBD, the maximum efficiency is around 97-97.4% at 4-5A. At 50-70V, the efficiency is higher than without SBD at more than 5A, but at 80V and 90V, the efficiency is lower than without SBD due to the loss caused by the capacitance between the terminal of SiC-SBD.

V. CONCLUSION

This study focused on two points in GaN-based half-bridge buck converters: the increase in the BSC voltage and the dead-time loss due to the large voltage drop during reverse conduction of the GaN HEMT. The BSC voltage can be maintained at an appropriate voltage regardless of dead time, frequency, or load current by connecting an anti-parallel diode to the low-side GaN device of the converter. The efficiency with the SBD was measured at various input voltages and load currents, and it was shown that connecting the SBD extends the range of high-efficiency operation.

ACKNOWLEDGMENT

I would like to thank Mr. Makoto Chiba, our technical staff, for the valuable discussions.

This work was supported by MEXT-Program for Creation of Innovative Core Technology for Power Electronics Grant Number JPJ009777.

REFERENCES

[1] M. Young, The Technical Writer's Handbook. Mill Valley, CA: University Science, 1989. J. Millán, P. Godignon, X. Perpiñà, A. Pérez-Tomás and J. Rebollo, "A Survey of Wide Bandgap Power Semiconductor Devices," in IEEE Transactions on Power Electronics, vol. 29, no. 5, pp. 2155-2163, May 2014, doi: 10.1109/TPEL.2013.2268900.

[2] J. P. Kozak et al., "Stability, Reliability, and Robustness of GaN Power Devices: A Review," in IEEE Transactions on Power Electronics, vol. 38, no. 7, pp. 8442-8471, July 2023, doi: 10.1109/TPEL.2023.3266365.

[3] H. Järvisalo, J. Korhonen, J. Rautio and P. Silventoinen, "Three-phase three-level GaN-ANPC inverter with a 1 MHz switching frequency," 2021 IEEE 8th Workshop on Wide Bandgap Power Devices and Applications (WiPDA), Redondo Beach, CA, USA, 2021, pp. 226-230, doi: 10.1109/WiPDA49284.2021.9645085..

[4] V. Ž. Lazarević and M. Vasić, "High-Frequency GaN-Based ANPC Three-Level Converter as a Low-Noise Arbitrary PWL Voltage Generator," in IEEE Journal of Emerging and Selected Topics in Power Electronics, vol. 10, no. 5, pp. 5997-6008, Oct. 2022, doi: 10.1109/JESTPE.2022.3161421.

[5] R. Hartwig, A. Hensler, T. Ellinger and C. Primas, "Reduced Parasitics Leading to a 99.2 % Efficient Single-Phase Nine-Level Inverter at a Switching Frequency of 800 kHz," 2021 IEEE Applied Power Electronics Conference and Exposition (APEC), Phoenix, AZ, USA, 2021, pp. 809-816, doi: 10.1109/APEC42165.2021.9487281.

[6] Y. Wu, M. Jacob-Mitos, M. L. Moore and S. Heikman, "A 97.8% Efficient GaN HEMT Boost Converter With 300-W Output Power at 1 MHz," in IEEE Electron Device Letters, vol. 29, no. 8, pp. 824-826, Aug. 2008, doi: 10.1109/LED.2008.2000921.

[7] Y. Xi, M. Chen, K. Nielson and R. Bell, "Optimization of the drive circuit for enhancement mode power GaN FETs in DC-DC converters," 2012 Twenty-Seventh Annual IEEE Applied Power Electronics Conference and Exposition (APEC), Orlando, FL, USA, 2012, pp. 2467-2471, doi: 10.1109/APEC.2012.6166168.

[8] W. Chen et al., "Impact of Parasitic Elements on Power Loss in GaN-based Low-voltage and High-current DC-DC Buck Converter," 2018 1st Workshop on Wide Bandgap Power Devices and Applications in Asia (WiPDA Asia), Xi'an, China, 2018, pp. 294-298, doi: 10.1109/WiPDAAsia.2018.8734534.

[9] M. Zdanowski and J. Rąbkowski, "Operation modes of the GaN HEMT in high-frequency half-bridge converter," 2016 Progress in Applied Electrical Engineering (PAEE), Koscielisko-Zakopane, Poland, 2016, pp. 1-6, doi: 10.1109/PAEE.2016.7605115.

[10] K. Yamaguchi, K. Katsura, T. Yamada and Y. Sato, "Criteria for Using Antiparallel SiC SBDs With SiC mosfets for SiC-Based Inverters," in IEEE Transactions on Power Electronics, vol. 35, no. 1, pp. 619-629, Jan. 2020, doi: 10.1109/TPEL.2019.2911988.

[11] E. A. Jones, F. F. Wang and D. Costinett, "Review of Commercial GaN Power Devices and GaN-Based Converter Design Challenges," in IEEE Journal of Emerging and Selected Topics in Power Electronics, vol. 4, no. 3, pp. 707-719, Sept. 2016, doi: 10.1109/JESTPE.2016.2582685.

[12] "GS61004B 100V Enhancement Mode GaN Transistor" GaN Systems Datasheet.

[13] "V10PW10C High Current Density Surface-Mount TMBS® (Trench MOS Barrier Schottky) Rectifier" Vishay General Semiconductor Datasheet.

[14] "IDD10SG60C SiC Schottky Diode" Infineon Technologies Datasheet.

[15] J. Roig and F. Bauwens, "Origin of Anomalous COSS Hysteresis in Resonant Converters With Superjunction FETs," in IEEE Transactions on Electron Devices, vol. 62, no. 9, pp. 3092-3094, Sept. 2015, doi: 10.1109/TED.2015.2455072.

[16] R. Hou, J. Lu and D. Chen, "Parasitic capacitance Eqoss loss mechanism, calculation, and measurement in hard-switching for GaN HEMTs," 2018 IEEE Applied Power Electronics Conference and Exposition (APEC), San Antonio, TX, USA, 2018, pp. 919-924, doi: 10.1109/APEC.2018.8341124.

Finite Control Set Model Predictive Control Based on In-Situ Junction Temperature for Reliability Enhancement of Power Converters

Jiale Zhou
ECE Department
University of North Carolina at Charlotte
Charlotte, United States
jzhou20@charlotte.edu

Ali Parsa Sirat
Design Center
East West Manufacturing
Asheville, United States
aparsasirat@ewmfg.com

Chondon Roy
ECE Department
University of North Carolina at Charlotte
Charlotte, United States
croy6@charlotte.edu

Qiang Mu
ECE Department
University of North Carolina at Charlotte
Charlotte, United States
qmu1@charlotte.edu

Zaheen Mustakin
ECE Department
University of North Carolina at Charlotte
Charlotte, United States
zmustaki@charlotte.edu

Luocheng Wang
ECE Department
University of North Carolina at Charlotte
Charlotte, United States
lwang45@charlotte.edu

Babak Parkhideh
ECE Department
University of North Carolina at Charlotte
Charlotte, United States
bparkhideh@charlotte.edu

Tiefu Zhao
ECE Department
University of North Carolina at Charlotte
Charlotte, United States
tzhao5@charlotte.edu

Abstract—In power converters, power semiconductor devices are often the components most prone to failure. As such, monitoring their health is crucial for enhancing the reliability of power electronics systems. In particular, the junction temperature of a power MOSFET can be used as an indicator of the stress experienced by the device. Therefore, it is possible to use the junction temperature to determine if the converter is operating within an acceptable operational range. The reliability of power semiconductors can be significantly improved when junction temperatures are maintained at lower levels. Based on the Arrhenius equation it is commonly accepted that a device's lifetime doubles for every 10 °C temperature reduction. To achieve this, a conventional finite control set model predictive control (FCS-MPC) method using a secondary problem formulation to reduce power loss and mitigate thermal stress through a thermal model is introduced. Based on this foundation, this paper proposes an FCS-MPC method based on in-situ junction temperature to reduce power loss and thermal stress, thus enhancing reliability of the power converters. First, the junction temperature of silicon carbide (SiC) devices is obtained by measuring their on-state resistance through double pulse tests. Subsequently, the relationship among junction temperature, on-state resistance, and current is described as an R_{dson}-T_j map. Finally, the in-situ junction temperature based FCS-MPC approach for power loss reduction and thermal stress mitigation has been presented. The effectiveness of the proposed solution has been verified through simulation results, demonstrating a 25% reduction in junction temperature compared to the traditional SPWM method.

Keywords— In-situ junction temperature measurement, finite control set model predictive control (FCS-MPC), reliability, on-state resistance, temperature sensitive electrical parameter (TSEP).

I. INTRODUCTION

Power semiconductors have been identified as one of the most vulnerable components in power electronic systems, and their thermal stresses impact the system's reliability. Consequently, reducing power loss and mitigating thermal stress in power semiconductor devices are significant steps toward enhancing the reliability of converters. Thermal stress before the failure of power semiconductor devices is primarily determined by factors such as junction temperature swing, maximum junction temperature, mean junction temperature, etc. [1], [3]. Low power loss and thermal stress indicate a lower junction temperature. As depicted in Fig. 1, the junction temperature originates from the internal die of either a power module or discrete power semiconductor device. Measuring the junction temperature directly is challenging due to the fully encapsulated enclosure. However, precise extraction and monitoring of the junction temperature serve as the foundation for power device loss calculation, lifetime prediction, health monitoring, and reliability assessment. In previous studies [2] and [5], authors utilized case temperature to estimate the junction temperature through a thermal model. This approach is typically employed under normal operating conditions. If a failure occurs or there is a change in operating conditions leading to a sudden alteration of the thermal model, the pre-established thermal model may exhibit significant deviations from the actual junction temperature. In other words, the real-time performance of the method for obtaining the junction temperature from the thermal model is unsatisfactory.

979-8-3503-3714-3/23 $31.00 © 2023 IEEE

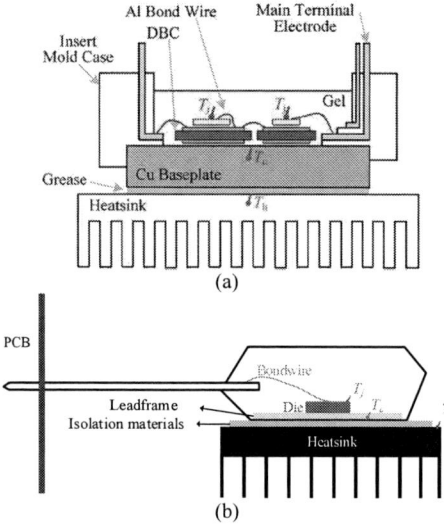

Fig. 1. The typical power semiconductor packages (a) power module [2] (b) TO-247 discrete device [4].

In [6], the concept of a temperature sensitive electrical parameter (TSEP) is introduced to indicate junction temperature. This method relies on using the power semiconductor device itself as a temperature sensing component, mapping its junction temperature information onto the electrical variables such as saturation voltage drop, turn-on delay time, short-circuit current, on-state resistance, current fall-time, etc. [6-8]. These electrical parameters serve as indicators of the thermal condition of power semiconductor devices [9]. Among the various existing methods, on-state resistance R_{dson} measurement appears to be particularly practical for the continuous operation mode of power converters. It can be utilized for either the direct measurement of T_j or a straightforward estimation of T_j, which helps in updating the online thermal parameters for monitoring and reliability purposes. The on-state resistance R_{dson} proves to be highly sensitive to temperature variations, and its value can fluctuate with changes in junction temperature T_j. The functional relationship between the on-state resistance R_{dson} and junction temperature T_j is simple and exhibits good linearity, making it easy to calibrate. Therefore, in this paper, the on-state resistance R_{dson} is used as an indicator of the junction temperature of power semiconductor devices.

By implementing in-situ junction temperature measurements, the operational health condition can be dynamically obtained and characterized. This leads to more accurate system modeling and more efficient control strategies for enhancing reliability. There are many method to reduce the power loss and thermal stress [2], [10], [11]. The traditional finite control set model predictive control (FCS-MPC) method, aimed at reducing power loss and thermal stress, relies on junction temperature calculated from thermal and loss models [2]. In this approach, power loss serves as the secondary objective function for power loss reduction and thermal stress mitigation. However, junction temperature calculation based on the thermal model is usually inaccurate due to practical degradation of power semiconductor devices or sudden changes in the thermal model. This inaccuracy can lead to erroneous

lifetime assessment and unreliable reliability enhancement methods. Therefore, in-situ junction temperature measurement is crucial for improving performance and accuracy in power loss reduction and thermal stress mitigation.

In this paper, an FCS-MPC method based on the in-situ junction temperature is proposed to limit the growth of junction temperature in power devices. The paper is organized as follows: Section II provides a brief introduction to the FCS-MPC method based on thermal modeling for power loss reduction and thermal stress mitigation. Moreover, FCS-MPC method based on in-situ junction temperature measurement is proposed. Furthermore, the physical offline relationship between junction temperature and on-state resistance of a SiC device using double pulse tests is described. Then, simulation results are presented in Section III. Finally, Section IV concludes the paper.

II. THE PRINCIPLE OF THE PROPOSED METHOD

A. The Conventional FCS-MPC Method Based on a Thermal Model

FCS-MPC offers the flexibility to achieve multiple control objectives simultaneously. In literature [2], FCS-MPC is introduced with a secondary problem formulation aimed at reducing power loss and thermal stress. In the control algorithm, the primary control objective is the power flow control, while the secondary control objective is the minimization of power loss, calculated using the power loss model of power semiconductor devices. Power loss encompasses both conduction loss and switching loss. The secondary problem formulation, featuring a similar l_2 norm-2 least square objective function for power loss, is presented as follows [2]:

$$J_s = \sum_{j=1}^{H} ||0 - E_{abc}[k+j]||_2^2 \qquad (1)$$

Where 0 represents the reference power loss, E_{abc} denotes the calculated power loss based on the loss model, and J_s represents the secondary objective value. Therefore, the integrated objective function, combining power flow control and power loss reduction, is presented as follows:

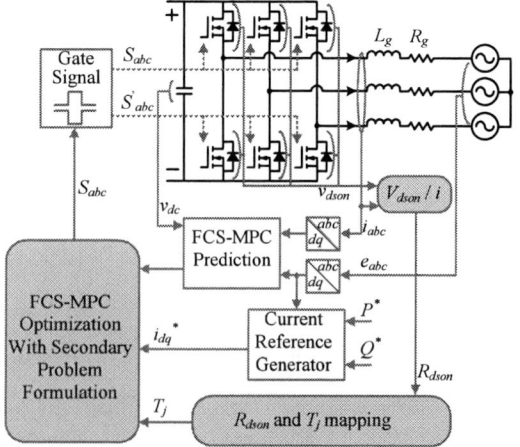

Fig. 2. FCS-MPC based on in-situ junction temperature control diagram.

979-8-3503-3714-3/23 $31.00 © 2023 IEEE

$$\min_{S_{abc}} \lambda_p \cdot J_p + \lambda_s \cdot J_s \qquad (2)$$

Where S_{abc} represents the switching state of power semiconductor devices, J_p stands for the primary objective value for power flow control, and λ_p and λ_s are the weighting factors for the primary and secondary objective functions, respectively. More detailed information about the secondary problem formulation for power loss reduction and thermal stress mitigation can be found in [2]. In this paper, λ_p is set to 1.

In [2], the calculation of power loss used as a secondary objective function is based on a thermal model. As discussed in Section I, the junction temperature obtained from the thermal model is not suitable for situations characterized by abrupt changes in operating conditions or thermal model alterations due to the degradation of power semiconductor devices. Therefore, in-situ junction temperature based FCS-MPC method is proposed.

B. The Proposed In-Situ Junction Temperature Based FCS-MPC Method

Considering the significance of real-time junction temperature extraction, voltage drop sensors are involved for junction temperature measurement. This subsection focuses on the proposed in-situ junction temperature control process. Fig. 2 illustrates the FCS-MPC based on in-situ junction temperature control diagram. The drain-to-source on voltage V_{dson} can be measured by V-ON voltage sensors [12], which are essential for the precise measurement of R_{dson}. When the power semiconductor device turns on, the voltage across its terminals

can be obtained. Subsequently, the on-state resistance can be determined using V_{dson} and the three-phase inductor current I. The R_{dson} value can be derived by dividing the drain-to-source on-state voltage by the flowing current. Both line LEM and switch Rogowski coil current sensors can be used for current measurement [13-17]. The LEM provides a more stable output, except when it experiences magnetic saturation or extremely high temperatures. The Rogowski coil can deliver accurate switch current measurements without being affected by saturation or temperature changes. However, in the context of R_{dson} calculation, having information about switching ringing is unnecessary. In other words, embedded switch current sensors can either obtain accurate current measurements for calculating R_{dson} or complicate the measurement system with complex electronic circuitry. To reduce complexity, LEM current sensors can be used to measure the inductor current, which remains continuous during the R_{dson} calculation.

Finally, the junction temperature obtained from R_{dson}-T_j map at different current levels will be incorporated into the secondary objective function for power loss calculations. R_{dson}-T_j map can be organized into a look-up table in the controller. This method of obtaining T_j ensures real-time T_j calculations for power loss estimation and can be applied to discrete power semiconductor devices in converter design.

C. R_{dson}-T_j Mapping

Estimating the junction temperature of the SiC MOSFET based on its on-state resistance value can be accomplished through either the voltage and current curves of the targeted device at various junction temperature levels or by measuring the actual physical on-state resistance at different junction temperatures using experimental procedures. However, the datasheet provides a sparse dataset on the relationship between R_{dson} and T_j, with significant temperature gaps, leading to

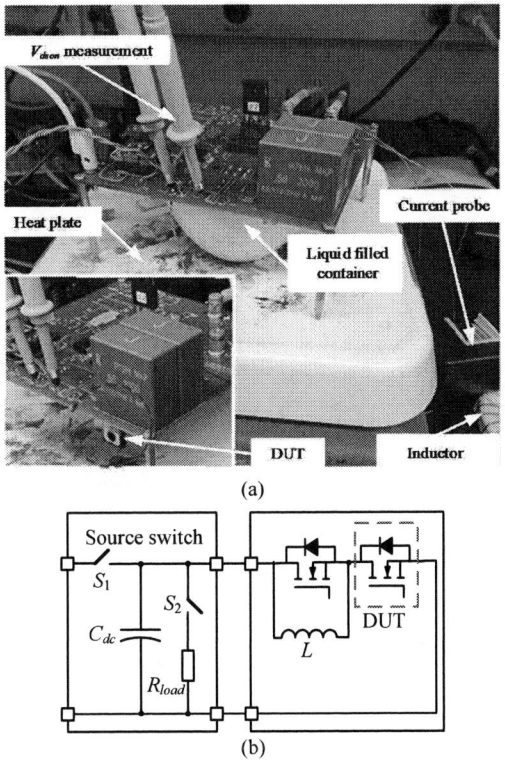

Fig. 3. Double pulse test platform (a) hardware test platform [19] (b) test schematic.

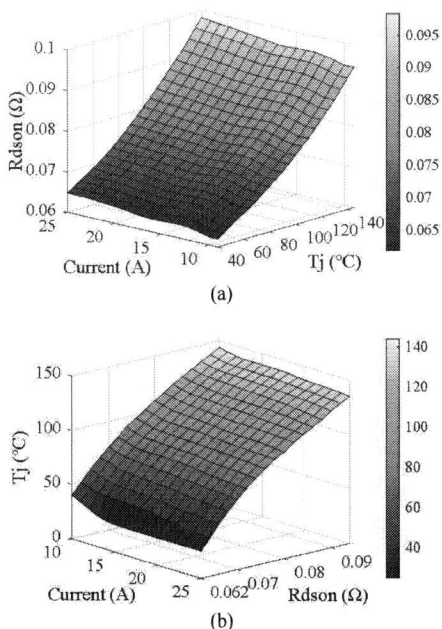

Fig. 4. The relationship among R_{dson}, T_j, and current (a) T_j-R_{dson} map (b) R_{dson}-T_j map.

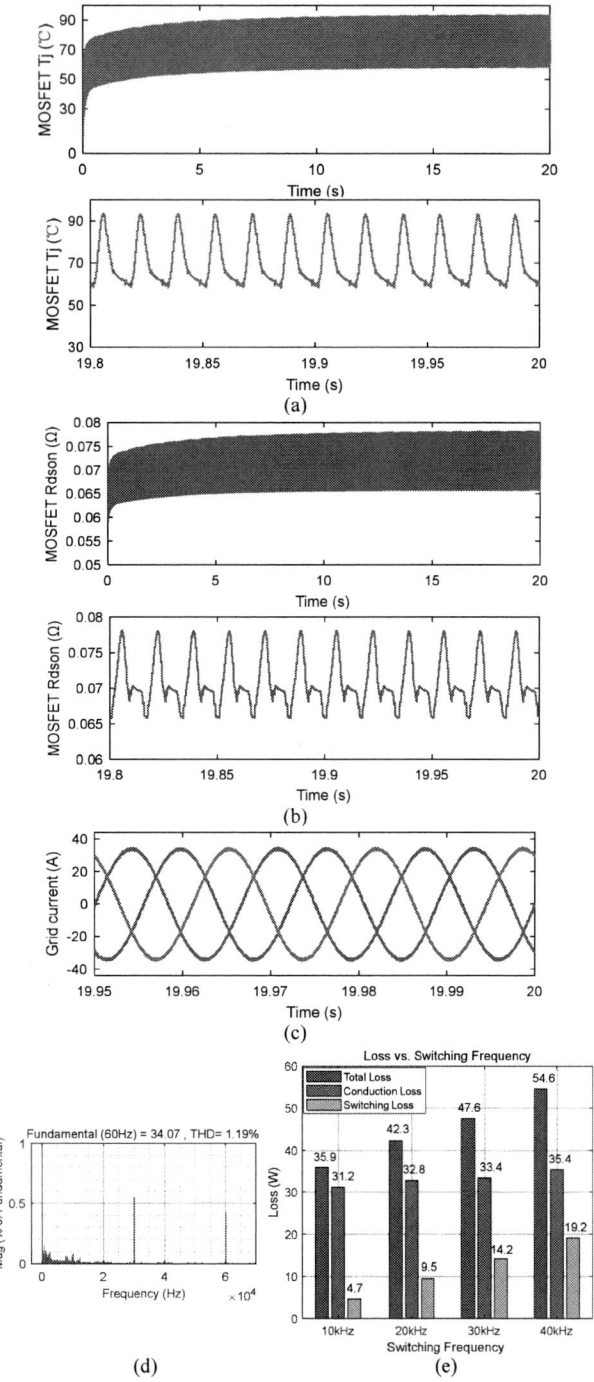

Fig. 5. PR controller based SPWM scheme (a) MOSFET junction temperature (b) MOSFET on-state resistance (c) Grid current (d) FFT analysis of grid current (e) Loss vs. switching frequency for each switch.

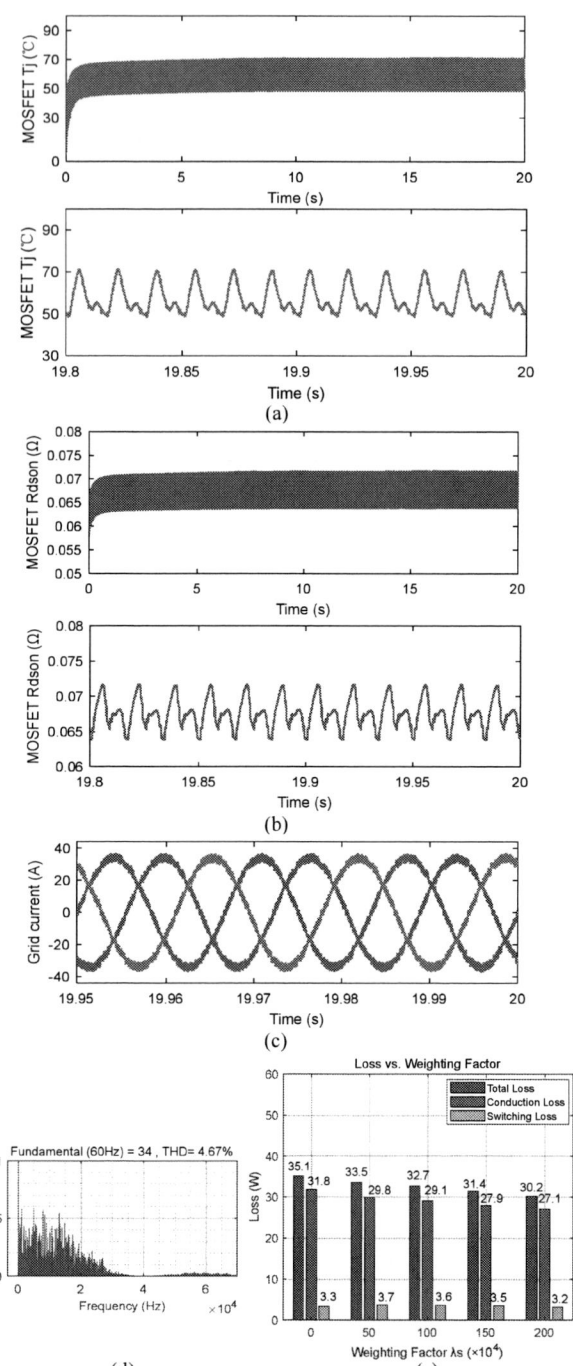

Fig. 6. In-situ junction temperature based FCS-MPC scheme (a) MOSFET junction temperature (b) MOSFET on-state resistance (c) Grid current (d) FFT analysis of grid current (e) Loss vs. weighting factor for each switch.

inaccurate junction temperature estimations derived from R_{dson}. Therefore, the relationship among R_{dson}, T_j, and current has been conducted through experiments in this paper. The relationship between the junction temperature and case temperature can be described as follows [21].

$$T_j = P_D Z_{jc}(t_p) + T_c \qquad (3)$$

Where T_j represents the junction temperature, P_D denotes power consumption, and T_c stands for the case temperature. $Z_{jc}(t_p)$ represents the dynamic thermal impedance from the case to the junction, which varies with the operation pulse width t_p.

Therefore, a simple R_{dson} reading procedure has been developed based on a short single-pulse current applied to the

TABLE I.	SIMULATION PARAMETERS	
Symbol	Parameters	Values
P	Rated power	20 kW
V_{dc}	DC-link voltage	1000 V
e_{abc}	Grid voltage	277/480 V
f	Grid frequency	60 Hz
i_{abc}	Grid current	24 A
L_g	Line inductance	3 mH
R_g	Line resistance	10 mΩ
MOSFET	Power semiconductor device	C2M0080120D
T_a	Air temperature	25 °C
$R_{\theta AH}$	Thermal resistance from air to heatsink	0.5 °C/W
$R_{\theta HC}$	Thermal resistance from heatsink to case	0.5 °C/W

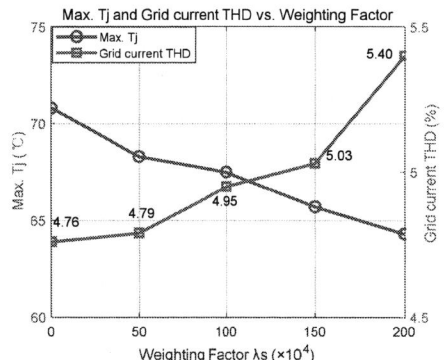

Fig. 7. The maximum Tj of MOSFET and grid current THD vs. weighting factor.

targeted device. According to (3), the junction temperature does not have sufficient time to significantly differ from the case temperature if the single-pulse current is less than 100 μs. Thus, by heating the case temperature of the power semiconductor device to a certain temperature using a heat plate, T_j can be assumed to be equal to the case temperature of the power semiconductor device.

Fig. 3 illustrates the double-pulse test platform for R_{dson} measurement at various junction temperature levels [18]. A heat plate is used to maintain the case temperature at a constant value. The targeted SiC power semiconductor device is the C2M0080120D from Wolfspeed [20].

The relationship among R_{dson}, T_j, and current is shown in Fig. 4. The test pulse current ranges from 9 A to 25 A, while the junction temperature varies from 40 °C to 145 °C. Fig. 4(a) presents the T_j-R_{dson} map, which shows the R_{dson} values obtained at different T_j and pulse current levels. It can be seen that R_{dson} increases with higher device junction temperatures and also rises with higher currents. Based on the data from Fig. 4(a), the R_{dson}-T_j map can be constructed, as shown in Fig. 4(b). Consequently, in-situ junction temperature can be estimated using the R_{dson}-T_j map.

III. SIMULATION RESULTS

To validate the effectiveness of the proposed in-situ junction temperature based FCS-MPC solution, simulation is conducted using MATLAB/Simulink and PLECS environments. The simulation parameters are detailed in Table I. The R_{dson}-T_j map obtained from Section II can be used to generate in-situ junction temperature in the simulation.

Performance comparisons between the proportional resonant controller based SPWM scheme and the proposed in-situ junction temperature based FCS-MPC method are presented in Fig. 5 and Fig. 6. The maximum junction temperature of MOSFET in the conventional SPWM scheme reaches 94 °C, whereas the proposed in-situ junction temperature based FCS-MPC scheme achieves a lower maximum junction temperature of 70 °C. Furthermore, the junction temperature swing in the proposed solution during steady state operation is 20 °C, nearly

29 °C smaller than that of the conventional SPWM method, as shown in Fig. 5(a) and Fig. 6(a). Fig. 5(b) and Fig. 6(b) illustrate the on-state resistance of the MOSFET for both the conventional SPWM method and the proposed scheme. It is evident that the proposed scheme exhibits lower R_{dson} than the traditional SPWM method during operation, leading to reduced junction temperature, as supported by the R_{dson}-T_j map from Fig. 4(b).

Fig. 6(c)(d) shows the grid current and its FFT analysis when the λ_s is set to $5*10^4$. Although the junction temperature is lower in the proposed solution, the THD of the grid current is higher compared to the conventional SPWM method. However, it still complies with IEEE Standard 519-2014, which stipulates that the THD of the grid current is supposed to be less than 5%. Fig. 5(e) shows the power loss versus switching frequency for each power semiconductor device in the conventional SPWM method. It can be seen that power loss increases with higher switching frequency. Fig. 6(e) shows the power loss versus weighting factor for each switch using the in-situ junction temperature based FCS-MPC scheme. It can be seen that power loss decreases as the weighting factor increases. In summary, the proposed solution effectively reduces power loss and junction temperature, thus enhancing the reliability of the power converters.

The maximum T_j of the MOSFET and grid current THD versus the weighting factor is shown in Fig. 7. As the weighting factor increases, the maximum junction temperature decreases, while the THD of the grid current increases.

IV. CONCLUSION

In this paper, a conventional thermal model based finite control set model predictive control (FCS-MPC) scheme is introduced. Based on this, an in-situ junction temperature based FCS-MPC scheme is proposed. Double pulse tests for on-state resistance measurement are conducted, and the relationship among the on-state resistance of power device, junction temperature, and current is depicted as an R_{dson}-T_j map. Subsequently, this R_{dson}-T_j map is used in the FCS-MPC solution to generate in-situ junction temperatures. Finally, the simulation results demonstrate significant benefits. The maximum junction temperature, when using the in-situ junction temperature based FCS-MPC, is reduced from 94 °C to 70 °C in comparison to the conventional SPWM method. Additionally, the THD of the grid current remains below 5%, aligning with the IEEE Standard

979-8-3503-3714-3/23 $31.00 © 2023 IEEE

519-2014. Moreover, the power loss can be reduced by using the proposed solution.

ACKNOWLEDGMENT

This project was funded through the North Carolina Renewable Ocean Energy Program. The authors would also like to thank the Energy Production and Infrastructure Center (EPIC) and Electrical and Computer Engineering Department at the University of North Carolina at Charlotte.

REFERENCES

[1] U. -M. Choi, F. Blaabjerg and K. -B. Lee, "Study and Handling Methods of Power IGBT Module Failures in Power Electronic Converter Systems," in *IEEE Transactions on Power Electronics*, vol. 30, no. 5, pp. 2517-2533, May 2015.

[2] L. Wang, J. He, T. Han, and T. Zhao, "Finite Control Set Model Predictive Control With Secondary Problem Formulation for Power Loss and Thermal Stress Reductions," in *IEEE Transactions on Industry Applications*, vol. 56, no. 4, pp. 4028-4039, July-Aug. 2020.

[3] R. Bayerer, T. Herrmann, T. Licht, J. Lutz, and M. Feller, "Model for Power Cycling lifetime of IGBT Modules - various factors influencing lifetime," *5th International Conference on Integrated Power Electronics Systems*, Nuremberg, Germany, 2008, pp. 1-6.

[4] H. Liu, T. Zhao, X. Xu, and J. Zhou, "Conduction Time Variation-Based Active Thermal Control Method for Si and SiC Hybrid Switch," 2022 *IEEE Energy Conversion Congress and Exposition* (ECCE), Detroit, MI, USA, 2022, pp. 1-5.

[5] L. Wang, T. Zhao, and J. He, "Centralized Thermal Stress Oriented Dispatch Strategy for Paralleled Grid-Connected Inverters Considering Mission Profiles," in *IEEE Open Journal of Power Electronics*, vol. 2, pp. 368-382, 2021.

[6] Y. Avenas, L. Dupont, and Z. Khatir, "Temperature Measurement of Power Semiconductor Devices by Thermo-Sensitive Electrical Parameters—A Review," in *IEEE Transactions on Power Electronics*, vol. 27, no. 6, pp. 3081-3092, June 2012.

[7] F. Yang, S. Pu, C. Xu, and B. Akin, "Turn-on Delay Based Real-Time Junction Temperature Measurement for SiC MOSFETs With Aging Compensation," in *IEEE Transactions on Power Electronics*, vol. 36, no. 2, pp. 1280-1294, Feb. 2021.

[8] W. Wei, W. Zhu, T. Liu, and G. Xu, "Junction Temperature Online Extraction Method for Power MOSFET by Current Fall Time," in *IEEE Transactions on Electron Devices*, vol. 69, no. 7, pp. 3811-3819, July 2022.

[9] H. Liu, J. Zhou, T. Zhao, and X. Xu, "Si IGBT and SiC MOSFET Hybrid Switch-Based Solid State Circuit Breaker for DC Applications," 2022 *IEEE Energy Conversion Congress and Exposition* (ECCE), Detroit, MI, USA, 2022, pp. 1-6.

[10] J. Zhou, H. Liu, T. Zhao, and X. Guo, "A Novel One-Dimension Space Vector Strategy for Multilevel Cascaded Inverters," 2023 *IEEE Applied Power Electronics Conference and Exposition* (APEC), Orlando, FL, USA, 2023, pp. 251-256.

[11] M. Andresen, K. Ma, G. Buticchi, J. Falck, F. Blaabjerg, and M. Liserre, "Junction Temperature Control for More Reliable Power Electronics," in *IEEE Transactions on Power Electronics*, vol. 33, no. 1, pp. 765-776, Jan. 2018.

[12] C. Roy, N. Kim, J. Gafford and B. Parkhideh, "On-State Voltage Measurement of High-Side Power Transistors in Three-Phase Four-Leg Inverter for In-Situ Prognostics," 2021 *IEEE Energy Conversion Congress and Exposition* (ECCE), Vancouver, BC, Canada, 2021, pp. 2770-2776, doi: 10.1109/ECCE47101.2021.9595082.

[13] A. Parsa Sirat, C. Roy, D. Evans, J. Gafford and B. Parkhideh, "In-Situ Ultrafast Sensing Techniques for Prognostics and Protection of SiC Devices," 2022 *IEEE 9th Workshop on Wide Bandgap Power Devices & Applications* (WiPDA), Redondo Beach, CA, USA, 2022, pp. 142-147.

[14] A. Parsa Sirat and B. Parkhideh, "Current Sensor Integration Issues with Wide-Bandgap Power Converters," *Sensors* (2023), 23(14), 6481.

[15] H. Niakan, A. P. Sirat, and B. Parkhideh, "A Novel Reset-Less Rogowski Switch-Current Sensor," in *IEEE Transactions on Power Electronics*, vol. 38, no. 4, pp. 4203-4206, April 2023.

[16] A. Parsa Sirat, H. Niakan, C. Roy, and B. Parkhideh, "Rogowski-Pair Sensor for High-Speed Switch Current Measurements without Reset Requirement," 2022 *IEEE Energy Conversion Congress and Exposition* (ECCE), Detroit, MI, USA, 2022, pp. 1-8.

[17] A. Parsa Sirat, H. Niakan, D. Evans, J. Gafford, and B. Parkhideh, "Ultra-Wideband Unidirectional Reset-Less Rogowski Coil Switch Current Sensor Topology for High-Frequency DC-DC Power Converters," 2023 *IEEE Applied Power Electronics Conference and Exposition* (APEC), Orlando, FL, USA, 2023, pp. 1662-1669.

[18] J. Zhou, H. Liu, T. Zhao, X. Xu, and Y. Wang, "SiC Bidirectional Solid-State Circuit Breaker with Soft-Start Function for Motor Control Center," 2023 *IEEE Applied Power Electronics Conference and Exposition* (APEC), Orlando, FL, USA, 2023, pp. 2307-2312.

[19] C. Roy, "In-Situ Instrumentation for Degradation Monitoring of Power Semiconductor Devices in Power Electronics Converters," Ph.D. dissertation, University of North Carolina at Charlotte, Charlotte, NC, USA, 2023.

[20] Wolfspeed, (2023). [Online] Available: https://assets.wolfspeed.com/uploads/2020/12/C2M0080120D.pdf

[21] Z. Dong, R. Ren, and F. Wang, "Evaluate I2t Capability of SiC MOSFETs in Solid State Circuit Breaker Applications," 2020 *IEEE Energy Conversion Congress and Exposition* (ECCE), Detroit, MI, USA, 2020.

979-8-3503-3714-3/23 $31.00 © 2023 IEEE

P-Type Doping Control of Magnetron Sputtered NiO for High Voltage UWBG Device Structures

Matthew A. Porter
Center for Power Electronics
Systems
Virginia Tech
Blacksburg, VA, USA
maporter@vt.edu

Yunwei Ma
Center for Power Electronics
Systems
Virginia Tech
Blacksburg, VA, USA
yunwei@vt.edu

Yuan Qin
Center for Power Electronics
Systems
Virginia Tech
Blacksburg, VA, USA
yuanqin@vt.edu

Yuhao Zhang
Center for Power Electronics
Systems
Virginia Tech
Blacksburg, VA
yhzhang@vt.edu

Abstract—A major challenge in the design and fabrication of power devices from ultra-wide bandgap (UWBG) materials is the lack of a native shallow acceptor dopant for most materials in that class. P-type regions in UWBG devices may alternatively be formed by the deposition of p-type wide bandgap materials such as nickel oxide (NiO) to form heterojunctions. This work examines the effectiveness of the modulation of the acceptor concentration (N_A) of magnetron sputtered NiO via the control of O_2 partial pressure during sputtering. NiO/n$^+$-Ga$_2$O$_3$ PN junctions are fabricated via sputtering using O_2/Ar percentage ratios ranging from 0% to 12.5%. The acceptor doping and acceptor level characteristics are studied via the dependence of PN junction capacitance upon voltage, temperature and frequency. It is found that the N_A can be controlled between 9×10^{17} cm^{-3} at 0% O_2/Ar to 2×10^{18} cm^{-3} at 12.5% O_2/Ar partial pressure. Studies of the capacitance of the diodes shows that the associated first ionization level of the Ni^{3+} vacancy thought to be responsible for the p-type doping of the NiO is 0.35 eV, with suggestion of a second ionization level at 0.54 eV seen in the 0% O_2/Ar sample. To demonstrate the feasibility of utilizing magnetron sputtered NiO in power devices, a lateral RESURF terminated Ga$_2$O$_3$ Schottky diode is fabricated. Charge balance to maximize breakdown voltage is demonstrated in the Schottky diode as a function of NiO RESURF thickness, showing the viability of NiO in high voltage power devices for field spreading regions such as RESURF layers and JTE structures.

Keywords—power electronics, power semiconductor devices, nickel oxide, gallium oxide, wide-bandgap, ultra-wide bandgap

I. INTRODUCTION

Ultra-wide bandgap semiconductors (UWBGs) such as β-Ga$_2$O$_3$ and AlN are promising candidates for next generation power device designs, due to an increase in critical electric field and implied reduction in potential device losses [1]. However, hole transport and p-type doping in UWBG materials remains a major obstacle to their successful use in power device designs. For example, no native acceptor dopant is currently known for β-Ga$_2$O$_3$, and the self-trapping of holes limits hole transport in that material [2], [3]. Beyond the lack of native dopants, advanced power device designs require a method of selective-area doping to create post-growth p-type regions in desired areas of the device [1]. Successful selective area p-type doping in many wide bandgap (WBG) and UWBG materials such as GaN via traditional methods such as implantation has been shown to

be challenging [4], [5], requiring high N_2 overpressure or pulsed thermal annealing for acceptor activation [6].

An alternative method of achieving effective selective-area p-type doping in UWBG power devices is through the deposition of WBG or UWBG materials which can be easily doped p-type. The transparent conductive oxide NiO is a naturally p-type material with a bandgap of 3.7 eV [3]. The native acceptor in NiO is believed to be due to Ni^{3+} charge states formed due to Ni vacancies in the material [7]. NiO deposited using conformal deposition techniques such as reactive magnetron sputtering and pulsed laser deposition has been demonstrated to maintain p-type conduction and doping after deposition [8]. In addition, high O_2 partial pressures during RF sputtering of NiO has been shown to modulate the effective hole density of the resulting NiO film [9], [10].

Recent work in incorporating NiO into power devices has been successful in using reactive magnetron sputtered NiO in β-Ga$_2$O$_3$ power PN diodes as both a junction termination extension (JTE) edge termination and hole injection layer, resulting in the achievement of breakdown voltages (V_{br}) of 2.5 kV with specific on-resistance ($R_{on,sp}$) of 5.9 mΩ.cm^2 [11]. Further work in [12] demonstrated avalanche capable NiO-Ga$_2$O$_3$ PN diodes with a V_{br} of 1.5 kV via an edge termination design using a combination of a high-k dielectric field plate and magnetron sputtered NiO JTE. NiO has also been utilized to form a charge-balance Ga$_2$O$_3$ with a V_{br} up to 10 kV and operational temperature of 200 °C [13]. In addition to forming planar PN junctions, NiO has also been used to form PN junction on 3D FinFET and trigate structures in GaN power devices [14]–[17].

This paper explores the variation of acceptor concentration (N_A) in reactive magnetron-sputtered NiO films via O_2 partial pressure control during the sputtering process. NiO films are deposited under varying O_2 partial pressure on a n$^+$-Ga$_2$O$_3$ substrate to form p-NiO/n$^+$-Ga$_2$O$_3$ diodes. The behavior of the capacitance of the fabricated diodes are studied as a function of voltage, frequency and temperature (C-V-T/f) to understand the dependence of the acceptor doping on O_2 partial pressure during sputtering, as well as the nature of the acceptor state giving rise to the doping. It is found that effective acceptor doping concentrations can be controlled between 9×10^{17} cm^{-3} to 2×10^{18} cm^{-3} utilizing the Ni^{3+} level believed to be responsible for the acceptor doping is a deep acceptor with ionization energy of 0.35 eV, with a second ionization level near 0.54 eV. Utilizing

---

This work is in part supported by National Science Foundation under Grants ECCS-2230412 and ECCS-2036915, Office of Naval Research monitored by Lynn Petersen (N00014-21-1-2183), and the Center for Power Electronics Systems Industry Consortium at Virginia Tech.

the studied NiO sputtering conditions, a p-NiO reduced-surface-field (RESURF) terminated lateral Ga₂O₃ Schottky diode similar to the one in [13], [18], [19] is fabricated. Charge balance as a function of p-NiO RESURF thickness is successfully achieved, demonstrating the potential for NiO as a p-type substitute for UWBG device applications.

II. DEVICE FABRICATION PROCESSES

A. P-NiO/n⁺-Ga₂O₃ Diode Fabrication

The p-NiO/n⁺-Ga₂O₃ diode design used for the characterization of NiO was fabricated via reactive magnetron sputtering of NiO on an (001) n⁺-Ga₂O₃ wafer with a doping level of 8.3×10^{18} cm⁻³ grown by Novel Crystal Technologies. Fig. 1(b) illustrates the fabrication process for the device. Ti/Au backside cathode contacts are formed via e-beam deposition on the n⁺-Ga₂O₃ wafer. p-NiO is then deposited via reactive magnetron sputtering. The sputtering process is described in more detail in Section III. Table 1 shows the four NiO sputtering conditions used to fabricate the diodes studied in this work. Recipe D (Ar/O₂ partial pressure ratio of 2:1) is used to deposit a p⁺-NiO layer for ohmic contact after deposition of a p-NiO film using recipes A, B or C. Ni/Au anode contacts are deposited on the p⁺-NiO via e-beam deposition.

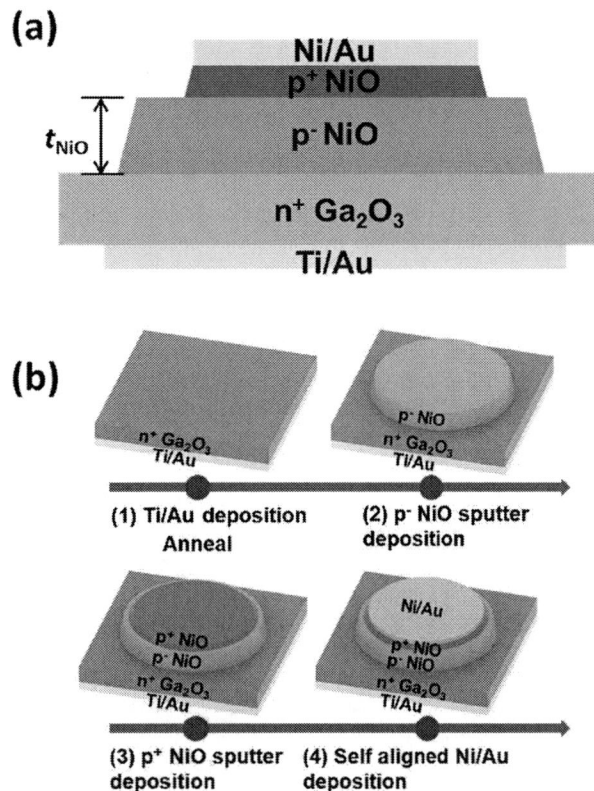

Fig. 1. P-NiO/n+Ga2O3 PN diode design and fabrication process. (a) illustrates the design of the diode, and (b) shows the fabrication process flow.

Table 1. NiO reactive magnetron sputtering conditions studied in this work.

Recipes	Ar flow (sccm)	O₂ flow (sccm)	Rate (nm/Hour)	Resistivity (Ω·m)	H_D (nm)
Recipe A	60	0	49	84	147/340/540
Recipe B	58	3	21	6.9	63/147/244
Recipe C	54	6.8	24	0.53	48/95/156
Recipe D	40	20	20	1.8×10^{-3}	N/A

B. P-NiO RESURF Terminated Lateral Ga₂O₃ Schottky Fabrication

The lateral Ga₂O₃ schottky diode used to study charge balance was fabricated on a Novel Crystal Technologies MBE-grown Ga₂O₃ n-/UID/n+ wafer. The doping level and thickness of the top n- channel layer were measured to be 2×10^{17} cm⁻³ over 70 nm via electrochemical C-V profiling. The UID layer was similarly measured to be doped at a concentration of 10^{16} cm⁻³ over 220 nm. Ni/Au Schottky contacts and Ti/Au ohmic contacts were formed via e-beam deposition at an anode-to-cathode separation of 17 μm. RESURF termination of the diode was accomplished prior to Schottky contact deposition via reactive magnetron sputtering of NiO of varying thickness using recipes A and B from Table 1. The length of the NiO RESURF region was 10 μm for all diodes tested.

III. NiO ACCEPTOR CHARACTERIZATION

The PN diodes described in Section II were used to characterize the dependence of NiO acceptor concentration on sputtering conditions. Three lots of diodes were fabricated using NiO sputtered via recipes A, B and C. Thickness of the p-NiO film was 48 nm for recipe C, 244 nm for recipe B and 540 nm for recipe A (Table 1). A Kurt. J. Lesker Inc. PVD-75 RF magnetron sputtering chamber was used to deposit the NiO. 99.9% purity NiO sputtering targets supplied by Kurt J. Lesker Inc. were utilized as the source for deposition. Ar and O₂ plasma was used with a total chamber pressure of 3 mTorr and RF power of 100 W.

Characterization of the diodes was performed using a temperature-controlled probe station and a Keithley 4200 parameter analyzer. Initial characterization of the resistivity of the deposited NiO films was performed via TLM test structures fabricated alongside the diodes under test. Table 1 shows the variation of NiO sheet resistance with varying Ar/O₂ partial pressure ratio. Sheet resistance decreases from 84 Ω·m to 0.53 Ω·m as Ar/O₂ partial pressure ratio increases from 0% (recipe A) to 12.5% (recipe C). Accounting for film thickness, this decrease in sheet resistance suggests an increasing hole concentration in the NiO film with increasing O₂ pressure during deposition.

Figs. 2 (a-c) show the initial C-V characterization as a function of sputtering recipe and device temperature. C-V measurements were performed at a test frequency of 5 kHz. The $1/C^2$ characteristics of the measured C-V characteristics show good linearity over all temperatures, with an intercept of

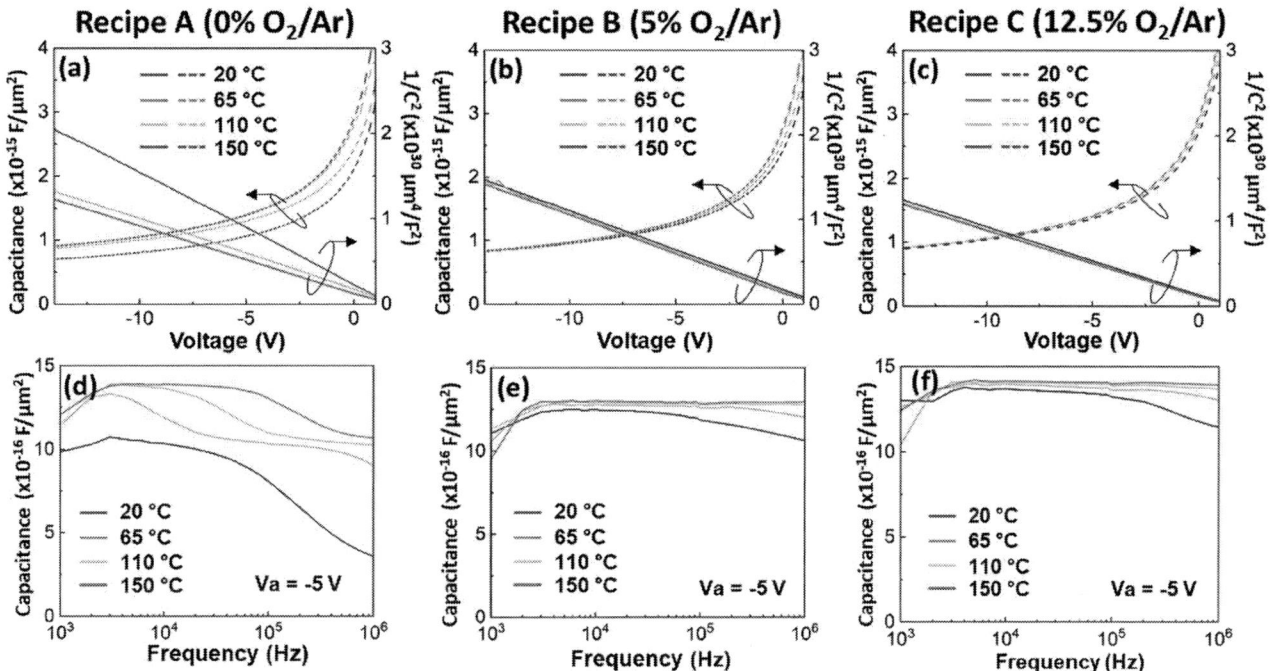

Fig. 2. C-V and C-f characterization of p-NiO/n⁺-Ga₂O₃ diodes versus temperature and NiO sputtering conditions. (a-c) show the C-V characterization as a function of temperature, and (d-f) show the temperature dependence of capacitance dispersion at -5 V reverse bias.

1.7 V corresponding to the built-in voltage (V_{bi}) of the p-NiO/n⁺-Ga₂O₃ junction. However, a large temperature dependence of the capacitance characteristics is observed for the 0% Ar/O2 partial pressure sputtering condition (recipe A). The N_A of the p-NiO layer is extracted using the measured $1/C^2$-V characteristics, correcting for the n⁺-Ga₂O₃ doping of 8.3×10^{18} cm⁻³ and assuming the relative permittivity of Ga₂O₃ and NiO to be 11.9 and 12.4, respectively [20][21]. The temperature dependence of the capacitance of the diode with NiO sputtered with a 0% Ar/O2 ratio (recipe A) resulted in a temperature dependent estimate of N_A. At 25 °C, the calculated doping of the 0% Ar/O2 NiO was 9.6×10^{17} cm⁻³ at 20 °C, and 1.58×10^{18} at 150 °C. The magnitude of N_A at 5 kHz of NiO sputtered with recipes B and C (5% and 12.5% O₂/Ar ratios) show little temperature dependence, and are calculated to be 1.5×10^{18} cm⁻³ and 1.9×10^{18} cm⁻³, respectively.

Figs. 2(d-f) show the measured C-f dispersion of the diodes as a function of sputtering recipe and device temperature at a reverse bias of -5 V. Frequency dispersion, with capacitance measured to decrease after an observed corner frequency, is found in diodes fabricated with NiO sputtered via all three recipes over the measured 5 kHz to 1 MHz range, with the most significant dispersion found in diodes sputtered with 0% O₂/Ar ratio (recipe A). For all recipes, the cutoff frequencies increase in magnitude as a function of temperature. For recipes B and C (5% and 12.5% O₂/Ar), the corner frequency moves beyond the measurable frequency range for temperatures greater than 65 °C. Two corner frequencies are visible in the C-f characteristics of the diodes with NiO sputtered with 0% Ar/O₂; both are visible

in the measured range at 65 °C and increase in magnitude with temperature.

The origin of the frequency dispersion in the measured C-f characteristics can be explained by a high activation energy of the acceptor level giving rise to the p-type doping in Ni-deficient NiO. To extract the acceptor activation energy, two methods were explored. TCAD simulation of the fabricated device structure was used to fit the measured C-f characteristics. Simulations were performed with an acceptor trap level giving rise to the p-type doping in the NiO layer. For the case of NiO deposited with 0% O₂/Ar (recipe A), two independent acceptor trap levels were assumed to achieve a good fit. Data fitting using the simulation output was used to extract the acceptor energy levels and doping concentrations necessary to fit the measured results.

Figs. 3 (a-c) show the results of the fitting of the simulation output to the measured C-f characteristics. The table in Fig. 3(d) shows the extracted acceptor characteristics from the fitting. For recipes B and C (5% and 12.5% O2/Ar), an acceptor ionization energy of 0.325 was found, and good agreement is found between the Na values obtained by the fit and the value extracted from C-V characterization at 5 kHz. For recipe A, the presence of two traps resulted in an extracted first acceptor ionization energy of 0.315 eV and second acceptor ionization energy of 0.44 eV. The acceptor ionization energy extracted for recipes B and C and the first ionization energy extracted for recipe A are close to value obtained via similar measurements on magnetron sputtered and PLD deposited NiO in [8], and are close to the

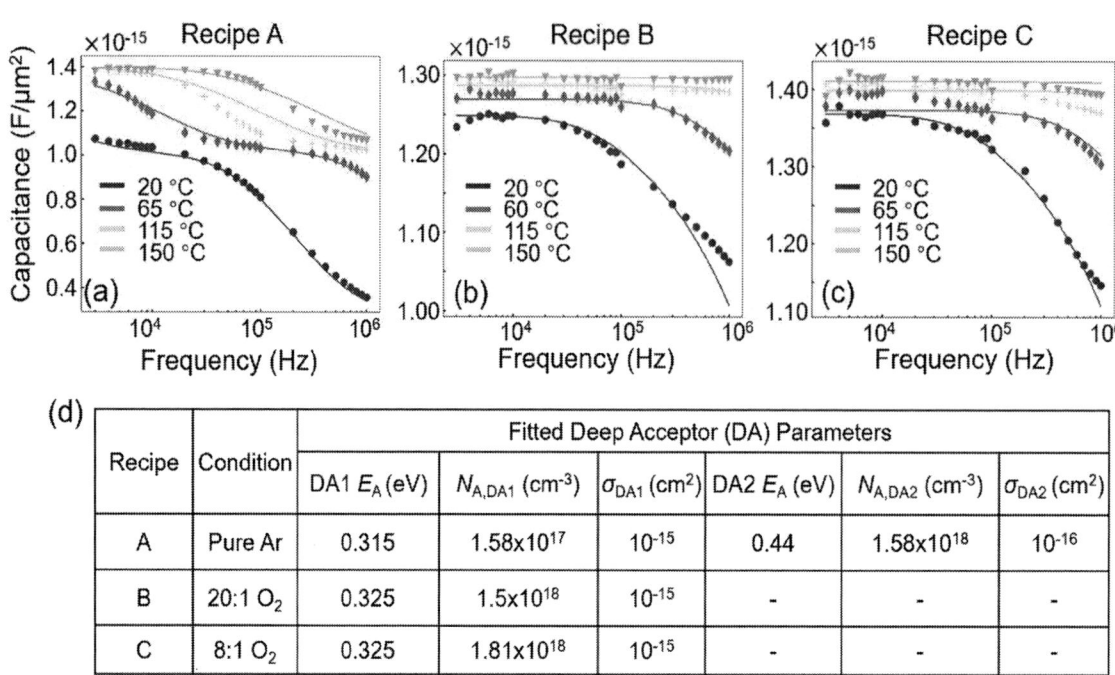

Recipe	Condition	Fitted Deep Acceptor (DA) Parameters					
		DA1 E_A (eV)	$N_{A,DA1}$ (cm^{-3})	σ_{DA1} (cm^2)	DA2 E_A (eV)	$N_{A,DA2}$ (cm^{-3})	σ_{DA2} (cm^2)
A	Pure Ar	0.315	1.58×10^{17}	10^{-15}	0.44	1.58×10^{18}	10^{-16}
B	20:1 O_2	0.325	1.5×10^{18}	10^{-15}	-	-	-
C	8:1 O_2	0.325	1.81×10^{18}	10^{-15}	-	-	-

Fig. 3. Fitted capacitance-frequency dispersion of p-NiO/n$^+$-Ga$_2$O$_3$ diodes at -5 V reverse bias versus temperature and sputtering conditions and TCAD extracted acceptor characteristics. (a-c) show the measured and fitted C-f dispersion characteristics as a function of temperature. The table in (d) shows the acceptor characteristics extracted via the TCAD fitting method.

theoretically expected value for the first ionization energy of the Ni^{3+} (0/-1) charge state associated with Ni vacancies in NiO. Similar theoretical calculations place the (-1/-2) Ni vacancy activation energy at 0.6 eV [22]. However, if the second acceptor energy extracted from the simulation is to be assigned to (-1/-2), then the extracted concentration of the second acceptor trap used to represent it should be twice that of the trap used to represent the (0/-1) transition. This discrepancy is a result of using multiple independent acceptor states to represent the varying charge state of a single physical acceptor, as the percentage of acceptors ionized in the (0/-1) and (-1/-2) states is dependent upon the local Fermi level and must sum to 100% [23].

To examine the nature of the acceptor level more closely in NiO films sputtered with 0% O$_2$/Ar partial pressure ratios, further measurements of the C-f dispersion at -5 V reverse bias were taken. Fig. 4 shows the results of C-f-T characterization of diodes fabricated with 0% O$_2$/Ar during NiO sputtering between 30 °C and 120 °C in 10 °C steps at -5V reverse bias. The temperature behavior of the first and second cutoff frequencies in the C-f dispersion can be clearly seen to be increasing. The acceptor activation energies are extracted from the dispersion measurements following the analysis of Schibli and Milnes for the C-f characteristics of a p-n$^+$ junction doped with a deep acceptor [24]. If the p-type acceptor is a deep level, then the corner frequency in the dispersion characteristics can be shown to be related to the acceptor activation energy by

$$\frac{\omega_c}{T} \propto \exp\left(\frac{E_a - E_v}{kT}\right) \quad (1)$$

In Equation (1), ω_c is the corner frequency, Ea-Ev is the acceptor activation energy relative to the valence band edge and T the temperature. The intersection of the measured C-f characteristics as a function of temperature with a capacitance value chosen near the visible corner frequency in the plot was used to extract the corner frequency. It is assumed that the analysis in [24] continues to hold for acceptors with multiple charge states. Using Equation 1, the extracted corner frequencies can be plotted in an Arrhenius plot to determine the acceptor activation energies. Fig. 4 (b-c) shows the Arrhenius plot associated with the two observed corner frequencies in the C-f-T characteristics of recipe A diodes. A first activation energy of 0.35±0.03 eV is found for the corner frequencies associated with the dispersion region marked by ω_{T1} in Fig 4 (a), while an activation energy of 0.54±0.08 eV is found for the corner frequencies associate with the region marked by ω_{T2}. These values are close to the Ni^{3+} (0/-1) activation energy found experimentally in [8] and to the theoretical value for the (-1/-2) activation energy predicted in [22]. These experimentally extracted activation energies also support, within the error of the measurement, the values found via the TCAD fitting method.

Both the experimentally extracted and TCAD fitted activation energies found allow us to conclude that the Ni^{3+} acceptor can become doubly ionized in sputtered NiO films using the Ni vacancy as the acceptor level. The band diagram shown in Fig. 4(d) illustrates the origin and consequences of this double ionization. Given that the Ni^{3+} can transition from a charge state of -1 to a charge state of -2, the depletion region in the NiO will contain two regions of charge: one with magnitude -qN_A, and another with charge -2q(N_A), the extent of which depends upon the positions $W_{d,A1}$ and $W_{d,A2}$ where the hole

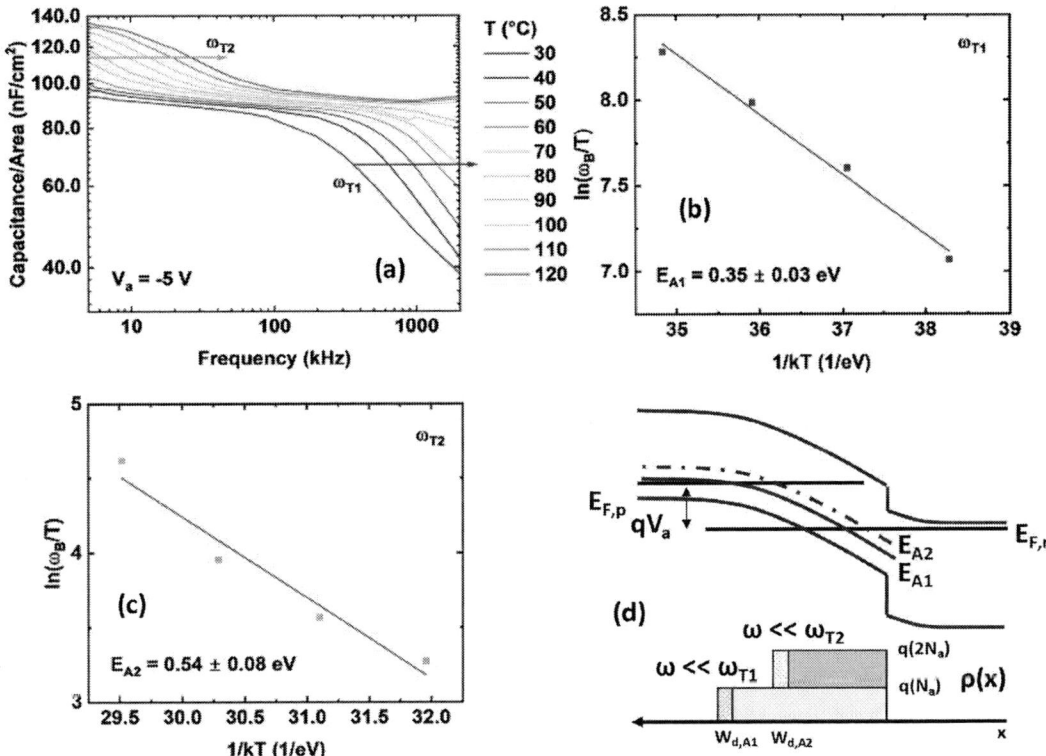

Fig. 4. Capacitance frequency dispersion at -5 V reverse bias for diodes fabricated with 0% O_2/Ar NiO. (b-c) show the extracted acceptor activation energies. (d) gives the band diagram for a p-NiO/n^+-Ga_2O_3 PN diode with the assumed energy levels, as well as the charge distribution in the depletion width.

quasi-Fermi level crosses the associated activation levels with the (0/-1) and (-1/-2) charge transitions. At low frequency, acceptors at the band edge can make the (0/-1) charge transition, and acceptors near the point at which the hole quasi-fermi level crosses the second ionization energy can make also make the (-1/-2) charge transition, with both contributing to the measured capacitance. As frequency increases, only the acceptors at the depletion region edge can make the (0/-1) charge transition under the AC perturbation, reducing the capacitance. At high enough frequency, the charge state of the deep acceptor cannot follow the applied AC bias, resulting in highly reduced capacitance. This result suggests that the physical acceptor concentration in Ni vacancy doped NiO can be extracted from PN C-V measurements at a frequency low enough such that only the singly ionized acceptors at the edge of the depletion region can respond to the AC perturbation, but not low enough such that the (-1/-2) level can respond.

IV. Charge Balance in NiO RESURF Terminations

Using the O_2/Ar sputtering conditions studied in Section III, reactive-magnetron sputtered NiO was used in a lateral RESURF Ga_2O_3 Schottky diode to demonstrate the viability of the material in power device applications. Lateral Schottky diodes with magnetron sputtered NiO p-type RESURF terminations were fabricated as described in Section II.B. The RESURF termination will maximize the measured breakdown

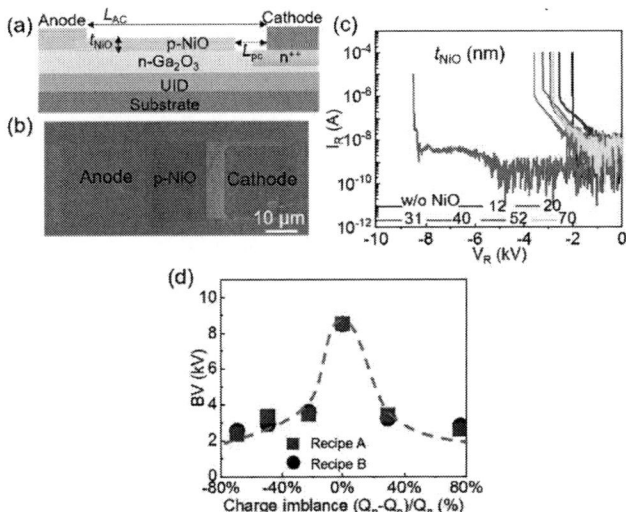

Fig. 5. p-NiO RESURF Ga_2O_3 Schottky diode device structure and breakdown voltage measurements as a function of NiO thickness. (a) and (b) show the device structure, (c) shows the measured breakdown as a function of NiO thickness for recipe A, and (d) plots Vbr vs. charge balance for both recipes.

voltage through the charge balance condition $\sigma_n L_{ac} = (L_{ac} - L_{pc}) t_{NiO} N_A$, where σ_n is the total doping dose of the n+ Ga_2O_3 channel, measured at 3.8×10^{12} cm^{-2} by electrochemical C-V profiling, L_{ac} is the anode-to-cathode separation of 17 um, and L_{pc} is the cathode-to-RESURF separation of 7 um. Recipes A and B (0% and 5% O_2/Ar) were used to deposit NiO RESURF layers of thickness varying between 12 nm and 70 nm for charge balance testing.

Fig. 5 shows the device structure, breakdown voltage measurements as a function of NiO thickness and temperature and breakdown voltage as a function of estimated charge imbalance. Diodes using a 0% O_2/Ar NiO RESURF layer reach an optimal breakdown voltage of 8 kV at a NiO thickness of 75 nm, while diodes using a 5% O_2/Ar NiO RESURF layer reach an optimum breakdown voltage of also around 8 kV at a NiO thickness of 40 nm. Based upon the charge balance condition at the optimum breakdown design, the estimated N_A value of the 0% O_2/Ar NiO RESURF film is 8.6×10^{17} cm^{-3} while the estimated N_A of the 5% $O2$/Ar NiO RESURF film is 1.6×10^{18} cm^{-3}. These results successfully demonstrate the feasibility of using reactive-magnetron sputtered NiO as a selective-area p-type dopant for an UWBG power device.

V. CONCLUSION

This work examines the effect of varying O_2 partial pressure during sputtering of NiO thin films. Through capacitance measurements on p-NiO/n$^+$-Ga_2O_3 diodes fabricated via magnetron sputtering, we have shown that the effective acceptor density in NiO deposited in this manner can be controlled from a magnitude of $\sim 9 \times 10^{17}$ cm^{-3} to 2×10^{18} cm^{-3} over an O_2/Ar partial pressure ratio range from 0% to 12.5%. Acceptor first and second ionization energies of 0.35 and 0.54 eV are extracted using the temperature dependent C-f dispersion of the fabricated diodes,, showing experimentally that the second charge state of the Ni^{3+} may occur in NiO films deposited via sputtering in the pure Ar atmosphere. A magnetron-sputtered NiO layer is used to successfully demonstrate charge balance in a lateral Ga_2O_3 Schottky diode, demonstrating the feasibility of the utilization of NiO for selective area doping in UWBG power devices. Further studies on the control of p-type doping in PLD deposited and Li-doped Ni films should be examined to further the understanding of the control of p-type doping in NiO for use in power device applications.

ACKNOWLEDGMENT

The authors thank the assistance and advice of Dr. Ming Xiao, Boyan Wang, Yifan Wang and Donald Leber at Virginia Tech, Joseph Spencer at NRL and Zhonghao Du at USC in assisting with device fabrication and technical support, as well as the collaboration with Silvaco on device simulation of our device structures.

REFERENCES

[1] Y. Zhang, F. Udrea, and H. Wang, "Multidimensional device architectures for efficient power electronics," *Nat Electron*, vol. 5, no. 11, Art. no. 11, Nov. 2022, doi: 10.1038/s41928-022-00860-5.

[2] J. L. Lyons, "A survey of acceptor dopants for β-Ga2O3," *Semicond. Sci. Technol.*, vol. 33, no. 5, p. 05LT02, Apr. 2018, doi: 10.1088/1361-6641/aaba98.

[3] J. A. Spencer, A. L. Mock, A. G. Jacobs, M. Schubert, Y. Zhang, and M. J. Tadjer, "A review of band structure and material properties of transparent conducting and semiconducting oxides: Ga 2 O 3 , Al 2 O 3 , In 2 O 3 , ZnO, SnO 2 , CdO, NiO, CuO, and Sc 2 O 3 ," *Applied Physics Reviews*, vol. 9, no. 1, p. 011315, Mar. 2022, doi: 10.1063/5.0078037.

[4] Y. Zhang et al., "Vertical GaN Junction Barrier Schottky Rectifiers by Selective Ion Implantation," *IEEE Electron Device Letters*, vol. 38, no. 8, pp. 1097–1100, Aug. 2017, doi: 10.1109/LED.2017.2720689.

[5] Y. Zhang and T. Palacios, "(Ultra)Wide-Bandgap Vertical Power FinFETs," *IEEE Transactions on Electron Devices*, vol. 67, no. 10, pp. 3960–3971, Oct. 2020, doi: 10.1109/TED.2020.3002880.

[6] J. D. Greenlee, B. N. Feigelson, T. J. Anderson, J. K. Hite, K. D. Hobart, and F. J. Kub, "Symmetric Multicycle Rapid Thermal Annealing: Enhanced Activation of Implanted Dopants in GaN," *ECS J. Solid State Sci. Technol.*, vol. 4, no. 9, p. P382, Aug. 2015, doi: 10.1149/2.0191509jss.

[7] R. Karsthof, A. M. Anton, F. Kremer, and M. Grundmann, "Nickel vacancy acceptor in nickel oxide: Doping beyond thermodynamic equilibrium," *Phys. Rev. Mater.*, vol. 4, no. 3, p. 034601, Mar. 2020, doi: 10.1103/PhysRevMaterials.4.034601.

[8] R. Karsthof, H. von Wenckstern, V. S. Olsen, and M. Grundmann, "Identification of LiNi and VNi acceptor levels in doped nickel oxide," *APL Materials*, vol. 8, no. 12, p. 121106, Dec. 2020, doi: 10.1063/5.0032102.

[9] M. Xiao et al., "First Demonstration of Vertical Superjunction Diode in GaN," in *2022 International Electron Devices Meeting (IEDM)*, Dec. 2022, p. 35.6.1-35.6.4. doi: 10.1109/IEDM45625.2022.10019405.

[10] X. Yin, Y. Guo, H. Xie, W. Que, and L. B. Kong, "Nickel Oxide as Efficient Hole Transport Materials for Perovskite Solar Cells," *Solar RRL*, vol. 3, no. 5, p. 1900001, 2019, doi: 10.1002/solr.201900001.

[11] B. Wang et al., "2.5 kV Vertical Ga2O3 Schottky Rectifier With Graded Junction Termination Extension," *IEEE Electron Device Letters*, vol. 44, no. 2, pp. 221–224, Feb. 2023, doi: 10.1109/LED.2022.3229222.

[12] F. Zhou et al., "An avalanche-and-surge robust ultrawide-bandgap heterojunction for power electronics," *Nat Commun*, vol. 14, no. 1, Art. no. 1, Jul. 2023, doi: 10.1038/s41467-023-40194-0.

[13] Y. Qin et al., "10-kV Ga2O3 Charge-Balance Schottky Rectifier Operational at 200 °C," *IEEE Electron Device Letters*, vol. 44, no. 8, pp. 1268–1271, Aug. 2023, doi: 10.1109/LED.2023.3287887.

[14] Y. Ma et al., "Tri-gate GaN junction HEMT," *Appl. Phys. Lett.*, vol. 117, no. 14, p. 143506, Oct. 2020, doi: 10.1063/5.0025351.

[15] Y. Ma, M. Xiao, Z. Du, H. Wang, and Y. Zhang, "Tri-Gate GaN Junction HEMTs: Physics and Performance Space," *IEEE Transactions on Electron Devices*, vol. 68, no. 10, pp. 4854–4861, Oct. 2021, doi: 10.1109/TED.2021.3103157.

[16] M. Xiao et al., "5 kV Multi-Channel AlGaN/GaN Power Schottky Barrier Diodes with Junction-Fin-Anode," in *2020 IEEE International Electron Devices Meeting (IEDM)*, Dec. 2020, p. 5.4.1-5.4.4. doi: 10.1109/IEDM13553.2020.9372025.

[17] Y. Zhang et al., "GaN FinFETs and trigate devices for power and RF applications: review and perspective," *Semicond. Sci. Technol.*, vol. 36, no. 5, p. 054001, Mar. 2021, doi: 10.1088/1361-6641/abde17.

[18] M. Xiao, Y. Ma, K. Liu, K. Cheng, and Y. Zhang, "10 kV, 39 mΩ·cm2 Multi-Channel AlGaN/GaN Schottky Barrier Diodes," *IEEE Electron Device Letters*, vol. 42, no. 6, pp. 808–811, Jun. 2021, doi: 10.1109/LED.2021.3076802.

[19] M. Xiao et al., "Multi-Channel Monolithic-Cascode HEMT (MC2-HEMT): A New GaN Power Switch up to 10 kV," in *2021 IEEE International Electron Devices Meeting (IEDM)*, Dec. 2021, p. 5.5.1-5.5.4. doi: 10.1109/IEDM19574.2021.9720714.

[20] K. V. Rao and A. Smakula, "Dielectric Properties of Cobalt Oxide, Nickel Oxide, and Their Mixed Crystals," *Journal of Applied Physics*, vol. 36, no. 6, pp. 2031–2038, Jul. 2004, doi: 10.1063/1.1714397.

[21] A. Fiedler, R. Schewski, Z. Galazka, and K. Irmscher, "Static Dielectric Constant of β-Ga2O3 Perpendicular to the Principal Planes (100), (010), and (001)," *ECS J. Solid State Sci. Technol.*, vol. 8, no. 7, p. Q3083, Mar. 2019, doi: 10.1149/2.0201907jss.

[22] H. D. Lee, B. Magyari-Köpe, and Y. Nishi, "Model of metallic filament formation and rupture in NiO for unipolar switching," *Phys. Rev. B*, vol. 81, no. 19, p. 193202, May 2010, doi: 10.1103/PhysRevB.81.193202.

[23] Milnes, A.G., *Deep Impurities in Semiconductors*. New York: John Wiley & Sons, Ltd, 1973.

[24] E. Schibli and A. G. Milnes, "Effects of deep impurities on n+p junction reverse-biased small-signal capacitance," *Solid-State Electronics*, vol. 11, no. 3, pp. 323–334, Mar. 1968, doi: 10.1016/0038-1101(68)90044-0.

Gate Lifetime of P-Gate GaN HEMT Under DC and Switching Overvoltage Stress

Bixuan Wang
Center for Power Electronics Systems (CPES)
Virginia Tech
Blacksburg, USA
bixuanwang@vt.edu

Ruizhe Zhang
CPES
Virginia Tech
Blacksburg, USA
rzzhang@vt.edu

Qihao Song
CPES
Virginia Tech
Blacksburg, USA
qihao95@vt.edu

Qiang Li
CPES
Virginia Tech
Blacksburg, USA
lqvt@vt.edu

Yuhao Zhang
CPES
Virginia Tech
Blacksburg, USA
yhzhang@vt.edu

Abstract—This paper investigates the gate lifetime of Schottky-type p-gate GaN HEMTs (SP-HEMTs) under positive gate voltage (V_G) stress. The V_G stress is implemented in both the DC bias and the switching at various frequencies (f_{sw}). For the switching stress, the impacts of two switching schemes, i.e., drain-and-source grounded (DSG) and inductive hard switching (HSW), on the gate lifetime are studied. The switching stress features the repetitive, resonant V_G ringing with an average dV_G/dt of 1~2 V/ns produced by a V_G-overshoot-generation circuit. It is found that, in either the DC or switching case, the gate lifetime shows a power-law relation with the V_G-stress magnitude and can be fitted by Weibull distribution. At f_{sw} up to 100 kHz, the switching lifetime is determined by the switching cycle numbers. The maximum allowable V_G values for a 10-year lifetime extracted under the DC bias and the 100-kHz DSG switching are found to be similar (~6 V). A higher V_G limit of ~10 V is revealed under the 100-kHz HSW condition. These results provide new insights of GaN HEMT gate lifetime under static and dynamic stresses and provide useful reference for the gate qualification.

Keywords—power electronics, GaN HEMT, gate reliability, lifetime, DC, spike, ringing, inductive power switching

I. INTRODUCTION

In recent years, gallium nitride (GaN) high-electron-mobility transistors (HEMTs) are gaining breakthrough adoptions in high-efficiency power conversions [1], [2]. Among various promising device architectures, Schottky-type p-gate GaN HEMT (SP-HEMT) has become popular in GaN device market [3], [4]. However, a reliability concern of GaN SP-HEMT is the narrow headroom in forward gate-to-source voltage (V_G) [4]. Many vendors suggest the ON-state operating V_G not to exceed 6~7 V from reliability considerations [5], [6]. Meanwhile, it is usually set as 5~6 V in practical applications to achieve a low on-resistance (R_{ON}). This produces a V_G window of only 1 V.

This narrow V_G window demands a thorough study of the gate overvoltage lifetime. Up to now, several methods have been applied to characterize the gate lifetime of commercial GaN SP-HEMTs, including the DC-bias stress, pulse I-V, and inductive switching circuit [7], [8]. In general, the characterization is desirable to be performed in the condition that best mimics the converter operations.

In practical converters, the gate overvoltage is usually produced due to the high slew rate and the parasitic inductance of the gate loop. Such gate ringing has a resonance profile and could occur with the switching in the drain-source loop. In the previous studies, the DC-V_G stress (referring as the constant-V_G stress in this paper) at V_G of 9~11 V has been adopted [9]–[12] to characterize the gate lifetime. Recently, the pulse-IV test with continuous square-wave V_G stress is used to mimic the ON/OFF operating V_G during switching [8], [13], [14]. However, the V_G profile in these two methods are different from the practical converter operations. Moreover, these tests are performed under the device drain-and-source-grounded (DSG) condition, albeit the device operating in high-voltage power switching [e.g., inductive hard switching (HSW)] in applications.

Recently, we developed a new circuit method that produces a resonance-like V_G overshoot with a pulse width down to 20 ns (dV_G/dt = 1~2 V/ns) during device's turn-ON transient, and the gate characterization can be performed under both DSG and HSW conditions in an inductive load converter [7], [15]. This work performs the gate lifetime characterization using this circuit method under the DSG and HSW conditions (referring as the switching stress in this paper), and compares the results with the gate lifetime under the DC stress. In switching characterizations, the impact of frequency (f_{sw}) on the switching lifetime, as well as the device degradation in the prolonged switching stress, are also studied. The max allowable V_G for a 10-year lifetime is extracted from all these three tests, and the results are compared.

II. TEST METHODS

A. DUT and Test Platforms

The device under test (DUT) in this work is the commercially available 650-V, 30-A GaN SP-HEMT [5]. Fig. 1(a) shows the DUT's schematic and the microscopic image of the gate region. Fig. 1(b) shows the DUT's static I_G-V_G characteristics with drain-to-source voltage (V_{DS}) of 0 V at 25 and 125 °C, measured using a Keysight B1505 Power Device Analyzer with a 0.1-A I_G compliance. Further measurement confirms that the DUT does not show degradation after this I_G-V_G characterization.

The schemes of the DC test and the switching tests under DSG and HSW are detailed as follows.

This work is supported in part by the National Science Foundation under the Grant ECCS-2202620 and in part by the Power Management Consortium of Center for Power Electronics Systems, Virginia Tech.

979-8-3503-3714-3/23 $31.00 © 2023 IEEE

1) DC Stress: The DUT is stressed under a constant gate-to-source overvoltage (V_G^{DC}) applied by B1505, with I_G evolution being tracked until the destructive gate failure, which is verified by the post-test characterizations. During the test, the DUT's drain and source are kept grounded, i.e., $V_{DS} = 0$ V. For statistical significance, 8 devices are stressed under each V_G^{DC} value (9, 9.5, and 10 V).

2) Switching Stress: The resonant V_G overshoot across DUT's gate and source (G-S) is generated by a carefully designed circuit, enabling power switching in DUT's drain-source (D-S) loop along with each overshoot. In this paper, the switching scheme in DUT's D-S loop is configured for DSG (for a direct comparison with results under DC stress), or HSW at 400-V bus voltage (V_{BUS}) with inductive load. The peak value of V_G overshoot [$V_{G(PK)}^{SW}$, representing the peak V_G stress in switching] as well as the pulse width (*PW*), can be modulated by circuit parameters. The *PW* is fixed at 20 ns in all tests in this work to best mimic common transient overshoots in practical operations. The lifetime is measured as the total switching time until failure, and is characterized at different $V_{G(PK)}^{SW}$ (15.5 to 17.5 V in DSG condition and 17.5 to 19.5 V in HSW condition), as well as different f_{SW} (10 kHz to 100 kHz). 10 devices are stressed in each condition.

The DUT's V_G and V_{DS} are sensed by a 1-GHz-bandwidth low-voltage passive probe (Tektronix TPP1000) and an 800-MHz-bandwidth high-voltage passive probe (Tektronix TPP0850), respectively, for fast-switching measurements.

The DUT's case temperature (T_C) is kept at 125 °C in all tests, heated by a power resistor attached to its top-side case (the DUT is top-side cooled that offers very low junction-to-case thermal resistance). The T_C is carefully monitored throughout the entire test, calibrated by a K-type thermocouple as well as an infrared camera. The stress is applied when DUT reaches thermal steady state in all tests.

B. Circuit Design for Switching Stress

Fig. 2 shows the exemplar circuit schematics for switching-stress test configured for 400-V inductive HSW condition. The circuit for switching stress in DSG condition shares the same overvoltage-generation design in the gate loop, but with the DUT's D-S keeps grounded. This gate-overvoltage-generation circuit produces a V_G overshoot, similar to how a V_{DS} overshoot is generated in the unclamped inductive switching (UIS) test [16]–[22].

The core methodology is to build up and store energy in an inductor (L_G) in the DUT's gate loop, with the L_G being charged by a voltage source ($V_{DD} = 0.5$ V in this work) while the switch (S_1, a 80-V GaN HEMT in this work [6]) is turned ON. When the S_1 is fast turned OFF, the energy in L_G creates a resonant overvoltage (V_{LG}) across the L_G. As the DUT's V_G becomes $V_{LG} + V_{DD}$, a resonant-like overvoltage is applied directly across the DUT's gate and source. The V_{LG} overshoot can be approximated as the resonance between L_G and an equivalent capacitance (the sum of the C_{ISS} of DUT and the C_{OSS} of S_1). Thus, the *PW* of the V_G overshoot, which is close to a half of the V_{LG}-overshoot resonance period, can be modulated directly by varying the L_G value. After L_G is determined, the $V_{G(PK)}^{SW}$ can be modulated by the energy stored in L_G, which is determined by the V_{DD} value and S_1 turn-ON time.

The HSW condition (Fig. 2) is realized by the hard turn ON by the V_G overshoot when exceeding DUT's threshold voltage (V_{TH}). In this work, the power loop is configured with a 400-V bus voltage (V_{BUS}), a load inductor (L_{LOAD}), a free-wheeling diode (*FWD*), and an inductor (L_P) to mimic power-loop parasitic inductance and suppress the high I_D-V_{DS} overlap (and thermal runaway). More circuit details are described in [7]. Fig. 3(a) shows the exemplar HSW test waveforms in a single switching cycle with $V_{G(PK)}^{SW} = 17.5$ V. Fig. 3(b) shows

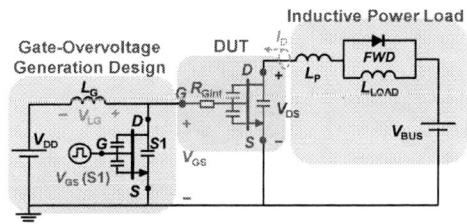

Fig. 2: Circuit schematics for switching stress in inductive hard switching (HSW) condition. Circuit for switching stress in the drain-source-grounded (DSG) condition shares the same overvoltage generation design in gate loop, but with the DUT's drain and source kept grounded.

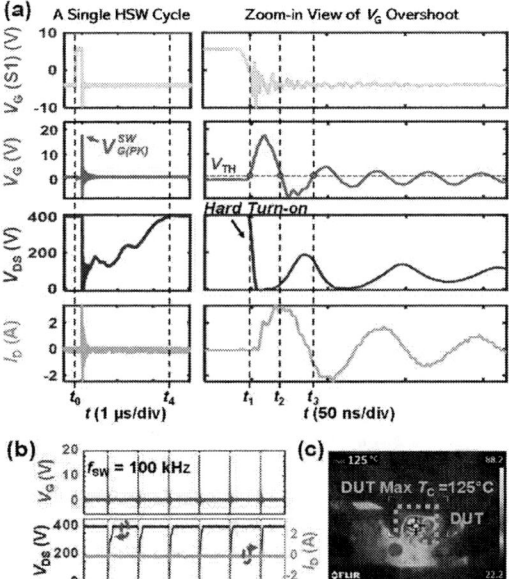

Fig. 3: Exemplar gate-overvoltage test waveforms in HSW condition: (a) (left) a whole switching cycle, (right) zoom-in view of V_G overshoot period; and (b) repetitive switching at f_{SW} = 100 kHz. (c) Thermal camera image showing calibrated DUT's case temperature (T_C) = 125 °C during switching.

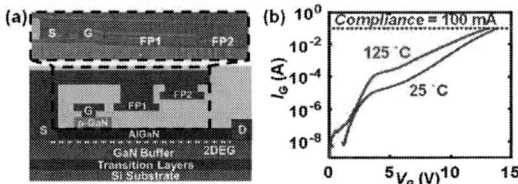

Fig. 1: (a) (Bottom) DUT schematics and (top) microscopic photo of gate region. (b) Static I_G-V_G characteristics at 25 and 125 °C.

979-8-3503-3714-3/23 $31.00 © 2023 IEEE

the exemplar 100-kHz continuous switching waveforms with DUT's temperature stabilizing at 125 °C [Fig. 3(c)]. The $V_{G(PK)}^{SW}$ in DSG condition (SW, DSG) shares the same characteristics as that in HSW.

This methodology essentially resembles the nature of the V_G ringing in practical converter operations, which is induced by the gate-loop parasitics during the device's turn-ON transients in power switching.

III. RESULTS AND DISCUSSIONS

In this paper, the gate failure is regarded as the DUT's non-recoverable loss of switching function, verified by post-test characterizations.

A. Failure Characteristics under DC Stress

Fig. 4(a) shows the statistical test waveforms of the I_G evolution during stress, under V_G^{DC} from 9 to 10 V. A sudden I_G increase suggests a complete or partial G-S shorting, which is confirmed by static I-V of the failed devices. Therefore, the lifetime is measured as the stressing time until the first increase in I_G is observed, as demonstrated in Fig. 4(b).

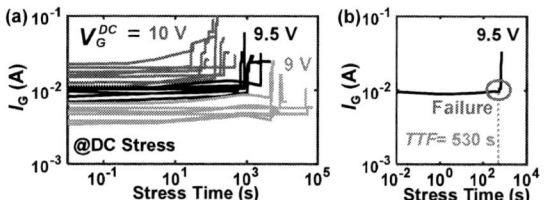

Fig. 4: (a) Gate lifetime (time to failure *TTF*) characteristics in DC gate-overvoltage test under different constant-V_G stress (V_G^{DC}) at 125 °C. (b) I_G sudden increase demonstrates a gate failure.

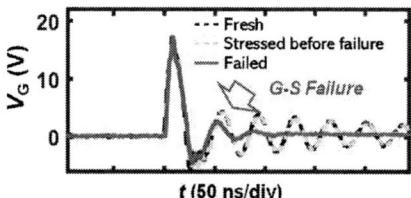

Fig. 5: DUT's V_G waveforms when the DUT is fresh, stressed before failure, and after failure. All waveforms are recorded during the consecutive gate-overvoltage switching test. The V_G waveform show similar characteristics in DSG and HSW conditions. The fast damping in V_G indicates a gate-to-source shorting.

B. Failure Characteristics under Switching Stress

Fig. 5 shows the typical V_G switching waveforms when the DUT is fresh, stressed before failure, and failed, all recorded from the consecutive switching in either DSG or HSW condition. In the failure waveform, a fast damping of the V_G ringing after $V_{G(PK)}^{SW}$ implies a G-S shorting. Post-test I-V confirms the destructive G-S failure and the functional D-S OFF-state blocking capability after the gate breakdown. Under switching stress, the lifetime is regarded as the total time until the first occurrence of fast damping in V_G ringing.

C. Lifetime Characteristics under DC and Switching

Under each stress condition, the failure probability (*P*) as a function of *TTF* can be obtained using the *TTF* data of the corresponding tested devices. Fig. 6 shows that, in each stress condition, the *TTF* can be fitted by Weibull distribution, with shape factor $\beta > 1$ in most cases, indicating a wear-out failure. The *TTF* (at *P* = 63% or 0.1%) extracted from the Weibull fitting show a power law relation with either the V_G^{DC} under DC stress or the $V_{G(PK)}^{SW}$ under switching stresses (Fig. 7), allowing us to predict the corresponding V_G boundary for a 10-year *TTF* from extrapolation, with the assumption that the *TTF* power law relation is also valid at lower stress value. In the DSG condition, the projected max $V_{G(PK)}^{SW}$ for a 10-year lifetime under 100-kHz switching stress (7.0 and 5.8 V for *P* = 63% and 0.1%, respectively) is similar to the projected max V_G^{DC} under DC stress, ~5 V lower than the max $V_{G(PK)}^{SW}$ under 100-kHz HSW. This suggests that 1) the DC test method can be used to project the max allowable $V_{G(PK)}^{SW}$ under 100-kHz switching in DSG condition, 2) the max allowable $V_{G(PK)}^{SW}$ under the HSW cannot be represented by the DC test, and the device has a higher gate overvoltage margin in HSW as compared to DSG.

Finally, the f_{SW} impact on the *TTF* in switching is studied. The characteristics of the switching-cycle-to-failure number (*SCTF#* = *TTF* × f_{SW}) at f_{SW} = 10 kHz and 100 kHz are compared (Fig. 8). The mean-*SCTF#* (*MSCTF#*) as well as Weibull distributions at the two f_{SW} are similar under the same 17.5-V $V_{G(PK)}^{SW}$ in DSG condition [Fig. 8(a)] and 19.5-V $V_{G(PK)}^{SW}$ in HSW condition [Fig. 8(a)], indicating that the switching *TTF* is dominated by the *SCTF#* instead of total time at f_{SW} in this range, implying a strong f_{SW} impact on *TTF*. Building on these findings, the max $V_{G(PK)}^{SW}$ for 10-year gate lifetime under switching stress at f_{SW} from 100 Hz to 100 kHz are predicted, revealing a higher gate overvoltage margin for 10-year lifetime at f_{SW} < 100 kHz than that projected by the DC test (Fig. 9). The gate lifetime in inductive hard switching is longer as compared to that in DSG, and it also increases with the decreased f_{SW}.

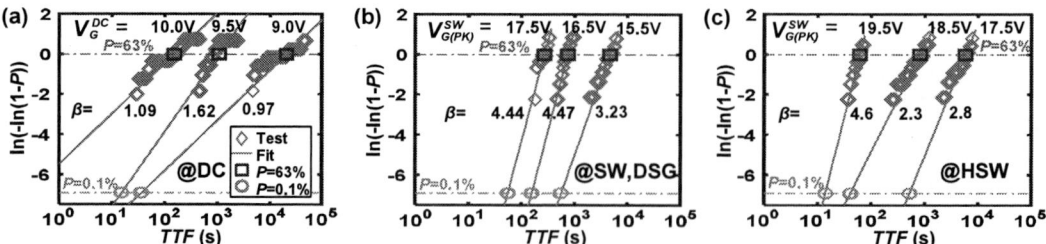

Fig. 6: Weibull distribution of the *TTF* at (a) different V_G^{DC} under DC stress, as well as different $V_{G(PK)}^{SW}$ under 100-kHz switching stress in (b) DSG and (c) HSW conditions. The *TTF* at failure probability *P* = 63% and 0.1% under each condition can be extracted from the corresponding fitting curve.

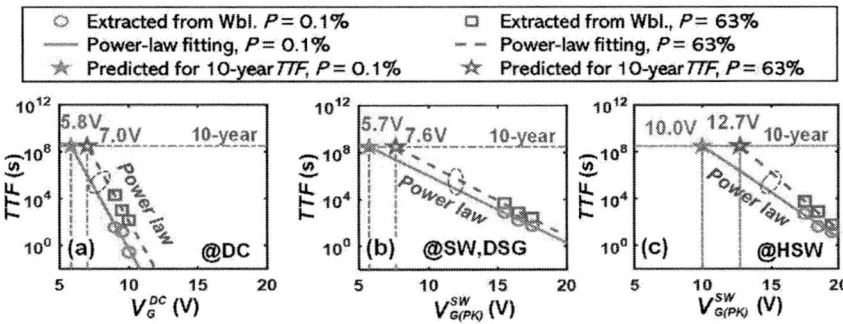

Fig. 7: For 10-year *TTF* at $P = 63\%$ and 0.1%, the (a) max V_G^{DC} under DC stress, as well as the max $V_{G(PK)}^{SW}$ under 100-kHz switching stress in (b) DSG and (c) HSW conditions are predicted based on power law extrapolation with the *TTF* extracted from the corresponding Weibull distribution.

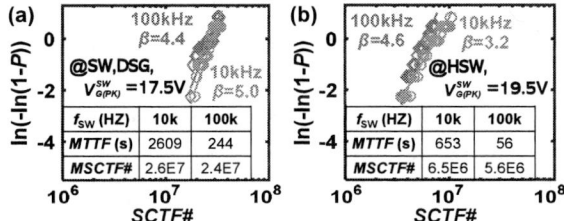

Fig. 8: (Top) Under switching stress, the Weibull distributions of the switching-cycle-to-failure number (*SCTF#* = *TTF* f*sw) at *f*sw = 10 k and 100 kHz are compared, under (a) $V_{G(PK)}^{SW}$= 17.5 V in DSG condition, and (b) $V_{G(PK)}^{SW}$= 19.5 V in HSW condition. (Bottom) The mean time to failure (*MTTF*) and the mean *SCTF#* (*MSCTF#*) of tested 10 DUTs in each stress condition. The *MSCTF#* show weak dependence on the *f*sw, indicating the strong impact of the *f*sw on the lifetime.

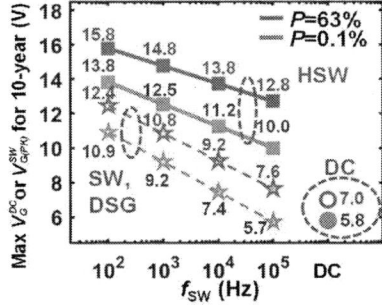

Fig. 9: For a 10-year lifetime at $P = 63\%$ and 0.1%, the predicted max V_G^{DC} under DC stress, as well as the max $V_{G(PK)}^{SW}$ under switching stress as a function of *f*sw (from 100 to 100 kHz), and with different switching schemes in device's power loop (DSG and HSW). The predicted $V_{G(PK)}^{SW}$ margin under 100-kHz switching in DSG condition is close to the V_G^{DC} under DC stress, and this margin increases at lower *f*sw. In HSW condition, the $V_{G(PK)}^{SW}$ margin is even higher.

IV. GATE DEGRADATION BEFORE FAILURE

The switching- lifetime tests suggest the GaN SP-HEMT has a higher gate overvoltage margin in the HSW as compared to the DSG with the same $V_{G(PK)}^{SW}$. As a verification, we measure the device characteristics intermittently during the prolonged DSG and HSW switching stress with the same $V_{G(PK)}^{SW}$. Here the intermittent characterization is performed by taking the device from the circuit board and measuring it on

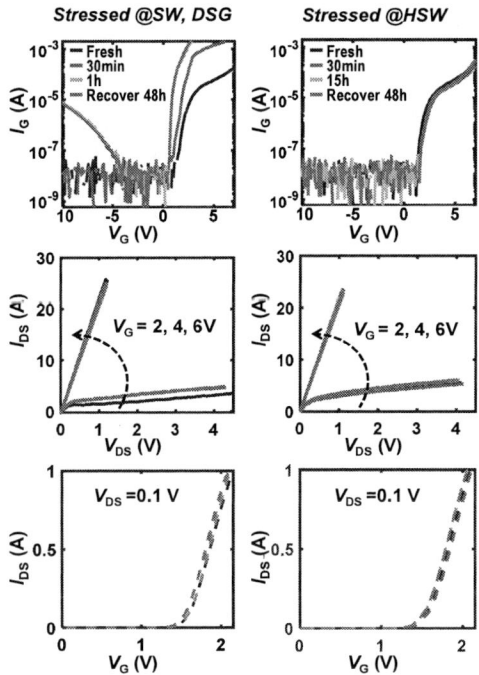

Fig. 10: Devices' gate leakage, R_{ON} and V_{TH} during prolonged switching stress at *f*sw = 100 kHz and $V_{G(PK)}^{SW}$ = 10V, in (left) DSG and (right) HSW conditions. Devices remain functional after stressed in each condition, while present enhanced gate reliability as well as on-resistance stability in HSW.

the curve tracer. Hence it is not an 'in-situ' measurement. The measured parametric shift is either permanent or unrecoverable in a few minutes (i.e., the time it takes to transfer the device from circuit board to curve tracer).

Fig. 10 compares the evolutions of devices' gate leakage, R_{ON} and V_{TH} under the prolonged stress in DSG (left) and HSW (right) conditions, both under 125 °C, *f*sw = 100 kHz and $V_{G(PK)}^{SW}$ = 10 V. In the DSG condition, an apparent increase in the forward gate current is observed after a 30-minute stress, indicating a degradation of the gate/p-GaN Schottky contact. After stressed for an hour, an increase is seen in the reverse gate current, indicating a subsequent degradation of the p-GaN/AlGaN/GaN PIN junction induced by the crowded electric field and p-GaN punch through after the initial degradation of Schottky contact. Such degradation behaviors are also reported elsewhere based on the electroluminescence

979-8-3503-3714-3/23 $31.00 © 2023 IEEE   261

emission [23] and the scanning-electron-microscope images of contacts condition [24]. Despite that the device remains functional after stress, these gate degradation are confirmed to be non-recoverable, thus posing a potential risk to device reliability. In HSW condition, all characteristics remain stable after the DUT is stressed for 15 hours, verifying a relieved voltage stressed across the gate compared with that in DSG condition. Note that the physical mechanism of this reduced stress under the HSW condition has been explained in our prior report [15].

V. CONCLUTION

This work compares the gate lifetime of GaN SP-HEMT under forward gate overvoltage tested under the DC bias and the repetitive V_G ringing switching. The switching tests are performed at different f_{SW} and with different switching schemes in device's power loop. For a 10-year lifetime, the predicted $V_{G(PK)}^{SW}$ margin under 100-kHz switching in DSG condition is close to the V_G^{DC} under the DC stress, and this margin increases at lower f_{SW}. Under the HSW condition, the $V_{G(PK)}^{SW}$ margin is even higher. This work suggests the insufficiency of using the DC test to project the gate overvoltage margin of GaN SP-HEMTs for power electronics applications, as the DC test results cannot capture the frequency dependence and the impact of main loop switching scheme. Switching-based test with a stress profile best mimicking the practical use conditions is essential for quantifying the gate reliability and lifetime of GaN HEMTs in power electronics applications.

REFERENCES

[1] Y. Zhang, F. Udrea, and H. Wang, "Multidimensional device architectures for efficient power electronics," *Nat. Electron.*, vol. 5, no. 11, Art. no. 11, Nov. 2022, doi: 10.1038/s41928-022-00860-5.

[2] N. Keshmiri, D. Wang, B. Agrawal, R. Hou, and A. Emadi, "Current Status and Future Trends of GaN HEMTs in Electrified Transportation," *IEEE Access*, vol. 8, pp. 70553–70571, 2020, doi: 10.1109/ACCESS.2020.2986972.

[3] K. Hoo Teo *et al.*, "Emerging GaN technologies for power, RF, digital, and quantum computing applications: Recent advances and prospects," *J. Appl. Phys.*, vol. 130, no. 16, p. 160902, Oct. 2021, doi: 10.1063/5.0061555.

[4] J. P. Kozak *et al.*, "Stability, Reliability, and Robustness of GaN Power Devices: A Review," *IEEE Trans. Power Electron.*, vol. 38, no. 7, pp. 8442–8471, Jul. 2023, doi: 10.1109/TPEL.2023.3266365.

[5] "GS66508T," GaN Systems. Accessed: Mar. 21, 2023. [Online]. Available: https://gansystems.com/gan-transistors/gs66508t/

[6] "EPC2214: Automotive 80 V, 47 A Enhancement-Mode GaN Power Transistor." Accessed: Mar. 21, 2023. [Online]. Available: https://epc-co.com/epc/products/gan-fets-and-ics/epc2214

[7] B. Wang *et al.*, "Gate Lifetime of P-Gate GaN HEMT in Inductive Power Switching," in *2023 35th International Symposium on Power Semiconductor Devices and ICs (ISPSD)*, May 2023, pp. 20–23. doi: 10.1109/ISPSD57135.2023.10147610.

[8] J. He, J. Wei, S. Yang, Y. Wang, K. Zhong, and K. J. Chen, "Frequency- and Temperature-Dependent Gate Reliability of Schottky-Type p -

GaN Gate HEMTs," *IEEE Trans. Electron Devices*, vol. 66, no. 8, pp. 3453–3458, 2019, doi: 10.1109/TED.2019.2924675.

[9] M. Ťapajna, O. Hilt, E. Bahat-Treidel, J. Würfl, and J. Kuzmík, "Gate Reliability Investigation in Normally-Off p-Type-GaN Cap/AlGaN/GaN HEMTs Under Forward Bias Stress," *IEEE Electron Device Lett.*, vol. 37, no. 4, pp. 385–388, Apr. 2016, doi: 10.1109/LED.2016.2535133.

[10] I. Rossetto *et al.*, "Time-Dependent Failure of GaN-on-Si Power HEMTs With p-GaN Gate," *IEEE Trans. Electron Devices*, vol. 63, no. 6, pp. 2334–2339, Jun. 2016, doi: 10.1109/TED.2016.2553721.

[11] S. Stoffels *et al.*, "Failure mode for p-GaN gates under forward gate stress with varying Mg concentration," in *2017 IEEE International Reliability Physics Symposium (IRPS)*, Apr. 2017, pp. 4B-4.1-4B-4.9. doi: 10.1109/IRPS.2017.7936310.

[12] A. Stockman *et al.*, "Gate Conduction Mechanisms and Lifetime Modeling of p-Gate AlGaN/GaN High-Electron-Mobility Transistors," *IEEE Trans. Electron Devices*, vol. 65, no. 12, pp. 5365–5372, Dec. 2018, doi: 10.1109/TED.2018.2877262.

[13] M. Millesimo *et al.*, "The Role of Frequency and Duty Cycle on the Gate Reliability of p-GaN HEMTs," *IEEE Electron Device Lett.*, vol. 43, no. 11, pp. 1846–1849, Nov. 2022, doi: 10.1109/LED.2022.3206610.

[14] M. Millesimo *et al.*, "Gate Reliability of p-GaN Power HEMTs Under Pulsed Stress Condition," in *2022 IEEE International Reliability Physics Symposium (IRPS)*, Mar. 2022, p. 10B.2-1-10B.2–6. doi: 10.1109/IRPS48227.2022.9764592.

[15] B. Wang *et al.*, "Dynamic Gate Breakdown of p-Gate GaN HEMTs in Inductive Power Switching," *IEEE Electron Device Lett.*, vol. 44, no. 2, pp. 217–220, Feb. 2023, doi: 10.1109/LED.2022.3227091.

[16] R. Zhang, J. P. Kozak, Q. Song, M. Xiao, J. Liu, and Y. Zhang, "Dynamic Breakdown Voltage of GaN Power HEMTs," in *2020 IEEE International Electron Devices Meeting (IEDM)*, 2020, p. 23.3.1-23.3.4. doi: 10.1109/IEDM13553.2020.9371904.

[17] R. Zhang, J. P. Kozak, M. Xiao, J. Liu, and Y. Zhang, "Surge-Energy and Overvoltage Ruggedness of P-Gate GaN HEMTs," *IEEE Trans. Power Electron.*, vol. 35, no. 12, pp. 13409–13419, Dec. 2020, doi: 10.1109/TPEL.2020.2993982.

[18] Q. Song, R. Zhang, Q. Li, and Y. Zhang, "Output Capacitance Loss of GaN HEMTs in Steady-State Switching," *IEEE Trans. Power Electron.*, pp. 1–10, 2023, doi: 10.1109/TPEL.2023.3279308.

[19] Q. Song, R. Zhang, J. P. Kozak, J. Liu, Q. Li, and Y. Zhang, "Robustness of Cascode GaN HEMTs in Unclamped Inductive Switching," *IEEE Trans. Power Electron.*, vol. 37, no. 4, pp. 4148–4160, Apr. 2022, doi: 10.1109/TPEL.2021.3122740.

[20] J. Liu *et al.*, "1.2-kV Vertical GaN Fin-JFETs: High-Temperature Characteristics and Avalanche Capability," *IEEE Trans. Electron Devices*, vol. 68, no. 4, pp. 2025–2032, Apr. 2021, doi: 10.1109/TED.2021.3059192.

[21] R. Zhang, Q. Song, Q. Li, and Y. Zhang, "Overvoltage Robustness of p-Gate GaN HEMTs in High Frequency Switching up to Megahertz," *IEEE Trans. Power Electron.*, vol. 38, no. 5, pp. 6063–6072, May 2023, doi: 10.1109/TPEL.2023.3237985.

[22] Q. Song, R. Zhang, Q. Li, and Y. Zhang, "Origin of Soft-Switching Output Capacitance Loss in Cascode GaN HEMTs at High Frequencies," *IEEE Trans. Power Electron.*, vol. 38, no. 11, pp. 13561–13566, Nov. 2023, doi: 10.1109/TPEL.2023.3299977.

[23] X. Tang *et al.*, "On the physics link between time-dependent gate breakdown and electroluminescence in Schottky-type p-GaN gate HEMTs," in *2022 IEEE 34th International Symposium on Power Semiconductor Devices and ICs (ISPSD)*, May 2022, pp. 57–60. doi: 10.1109/ISPSD49238.2022.9813615.

[24] A. Mehta, H. Shichijo, J. Joh, C. Suh, and M. Kim, "Degradation and Failure Mechanism of p-GaN Gate E-Mode GaN HEMTs," *ECS Trans.*, vol. 112, no. 2, p. 9, Sep. 2023, doi: 10.1149/11202.0009ecst.

Fabrication AlGaN/GaN Fin-HEMTs With Hexagon Nano-scale Fin Channel

Yu-Hsuan Lu
Graduate School of Advanced Technology
National Taiwan University
Taipei, Taiwan
d11k44005@ntu.edu.tw

Yu-Cheng Chang
Graduate Institute of Photonics and Optoelectronics
National Taiwan University
Taipei, Taiwan
r09941153@ntu.edu.tw

Wei-Ju Lu
Graduate Institute of Photonics and Optoelectronics
National Taiwan University
Taipei, Taiwan
r10941005@ntu.edu.tw

Feng-Ting Lin
Graduate Institute of Electronics Engineering
National Taiwan University
Taipei, Taiwan
r10943072@ntu.edu.tw

Bo-Hsun Xu
Graduate Institute of Photonics and Optoelectronics
National Taiwan University
Taipei, Taiwan
r10941150@ntu.edu.tw

Chao-Hsin Wu
Graduate School of Advanced Technology
National Taiwan University
Taipei, Taiwan
chaohsinwu@ntu.edu.tw

Abstract—This manuscript briefly presents the fabrication and characterization of AlGaN/GaN Fin-High-Electron-Mobility Transistors (Fin-HEMTs). The hexagon nanoscale fin channel was fabricated using TMAH wet etching solution, which reduced the surface roughness and shrunk the fin width (W_{fin}) to 153nm. The device with fin channel structure exhibits a maximum current density of 1285.01 mA/mm and improved thermal current density reduction from 27% to 13%. Moreover, at a gate bias of 0.5V and a drain bias of 6V, the cutoff frequency (f_T) and the maximum oscillation frequency (f_{max}) are 2.082 GHz and 6.472 GHz, respectively.

Keywords—*Fin-HEMTs, hexagon, cutoff frequency, maximum oscillation frequency*

I. INTRODUCTION

Over the past few years, the wireless communication market has shifted towards developing high bandwidth and low latency channels to meet the requirement of 5G communication systems. The gate length of transistors has also continued to shrink for improved high-frequency performances. However, while the gate length has shrunk to the nanometer scale, undesirable short channel effects hamper the device characteristics [1]. In order to overcome the short channel effect, a fin channel structure was applying to high electron mobility transistors. This structure can effectively enhance the gate control ability and minimize the bias dependence on the transconductance value (g_m) and f_T/f_{max} [2]. The improved high-frequency responses with excellent linearity make these devices well-suited for broadband radio frequency amplifiers, and this structure is expected to become a must have feature of the next generation of high electron mobility transistors.

II. DEVICE FABRICATION

The device fabrication starts from fin channel etching, a 100 nm of SiN hard mask layer was deposited, and a fin length (L_{fin}) of 1μm a fin width (W_{fin}) of 300 nm were patterned by electron-beam lithography (EBL) and defined through the inductively coupled plasma reactiveion etching (ICP-RIE). Subsequently, the wet etching treatment is performed at 70 °C for 10 minutes using 10% concentration tetramethylammonium hydroxide (TMAH) solution to repair the sidewall defects and narrow the fin width (W_{fin}). The fin channels were placed along the epi-taxial m-plane to minimize the irregular edges and re-duce channel skewing [3-5]. Fig. 1 shows the AFM images before and after TMAH wet etching process, with the channel placed along epitaxial m-plane, the different wet etching rate shaped the etching trench to hexagon form. The TMAH wet etching successfully shrunk the fin width (W_{fin}) from 300 nm to 153 nm which could improve the gate control ability. After fin channel formation, the hard mask was lifted off by BOE solution and the mesa isolation is done by Cl_2/BCl_3 plasma. Then the source/drain metal deposition with Ti/Al/Ni/Au, followed by

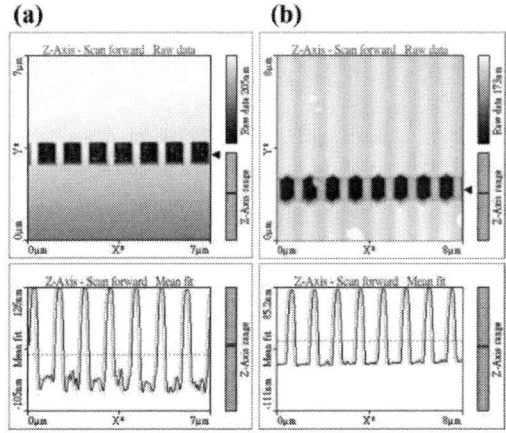

Fig. 1 AFM image of the fin channel array (a) before and (b) after TMAH wet etching process.

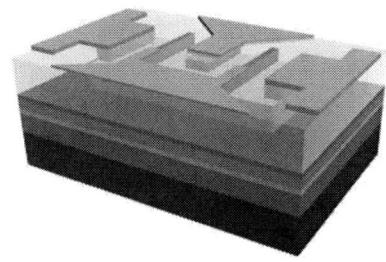

Fig. 2 Schematic structure of the double-finger AlGaN/GaN HEMTs with GSG pad.

rapid thermal annealing (RTA) at 925 °C for 60s in a N_2 furnace to form Ohmic contacts. The gate is defined by mask aligner with Ni/Au metallization which deposited directly on the fin channel without any insulating dielectrics to form the Schottky contact on all sides of the fin channel. Finally, the standard back-end process includes silicon dioxide (SiO_2) planarization and coplanar waveguide metallization for RF GSG probing. The schematic structure of the double-finger AlGaN/GaN HEMTs is shown in Fig. 2.

III. RESULTS AND DISCUSSION

A. DC Measurement

The DC transfer characteristics (I_d-V_g) of planar HEMT and Fin-HEMT are shown in Fig. 3. Two devices were fabricated on the same SiC substrates, and both devices share device dimensions of L_G= 2 μm, W_G= 2×50 μm, and L_{GD}/L_{GS}= 5/5 μm. For Fin-HEMTs, the DC performance is normalized to the effective gate width (W_{eff}), which can be calculated by W_{eff} = W_g×(W_{fin}/W_{period}). The fin period is consisting of one fin body and one side of the etched region. The effective gate width (W_{eff}) in this part is (2×50 μm)×(153 nm/1 μm) = 15.3 μm. For planar HEMT, the threshold voltage (Vth) is -3.44V, and the maximum current density is 558.70 mA/mm at V_{ov}= 5 V and V_d= 6 V. On the other hand, for Fin-HEMT the threshold voltage (Vth) is -2.72V, and the maximum current density is 1285.01 mA/mm at V_{ov}= 5 V and V_d= 6 V after normalized by the effective gate width. The threshold voltage (V_{th}) could shifts to the positive direction by the fin structure and keep shifting while narrowing down the fin width. The main reasons for the threshold voltage (V_{th}) shift are the increase in the depletion ratio of the gate sidewall and the decrease in the piezoelectric polarization effect of the AlGaN barrier layer due to stress relaxation [6]. However, the reverse gate leakage current of Fin-HEMT is about one order of magnitude larger than that of planar HEMT which can be attributed to leakage from the sidewall gate metal. At forward biased, the gate leakage current of Fin-HEMT increases compared rapidly due to the electron transport to the sidewall gate metal. When the plasma etching method was used to form a fin-nanochannel array, a larger subthreshold swing was induced by plasma etching damage, and a higher gate leakage current was caused by the resulting trap-assisted tunneling effect. As the results, the treatment of TMAH solution is crucial for ameliorated the sidewall defects induced by plasma etching damage.

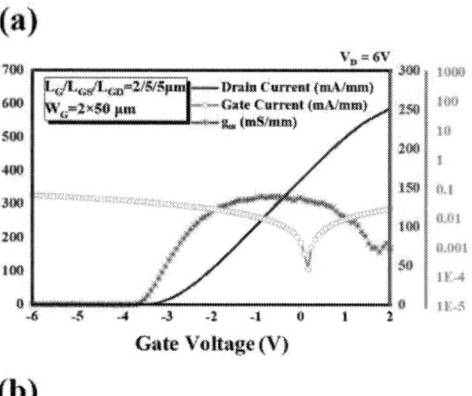

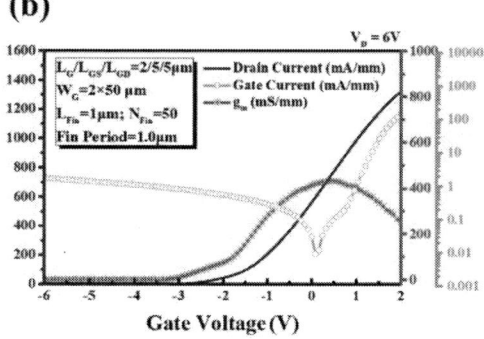

Fig. 3 DC Transfer Characteristic at V_D=6V of the device (a) Planar HEMT and (b) Fin-HEMT.

B. Thermal Analysis

The DC output characteristics (Id-Vd) under different temperature of planar HEMT and Fin-HEMT are shown in Fig.

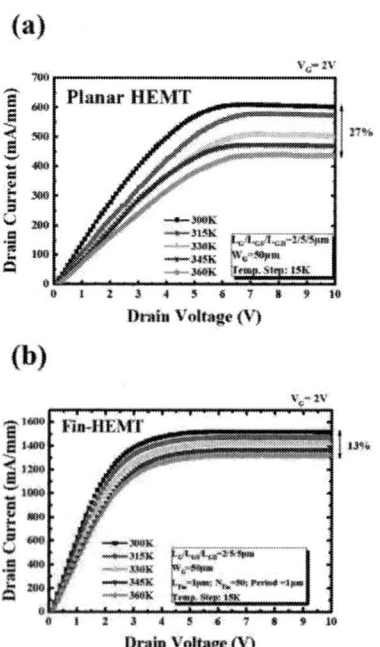

Fig. 4 DC Output Characteristic at V_G=2V of the device under different ambient temperature (a) Planar HEMT and (b) Fin-HEMT.

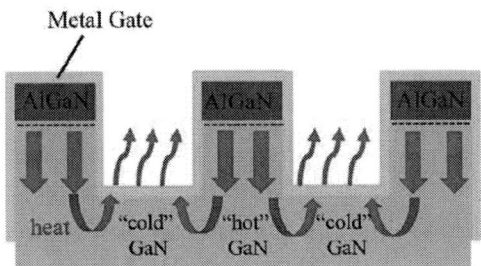

Fig. 5 The assumption heat dissipation path for AlGaN/GaN Fin-HEMTs.

4. In this section, we choose single-finger gate device and both devices share device dimensions of L_G=2 μm, W_G=50 μm, and L_{GD}/L_{GS}= 5/5 μm. During the DC measurement, a TEC-controller was used to control the ambient temperature from 300K to 360K with every 15K interval. Owing to High-Electron-Mobility Transistors (HEMTs) do not generate carriers through doping, the impact of ionized impurity scattering between the free dopant atoms is relatively low [7], whereas phonon scattering has a greater impact on carrier mobility. When the oscillation between atoms becomes more severe with the increase of temperature, the carrier mobility will decrease and attenuate the device current density. In order to characterize the device degradation by temperature more intuitive, we calculate the current density reduction (CDR) expressed as:

$$CDR = \left| \frac{I_{D@85°C} - I_{D@25°C}}{I_{D@25°C}} \right| \times 100\% \quad (1)$$

According to Fig. 4, the on-resistance increases as the temperature increases due to the mobility degradation in the 2DEG channel. However, the fin channel structure provided additional heat dissipation path between each channel shown in Fig. 5. Compared to the planar structure, Fin-HMETs successfully lower the thermal current density reduction from 27% to 13%.

C. RF Measurement

Afterward, the S-parameters of the device are characterized for a frequency range from 0.04 to 40 GHz using an Agilent N5225A PNA network analyzer. The AlGaN/GaN Fin-HMETs device investigated in this session has a dimension of L_G= 2 μm, W_G= 2×50 μm, and L_{GD}/L_{GS}= 5/5 μm. On-wafer open and short calibration is implemented to de-embed parasitic capacitances and inductances of metal pad during the measurement. Fig. 6 shows the de-embedded small-signal short-circuit current gain |h21|, unilateral gain U and maximum stable gain (MSG) of the device when biased at V_G = 0.5 V and V_D = 6 V. After subtracting the parasitic effects, the de-

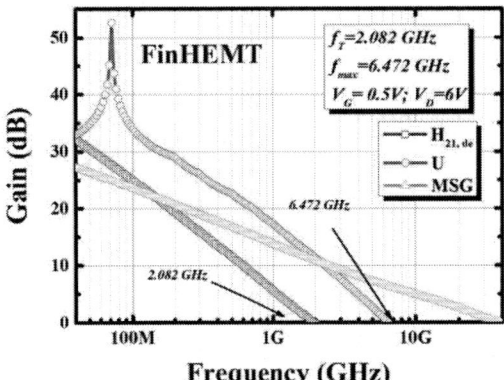

Fig. 6 Small-signal RF performance of Fin-HEMT at a quiescent bias of V_G=0.5V, V_D=6V after de-embedding.

embedded f_T and f_{max} is 2.082 GHz and 6.472 GHz, respectively, which are extracted by extrapolating using the slope of −20 dB/dec. The high-frequency performance could be improved when shrinking gate length (L_g) or adjusting fin geometry [8].

ACKNOWLEDGMENT

This work is supported by the program: NCSIST-111HT942003 through the National Chung-Shan Institute of Science & Technology.

REFERENCES

[1] T. Dutta et al., "Origins of the short channel effects increase in III-V nMOSFET technologies," 2012 13th International Conference on Ultimate Integration on Silicon (ULIS), pp. 25-28, doi: 10.1109/ULIS.2012.6193348, 2012

[2] E. Ture et al., "GaN-Based Tri-Gate High Electron Mobility Transistors," 2018

[3] N.A. Taradeh et al., "Characterization of m-GaN and a-GaN Crystallographic Planes after Being Chemically Etched in TMAH Solution, "MDPI Energies, 2020

[4] J. He, M. Feng et al., "On-wafer fabrication of cavity mirrors for InGaN-based laser diode grown on Si," Science Reports, 2018

[5] Im, Ki Sik et al., "Fabrication of AlGaN/GaN Ω-shaped nanowire fin-shaped FETs by a top-down approach, " Ap-plied Physics Express. 8. 066501. 10.7567/APEX.8.066501, 2015

[6] Zhang, Meng et al. "Influence of Fin Configuration on the Characteristics of AlGaN/GaN Fin-HEMTs." IEEE Transac-tions on Electron Devices 65: 1745-1752, 2018

[7] Bagnall et al., "Device-level thermal analysis of GaN-based electronics." 2013

[8] H. -S. Zhang et al., "Influence of Different Fin Configura-tions on Small-Signal Performance and Linearity for Al-GaN/GaN Fin-HEMTs," in IEEE Transactions on Electron Devices, vol. 66, no. 8, pp. 3302-3309, 2019

Short-Circuit Ruggedness Characterization of State-of-the-Art 3.3 kV SiC MOSFETs

Yizhou Cong[1], Peiwen Jiang[1], Ke Wang[1], Pengyu Fu[1], Jin Wang[1]
Ashish Kumar[2], Kraig Olejniczak[2]
[1]Center for High Performance Power Electronics (CHPPE), The Ohio State University, Columbus, OH
[2]Wolfspeed Inc., Fayetteville, AR

Abstract—This paper presents the short-circuit evaluation of state-of-the-art 3.3 kV silicon carbide (SiC) metal–oxide–semiconductor field-effect transistors (MOSFETs) from Wolfspeed, encompassing discrete devices rated at 40 A and power modules rated at 800 A. The SiC MOSFET degradation under repetitive short-circuit tests is evaluated by comparing initial and 250-cycle interval censored static parameters (up to 1,000 cycles). The threshold voltage and on-state resistance variations can be observed. Destructive tests on discrete devices and power modules reveal a maximum short circuit withstand time of 3 μs at 2 kV.

Index Terms—Short circuit, MOSFET, silicon carbide (SiC), device degradation, short circuit withstand time, medium voltage

I. INTRODUCTION

Silicon carbide (SiC) metal–oxide–semiconductor field-effect transistors (MOSFETs) offer significant advantages with greater switching speed, higher switching frequency, and enhanced maximum operating temperature compared to their Si counterparts. These advantages, combined with a drive to simplify control by using fewer stacked devices, position medium-voltage (MV) SiC MOSFETs as a promising choice for next-generation MV system applications, including grid-tie converters, solid-state transformers, solid-state circuit breakers, and traction inverters. However, a notable hurdle for SiC device adoption is their diminished short-circuit (SC) ruggedness compared with Si IGBTs. This is attributed to their shorter channel length and increased saturation current in a comparable die size to Si IGBTs.

The degradation of static/dynamic parameters and short circuit withstand time (SCWT) are two aspects to evaluate the device SC ruggedness. To analyze device parameter degradation, the repetitive SC test is employed to accelerate the device degradation and emulate real-world scenarios such as the inrush current during motor start-up. Static parameters encompass threshold voltage, on-state resistance, gate leakage current, and drain leakage current, as highlighted in various studies [1]–[4]. Meanwhile, dynamic parameters feature gate charge, switching time, and switching energy

Effort sponsored by the U.S. Government under OTA Project AMTC-19-07-001 under Prime Contract W9124P-19-9-0001. The U.S. Government is authorized to reproduce and distribute reprints for Governmental purposes notwithstanding any copyright notation thereon. The views and conclusions contained herein are those of the authors and should not be interpreted as necessarily representing the official policies or endorsements, either expressed or implied, of the U.S. Government.

[4], [5]. Predominantly, the quality of the SiC/SiO2 interface and the degradation of the gate dielectric are identified as primary causes of device deterioration in these publications. The SCWT is derived from destructive SC tests. As SiC semiconductor fabrication technology advances, there has been a surge in MV SiC MOSFETs entering the market in recent years. A concise overview of discrete SiC MOSFETs' SCWT is provided in [6]. Their SCWT lags behind Si IGBTs, which commonly achieves 10 μs [7]. Thus, the SCWT of SiC MOSFETs, including discrete devices and power modules, is crucial to gate driver designers to determine a reasonable fault response time. Despite the emergence of MV SiC MOSFETs, their availability in the market remains limited, with SCWT data for these devices, particularly for MV SiC power modules, seldom disclosed.

Earlier research, as cited in [8], explored device degradation after 100 cycles. In contrast, this paper examines four devices over 1,000 SC cycles, offering a detailed insight into degradation patterns. Additionally, the SCWT for SiC power MOSFETs is presented in the paper, particularly the SCWT for high-current power modules, bridging the knowledge gap in device protection design.

The paper is organized as follows. Section II introduces the high current SC setup and test procedures for repetitive and destructive SC tests. Section III details the submodule degradation under repetitive SC tests with static evaluation results. Section IV presents destructive SC test results from two discrete devices under 2 kV and 2.5 kV bus voltage and a power module at a 2kV dc bus voltage. Finally, Section V concludes the paper.

II. SHORT-CIRCUIT EXPERIMENTAL SETUP AND PROCEDURES

The SiC MOSFETs under investigation, showcased in Fig. 1, come in two distinct packages. The power module, nominally rated at 800 A (populated with 20 dies) at room temperature, is evaluated with both repetitive and destructive SC tests. The discrete device, nominally rated at 40 A at room temperature, is exclusively subjected to destructive SC tests. Notably, all these SiC devices utilize the same SiC MOSFET die.

979-8-3503-3714-3/23 $31.00 © 2023 IEEE

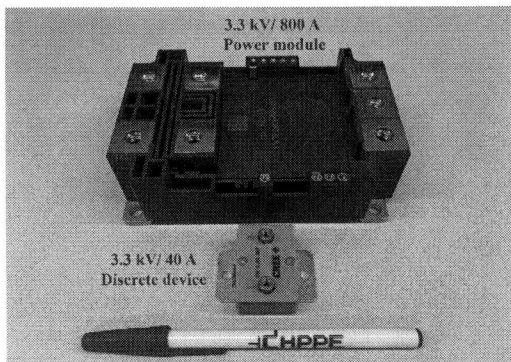

Fig. 1: Investigated SiC MOSFETs with two kinds of packages

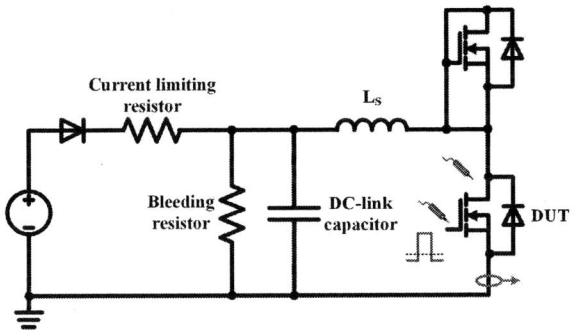

Fig. 2: Short-circuit test circuit diagram

A. High-current short-circuit test setup

An Agilent B1505A curve tracer is employed to gauge the device's static characteristics, and a customized SC test setup is built as illustrated in Fig. 2. Within this figure, the lower device of the 3.3-kV power module serves as the device under test (DUT). To safeguard the upper device from potential harm, its Drain, Gate, and Source are interconnected.

To protect the power source from potential damage from high output currents caused by the DUT failing to shut off during a SC event, a 10-kΩ current-limiting resistor is placed in series with the power source. This resistor can also decouple the power source during such events. The DC-link capacitor bank has four identical 50-μF capacitors connected in parallel to provide sufficient SC current. Given the power module's size and anticipated high SC current, no PCB busbar or nearby decoupling capacitor is provided. Instead, a copper busbar is directly affixed to the power module as depicted in Fig. 3.

The power loop stray inductance (Ls) mainly comprises the equivalent series inductance (ESL) of the dc-link capacitors and the stray inductance introduced by the copper busbar. This Ls contributes to an overshoot in the drain-to-source voltage (Vds) during turn-off transients. As such, it is crucial to choose a dc-link capacitor with minimal ESL and meticulously design the busbar to decrease stray inductance. Finite-element simulations conducted on the copper busbar indicate a total inductance of 19.8 nH. In Fig. 3, the copper busbar is insulated by Kapton tape, and a 3-D printed insulating layer is placed between the positive and negative busbars to enhance insulation.

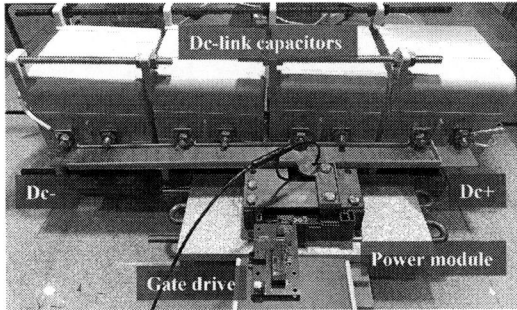

Fig. 3: Short-circuit test setup for power module

The customized gate driver is simplified with only fundamental turn-on and turn-off functionalities without any auxiliary functions. It operates at gate voltages of +15 V and -5 V. For the power module encompassing 20 dies, the turn-on resistance is set at 0.5 Ω, while the turn-off resistance is at 11 Ω. This configuration ensures a soft turn-off and emulates the de-saturation status of a normal gate driver with an integrated SC protection function. Appropriate gate resistance scaling is applied based on the number of dies in the device.

B. Short-circuit test procedures

For the repetitive SC test, 1000 SC cycles in total are applied to the power module with a 2 kV dc bus voltage. Static characterizations are performed every 250 cycles to measure the device threshold voltage, on-state resistance, gate leakage current, and drain leakage current. For each cycle, the DUT is turned on for 1-μs and waits for 1 second before the next cycle so that the heat generated during the SC event can be fully dissipated and the junction temperature can remain at the ambient temperature (25 °C) before the next cycle.

The destructive SC test is conducted with a 2 kV dc bus voltage. The pulse width starts from 1 μs and increases gradually until the device fails. The SCWT and its corresponding SC energy can be acquired from the experimental waveforms.

III. DEGRADATION UNDER REPETITIVE SHORT-CIRCUIT TESTS

Five power modules with twenty dies populated (PM 1 to PM 5) are tested for device degradation. Test results of the first cycle and the 1000th cycle of PM 2 are displayed in Fig. 4. After 1000 cycles, both the peak current and overshoot voltage exhibit almost no variations from the first cycle. The computed SC energy is approximately 11.4 J as determined by the following equation:

$$E_{sc} = \int_0^{t_{sc}} V_{ds} I_{ds} \, dt \qquad (1)$$

After every 250 cycles, a static characterization is conducted to assess the device degradation. The threshold voltage is tested with identical Vds and gate-to-source (Vgs) voltages. It is measured when the drain-to-source current (Ids) is 20 mA. Initial threshold voltages of all investigated power modules are listed in Table. I. Threshold voltage variations are illustrated in Fig. 5. According to the test results, almost all modules show

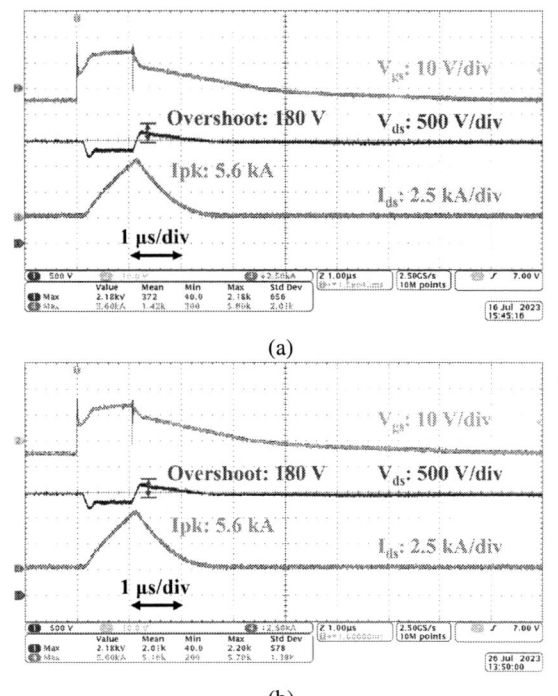

(a)

(b)

Fig. 4: Repetitive short-circuit test results of PM 2: (a) The first cycle; (b) The 1000th cycle

negative threshold voltage drift under repetitive SC tests. The maximum negative threshold voltage drift is -0.14 V from PM 1. The negative voltage drift can stem from holes trapped in the gate oxide under the condition of relatively short pulse width, low gate voltage, or low SC energy [1]–[3]. In the meanwhile, PM 2 exhibits a minor positive threshold voltage drift. This can be caused by the electrons trapped in the gate oxide under an electric field with opposite direction [2], [3].

The on-state resistance, measured with a Vgs of 15 V and

an Ids of 120 A, can be found in Table. II. The variation of the on-state resistance is normalized based on their initial values and presented in Fig. 6. The results illustrate that the on-resistance generally shows a slight downward trend while only PM 1 presents an upward trend with a maximum variation of +3.1%. The on-resistance variation can be attributed to the combined effects of gate oxide degradation mentioned in the last paragraph and package aging. The device on-resistance mainly includes channel resistance, accumulation layer resistance, JFET region resistance, drift region resistance, and substrate resistance. Both channel resistance and accumulation layer resistance would decrease as the threshold voltage decreases [9], which can be the cause of the reduced on-resistance. The package aging under repetitive SC tests, such as potential lift-off or crack of bond wires and delamination or voids of the solder joints [10], may cause on-resistance to rise. Nevertheless, with the existing results from Fig. 6 and considering the trend of threshold voltages of these devices, these variations have not indicated obvious degradation either from the device or the package.

The gate leakage and drain leakage currents are also measured for all five modules. Waveforms of each kind of parameter measured after every 250 cycles are almost overlapped, which is observed from the results of all five modules. These overlapped waveforms indicate that the device does not have apparent degradation after 1000 SC cycles.

IV. SURVIVABILITY UNDER DESTRUCTIVE SHORT-CIRCUIT TESTS

Destructive tests are initially performed on discrete devices at 2 kV and 2.5 kV, representing the typical operating voltage for 3.3 kV devices. The destructive test results on discrete devices are presented in Fig. 7 and Fig. 8, the SC current reaches its peak value and then reduces gradually with the pulse width. This behavior results from the combined effect

TABLE I: Initial threshold voltages

PM No.	1	2	3	4	5
Vth (V)	2.91	3.18	2.83	3.13	3.15

TABLE II: Initial on-state resistances

PM No.	1	2	3	4	5
Rdson (mΩ)	2.92	2.91	3.25	3.18	3.05

Fig. 5: Threshold voltage shift

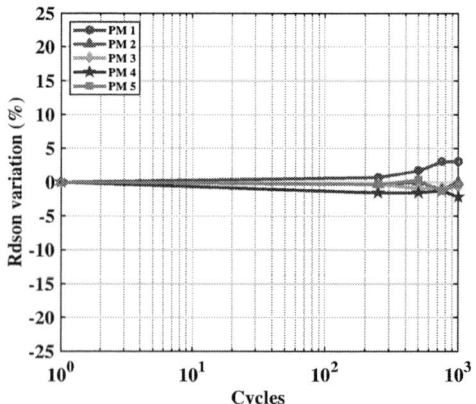

Fig. 6: On-state resistance variation

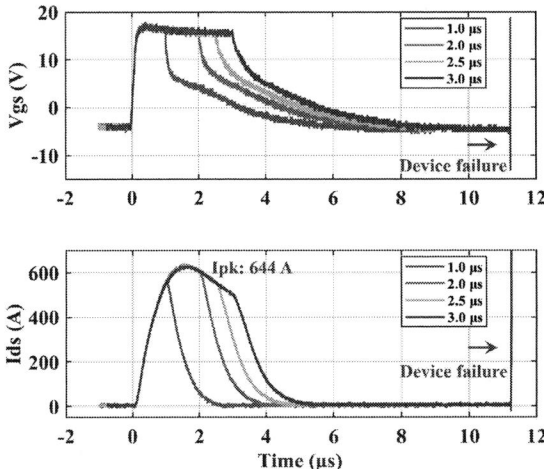

Fig. 7: Short-circuit behavior of discrete device at 2 kV

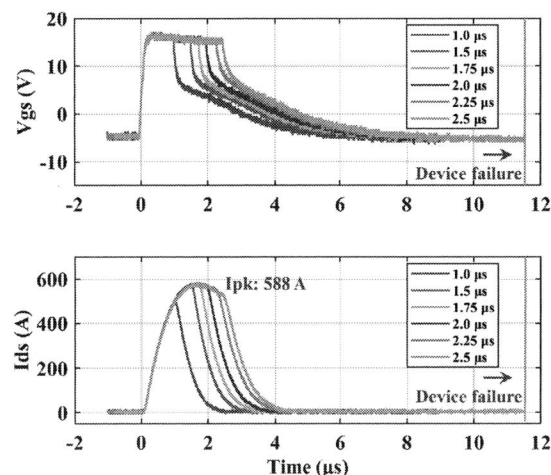

Fig. 8: Short-circuit behavior of discrete device at 2.5 kV

of the rise of the carrier number and the rise and subsequent reduction of electron mobility with temperature [11], [12]. The difference of saturation currents between the two devices can be attributed to their distinct threshold voltages according to the equation [9]:

$$I_{d,sat} = \frac{ZC_{ox}\mu_{inv}}{2L_{ch}}(V_{gs} - V_{th})^2, \quad (2)$$

where Z is the channel width orthogonal to the cross-section view of the device, C_{ox} is the specific gate oxide capacitance, μ_{ch} is the inversion layer mobility, L_{ch} is the channel length, and V_{th} is the threshold voltage. These two devices share the same design parameters from (2) but the threshold voltages of these two devices are 2.74 V and 3.01 V respectively at 1 mA, which may cause the saturation current difference.

According to the test results shown in Fig. 7 and Fig. 8, all current waveforms of each device overlap indicating the device does not experience considerable saturation current degradation under the destructive SC tests. Based on the test findings, the discrete device's SCWT measures 2.5 μs at 2 kV and 2.25 μs at 2.5 kV, with respective SC energies of 3.11 J and 3.12 J. The SCWT can be longer with lower gate resistance for higher turn-off speed. In all test conditions, the device can be successfully turned off during the SC event. However, they both failed at around 7 μs after the SC event, and their critical energy can be obtained to be 3.51 J and 3.32 J. Due to the feature of the failure waveform, the device failure can be caused by thermal runaway of potential hot pots [13] or melted surface metalization on the device source [14].

For the power module destructive SC tests, the DUT with 20 dies is tested at 2 kV and the test results are presented in Fig. 9. The SCWT can be determined to be 3.0 μs with the SC energy of 55.71 J. Also, the critical energy can be derived to be 69.77 J with 3.5 μs.

Fig. 10 depicts the short-circuit energy from discrete devices at 2 kV and 2.5 kV bus voltage and the short-circuit energy per die of the power module at 2 kV. The per-die short-circuit

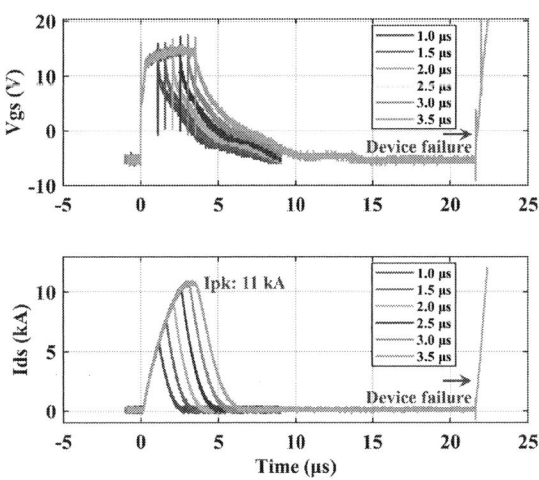

Fig. 9: Short-circuit behavior of power module at 2 kV

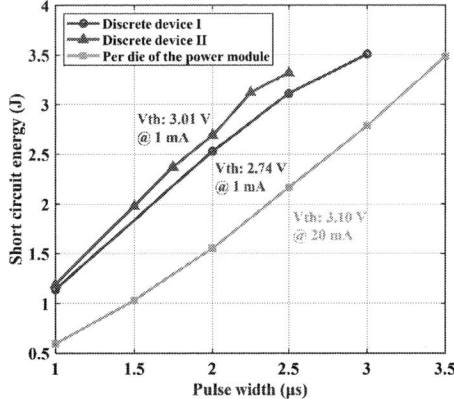

Fig. 10: Short-circuit energies of discrete devices and power module with different pulse widths at 2 kV

TABLE III: Destructive short-circuit test results

DUT	Bus voltage (kV)	SCWT (μs)	E_{sc} at SCWT (J)
Discrete device I	2.0	2.5	3.11
Discrete device II	2.5	2.25	3.12
Power Module I	2.0	3.0	55.71

energy is equal to the power module short-circuit energy divided by 20 to compare with discrete devices. The last point of each waveform is the critical short-circuit energy to cause device failure. The averaged short-circuit energy from the power module is lower than the discrete device at the same pulse width due to the higher threshold voltage. However, the critical short-circuit energies are almost identical. This implies that with suitable packaging the power module SC survivability would not suffer from derating compared with discrete devices.

V. CONCLUSION

The SC ruggedness evaluation of state-of-the-art 3.3 kV SiC MOSFETs is presented in the paper. Repetitive SC tests reveal a maximum threshold voltage shift of -0.14 V and a +3.1% peak variation in on-resistance after 1,000 cycles. The SCWT and corresponding SC energy obtained from destructive SC tests are summarized in Table. III. At 2 kV, the discrete device can survive 2.5-μs pulse width while the power module can withstand 3-μs pulse width. The longer SCWT from power module compared with the discrete device can be explained by the difference in threshold voltage.

ACKNOWLEDGMENT

The work presented in this paper is sponsored by Wolfspeed. The team appreciates test samples and technical inputs provided by the Wolfspeed team throughout the project.

REFERENCES

[1] X. Zhou, H. Su, Y. Wang, R. Yue, G. Dai, and J. Li, "Investigations on the Degradation of 1.2-kV 4H-SiC MOSFETs Under Repetitive Short-Circuit Tests," *IEEE Transactions on Electron Devices*, vol. 63, no. 11, pp. 4346–4351, 2016.

[2] J. Sun, J. Wei, Z. Zheng, Y. Wang, and K. J. Chen, "Short Circuit Capability and Short Circuit Induced V_{TH} Instability of a 1.2-kV SiC Power MOSFET," *IEEE Journal of Emerging and Selected Topics in Power Electronics*, vol. 7, no. 3, pp. 1539–1546, 2019.

[3] Y. Li, X. Zhou, Y. Zhao, Y. Jia, D. Hu, Y. Wu, L. Zhang, Z. Chen, and A. Q. Huang, "Gate Bias Dependence of VTH Degradation in Planar and Trench SiC MOSFETs Under Repetitive Short Circuit Tests," *IEEE Transactions on Electron Devices*, vol. 69, no. 5, pp. 2521–2527, 2022.

[4] R. Yu, S. Jahdi, P. Mellor, L. Liu, J. Yang, C. Shen, O. Alatise, and J. Ortiz-Gonzalez, "Degradation Analysis of Planar, Symmetrical and Asymmetrical Trench SiC MOSFETs Under Repetitive Short Circuit Impulses," *IEEE Transactions on Power Electronics*, vol. 38, no. 9, pp. 10 933–10 946, 2023.

[5] J. Wei, S. Liu, L. Yang, J. Fang, T. Li, S. Li, and W. Sun, "Comprehensive Analysis of Electrical Parameters Degradations for SiC Power MOSFETs Under Repetitive Short-Circuit Stress," *IEEE Transactions on Electron Devices*, vol. 65, no. 12, pp. 5440–5447, 2018.

[6] D. Xing, X. Lyu, J. Liu, C. Xie, A. Agarwal, and J. Wang, "3300-V SiC MOSFET Short-Circuit Reliability and Protection," in *2021 IEEE Applied Power Electronics Conference and Exposition (APEC)*, 2021, pp. 1262–1266.

[7] R. Chokhawala, J. Catt, and L. Kiraly, "A discussion on IGBT short-circuit behavior and fault protection schemes," *IEEE Transactions on Industry Applications*, vol. 31, no. 2, pp. 256–263, 1995.

[8] K. Wang, Y. Cong, P. Fu, X. Li, Q. Cheng, B. Hu, J. Wang, A. Kumar, K. Olejniczak, D. Pelletier, Z. Cole, A. Deshpande, and A. Goyal, "Short-Circuit Ruggedness and Partial Discharge Evaluation of a 3.3 kV SiC MOSFET Power Module," in *2022 IEEE 9th Workshop on Wide Bandgap Power Devices Applications (WiPDA)*, 2022, pp. 154–158.

[9] B. J. Baliga, *Fundamentals of power semiconductor devices*. Springer Science & Business Media, 2010.

[10] S. Dusmez, H. Duran, and B. Akin, "Remaining Useful Lifetime Estimation for Thermally Stressed Power MOSFETs Based on on-State Resistance Variation," *IEEE Transactions on Industry Applications*, vol. 52, no. 3, pp. 2554–2563, 2016.

[11] A. Pérez-Tomás, P. Brosselard, P. Godignon, J. Millán, N. Mestres, M. R. Jennings, J. A. Covington, and P. A. Mawby, "Field-effect Mobility Temperature Modeling of 4H-SiC Metal-oxide-semiconductor Transistors," *Journal of Applied Physics*, vol. 100, no. 11, p. 114508, 12 2006.

[12] P. Gaubert, A. Teramoto, W. Cheng, and T. Ohmi, "Relation Between the Mobility, $1/f$ Noise, and Channel Direction in MOSFETs Fabricated on (100) and (110) Silicon-Oriented Wafers," *IEEE Transactions on Electron Devices*, vol. 57, no. 7, pp. 1597–1607, 2010.

[13] G. Romano, A. Fayyaz, M. Riccio, L. Maresca, G. Breglio, A. Castellazzi, and A. Irace, "A Comprehensive Study of Short-Circuit Ruggedness of Silicon Carbide Power MOSFETs," *IEEE Journal of Emerging and Selected Topics in Power Electronics*, vol. 4, no. 3, pp. 978–987, 2016.

[14] K. Han, A. Kanale, B. J. Baliga, B. Ballard, A. Morgan, and D. C. Hopkins, "New Short Circuit Failure Mechanism for 1.2kV 4H-SiC MOSFETs and JBSFETs," in *2018 IEEE 6th Workshop on Wide Bandgap Power Devices and Applications (WiPDA)*, 2018, pp. 108–113.

AUTHOR INDEX

Agamy, Mohammed...201
Agarwal, Anant K. 38, 80, 84, 88
Aghdam, Sima Azizi..201
Allerman, Andrew A. 42
Allioua, Abdelmoumin..58
Alonso, Corinne...46
Anderson, Travis J. 109
Arriola, Emmanuel 99
Bai, Hua...219, 225
Barr, Nathaniel..142
Barua, Himel...74
Basler, Michael...21
Beczkowski, Szymon..25
Bhattacharya, Monikuntala.................38, 80, 84, 88
Binder, Andrew T..42
Bisi, Davide 11
Bobde, Madhur..154
Bosch, Michael..207
Boshkovski, Filip..99
Boutry, Arthur...138
Breidenstein, Daniel 15
Castagna, M. E..64
Chang, Yu-Cheng..263
Chatterjee, Urmimala..52
Chatty, Kiran...154
Chen, Ruirui 174, 219, 225
Chen, Su-Wen...142
Chen, Yuzhi...125
Chen, Zeyu..148
Chini, A...64
Chowdhury, Shajjad..74
Christou, Aristos...169
Cioni, M..64
Cong, Yizhou...266
Cruse, Bill...11
Deboer, Skylar..159
Dieckerhoff, Sibylle...5
Driesen, Johan..52
Dudley, Michael..148
Dürbaum, Thomas 15
Ebrahimian, Armin 31, 131
Fang, Jiayue..119
Faruque, Asif..213
Ferretti, Jacopo 179
Flicker, Jack..163
Frey, Dick...142
Fu, Pengyu..266
Gallagher, James C. 109

Gao, Zihan......................................174, 219, 225
Gendron, Amaury.. 142
Geng, Xiaomeng 5
Georgiev, Daniel G.. 109
Gill, Lee 163
Giorgino, G. 64
Glaser, Caleb 42
Goodnick, Stephen 163
Govaerts, George.. 52
Griepentrog, Gerd 58
Gu, Albert.. 142
Gudino, Natalia... 169
Gupta, Geetak...11
Hassan, Ayesha... 213
Hilt, Oliver 5
Hobart, Karl D... 109
Hontz, Michael R... 109
Huang, Yulu..11
Iucolano, F... 64
Jacobs, Alan G... 109
Jang, Seung Yup..148, 159
Jiang, Peiwen 266
Jimenez, Sergio 138
Jin, Michael...38, 80, 84, 88
Jørgensen, Asger Bjørn 25
Kallfass, Ingmar..190, 207
Kaplar, Robert J.. 42
Kaplar, Robert 163
Kato, Yosuke 240
Ke, Chao-Yang 69
Ker, Ming-Dou 69
Khan, Waqar A. 31
Khanna, Raghav 109
Kim, Dongyoung..148, 159
Koch, Dominik..190, 207
Koehler, Andrew D. 109
Kohlhepp, Benedikt.. 15
Krause, David.. 58
Kumar, Ashish 266
Kuring, Carsten 5
Kuroi, Takashi... 185
Lal, Rakesh 11
Lemmon, Andrew.. 138
Li, Chi... 125
Li, Dingrui..174, 219, 225
Li, Hu..119
Li, Qiang...236, 258
Li, Zhongda 201

Lin, Feng-Ting 263
Lin, Min 219, 225
Liu, Zhan 154
Losee, Peter 201
Lu, Albert 231
Lu, Guo-Quan 1, 99
Lu, Shengchang 1
Lu, Wei-Ju 263
Lu, Yu-Hsuan 263
Luongo, G. 64
Lynch, Justin 1, 185
Ma, Wenjie 119
Ma, Yunwei 251
Mancini, Stephen A 148
Mantooth, H. Alan 213
Marcault, Emmanuel 46
Marletta, G. 64
Martinez, Wilmar 52
Meyer, Dennis 104
Miccoli, C. 64
Michaels, Alan 163
Mishra, Umesh 11
Mönch, Stefan 21
Morgan, Adam J 159
Morgan, Adam 1
Moschetti, M. 64
Mu, Qiang 245
Munk-Nielsen, Stig 25
Mustakin, Zaheen 245
Natori, Kenji 240
Nelson, Tolen 109
Neufeld, Carl 11
Ngo, Khai D. T. 1
Niakan, Hossein 195
Nicholas, Carl 1, 99
Nuzzo, Jeremy 207
Olejniczak, Kraig 266
Ozpineci, Burak 74
Pandey, Prakash 109
Pang, Xiaohu 119
Parkhideh, Babak 195, 245
Porter, Matthew A 251
Porter, Matthew 231
Qian, Jiashu 38, 80, 84, 88
Qin, Yuan 251
Quay, Rüdiger 21
Raghothamachar, Balaji 148
Rahman, Mohammad Dehan 114
Reiner, Richard 21
Roche, Ludovic 46
Roy, Chondon 245
Rueß, Manuel 190, 207

Rummel, Brian D. 42
Sabzevari, Seyed Iman Hosseini 31, 131
Sakamoto, Kazuma 240
Salemi, Arash 154
Sangiorgi, Enrico 179
Sato, Yukihiko 240
Schiapparelli, Giacomo-Piero 179
Schoch, Benjamin 207
Sdrulla, Dumitru 142
Setera, Brett 169
Shahabi, Ali 104
Sheridan, David C. 154
Shi, Limeng 38, 80, 84, 88
Shimbori, Atsushi 80, 84, 88
Shoemaker, Jonah 163
Sirat, Ali Parsa 195, 245
Smith, Peter 11
Song, Qihao 231, 236, 258
Song, Xiaoqing 114
Starr, Linda 104
Steinfeldt, Jeffrey 42
Sung, Woongje 1, 148, 159, 185
Takata, Tetsuya 142
Tallarico, Andrea Natale 179
Tarmoom, Ehab 104
Tokoro, Nobuhiro 185
Tolbert, Leon M. 219, 225
Trémouilles, David 46
Tringali, C. 64
Wada, Ryota 185
Waltereit, Patrick 21
Wang, Bixuan 258
Wang, Fred 174, 219, 225
Wang, Jin 266
Wang, Ke 266
Wang, Luocheng 245
Wang, Yifan 231
Wang, Yuyang 213
Weise, Nathan 31, 131
White, Marvin H. 38, 80, 84, 88
Wieczorek, Nick 5
Wilkins, Jon 74
Wolf, Mihaela 5
Wong, Hiu Yung 231
Wrana, Dominik 207
Wu, Chao-Hsin 263
Wu, Yifan 125
Xiao, Ming 231
Xu, Bo-Hsun 263
Xu, Zhuxian 80, 84, 88
Yang, Fei 92
Yates, Luke 42

Yee, Nathan	231
Yin, Shan	119
Yu, Hengyu	38, 80, 84, 88
Yu, Sheng-Yang	92
Zäch, Mike	25
Zhang, Jin	154
Zhang, Ruizhe	231, 258
Zhang, Xuning	104
Zhang, Yuhao	231, 236, 251, 258
Zhang, Zichen	1, 99
Zhao, Tiefu	245
Zheng, Zedong	125
Zhou, Jiale	245
Zhu, Lisi	154
Zingariello, Andrea	58

IEEE
445 Hoes Lane
Piscataway, NJ 08854-4141

ISBN 979-8-3503-3714-3